T0344937

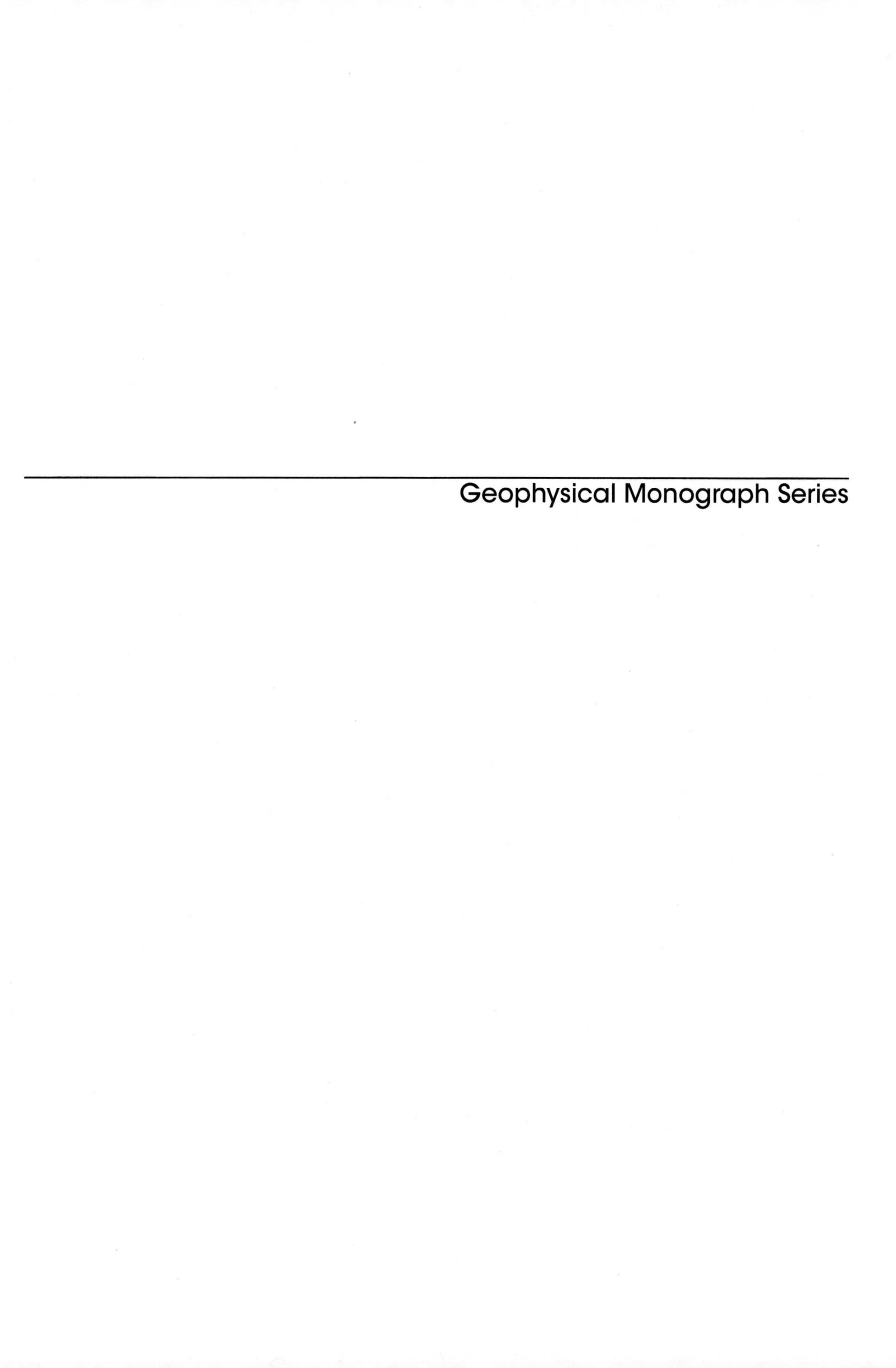
Geophysical Monograph Series

175 A Continental Plate Boundary: Tectonics at South Island, New Zealand *David Okaya, Tim Stem, and Fred Davey (Eds.)*

176 Exploring Venus as a Terrestrial Planet *Larry W. Esposito, Ellen R. Stofan, and Thomas E. Cravens (Eds.)*

177 Ocean Modeling in an Eddying Regime *Matthew Hecht and Hiroyasu Hasumi (Eds.)*

178 Magma to Microbe: Modeling Hydrothermal Processes at Oceanic Spreading Centers *Robert P. Lowell, Jeffrey S. Seewald, Anna Metaxas, and Michael R. Perfit (Eds.)*

179 Active Tectonics and Seismic Potential of Alaska *Jeffrey T. Freymueller, Peter J. Haeussler, Robert L. Wesson, and Göran Ekström (Eds.)*

180 Arctic Sea Ice Decline: Observations, Projections, Mechanisms, and Implications *Eric T. DeWeaver, Cecilia M. Bitz, and L.-Bruno Tremblay (Eds.)*

181 Midlatitude Ionospheric Dynamics and Disturbances *Paul M. Kintner, Jr., Anthea J. Coster, Tim Fuller-Rowell, Anthony J. Mannucci, Michael Mendillo, and Roderick Heelis (Eds.)*

182 The Stromboli Volcano: An Integrated Study of the 2002–2003 Eruption *Sonia Calvari, Salvatore Inguaggiato, Giuseppe Puglisi, Maurizio Ripepe, and Mauro Rosi (Eds.)*

183 Carbon Sequestration and Its Role in the Global Carbon Cycle *Brian J. McPherson and Eric T. Sundquist (Eds.)*

184 Carbon Cycling in Northern Peatlands *Andrew J. Baird, Lisa R. Belyea, Xavier Comas, A. S. Reeve, and Lee D. Slater (Eds.)*

185 Indian Ocean Biogeochemical Processes and Ecological Variability *Jerry D. Wiggert, Raleigh R. Hood, S. Wajih A. Naqvi, Kenneth H. Brink, and Sharon L. Smith (Eds.)*

186 Amazonia and Global Change *Michael Keller, Mercedes Bustamante, John Gash, and Pedro Silva Dias (Eds.)*

187 Surface Ocean–Lower Atmosphere Processes *Corinne Le Quèrè and Eric S. Saltzman (Eds.)*

188 Diversity of Hydrothermal Systems on Slow Spreading Ocean Ridges *Peter A. Rona, Colin W. Devey, Jérôme Dyment, and Bramley J. Murton (Eds.)*

189 Climate Dynamics: Why Does Climate Vary? *De-Zheng Sun and Frank Bryan (Eds.)*

190 The Stratosphere: Dynamics, Transport, and Chemistry *L. M. Polvani, A. H. Sobel, and D. W. Waugh (Eds.)*

191 Rainfall: State of the Science *Firat Y. Testik and Mekonnen Gebremichael (Eds.)*

192 Antarctic Subglacial Aquatic Environments *Martin J. Siegert, Mahlon C. Kennicut II, and Robert A. Bindschadler*

193 Abrupt Climate Change: Mechanisms, Patterns, and Impacts *Harunur Rashid, Leonid Polyak, and Ellen Mosley-Thompson (Eds.)*

194 Stream Restoration in Dynamic Fluvial Systems: Scientific Approaches, Analyses, and Tools *Andrew Simon, Sean J. Bennett, and Janine M. Castro (Eds.)*

195 Monitoring and Modeling the Deepwater Horizon Oil Spill: A Record-Breaking Enterprise *Yonggang Liu, Amy MacFadyen, Zhen-Gang Ji, and Robert H. Weisberg (Eds.)*

196 Extreme Events and Natural Hazards: The Complexity Perspective *A. Surjalal Sharma, Armin Bunde, Vijay P. Dimri, and Daniel N. Baker (Eds.)*

197 Auroral Phenomenology and Magnetospheric Processes: Earth and Other Planets *Andreas Keiling, Eric Donovan, Fran Bagenal, and Tomas Karlsson (Eds.)*

198 Climates, Landscapes, and Civilizations *Liviu Giosan, Dorian Q. Fuller, Kathleen Nicoll, Rowan K. Flad, and Peter D. Clift (Eds.)*

199 Dynamics of the Earth's Radiation Belts and Inner Magnetosphere *Danny Summers, Ian R. Mann, Daniel N. Baker, and Michael Schulz (Eds.)*

200 Lagrangian Modeling of the Atmosphere *John Lin (Ed.)*

201 Modeling the Ionosphere-Thermosphere *Jospeh D. Huba, Robert W. Schunk, and George V Khazanov (Eds.)*

202 The Mediterranean Sea: Temporal Variability and Spatial Patterns *Gian Luca Eusebi Borzelli, Miroslav Gacic, Piero Lionello, and Paola Malanotte-Rizzoli (Eds.)*

203 Future Earth - Advancing Civic Understanding of the Anthropocene *Diana Dalbotten, Gillian Roehrig, and Patrick Hamilton (Eds.)*

204 The Galápagos: A Natural Laboratory for the Earth Sciences *Karen S. Harpp, Eric Mittelstaedt, Noémi d'Ozouville, and David W. Graham (Eds.)*

205 Modeling Atmospheric and Oceanic Flows: Insightsfrom Laboratory Experiments and Numerical Simulations *Thomas von Larcher and Paul D. Williams (Eds.)*

206 Remote Sensing of the Terrestrial Water Cycle *Venkat Lakshmi (Eds.)*

207 Magnetotails in the Solar System *Andreas Keiling, Caitríona Jackman, and Peter Delamere (Eds.)*

208 Hawaiian Volcanoes: From Source to Surface *Rebecca Carey, Valerie Cayol, Michael Poland, and Dominique Weis (Eds.)*

209 Sea Ice: Physics, Mechanics, and Remote Sensing *Mohammed Shokr and Nirmal Sinha (Eds.)*

210 Fluid Dynamics in Complex Fractured-Porous Systems *Boris Faybishenko, Sally M. Benson, and John E. Gale (Eds.)*

211 Subduction Dynamics: From Mantle Flow to Mega Disasters *Gabriele Morra, David A. Yuen, Scott King, Sang Mook Lee, and Seth Stein (Eds.)*

212 The Early Earth: Accretion and Differentiation *James Badro and Michael Walter (Eds.)*

213 Global Vegetation Dynamics: Concepts and Applications in the MC1 Model *Dominique Bachelet and David Turner (Eds.)*

214 Extreme Events: Observations, Modeling and Economics *Mario Chavez, Michael Ghil, and Jaime Urrutia-Fucugauchi (Eds.)*

215 Auroral Dynamics and Space Weather *Yongliang Zhang and Larry Paxton (Eds.)*

216 Low-Frequency Waves in Space Plasmas *Andreas Keiling, Dong-Hun Lee, and Valery Nakariakov (Eds.)*

217 Deep Earth: Physics and Chemistry of the Lower Mantle and Core *Hidenori Terasaki and Rebecca A. Fischer (Eds.)*

218 Integrated Imaging of the Earth: Theory and Applications *Max Moorkamp, Peter G. Lelievre, Niklas Linde, and Amir Khan (Eds.)*

219 Plate Boundaries and Natural Hazards *Joao Duarte and Wouter Schellart (Eds.)*

220 Ionospheric Space Weather: Longitude and Hemispheric Dependences and Lower Atmosphere Forcing *Timothy Fuller-Rowell, Endawoke Yizengaw, Patricia H. Doherty, and Sunanda Basu (Eds.)*

221 Terrestrial Water Cycle and Climate Change: Natural and Human-Induced Impacts *Qiuhong Tang and Taikan Oki (Eds.)*

222 Magnetosphere-Ionosphere Coupling in the Solar System *Charles R. Chappell, Robert W. Schunk, Peter M. Banks, James L. Burch, and Richard M. Thorne (Eds.)*

Geophysical Monograph 223

Natural Hazard Uncertainty Assessment

Modeling and Decision Support

Karin Riley
Peter Webley
Matthew Thompson
Editors

This Work is a copublication of the American Geophysical Union and John Wiley and Sons, Inc.

WILEY

This Work is a copublication of the American Geophysical Union and John Wiley & Sons, Inc.

Published under the aegis of the AGU Publications Committee

Published by John Wiley & Sons, Inc., Hoboken, New Jersey
Published simultaneously in Canada

Library of Congress Cataloging-in-Publication Data

ISBN: 978-1-119-02786-7

Cover images:
Fire image on left: A crown fire moves through a forest. Because of their high intensity and flame lengths, crown fires are difficult to suppress and pose a hazard to many highly valued resources.
Map on right: The model FSPro provides probabilistic outputs that predict fire spread over a period of 1–4 weeks.
Volcano image on right: Activity Heat - Temperature Motion Adventure Bizarre Exploration Nature Horizontal Outdoors Exploding Ecuador Red Dark Fire Steam Mountain Volcano Sunset Night Snow Fumarole Lava Earthquake Backgrounds Erupting Geology Hell Glowing Photography Vulcanology Flowing Tungurahua Volcano strombolian

Printed in the United States of America

10 9 8 7 6 5 4 3 2 1

CONTENTS

CONTRIBUTORS

J. Abatzoglou
University of Idaho, Moscow, Idaho, USA

Eric Ross Anderson
Earth System Science Center, University of Alabama in Huntsville, Huntsville, Alabama, USA

Delia Arnold
Central Institute for Meteorology and Geodynamics (ZAMG), Vienna, Austria

D. Bachelet
Conservation Biology Institute, Corvallis, Oregon, and Oregon State University, USA

M. Borga
Department of Land, Environment, Agriculture and Forestry, University of Padova, Legnaro, Italy

Marcus Bursik
Department of Geology, University at Buffalo, SUNY, Buffalo, New York, USA

Lizeth Caballero
Facultad de Ciencias/Instituto de Geología, UNAM, Mexico City, Mexico

David Calkin
Rocky Mountain Research Station, US Forest Service, Missoula, Montana, USA

Lucia Capra
Centro de Geociencias, UNAM, Querétaro, Mexico

Jaime A. Collazo
Department of Applied Ecology, North Carolina State University, Raleigh, North Carolina, and US Geological Survey, North Carolina Cooperative Fish and Wildlife Research Unit, North Carolina State University, Raleigh, North Carolina, USA

Antonio Costa
Istituto Nazionale di Geofisica e Vulcanologia, INGV, Sezione di Bologna, Bologna, Italy

Jennifer Costanza
Department of Forestry and Environmental Resources, North Carolina State University, Raleigh, North Carolina, USA

Jonathan Dehn
Geophysical Institute, University of Alaska Fairbanks, Fairbanks, Alaska, and Volcanic Ash Detection, Avoidance and Preparedness for Transportation (V-ADAPT), Inc., Fairbanks, Alaska, USA

Mauro Antonio Di Vito
Istituto Nazionale di Geofisica e Vulcanologia, INGV, Osservatorio Vesuviano, Napoli, Italy

Luke T. Ellison
Climate and Radiation Laboratory, NASA Goddard Space Flight Center, Greenbelt, Maryland, and Science Systems and Applications Inc., Lanham, Maryland, USA

K. Ferschweiler
Conservation Biology Institute, Corvallis, Oregon, USA

Xiang Gao
Massachusetts Institute of Technology, Cambridge, Massachusetts, USA

Gernot Geppert
Universität Hamburg, Hamburg, Germany

Wenyu Gong
Geophysical Institute, University of Alaska Fairbanks, Fairbanks, Alaska, USA

Robert E. Griffin
Atmospheric Science Department, University of Alabama in Huntsville, Huntsville, Alabama, USA

Michael Hand
Rocky Mountain Research Station, US Forest Service, Missoula, Montana, USA

Jessica R. Haas
Rocky Mountain Research Station, US Forest Service, Missoula, Montana, USA

Kevin D. Hyde
University of Wyoming, Laramie, Wyoming, USA

Charles Ichoku
Climate and Radiation Laboratory, NASA Goddard Space Flight Center, Greenbelt, Maryland, USA

Daniel E. Irwin
NASA Marshall Space Flight Center, Huntsville, Alabama, USA

Matthew D. Jones
Center for Computational Research, University at Buffalo, SUNY, Buffalo, New York, USA

Johannes W. Kaiser
Max Planck Institute for Chemistry, Mainz, Germany

Su Young Kang
Department of Geological Sciences, Pusan National University, Busan, Korea

Maureen C. Kennedy
University of Washington, School of Interdisciplinary Arts and Sciences, Division of Sciences and Mathematics, Tacoma, Washington, USA

Kwang-Hee Kim
Department of Geological Sciences, Pusan National University, Busan, Korea

Nina Iren Kristiansen
Norwegian Institute for Air Research (NILU), Kjeller, Norway

Erin Leidy
Charles Stark Draper Laboratory, Cambridge, Massachusetts, and Massachusetts Institute of Technology, Cambridge, Massachusetts, USA

Yannick Le Page
Pacific Northwest National Laboratory, Joint Global Change Research Institute, University of Maryland, College Park, Maryland, USA

Zhong Lu
Southern Methodist University, Dallas, Texas, USA

Reza Madankan
Department of Mechanical and Aerospace Engineering, University at Buffalo, SUNY, Buffalo, New York, USA

Natasha Markuzon
Charles Stark Draper Laboratory, Cambridge, Massachusetts, USA

F. Marra
Department of Geography, Hebrew University of Jerusalem, Israel

Damien Martin
School of Physics and Centre for Climate and Air Pollution Studies, National University of Ireland Galway (NUIG), Galway, Ireland

Warner Marzocchi
Istituto Nazionale di Geofisica e Vulcanologia, INGV, Sezione di Roma1, Roma, Italy

Christian Maurer
Central Institute for Meteorology and Geodynamics (ZAMG), Vienna, Austria

Donald McKenzie
Pacific Wildland Fire Sciences Laboratory, Pacific Northwest Research Station, US Forest Service, Seattle, Washington, USA

Alexa McKerrow
Department of Applied Ecology, North Carolina State University, Raleigh, North Carolina, and Core Science Analytics, Synthesis & Libraries, US Geological Survey, Raleigh, North Carolina, USA

Franz Meyer
Geophysical Institute, University of Alaska Fairbanks, Fairbanks, Alaska, USA

E. I. Nikolopoulos
Department of Land, Environment, Agriculture and Forestry, University of Padova, Legnaro, Italy

Colin O'Dowd
School of Physics and Centre for Climate and Air Pollution Studies, National University of Ireland Galway (NUIG), Galway, Ireland

Krishna Pacifici
Department of Forestry and Environmental Resources, Program in Fisheries, Wildlife, and Conservation Biology, North Carolina State University, Raleigh, North Carolina, USA

Abani Patra
Department of Mechanical and Aerospace Engineering, University at Buffalo, SUNY, Buffalo, New York, USA

E. Bruce Pitman
Department of Mathematics, University at Buffalo, SUNY, Buffalo, New York, USA

Solene Pouget
Department of Geology, University at Buffalo, SUNY, Buffalo, New York, USA

Razvan Rădulescu
School of Physics and Centre for Climate and Air Pollution Studies, National University of Ireland Galway (NUIG), Galway, Ireland

Florian Rauser
Max Planck Institute for Meteorology, Hamburg, Germany

John Regan
Charles Stark Draper Laboratory, Cambridge, Massachusetts, USA

Brian Reich
Department of Statistics, North Carolina State University, Raleigh, North Carolina, USA

Karin Riley
Rocky Mountain Research Station, US Forest Service, Missoula, Montana, and Numerical Terradynamic Simulation Group, College of Forestry and Conservation, University of Montana, Missoula, Montana, USA

Laura Sandri
Istituto Nazionale di Geofisica e Vulcanologia, INGV, Sezione di Bologna, Bologna, Italy

Adam Schlosser
Massachusetts Institute of Technology, Cambridge, Massachusetts, USA

Joe H. Scott
Pyrologix LLC, Missoula, Montana, USA

T. Sheehan
Conservation Biology Institute, Corvallis, Oregon, USA

Tarung Singh
Department of Mechanical and Aerospace Engineering, University at Buffalo, SUNY, Buffalo, New York, USA

Puneet Singla
Department of Mechanical and Aerospace Engineering, University at Buffalo, SUNY, Buffalo, New York, USA

Catherine Slesnick
Charles Stark Draper Laboratory, Cambridge, Massachusetts, USA

Mikhail Sofiev
Finnish Meteorological Institute (FMI), Helsinki, Finland

Kerstin Stebel
Norwegian Institute for Air Research (NILU), Kjeller, Norway

E. Ramona Stefanescu
Department of Mechanical and Aerospace Engineering, University at Buffalo, SUNY, Buffalo, New York, USA

Andreas Stohl
Norwegian Institute for Air Research (NILU), Kjeller, Norway

Cathelijne Stoof
Wageningen University, Wageningen, Netherlands

Robert Sulpizio
Dipartimento di Scienze della Terra e Geoambientali, Università di Bari, Bari, Italy, and Istituto per la Dinamica dei Processi Ambientali, Consiglio Nazionale delle Ricerche, IDPA-CNR, Milano, Italy

Adam J. Terando
US Geological Survey, Southeast Climate Science Center, Raleigh, North Carolina, and Department of Applied Ecology, North Carolina State University, Raleigh, North Carolina, USA

Matthew Thompson
Rocky Mountain Research Station, US Forest Service, Missoula, Montana, USA

Pablo Tierz
Istituto Nazionale di Geofisica e Vulcanologia, INGV, Sezione di Bologna, Bologna, Italy

Anne Tillery
US Geological Survey, Albuquerque, New Mexico, USA

Rosario Vázquez
Centro de Geociencias, UNAM, Querétaro, Mexico

Julius Vira
Finnish Meteorological Institute (FMI), Helsinki, Finland

Jun Wang
Department of Earth and Atmospheric Sciences, University of Nebraska, Lincoln, Nebraska, USA, *and Now at* Center for Global and Regional Environmental Research, and Department of Chemical and Biochemical Engineering, University of Iowa, Iowa City, Iowa, USA

Jord J. Warmink
Department of Water Engineering and Management,
University of Twente, Enschede, Netherlands

Peter Webley
Geophysical Institute, University of Alaska Fairbanks,
Fairbanks, Alaska, and Volcanic Ash Detection,
Avoidance and Preparedness for Transportation
(V-ADAPT), Inc., Fairbanks, Alaska, USA

Gerhard Wotawa
Central Institute for Meteorology and Geodynamics
(ZAMG), Vienna, Austria

Yun Yue
Department of Earth and Atmospheric Sciences,
University of Nebraska, Lincoln,
Nebraska, USA

Lucia Zaccarelli
Istituto Nazionale di Geofisica e Vulcanologia,
INGV, Sezione di Bologna,
Bologna, Italy

1

Uncertainty in Natural Hazards, Modeling and Decision Support: An Introduction to This Volume

Karin Riley,[1] Matthew Thompson,[1] Peter Webley,[2] and Kevin D. Hyde[3]

1.1. INTRODUCTION

Modeling has been used to characterize and map natural hazards and hazard susceptibility for decades. Uncertainties are pervasive in natural hazards analysis, including a limited ability to predict where and when extreme events will occur, with what consequences, and driven by what contributing factors. Modeling efforts are challenged by the intrinsic variability of natural and human systems, missing or erroneous data, parametric uncertainty, model-based or structural uncertainty, and knowledge gaps, among other factors. Further, scientists and engineers must translate these uncertainties to inform policy decision making, which entails its own set of uncertainties regarding valuation, understanding limitations, societal preferences, and cost-benefit analysis. Thus, it is crucial to develop robust and meaningful approaches to characterize and communicate uncertainties.

Only recently have researchers begun to systematically characterize and quantify uncertainty in the modeling of natural hazards. Many factors drive the emergence of these capabilities, such as technological advances through increased computational power and conceptual development of the nature and complexity of uncertainty. These advances, along with increased sophistication in uncertainty analysis and modeling, are currently enabling the use of probabilistic simulation modeling, new methods that use observational data to constrain the modeling approaches used, and other quantitative techniques in the subdisciplines of natural hazards. In turn, these advances are allowing assessments of uncertainty that may not have been possible in the past.

Given the expanding vulnerability of human populations and natural systems, management professionals are ever more frequently called upon to apply natural hazard modeling in decision support. When scientists enter into predictive services, they share professional, moral, legal, and ethical responsibilities to account for the uncertainties inherent in predictions. Where hazard predictions are flawed, limited resources may be unjustifiably be spent in the wrong locations, property may be lost, already stressed ecosystems may be critically damaged, and potentially avoidable loss of human life may occur. These essential concerns for reliable decision support compel thorough characterization of the uncertainties inherent in predictive models.

1.2. ORIGINS AND OBJECTIVES OF THIS VOLUME

This volume is an outcome of the 2013 American Geophysical Union (AGU) Fall Meeting session entitled "Uncertainty in Natural Hazard Assessment: Volcanoes, Earthquakes, Wildfires, and Weather Phenomena," which was a combination of two AGU Focus Group Sections: Natural Hazards and Volcanology/Geochemistry/Petrology. The session was inspired in part by the AGU SWIRL program, which encourages interdisciplinary research. In 2013, the SWIRL program offered a theme "Characterizing Uncertainty." In the session, researchers from volcanology, wildfire, landslide analysis, and other fields were brought together to compare results in characterizing uncertainties and developing methods for spatial and temporal understanding of event probability.

[1] *Rocky Mountain Research Station, US Forest Service, Missoula, Montana, USA*

[2] *Geophysical Institute, University of Alaska Fairbanks, Fairbanks, Alaska, USA*

[3] *University of Wyoming, Laramie, Wyoming, USA*

Natural Hazard Uncertainty Assessment: Modeling and Decision Support, Geophysical Monograph 223, First Edition.
Edited by Karin Riley, Peter Webley, and Matthew Thompson.

This monograph focuses largely on the work presented at this AGU session, as well as other presentations from across the 2013 AGU fall meeting that had a focus associated with the AGU SWIRL theme, "Characterizing Uncertainty."

The principal objectives of this monograph are to provide breadth in terms of the types of natural hazards analyzed, to provide depth of analysis for each type of natural hazard in terms of varying disciplinary perspectives, and to examine emerging techniques in detail. As a result, the volume is largely application focused and targeted, with an emphasis on assorted tools and techniques to address various sources of uncertainty. An additional emphasis area includes analyzing the impacts of climate change on natural hazard processes and outcomes. We chose studies from various continents to highlight the global relevance of this work in mitigating hazards to human life and other natural and socioeconomic values at risk. In assembling studies across types of natural hazards, we illuminate methodologies that currently cross subdisciplines, and identify possibilities for novel applications of current methodologies in new disciplines.

To our knowledge, this volume is unique in that it brings together scientists from across the full breadth of the AGU scientific community, including those in real-time analysis of natural hazards and those in the natural science research community. Taken together, the chapters provide documentation of the common themes that cross these disciplines, allowing members of the AGU and broader natural hazards communities to learn from each other and build a more connected network.

We hope this will be a useful resource for those interested in current work on uncertainty classification and quantification and that it will encourage information exchange regarding characterization of uncertainty across disciplines in the natural and social sciences and will generally benefit the wider scientific community. While the work does not exhaustively address every possible type of hazard or analysis method, it provides a survey of emerging techniques in assessment of uncertainty in natural hazard modeling, and is a starting point for application of novel techniques across disciplines.

1.3. STRUCTURE

The remainder of this chapter introduces the contents of each part and chapter, and then distills emergent themes for techniques and perspectives that span the range of natural hazards studied. The monograph is composed of three main parts: (1) Uncertainty, Communication, and Decision Support (4 chapters); (2) Geological Hazards (7 chapters); and (3) Biophysical and Climatic Hazards (10 chapters). Specific types of natural hazards analyzed include volcanoes, earthquakes, landslides, wildfires, storms, and nested disturbance events such as postfire debris flows.

1.3.1. Part I: Uncertainty, Communication, and Decision Support

Here we provide a broad, cross-disciplinary overview of issues relating to uncertainty characterization, uncertainty communication, and decision support. Whereas most chapters in the subsequent two sections address specific quantitative analysis and modeling techniques, we begin with more qualitative concerns. We address questions related to various facets of uncertainty, introduce some basic tenets of uncertainty analysis, discuss challenges of clear communication across disciplines, and contemplate the role of uncertainty assessment in decision processes as well as at the science-policy interface.

In Chapter 2, Thompson and Warmink provide an overarching framework for identifying and classifying uncertainties. While they focus on uncertainty analysis in the context of modeling, the basic framework can be expanded to consider sources of uncertainty across the stages of decision making and risk management. While other typologies and frameworks exist and may be more suitable to a specific domain, the main point is the importance of beginning with the transparent and systematic identification of uncertainties to guide subsequent modeling and decision processes.

In Chapter 3, Rauser and Geppert focus on the problem of communicating uncertainty between disciplines. The authors bring to bear perspectives from the Earth system science community, leveraging insights from a series of workshops and conferences focused on understanding and interpreting uncertainty. Like natural hazards analysis, the field of Earth system science integrates a wide range of scientific disciplines, and so lessons on developing a common language of uncertainty across disciplines are highly relevant. As with *Thompson and Warmink* [Chapter 2, this volume], the authors stress the importance of being clear and explicit regarding the types and characteristics of uncertainties faced.

Last, *Thompson et al.* [Chapter 4, this volume] and *Webley* [Chapter 5, this volume] provide examples of operational decision support systems that incorporate uncertainty and probability. Thompson and coauthors focus on the context of wildfire incident management, and discuss the use of stochastic fire simulation to generate probabilistic information on possible fire spread, how this information can facilitate strategic and tactical decision making, and future directions for risk-based wildfire decision support. Webley focuses on the context of volcanic-ash cloud dispersal, and discusses the different types of uncertainties that play a role in assessing volcanic-ash hazard, and how the research community is working with operational volcanic-ash advisory groups to improve decision making and the application of probabilistic modeling in the real-time hazard assessment and development of ash advisories for the aviation community.

In summary, this opening section introduces a framework for identifying and classifying uncertainty, presents insights on communicating uncertainty across disciplines, and illustrates two current examples of operational decision support systems that incorporate uncertainty and probability. Consider using this section as a lens through which to view subsequent sections: for example, examine the types of uncertainties the authors address, how the authors characterize the uncertainties, how they describe and communicate the uncertainties, how they tailor their analysis to match the uncertainties faced, and the types of decisions the analyses might support.

1.3.2. Part II: Geological Hazards

The type, size, and magnitude of hazards from geological processes are highly variable. Therefore, operational organizations as well as research scientists need to be able to classify the uncertainty and quantify the potential range of possible scenarios for hazard magnitude and timing. Being able to quantify the uncertainty can then lead to increasing confidence in the assessment of hazards and reducing the risk of exposure to hazards. The chapters in this section cover work currently occurring in uncertainty quantification of volcanic, earthquake, and landslide processes. Assessment of geological hazards is further complicated by the fact that they are sometimes nested. For example, a volcanic eruption may spawn lahars. Chapters in this section address topics regarding natural hazard patterns in both space and time, the role of physical and probabilistic analyses for forecasting and risk assessment, and novel methods for event early warning and response.

Webley et al. [Chapter 6, this volume] focus on developing a volcanic-ash dispersion modeling framework that accounts for uncertainties in the initial source parameters and variability in the numerical weather prediction (NWP) data used for the ash dispersal. The authors illustrate that in building a probabilistic approach, where a one-dimensional plume model is coupled to a Lagrangian ash-dispersion model, the uncertainty in the downwind ash concentrations can be quantified. Outputs include estimates of the mean ash concentrations, column mass loading, and ash fallout, along with the probability of mass loading or concentration exceeding a defined threshold. Comparison of their probabilistic modeling with observational data further constrains the uncertainties. The authors identify the need for new research projects to work with the end users to ensure that products are developed for transition to the operational environment.

Gong et al. [Chapter 7, this volume] highlight the uncertainties in estimating the magma source beneath volcanoes using spaceborne Interferometric Synthetic Aperture Radar (InSAR) measurements. The authors use InSAR data to estimate the volcanic source parameters, and subsequently illustrate how the accuracy in the inversion method is influenced by radar phase measurements. By using a Mogi source model approach, they discuss how different components of the InSAR deformation measurements, such as topography, orbital location, decorrelation, and tropospheric variability, can impact the estimation of the volcanic source parameter, such as magma storage depth and change in volume with time. When several parameters can be constrained, such as magma compressibility and topographical variability, the accuracy of estimates of magma depth and volume over time can be improved and increase our understanding of the volcanic system.

Kristiansen et al. [Chapter 8, this volume] focus on uncertainties that exist in volcanic emission clouds, including both ash and sulfur dioxide concentrations. Observational data are used to constrain the source terms using inversion modeling approaches, data assimilation, and ensemble modeling (consisting of inputs, numerical weather prediction [NWP], and multimodel ensemble approaches). One eruption, Grimsvötn volcano in 2011, is used as a case study to illustrate how an integrated approach that couples modeling with observations and compares multiple dispersion models can reduce uncertainties in the downwind volcanic emissions and increase confidence in forecasts for use in real-time hazard assessment.

Tierz et al. [Chapter 9, this volume] continue the focus on volcanoes, assessing the uncertainty in pyroclastic density currents (PDCs) through simulated modeling. A Monte Carlo modeling approach is applied to a study site on Mt. Vesuvius, Italy, to assess which parameters have the greatest influence on the simulated PDCs. The analysis specifies the different uncertainties that exist in modeling PDCs and quantifies their impact on the PDC simulations and predictability. Results demonstrate that the theoretical uncertainties in the Monte Carlo modeling outweigh, by up to a factor of 100, the uncertainties in the initial observations that drive the model.

Kang and Kim [Chapter 10, this volume] estimate losses from an earthquake using a site classification map, and demonstrate how improved knowledge of local site conditions can reduce uncertainty in predicting losses from future earthquakes. The authors present a new earthquake hazard classification map for different regions in South Korea, which enables them to better constrain the impacts of the underlying soil and ground structure to local buildings, thus producing improved estimates of potential loss from different earthquake scenarios. Impact to infrastructure ranging from residential buildings to essential facilities such as hospitals, schools, and fire stations is estimated. The authors discuss how the improved site classification map could be used in decision making and provide more reliable estimates of earthquake loss in developing emergency plans.

Anderson et al. [Chapter 11, this volume] assess how different preprocessing techniques applied to digital elevation models (DEMs) could influence the delineation of debris flow inundation hazard zones. Results show that use of globally applicable DEMs and specific preprocessing techniques can impact the accuracy of the extent of the debris flows and reinforces need for DEMs with higher spatial resolution. Errors in the processed global DEMs propagate into the lahar modeling, leading to inaccurate debris flow maps, and thus reduce confidence in the modeling needed for critical decision making. The authors propose that continued conversations with the end users of the modeling are needed so end users can better understand the limitations of the modeling and potential errors in the debris flow maps.

The final chapter in the section comes from *Caballero et al.* [Chapter 12, this volume] who focus on evaluating lahar simulation flow modeling for two active volcanoes in Mexico. The authors analyze the impact of input parameter selection on the model's capability to match the observations of a real world lahar. While the results illustrate that the approach is an excellent tool for lahar modeling, several inputs (such as input hydrograph and rheologic coefficients) can have a significant impact on the spatial and temporal accuracy of the simulations as well as the predicted magnitude of the lahar. Retrospective analysis of well-studied volcanoes can be used to constrain the uncertainties in these input parameters, but there is a need to specifically improve the rheologic coefficient measurements or at least better understand how the variability impacts the modeling results.

In summary, the chapters in this section focus on prediction of volcanic ash clouds, using deformation to estimate changing volcanic magma sources, earthquake loss estimations, modeling of pyroclastic density currents, lahars, and debris flows. Each chapter highlights the uncertainties that can impact modeling of these geologic hazards and the need for observational data to both constrain the uncertainties and potentially be used with inversion methods to initialize future simulations.

1.3.3. Part III: Biophysical and Climatic Hazards

This part focuses on advancements in uncertainty and risk assessment for natural hazards driven by biological, physical, and climatic factors. Similar to the previous section, chapters address a variety of topical issues related to understanding and forecasting hazards across spatiotemporal scales, germane to both research and management communities. Methods range from ensemble forecasting to scenario analysis to formal quantification of parameter uncertainty, among others. A key theme in this section is the consideration of future climatic conditions and their relationship to natural hazard processes.

In the first chapter, *Riley and Thompson* [Chapter 13, this volume] systematically identify and classify model-based uncertainties in current wildfire modeling approaches, in order to contribute to understanding, evaluation, and effective decision-making support. For each source of uncertainty identified, their analysis characterizes the nature (limited knowledge or variability), where it manifests in the modeling process, and level on a scale from total determinism to total ignorance. Uncertainty compounds and magnifies as the time frame of the modeling effort increases from the incident level to the 10 yr planning period to a 50 yr period, during which climate change must be incorporated into analyses.

Ichoku et al. [Chapter 14, this volume] evaluate the implications of measuring emissions from fires using satellite imagery of different resolutions from various remote sensing platforms. Their methodology includes a literature review and meta-analysis of the uncertainty ranges of various fire and smoke variables derived from satellite imagery, including area burned, flaming versus smoldering combustion, and smoke constituents. Findings indicate that as satellite resolution decreases, uncertainty increases. The authors note that most of the variables are observed at suboptimal spatial and temporal resolutions, since the majority of fires are smaller than the spatial resolution of the satellites, resulting in inaccuracy in estimation of burned area and fire radiative power. Discrepancies are smaller where satellite observations are more complete. As a result of this study, the authors recommend further research that combines ground-based, airborne, and satellite measurements with modeling in order to reduce uncertainty.

Kennedy and McKenzie [Chapter 15, this volume] couple a regional GIS-based hydroecologic simulation system with a new fire model, at a level of aggregation and process detail commensurate with the inputs. The new fire model (WMFire_beta) expands the exogenously constrained dynamic percolation (ECDF) model by varying the probability of fire spread from a burning cell to each of its orthogonal neighbors based on vegetation, weather, and topographic parameters. The authors utilize fractal dimension (complexity of the fire perimeter) and lacunarity (measure of unburned space within the fire perimeter) to assess which combinations of the model parameters produced a run with similar characteristics to the Tripod Fire in Washington state. Findings indicate that the model is not sensitive to fuel moisture, meaning that the equation for assigning it was not sufficient to capture the role of fuel moisture. This methodology enables the authors to falsify or verify components of model structure, and suggests development of a new representation that would improve the model and reduce uncertainty.

Terando et al. [Chapter 16, this volume] project changes in the frequency of extreme monthly area burned by wildfires for a study area in coastal Georgia at the end of the 21st century. A statistical model based on aggregated monthly area burned from 1966 to 2010 is used to predict the number of months with extreme area burned under future climate conditions. Uncertainty in future climate is addressed by the use of ensemble datasets, which weight the contribution of each general circulation model (GCM) based on performance during the recent historic period. Sources of uncertainty include variation in outputs from GCMs, the effects of different methods for weighting GCM outputs in ensemble models, sparse observations of months with extreme area burned, possible future changes in fuel characteristics over time, and changes in fire suppression actions. By the end of the 21st century, the model indicates increased probability of more frequent months with extreme area burned, likely due to longer and hotter wildfire seasons. However, 95% projection intervals for the three emissions scenarios all span zero. The authors conclude that while there is large uncertainty in these projections, the results give a more informative depiction of the current state of knowledge, and suggest that the increase in the projection of number of months with extreme area burned indicates the need for future large damages to be considered in risk assessments.

Bachelet et al. [Chapter 17, this volume] utilize the dynamic global vegetation model MC2 with fire enabled in order to simulate vegetation distribution and carbon storage under a suite of climate futures from the Coupled Model Intercomparison Project 5 (CMIP5), including two greenhouse gas concentration trajectories and the outputs of 20 general circulation models. All models predict a warmer future (although the magnitude varies), but large differences exist in magnitude and seasonality of precipitation. Large shifts in vegetation toward warmer types are predicted by the results (e.g., temperate to subtropical forest), with the shifts sometimes being rapid because they are driven by fire. Some results are not intuitive, for example, area burned was episodically larger under the lower greenhouse gas emission trajectory (RCP 4.5 versus 8.5) because the milder climate promotes fuel buildup under lower drought stress, leading to subsequently larger area burned. Uncertainties in projected area burned and areal extent of vegetation type are large across general circulation models and greenhouse gas trajectories (for example, change in areal extent of deciduous forest ranged from −95% to 1453% from the time periods 1972–2000 to 2071–2100).

Le Page [Chapter 18, this volume] examines the sensitivity of fires to climate, vegetation, and anthropogenic variables in the Human-Earth System FIRE (HESFIRE) model, a global fire model, which runs at 1 degree grid resolution. Because HESFIRE includes a suite of variables including climate (e.g., ignition probability from lightning, relative humidity) and human variables (e.g., ignition probability from land use, fragmentation), it has the potential to project fire activity under future climate or societal scenarios. In this study, a set of model parameters were varied in order to evaluate model sensitivity. In addition, the study evaluates the sensitivity of the model outputs to alternative input data (two land cover datasets and two climate datasets). Results indicate that the model is most sensitive to fuel limitation in arid and semiarid ecosystems, with sensitivity to landscape fragmentation being dominant in most grasslands and savannahs. Use of alternative climate and land cover datasets produces changes in projected area burned as large as 2.1 times. Le Page concludes that model evaluations should include sensitivity analyses, as well as investigations of how models represent fundamental aspects of fire ecology, in order to characterize model performance and uncertainties.

Hyde et al. [Chapter 19, this volume] review the current status of debris flow prediction following wildfires, and present a conceptual model of the general sequence of conditions and processes leading to these hazardous events. Six components constitute the postfire debris-flow hazard cascade: biophysical setting, fire processes, fire effects, rainfall, debris flow, and values at risk. Current knowledge and predictive capabilities vary between these components, and no single model or prediction approach exists with capacity to link the sequence of events in the postfire debris-flow hazard cascade. Defining and quantifying uncertainties in predicting postfire debris flows requires addressing knowledge gaps, resolving process contradictions, and conducting new research to develop a comprehensive prediction system.

Haas et al. [Chapter 20, this volume] couple two wildland fire models with a debris flow prediction model to assess which watersheds on a landscape in New Mexico, USA, are most susceptible to a combination of moderate-to-high severity fire followed by debris flows. The methodology allows for prefire estimation of the probability of a postfire debris flow based on a storm of a set of certain recurrence intervals and a set of simulated fire events. A primary innovation of the approach is an improved ability to capture variability surrounding the size, shape, and location of fire perimeters with respect to watershed boundaries. The authors note that identifying watersheds with highest probability and volume of postfire debris flows could assist land managers in evaluating potential mitigation measures such as fuel reduction treatments or retention dams.

Nikolopoulos et al. (Chapter 21) investigate implications of using thresholds in storm intensity and duration recorded at rain gauges to predict debris flows, which often occur at remote locations away from the sparse network of rain gauges. The study utilizes radar data of 10 storms in northern Italy that spawned 82 debris flows. Several different spatial interpolation methods were used to estimate rainfall at remote debris flow locations. All methods underpredict rainfall at debris flow locations, due in part to the localized nature of the rainfall, and in part to statistical properties of interpolation methods. In this study, uncertainty results largely from sparse data, and the authors suggest that uncertainty in predicting debris flows might be reduced by using radar data to refine models that predict debris flow occurrence.

Markuzon et al. [Chapter 22, this volume] address the limitations of using precipitation measurements to predict landslides and test the effect of using a combination of precipitation and longer-term atmospheric conditions (i.e., temperature, atmospheric pressure, and winds) on landslide predictions. Current methods rely on often limited and faulty precipitation data and coarse-scale precipitation estimates, such as those derived from general circulation models, that result in smoothed rainfall estimates and underprediction of landslide events. The need to forecast landslide probability under different climate change scenarios requires new methods to overcome the effects of uncertainty in precipitation measurements and estimations on the accuracy of landslide predictions. The authors demonstrate that a combination of antecedent and concurrent weather conditions effectively detects and predicts landslide activity and therefore can be used to estimate changes in landslide activity relative to changing climate patterns.

In summary, the chapters in this section address an assortment of biophysical and climatic hazards, ranging from wildfire to precipitation-induced landslides to coupled hazards such as postfire debris flows. Chapters acknowledge uncertainty present in natural hazard prediction for the current time period, often due to sparse data, with uncertainty compounding as climate change is expected to produce alterations in vegetation and disturbance regimes. The capability to assess uncertainty varies across different disciplines based on the current state of knowledge. Collectively, this section reveals knowledge gaps in modeling of biophysical and climatic hazards.

1.4. A SYNTHESIS: LEARNING FROM THIS MONOGRAPH

This overview of the chapters illustrates how natural hazard sciences and modeling efforts vary across multiple dimensions in terms of data availability, sufficiency, and spatiotemporal scale, the relative amount and cumulative expertise of scientists working in each field, and the state of the science in quantitative uncertainty assessment. Chapters vary widely in content and focus, as well as how each set of authors characterizes, quantifies, and assesses uncertainty. The intended applications of individual chapters also vary, ranging from informing future research [e.g., *Ichoku et al.*, Chapter 14, this volume] to informing decision making and land management [e.g., *Haas et al.*, Chapter 20, this volume] among others. These observations speak as strongly as anything to the state of uncertainty science: assessment of uncertainty is robust in some areas and arguably nascent or even nonexistent in others. As *Hyde et al.* [Chapter 19, this volume] offered in their assessment of postfire debris flow hazards, where information and methods are not consistent, there can be no comprehensive assessment of uncertainty. Synthesis of these similarities and differences in scope and state of the science across chapters in this monograph highlights important synergies and opportunities for cross-hazard collaboration and learning. See Tables 1.1 and 1.2 for summaries of themes, techniques, and methods.

In compiling this volume, we learned that systematic identification of sources of uncertainty is a research endeavor in its own right; we present a framework that can be used in natural hazards [*Thompson and Warmink*, Chapter 2, this volume], and give an example for wildfire modeling [*Riley and Thompson*, Chapter 13, this volume]. Because of the resources required to perform systematic identification of uncertainty, the effort to do so can be prohibitive to researchers in the natural hazards, since limited resources are often available to produce natural hazard prediction models and outputs. In addition, once sources of uncertainty have been identified, quantifying each often requires further research. Because quantification of many sources of uncertainty has not been undertaken in many disciplines, it's challenging at this point for researchers to assess how the combination of multiple uncertainties might affect their model projections. As the body of knowledge on uncertainty grows, this task should become easier. We hope this monograph is a step toward accomplishing that goal.

Uncertainty can be broadly classified into two natures: knowledge and variability. Knowledge uncertainty can be reduced by further research, for example, by improving input data [e.g., *Webley et al.*, Chapter 6, this volume and *Kristiansen et al.*, Chapter 8, this volume]. Variability uncertainty is based on inherent variability in a system, and can't be eliminated, with weather and future climate being a recurring example [e.g., *Bachelet et al.*, Chapter 17, this volume]. Uncertainty manifests at different locations in the modeling process, for example in inputs or model structure. Understanding the nature and location of uncertainty can help researchers choose methods for addressing uncertainty.

We observe that the language used to describe uncertainty is still nascent. The term "uncertainty" itself is often used somewhat generically, without an attempt to define or classify it. Throughout the monograph, the use of terms (for example "structural uncertainty," which appears in both

Table 1.1 Emergent Themes Based on Synthesis of This Volume

Uncertainty is often due to sparse data, for example, due to sparse rain gauge locations [*Nikolopoulos et al.*, Chapter 21, this volume], fire records from documentary or satellite sources [*Terando et al.*, Chapter 16, this volume; *Ichoku et al.*, Chapter 14, this volume], or DEMs [*Anderson et al.*, Chapter 11, this volume]
When predicting a phenomenon, uncertainty can often be fairly well constrained in some aspects but not in others, due to gaps in data and knowledge, as in post-wildfire debris flow prediction [*Hyde et al.*, Chapter 19, this volume]
Uncertainty can stem from future events, for example, climate change effects on vegetation and fire [*Le Page*, Chapter 18, this volume; *Bachelet et al.*, Chapter 17, this volume], climate change effects on area burned [*Terando et al.*, Chapter 16, this volume], post-fire debris flows on a pre-fire landscape [*Haas et al.*, Chapter 20, this volume], uncertainty in fire extent for a given incident due to upcoming weather [*Thompson et al.*, Chapter 4, this volume]
Uncertainty is often driven by a combination of factors, necessitating broad uncertainty analysis of a phenomenon as a first step. Several chapters focused on such broad assessments, including fire modeling [*Riley and Thompson*, Chapter 13, this volume], post-fire debris flows [*Hyde et al.*, Chapter 19, this volume], and volcanic ash clouds [*Webley et al.*, Chapter 6, this volume; *Kristiansen et al.*, Chapter 8, this volume].
Assessing model complexity is important; when linking models, complexity should be proportional to other models. Uncertainty should be possible to assess, which is more difficult with complex models [*Kennedy and McKenzie*, Chapter 15, this volume].
Uncertainty can derive from scale of measurement relative to temporal and spatial scale of phenomena, for example in satellite measurements of fire extent and emissions [*Ichoku et al.*, Chapter 14, this volume] and DEM spatial resolution for debris flow prediction [*Anderson et al.*, Chapter 11, this volume].
Uncertainty is often introduced by input data, and it's important to address the sensitivity of the model to uncertainty in inputs. Many chapters mentioned this, and some addressed it quantitatively [*Terando et al.*, Chapter 16, this volume; *Le Page*, Chapter 18, this volume; *Webley et al.*, Chapter 6, this volume; *Anderson et al.*, Chapter 11, this volume].
Uncertainty can also exist in the model structure, which can be tested by field validation as in *Kennedy and McKenzie* [Chapter 15, this volume] or by varying model coefficients, as in *Le Page* [Chapter 18, this volume].
Observations can be used to constrain uncertainties, and to initialize future modeling, as in *Webley et al.* [Chapter 6, this volume] and *Kristiansen et al.* [Chapter 8, this volume].
Assessment of uncertainty in natural hazards is applicable across spatiotemporal scales. Spatial extents of chapters ranged from study areas in specific locales, such as *Terando et al.* [Chapter 16, this volume] in coastal Georgia, USA, up to global scale, as in *Le Page* [Chapter 18, this volume]. Temporal extents also varied widely, from that of a single natural hazard, such as a single wildfire incident as in *Thompson et al.* [Chapter 4, this volume], up to the expected effects of climate change, as in *Bachelet et al.*'s [Chapter 17, this volume] examination of vegetation and wildfire regimes.
Uncertainty compounds as spatial and temporal scales expand, as in wildfire modeling [*Riley and Thompson*, Chapter 13, this volume].

Table 1.2 Selected Techniques and Methods for Handling Uncertainty in the Natural Hazards

Statistical models, e.g. *Terando et al.* [Chapter 16, this volume]
Monte Carlo simulation, e.g. *Gong et al.* [Chapter 7, this volume], *Tierz et al.* [Chapter 9, this volume]
Process models and dynamic global vegetation models, e.g. *Bachelet et al.* [Chapter 17, this volume]
Spatial techniques
Hazard propagation from cell to cell, e.g. *Kennedy and McKenzie* [Chapter 15, this volume] and *Haas et al.* [Chapter 20, this volume]
Interpolation, e.g. *Nikolopoulos et al.* [Chapter 21, this volume]
Use of remotely sensed data
Satellite imagery, e.g. *Ichoku et al.* [Chapter 14, this volume]
Radar data, e.g. *Nikolopoulos et al.* [Chapter 21, this volume]
Sensitivity testing as a means to quantify uncertainty, especially where uncertainty is high due to large spatiotemporal scales, e.g. *Le Page* [Chapter 18, this volume], *Kang and Kim* [Chapter 10, this volume], and *Anderson et al.* [Chapter 11, this volume]
Probabilistic outputs as a way to convey uncertainty, as in fire models in *Haas et al.* [Chapter 20, this volume] and *Thompson et al.* [Chapter 4, this volume], and volcanic ash clouds as in *Webley et al.* [Chapter 6, this volume]. In fact, probabilistic outputs from fire models and volcanic ash cloud models were markedly similar, raising the question of whether similar techniques might be applicable in other disciplines where they are not yet used.
Scenario planning is a useful approach where uncertainty is high, as for precipitation delivery from storms [*Haas et al.*, Chapter 20, this volume], future climate (e.g. *Terando et al.* [Chapter 16, this volume]), earthquakes (e.g. *Kang and Kim* [Chapter 10, this volume]), and topographic mapping for debris flow generation [*Anderson et al.*, Chapter 11, this volume].
Integrated approaches, for example, comparison of statistical modeling and observational data, e.g. *Webley et al.* [Chapter 6, this volume], *Gong et al.* [Chapter 7, this volume], *Kristiansen et al.* [Chapter 8, this volume], and *Anderson et al.* [Chapter 11, this volume].

Terando et al. [Chapter 16, this volume] and *Riley and Thompson* [Chapter 13, this volume], varies across chapters. We have not attempted to resolve the use of terms, but instead find that it emphasizes the current state of uncertainty sciences, and articulately argues for the need for defining a common terminology, as proposed by *Rauser and Geppert* [Chapter 3, this volume].

This monograph identifies the need for techniques to transition from the research domain to the operational domain for effective decision making, as well as the need to involve end users in developing probabilistic approaches [*Webley et al.*, Chapter 6, this volume]. There is also the need to open discussions between the research community and end users in order to foster understanding of the limitations of the methods used so the data can be applied to make more informed decisions [*Anderson et al.*, Chapter 11, this volume] and to ensure that new approaches can transition into operations. The objective of such collaborative efforts will be to improve confidence in the interpretation of final model simulations and the application of model results for improved decision support.

In closing, we recommend several directions for future work. Uncertainty assessment would benefit from increased attention to systematic identification, classification, and evaluation of uncertainties. In addition, need exists for increased emphasis on clear communication of uncertainties, their impact on modeling efforts, and their impact on decision processes. As natural hazard and modeling efforts become increasingly interdisciplinary, an emphasis on targeting common language and understanding across disciplines becomes necessary. With a number of techniques now available to researchers, uncertainty can be constrained and the confidence in modeling of natural hazards increased. The concept of value of information can be brought to bear to inform decisions across contexts, ranging from determining the merit of investing in long-term research and monitoring to postponing time-pressed decisions to gather more information. This process will entail linking the field of uncertainty analysis with tools and concepts of decision analysis to characterize how and whether reduced uncertainty might influence decisions and outcomes. Last, and related to decision support, future work could consider involvement of end users for real-time hazard assessment during probabilistic workflow development, which will ensure end users can confidently apply new tools and understand derived products.

1.5. CONCLUSION

This volume arguably presents evidence that there is not yet a comprehensive recognition of the need for thorough uncertainty assessment nor consistent approaches to conduct these assessments. Yet, the need for uncertainty assessment has never been greater. Some natural hazards, such as wildfire, are on the rise, and the effect of others, such as landslides, is growing due to expanding human populations. The combined effect may strain already impaired natural resources. Inherent in the predictive services are professional, moral, legal, and ethical responsibilities to account for uncertainties inherent in predictions. The consequences of inaccurate predictions can be high: limited resources may be spent in the wrong locations, property may be lost, and human casualties can occur. We therefore advocate for a coordinated development of the science and practice of uncertainty assessment.

ACKNOWLEDGMENTS

We are grateful for the editorial staff at Wiley Books, including in particular our editor Dr. Rituparna Bose and editorial assistant Mary Grace Hammond. We appreciate the guidance of our Editorial Review Board, including Dr. Rebecca Bendick, Professor of Geosciences, University of Montana; Dr. Anna Klene, Professor of Geography, University of Montana; Dr. Ulrich Kamp, Professor of Geography, University of Montana; and Dr. Helen Dacre, Senior Lecturer, University of Reading. We also thank our anonymous reviewers for their expertise.

Part I
Uncertainty, Communication, and Decision Support

Matthew Thompson
Editor-in-Chief

2

Natural Hazard Modeling and Uncertainty Analysis

Matthew Thompson[1] and Jord J. Warmink[2]

Essentially, all models are wrong, but some are useful.

George E. P. Box

ABSTRACT

Modeling can play a critical role in assessing and mitigating risks posed by natural hazards. These modeling efforts generally aim to characterize the occurrence, intensity, and potential consequences of natural hazards. Uncertainties surrounding the modeling process can have important implications for the development, application, evaluation, and interpretation of models. In this chapter, we focus on the analysis of model-based uncertainties faced in natural hazard modeling and decision support. Uncertainty analysis can help modelers and analysts select appropriate modeling techniques. Further, uncertainty analysis can ensure decision processes are informed and transparent, and can help decision makers define their confidence in model results and evaluate the utility of investing in reducing uncertainty, where feasible. We introduce a framework for identifying and classifying uncertainties, and then provide practical guidance for implementing that framework. We review terminology and offer examples of application to natural hazard modeling, culminating in an abbreviated illustration of uncertainty analysis in the context of wildfire and debris flow modeling. The objective of this brief review is to help readers understand the basics of applied uncertainty theory and its relation to natural hazard modeling and risk assessment.

2.1. INTRODUCTION

Natural hazards can have devastating consequences including the loss of human life and significant socioeconomic and ecological costs. Natural hazards may be isolated events or they may be linked with cascading effects, for instance, debris flows after volcanic eruptions or wildfires. Although often destructive, these hazards are the result of natural processes with a range of potential environmental benefits as well (e.g., groundwater recharge after a flood). It is therefore important for society to be able to better understand, forecast, and balance the risks posed by natural hazards, in order to prepare for and mitigate those risks.

Broadly speaking, risk mitigation strategies can target either the natural hazard itself or the potential consequences. With respect to the former, reducing the likelihood or intensity of the hazard itself is only a feasible option in select cases, as in the case of wildfires, through preventing human-caused ignitions, manipulating fuel conditions, and increasing firefighting suppression capacity. With respect to the latter, reducing vulnerability is a more universally applicable mitigation strategy, which entails both reducing exposure through, for example, zoning to restrict development in hazard-prone areas and reducing susceptibility to loss through construction practices.

[1] *Rocky Mountain Research Station, US Forest Service, Missoula, Montana, USA*

[2] *Department of Water Engineering and Management, University of Twente, Enschede, Netherlands*

Natural Hazard Uncertainty Assessment: Modeling and Decision Support, Geophysical Monograph 223, First Edition.
Edited by Karin Riley, Peter Webley, and Matthew Thompson.

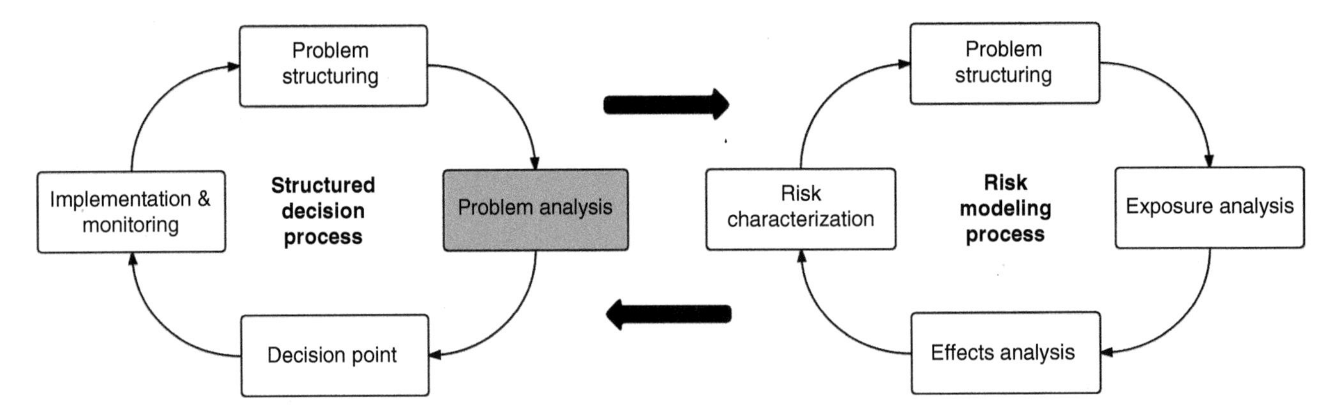

Figure 2.1 The four primary stages of a structured decision-making process and their relation to the four primary stages of a risk modeling process. Figure modified from *Marcot et al.* [2012], *Ascough II et al.* [2008], and *US Environmental Protection Agency* [1992].

The implementation of actions to manage risks from natural hazards begins with a decision process. The decision process may be formal or informal, and can span a range of decision makers from regulatory agencies to individual homeowners. Modeling can play a critical role in informing these decisions.

Figure 2.1 illustrates a generalized risk management process, and highlights the role of risk modeling in informing decision processes. The decision process has four primary stages: (1) problem structuring, (2) problem analysis, (3) decision point, and (4) implementation and monitoring [*Marcot et al.*, 2012]. In the first stage, the problem context is framed, relevant natural hazards are identified, and objectives and evaluation criteria are defined. In the second stage, risk management options are defined and evaluated, key uncertainties are identified, and potential trade-offs analyzed. In the third stage, a decision for a particular course of action is reached, and, in the last stage, the decision is implemented and monitoring actions may be undertaken.

We highlight the problem analysis stage because it entails the principal natural hazard and risk modeling components and provides the informational basis for evaluating consequences and assessing trade-offs to support decisions. However, uncertainty arises in all stages of the risk management and modeling process and the presented tools are to a large extent also applicable across other stages.

The risk modeling process similarly has four primary stages: (1) problem structuring, (2) exposure analysis, (3) effects analysis, and (4) risk characterization [*U.S. Environmental Protection Agency*, 1992; *Thompson et al.*, 2015]. In the first stage, the modeling objectives, scope of analysis, and assessment endpoints are identified, as are the salient characteristics of the natural hazards being analyzed. Exposure analysis, the second stage, examines the likelihood, intensity, and potential interaction of natural hazards with values at risk. Effects analysis next examines potential consequences as a function of exposure levels, often depicted with dose-response curves. In the risk characterization stage, results are synthesized to provide useful information for the decision process. Implicit in the risk modeling process depicted in Figure 2.1 are the steps of collecting and processing data, developing the conceptual model(s), selecting and applying the model(s), and calibrating and validating results.

Natural hazard modeling efforts generally aim to characterize the occurrence, intensity, and potential consequences of natural hazards. The field is wide ranging and involves a multitude of disciplines including risk analysis, statistics, engineering, and the natural sciences. Part of the reason the field is so broad is that characteristics of natural hazards themselves are broad, in terms of the relevant spatial and temporal scales of analysis, the underlying natural and anthropogenic processes driving hazard dynamics, and the degree of control humans have over those processes. Key modeling questions often relate to the location, timing, duration, and magnitude of hazardous events, as well as their causal pathways, cascading effects, and potential feedbacks on future hazard and risk. A key feature of natural hazard modeling is the reliance on probabilistic and integrated environmental modeling techniques.

Regardless of their scope and complexity, models are still fundamentally an abstraction of reality. This abstraction can have important implications for how models are developed, applied, evaluated, and interpreted. Principal among these concerns are uncertainties surrounding model inputs, the modeling process, and model outputs. Unaddressed or overlooked uncertainties can ultimately lead to ill-informed and inefficient decisions, in the worst case leading to increased hazard and/or vulnerability. Resultantly, an essential component of effective hazard and risk management is the analysis of uncertainties.

As we will describe in this chapter, there are a number of attributes with which uncertainty can be characterized. One important question is where uncertainties originate (i.e., from measurement error or knowledge gaps or modeling approximations or intrinsic system variability). A related question is whether the uncertainty is in some sense reducible through additional research and data collection; intrinsic system variability is considered irreducible [*Rougier et al.*, 2013]. Having a solid understanding of model-based uncertainties is important for a number of reasons. First, decision makers are able to define their level of confidence in model outputs and as a result decision processes are more informed and transparent. Second, decision makers can assess the degree to which uncertainty may affect choice of the best course of action and estimate the value of additional information. Third, decision makers can evaluate options for reducing uncertainty. Where the value of additional information is high, and where this information can be obtained (i.e., the uncertainty is reducible), then investing in additional research and monitoring or adopting an adaptive management approach may be warranted [*Thompson et al.*, 2013]. In turn, knowledge gained from monitoring may be used to update and inform modeling efforts, or could result in a reframing of how the problem is understood and a change in management strategy. Of course, not all forms of uncertainty are reducible, and attempting to reduce all forms of uncertainty may be an inefficient use of resources. It is therefore necessary to systematically assess model-based uncertainties.

In this chapter, we review concepts related to the identification, classification, and evaluation of uncertainties faced in natural hazard modeling and decision support. Our primary objectives are to introduce a formalized framework for analyzing uncertainties, and to provide practical guidance for implementing that framework. We introduce a typology to categorically describe sources of uncertainty along three dimensions, present an "uncertainty matrix" as a graphical tool to illustrate the essential features of the typology, and present a decision tree to facilitate proper application of the uncertainty matrix. We hope this chapter helps readers not just understand but also see how to actually apply uncertainty theory. Throughout our chapter, we build from the broader literature of environmental modeling and risk assessment, in particular from the work of *Walker et al.* [2003], *Refsgaard et al.* [2007], *Ascough II et al.* [2008], *Maier et al.* [2008], *Kwakkel et al.* [2010], *Warmink et al.* [2010], *Skinner et al.* [2014a], and *Skinner et al.* [2014b].

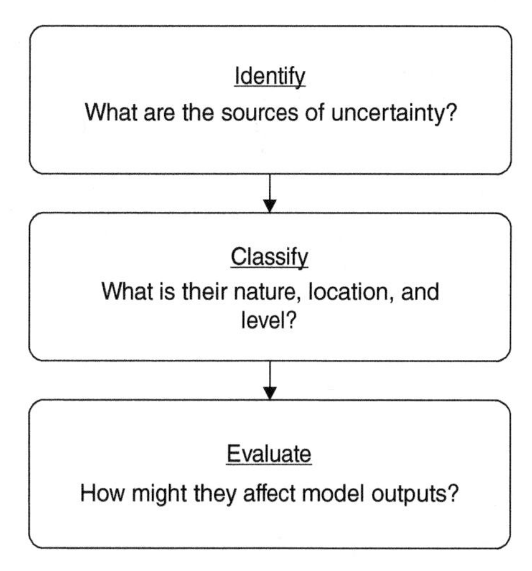

Figure 2.2 Conceptual overview of a three-step process for uncertainty analysis.

2.2. IDENTIFYING AND CLASSIFYING UNCERTAINTIES

Figure 2.2 provides a generalized overview of the steps of uncertainty analysis. The development and evaluation of modeling approaches is iterative in nature and premised on transparently identifying, classifying, and evaluating how uncertainties may influence model results and ultimately decision processes. Uncertainty analysis begins with the identification of potential sources of uncertainty. Having a clear, systematic, and consistent approach to identifying uncertainties can help modelers and analysts identify salient uncertainties. By identifying up front the sources of uncertainties faced, modelers can identify approaches and techniques that might be most suited to the problem at hand. In turn, model evaluation can help identify uncertainties that may be introduced due to the structure and technical implementation of the particular model(s) chosen. Later in this chapter, we will return to the selection of appropriate techniques to evaluate uncertainty.

The identification and classification of uncertainties is often driven by the experience and best judgment of modelers and analysts. An uncertainty typology can be a particularly useful tool to help identify, define, and communicate the important features of uncertainties faced within the specific modeling context. Typologies can help modelers and analysts better understand and differentiate uncertainties faced in the modeling process, and do so in a systematic and consistent fashion. It is critical that typologies are complete and consistent to avoid generation of misleading hazard and risk assessments. *Walker et al.* [2003] classifies uncertainty along three dimensions: (1) the *nature* of the uncertainty (i.e., the underlying cause of how the uncertainty came to exist); (2) the *location* of

Table 2.1 Definitions and Examples of the Nature Dimension of Uncertainty

Natures	Definitions	Examples
Linguistic	Ambiguity, vagueness, contextual dependency, evolving definitions	Definitions and conceptions of "sustainable" and "resilient"
Knowledge	Limitations of scientific understanding (reducible); also called epistemic	Knowledge gaps in understanding of the processes driving volcanic ash dispersal
Variability	Inherent variability of natural and human systems (irreducible); also called aleatory	Weather patterns driving fires and floods
Decision	Social cost-benefit analysis; unknown or inconsistent preferences	How to value a human life

Source: Modified from *Ascough et al.* [2008].

Table 2.2 Definitions and Examples of the Location Dimension of Uncertainty

Locations	Definitions	Examples
Context	Assumptions and choices underlying the modeling process	Spatiotemporal scope of analysis
Input	Data to define or describe relevant characteristics for a specific model run	Measurement error
Model structure	Relationships between variables or model components and underlying system	Relying on an empirical rather than a process-based model
Model technical	Technical and numerical aspects related to algorithmic and software implementation	Trade-offs between resolution and processing time
Parameters	A priori determined values invariant within chosen context and algorithmic representation	Stress drop parameter in earthquake modeling

Source: Modified from *Warmink et al.* [2010].

uncertainty in the modeling or decision process; and (3) the *level* of the uncertainty, along the spectrum from total determinism to total ignorance. It is important to note that there could be multiple classification schemas for each dimension [*Skinner et al.*, 2014a], which may be more or less applicable depending upon the specific context.

To begin, we borrow from *Ascough et al.* [2008], who define four main natures of uncertainty: linguistic, knowledge, variability, and decision (Table 2.1). Linguistic uncertainty relates to ambiguity, vagueness, and contextual dependency of terminology. In fact, the field of uncertainty analysis itself has struggled with using a common lexicon for characterizing uncertainties across scientific disciplines [*Romanowicz and Macdonald*, 2005; *Rauser and Geppert*, Chapter 3, this volume]. Variability (or aleatory) uncertainty is an attribute of reality and refers to the inherent randomness of the natural system and by definition cannot be reduced. It is also referred to as objective uncertainty, external uncertainty, stochastic uncertainty, or random uncertainty [*van Asselt and Rotmans* 2002]. Climate change or weather predictions that drive natural hazards are examples of variability uncertainties. Knowledge (or epistemic) uncertainty refers to the limitation of our knowledge. It can be reduced by improved system understanding due to scientific research or acquiring more data. An example is the main process that is responsible for the dispersion of volcanic ash after an eruption. Scientific research may answer this question, thereby reducing the uncertainty. Decision uncertainty enters the decision-making process after the estimation of risk has been generated. It deals with controversy about valuing social objectives, such as the value of a human life. Decision uncertainty can also refer to ambiguity or multiple equally valid frames of reference [*De Wulf et al.*, 2005], where no single truth exists.

How the location dimension is classified will very much depend on the scope of the uncertainty analysis. As described earlier, if the scope extends across the entire decision process then locations could include uncertainties related to how problems are defined and framed through to logistics of implementation and adaptive management. *Warmink et al.* [2010] define five main locations of model-based uncertainty: context, input, model structure, model technical, and parameters (Table 2.2). Context uncertainty refers to the underlying assumptions of the model, which are choices often made during the selection of a certain type of model. For example, using a two-dimensional or three-dimensional model or using a global circulation model or a regional model to predict weather patterns. Input uncertainties refer to the data that describe the modeling domain and time or space dependent driving forces (e.g., solar radiation). Uncertainties in these data may be caused by measurement errors. Model structure uncertainty refers to the processes in the model that describe the system relations. Using an empirical relation instead of a process-based description may cause model structure uncertainties. Model technical

Table 2.3 Definitions and Examples of the Level Dimension of Uncertainty

Levels	Definitions	Examples
Statistical	Possible to characterize uncertainties probabilistically	Hazard occurrence probability
Scenario	Possible to characterize uncertain outcomes but not their respective probabilities	Future global emissions and climate scenarios
Recognized ignorance	Impossible to outline different possibilities or their outcomes	No-analog vegetation shifts under climate change

Source: Modified from *Walker et al.* [2003] and *Skinner et al.* [2014a].

Table 2.4 Stylized Uncertainty Matrix Illustrating How Sources of Uncertainty Can Be Analyzed According to Each Dimension: Nature, Location, Level

	Nature				Location					Level		
Source	L	K	V	D	C	I	MS	MT	P	St	Sc	RI
Definition of forest resiliency	X				X						X	
Frequency of natural and human-caused ignitions			X			X				X		
Equations predicting fire spread and intensity		X					X				X	
Role of ash in post-fire debris flow initiation		X			X							X
Discretized landscape		X						X			X	
Value of assets and resources impacted by debris flow				X		X				X		

Note: L = linguistic, K = knowledge, V = variability, D = decision, C = Context, I = Input, MS = Model structure, MT = Model technical, P = Parameter, St = Statistical, Sc = Scenario, RI = Recognized ignorance.

uncertainties arise in the technical and numerical implementation. Finally, parameter uncertainties refer to the constants in the model that can have a physical or empirical background. Uncertainty in parameters can be related to model calibration.

Last, the level dimension of uncertainty reflects the variety of distinct levels of knowledge, and is generally broken down according to degree of confidence in probabilities and outcomes. However, these concepts originate from broader risk analysis principles focused on hazardous events, and may be difficult to directly translate to specific analysis of a given source of uncertainty depending upon its nature and location. Thus, the level of any given uncertainty can be highly context dependent. Here we borrow from *Walker et al.* [2003] and *Skinner et al.* [2014a], who define three main levels of uncertainty: statistical, scenario, and recognized ignorance (Table 2.3). Determinism is omitted because there is no uncertainty; total ignorance is similarly omitted since it isn't possible to identify and classify what isn't known.

Table 2.4 provides an abbreviated uncertainty matrix wherein each identified source of uncertainty is classified according to the three dimensions. Our intent is not to comprehensively enumerate all potential sources of uncertainty or all potential combinations of nature/location/level, but rather to illustrate how an uncertainty matrix can be developed. As an example we focus on modeling efforts that assess the potential for wildfires and the subsequent threat of postfire debris flows, the results of which can ultimately inform forest management and risk mitigation planning [e.g., *Tillery et al.*, 2014; *Haas et al.*, Chapter 20, this volume). Beginning with the top row, linguistic uncertainty regarding alternative definitions of forest resiliency could lead to different evaluation criteria and assumptions driving model selection (location = context). The second row indicates that the frequency of lightning-caused ignitions, an input to fire-prediction models, is subject to natural variability that can be characterized statistically. The third row indicates knowledge gaps in the structure of models that predict fire spread and intensity (level = scenario). There are similarly knowledge gaps regarding the role of ash in postfire debris flow initiation (fourth row), which is very poorly understood (level = recognized ignorance), and which likely influences model assumptions (location = context). Next, variability in how vegetation recovers between the fire and storm event can influence calculations of debris flow likelihood. Although these dynamics can be modeled directly, the rate of recovery can also be used as a parameter in longer-term modeling efforts [e.g., *Jones et al.*, 2014] that may take on a range of values.

The next row identifies that discretized representations of the landscape can represent a form of knowledge uncertainty relating to technical model implementation, whose influence can often only be discerned through running the model with different configurations (level = scenario). Last, when quantifying the socioeconomic and ecological consequences of postfire debris flows, an input to risk assessment calculations, how assets and resources are valued is a source of decision uncertainty that can often be characterized statistically through econometric and related techniques. More detailed descriptions and classifications of uncertainties faced in fire and debris flow modeling are available in *Riley and Thompson* (Chapter 13, this volume) and *Hyde et al.* (Chapter 19, this volume).

Although the final outcome may appear simple, the actual population of such a table can be a complex and challenging endeavor. In practice, even experienced analysts and modelers may be unable to identify and classify the entire universe of possible uncertainties for any given context. Nevertheless the generation and evaluation of uncertainty matrices reflect best practices in modeling and uncertainty assessment.

2.3. GUIDANCE FOR IDENTIFYING AND CLASSIFYING UNCERTAINTIES

It is imperative to describe each uncertainty accurately so that it can be uniquely identified across all three dimensions. Ideally, the identification of uncertainties results in a list of unique and complementary uncertainties. Complementary implies that the uncertainties do not overlap, which may result in overestimating of the uncertainty. Unique implies that the uncertainties are comparable, which can be essential to decision making. *Warmink et al.* [2010] defined three decision trees to aid the population of the uncertainty matrix; we modified these decision trees to match our dimensions of uncertainty.

Figure 2.3 presents the decision trees to facilitate identification of the three dimensions of uncertainty. In uncertainty identification practice, the borders of the classes prove to be difficult leading to discussion about the exact classification of an uncertainty. The decision trees are based on strict definitions of the individual classes in the matrix and help to clarify the classification criteria. To identify an uncertainty, we start at the left (nature) tree and try to answer the questions. After following all three trees, the uncertainty has a nature, location, and level and can be uniquely classified.

Each uncertainty is well defined if it fits into a single class in the uncertainty matrix, so it belongs to only one nature, one location, and one dimension. For instance, an uncertainty should never be located in the context and in the model structure, in which case it is poorly defined. In the decision trees, this implies that we need to be able to answer all questions. If we cannot answer a question, this means that the uncertainty is not well defined and needs to be specified by describing it more accurately. One possibility to better specify an uncertainty is to unravel it into two (or more) separate uncertainties, for example one uncertainty in the model structure and one uncertainty in the model context. Then for both uncertainties the identification starts again. This iterative process ultimately results in a list of unique and complementary uncertainties.

Ideally, the list of uncertainties is complete after the identification process. In practice it may not be possible reach this ideal situation, because we will never be able to cover all uncertainties that influence the model outcomes. Expert elicitation methods in combination with the decision trees, however, can likely increase the number of identified uncertainties. Comprehensive and thorough discussion of all possible sources of uncertainty can encourage experts to look beyond their first thoughts and more deeply consider model-based uncertainties including implicit assumptions and other contextual factors. A structured identification of uncertainties results in a better overview of uncertainties, which is an essential first step in uncertainty management.

2.4. TECHNIQUES FOR EVALUATING UNCERTAINTY

There is a rich set of modeling frameworks and uncertainty evaluation techniques that can be used throughout modeling and decision processes [*Matott et al.*, 2009; *Bastin et al.*, 2013]. These techniques can be used to assess potential uncertainty propagation and uncertainty in model outputs, and more generally to identify appropriate modeling approaches given the characteristics of uncertainties faced. *Refsgaard et al.* [2007] identify five groups of uncertainty analysis methodologies that differ according to purpose of use: (1) preliminary identification and characterization of sources of uncertainty, (2) assessment of levels of uncertainty, (3) analysis of uncertainty propagation through models, (4) ranking sources of uncertainty, and (5) reduction of uncertainty. A comprehensive review of all possible techniques along with a mapping to all possible combinations of three-dimensional uncertainty classifications is beyond the scope of this chapter. For instance, to assess the effect of an uncertainty due to model structure with a knowledge nature and scenario level, we can use the scenario analysis technique. However, sensitivity analysis [e.g., *Van der Perk*, 1997] or Monte-Carlo-based methods [e.g., *Pappenberger et al.*, 2006] are also applicable. Even expert elicitation might be used to quantify the uncertainty due to model structure error [*Warmink et al.*, 2011].

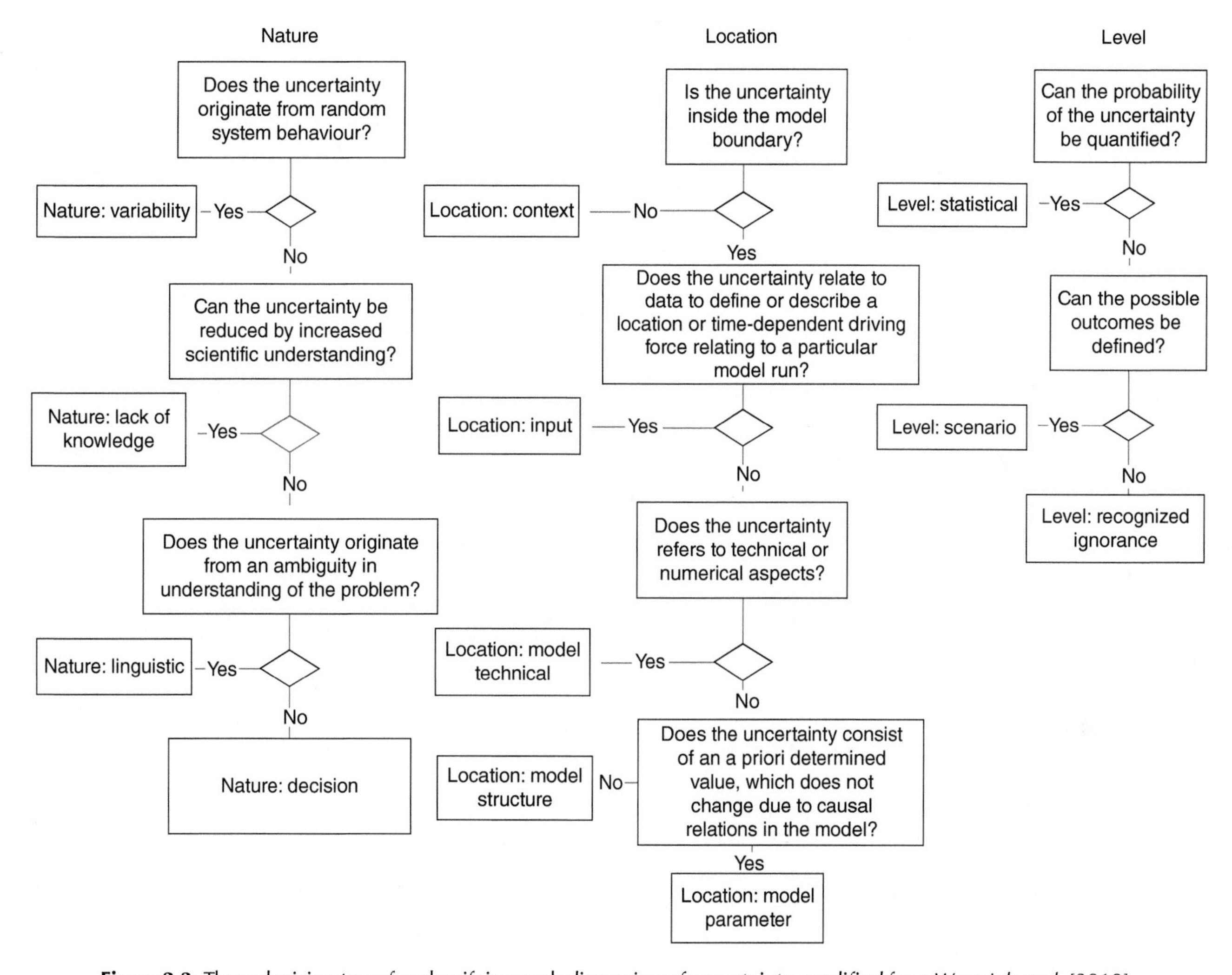

Figure 2.3 Three decision trees for classifying each dimension of uncertainty, modified from *Warmink et al.* [2010].

The advantages of using Monte-Carlo-based methods are that it is objective and uncertainties are quantified. Disadvantages are that the data acquisition and modeling process are time consuming and that only statistical uncertainties are considered. Expert elicitation on the other hand is relatively quick and provides a more comprehensive overview of the uncertainties, because it analyzes all levels and natures of uncertainty. However, its disadvantage is that it is more subjective. Another example is an uncertainty in the model context, where the decision of which input to use to predict a certain risk was not agreed upon. This uncertainty may have a decision nature and a level of (recognized) ignorance. Techniques to manage this uncertainty include scenario analysis [*Refsgaard et al.*, 2007], multicriteria decision analysis, or approaches toward resolving conflicting views [*Brugnach et al.*, 2008].

For each class in the uncertainty matrix, many techniques exist. In general terms, sensitivity analysis, scenario analysis, Monte Carlo simulation, fuzzy logic, expert judgment elicitation, Bayesian belief networks, multicriteria decision analysis, and combinations thereof, are common approaches. The most suitable technique in the end depends on the amount of available data and the required level of detail. More information on these techniques and other approaches to evaluating uncertainty can be found in *van der Sluijs et al.* [2005], *Refsgaard et al.* [2007], *Matott et al.* [2009], *Bastin et al.* [2013], *Thompson et al.* [2013], and *Skinner et al.* [2014a,b]. Here we focused on the identification and classification of uncertainties, using the uncertainty typology and uncertainty matrix as critical initial steps that can offer guidance for the appropriate evaluation and management of uncertainty. Starting with clarity in modeling context and objectives along with a firm understanding of uncertainties faced can go a long way toward selection of appropriate approaches. The selection of a specific uncertainty evaluation technique should link directly with the specific uncertainty, and should incorporate all three dimensions of that uncertainty.

2.5. DISCUSSION

In this chapter, we focused on the application of uncertainty analysis methods and tools to the context of natural hazard modeling. In particular, we introduced three key tools, the uncertainty typology (the three dimensions of uncertainty), the uncertainty matrix (a graphical overview of the essential features of uncertainty), and decision trees (guidance for populating the typology and matrix) and described their relation to identification, classification, and evaluation of uncertainties. We mentioned some of the more common techniques and gave direction as to how to select one to use in the development as well as evaluation of natural hazard modeling.

It is important to recognize that model-based uncertainties are not necessarily the most salient or significant impediment to efficient hazard and risk mitigation. Technical assessments of uncertainty, although necessary, may be insufficient when considering the broader context in which decisions are made [*Brown*, 2010]. That is, uncertainties can influence all stages of the decision process, and, depending upon the context it may be important to comprehensively analyze their characteristics and potential consequences. Uncertainty analyses that focus on the entire decision-making process will necessarily entail a broader set of uncertainties related to human communication, perceptions, and preferences, and may entail a different set of approaches to addressing the respective uncertainties [*Maier et al.*, 2008].

Uncertainty analysis aids in development of a common understanding of modeling efforts, can serve as a starting point for modeling processes, and can clarify the role of modeling in broader decision processes. The frameworks we introduced here for the three dimensions of uncertainty are not intended to be universal, and every context will need to be evaluated in its own right. Identifying and classifying uncertainties facilitates communication between stakeholders, scientists/analysts, and decision makers, and helps prioritize effective ways of managing uncertainties [*Gabbert et al.*, 2010]. Clear communication can become even more important when modeling efforts are interdisciplinary [*Rauser and Geppert,* Chapter 3, this volume], and span modeling domains from hazard likelihood and intensity to consequences and mitigation opportunities.

REFERENCES

Ascough, J. C., II, H. R. Maier, J. K. Ravalico, and M. W. Strudley (2008), Future research challenges for incorporation of uncertainty in environmental and ecological decision-making, *Ecol. Mod.*, *219*(3–4), 383–399.

Bastin, L., D. Cornford, R. Jones, G. B. Heuvelink, E. Pebesma, C. Stasch, S. Nativi, P. Mazzetti, and M. Williams (2013), Managing uncertainty in integrated environmental modelling: The UncertWeb framework, *Environ. Mod. Soft.*, *39*, 116–134.

Brown, J. D. (2010), Prospects for the open treatment of uncertainty in environmental research, *Prog. Phys. Geog.*, *34*(1), 75–100.

Brugnach, M., A. Dewulf, C. Pahl-Wostl, and T. Taillieu (2008), Toward a relational concept of uncertainty: about knowing too little, knowing too differently, and accepting not to know, *Ecol. Soc.*, *13*(2), 30; [online] http://www.ecologyandsociety.org/vol13/iss2/art30/.

Brugnach, M., A. Dewulf, H. J. Henriksen, and P. van der Keur (2011), More is not always better: Coping with ambiguity in natural resources management, *J. Environ. Man.*, *92*(1), 78–84.

DeWulf, A., M. Craps, R. Bouwen, T. Taillieu, and C. Pahl-Wostl (2005), Integrated management of natural resources: Dealing with ambiguous issues, multiple actors and diverging frames, *Water Sci. Tech.*, *52*(6), 115–124.

Gabbert, S., M. van Ittersum, C. Kroeze, S. Stalpers, F. Ewert, and J. A. Olsson (2010), Uncertainty analysis in integrated assessment: the users' perspective, *Reg. Environ. Change*, *10*(2), 131–143.

Jones, O. D., P. Nyman, and G. J. Sheridan (2014), Modelling the effects of fire and rainfall regimes on extreme erosion events in forested landscapes, *Stochas. Environ. Res. Risk Assess.*, *28*, 2015–2025.

Kwakkel, J. H., W. E. Walker, and V.A.W.J. Marchau (2010), Classifying and communicating uncertainties in model-based policy analysis, *Int. J. Tech., Pol. Man.*, *10*(4), 299–315.

Maier, H. R., J. C. Ascough II, M. Wattenbach, C. S. Renschler, W. B. Labiosa, and J. K. Ravalico (2008), Uncertainty in environmental decision making: Issues, challenges, and future directions, in *Environmental Modelling, Software and Decision Support: State of the Art and New Perspectives*, edited by A. J. Jakeman, A. A. Voinov, A. E. Rizzoli, and S. H. Chen, Elsevier, Amsterdam.

Marcot, B. G., M. P. Thompson, M. C. Runge, F. R. Thompson, S. McNulty, D. Cleaves, M. Tomosy, L. A. Fisher, and A. Bliss (2012), Recent advances in applying decision science to managing national forests, *For. Ecol. Man.*, *285*, 123–132.

Matott, L. S., J. E. Babendreier, and S. T. Purucker (2009), Evaluating uncertainty in integrated environmental models, a review of concepts and tools, *Water Resour. Res.*, *45*, W06421

Pappenberger, F., P. Matgen, K. J. Beven, J. -B. Henry, L. Pfister, and P. de Fraipont (2006), Influence of uncertain boundary conditions and model structure on flood inundation predictions, *Adv. Water Resour.*, *29*(10), 1430–1449; doi:10.1016/j.advwatres.2005.11.012.

Refsgaard, J. C., J. P. van der Sluijs, A. L. Højberg, and P. A. Vanrolleghem (2007), Uncertainty in the environmental modelling process, A framework and guidance, *Environ. Mod. Soft.*, *22*, 1543–1556.

Romanowicz, R., and R. Macdonald (2005), Modelling uncertainty and variability in environmental systems, *Acta Geophysica Pol.*, *53*, 401–417.

Rougier, J., S. Sparks, and L. J. Hill, eds. (2013), *Risk and Uncertainty Assessment for Natural Hazards*, Cambridge University Press.

Skinner, D. J., S. A. Rocks, and S. J. Pollard (2014a), A review of uncertainty in environmental risk: characterising potential natures, locations and levels, *J. Risk Res.*, *17*(2), 195–219.

Skinner, D. J., S. A. Rocks, S. J. Pollard, and G. H. Drew (2014b), Identifying uncertainty in environmental risk assessments: The development of a novel typology and its implications for risk characterization, *Human and Ecological Risk Assessment: An International Journal*, *20*(3), 607–640.

Thompson, M. P., B. G. Marcot, F. R. Thompson, S. McNulty, L. A. Fisher, M. C. Runge, D. Cleaves, and M. Tomosy (2013), The science of decision making: Applications for sustainable forest and grassland management in the National Forest System, Gen. Tech. Rep. WO-GTR-88. Washington, DC: U.S. Department of Agriculture, Forest Service.

Thompson, M. P., J. R. Haas, J. W. Gilbertson-Day, J. H. Scott, P. Langowski, E. Bowne, and D. E. Calkin (2015), Development and application of a geospatial wildfire exposure and risk calculation tool, *Environ. Mod. Soft.*, *63*, 61–72.

Tillery, A. C., J. R. Haas, L. W. Miller, J. H. Scott, and M. P. Thompson (2014), Potential Postwildfire Debris-Flow Hazards: A Prewildfire Evaluation for the Sandia and Manzano Mountains and Surrounding Areas, Central New Mexico, US Geological Survey Scientific Investigations Report 2014-5161, http://dx.doi.org/10.3133/sir20145161.

U.S. Environmental Protection Agency (1992), Framework for Ecological Risk Assessment, EPA/630/R-92/001; available online at: http://www2.epa.gov/sites/production/files/2014-11/documents/framework_eco_assessment.pdf.

Van der Perk, M. (1997), Effect of model structure on the accuracy and uncertainty of results from water quality models, *Hydrolog. Processes* *11*(3), 227–239; doi: 10.1002/(SICI)1099-1085(19970315)11:3<227::AID-HYP440>3.0.CO;2-#.

van der Sluijs, J. P., M. Craye, S. Funtowicz, P. Kloprogge, J. Ravetz, and J. Risbey (2005), Combining quantitative and qualitative measures of uncertainty in model-based environmental assessment: The NUSAP system, *Risk Anal.*, *25*(2), 481–492.

van Asselt, M. B., and J. Rotmans (2002), Uncertainty in integrated assessment modelling, *Clim. Change*, *54*(1–2), 75–105.

Walker, W. E., P. Harremoës, J. Rotmans, J. P. van der Sluijs, M. B. A. van Asselt, P. Janssen, and M. P. Krayer von Krauss (2003), Defining uncertainty: A conceptual basis for uncertainty management in model-based decision support, *Integrat. Assess.*, *4*(1), 5–17.

Warmink, J., J. Janssen, M. J. Booij, and M. S. Krol (2010), Identification and classification of uncertainties in the application of environmental models, *Environ. Mod. Soft.*, *25*(12), 1518–1527.

Warmink, J. J., H. Van der Klis, M. J. Booij, and S. J. M. H. Hulscher (2011), Identification and quantification of uncertainties in a hydrodynamic river model using expert opinions, *J. Water Res. Man.*, *25*(2), 601–622; doi: 10.1007/s11269-010-9716-7.

3

Understanding Uncertainty as a Key Interdisciplinary Problem in Earth System Science

Florian Rauser[1] and Gernot Geppert[2]

ABSTRACT

The emerging field of Earth system science integrates strongly disciplinary sciences with a multitude of new interdisciplinary fields in the joined attempt to understand the interactions and properties of the Earth system as a whole. Cross-disciplinary language problems are a natural consequence as every discipline brings with it a multitude of definitions, historical connotations, and interpretations of scientific language. In an attempt to help remedy this situation for Earth system science, the Young Earth System Science community (YESS) has worked in recent years on creating a description of what is meant when we talk of uncertainty in the Earth system sciences. The difference in methodology and quality of what different scientists mean when they refer to uncertainty is large, and in a series of workshops in 2012 and 2014 and an international conference in 2013, YESS has tried to bridge the disciplinary gaps a little. We believe that a consistent uncertainty language treatment within our field of research is a first step to a concrete representation of uncertainty in all spheres of scientific communication, from the disciplinary communication to the communication with stakeholders and the public.

The Earth system science enterprise aims to include all scientific disciplines that are necessary to describe the Earth system and its evolution on a variety of spatial and temporal scales. Depending on those scales, the number of disciplines changes and there is no uniform definition of which disciplines are part of Earth system science. Here, we define all disciplines that are required to describe the evolution of the Earth in the past and the future as part of the Earth system. These disciplines include the natural sciences as well as these social sciences that are concerned with human behavior as a cause or as an effect of environmental phenomena. This is important as for some temporal and spatial scales, the human dimension does not play too much of a role and the problem of uncertainty quantification and communication becomes one-way oriented: from science to the public audience. In our view of Earth system science, this view does not hold, as we are explicitly discussing subsystems of the Earth system, such as climate, that are affected by humans but at the same time affect humans themselves. For the sake of simplicity, we omit the discussion of longer timescales than centuries in this chapter, even though many of the discussed concepts here can also be applied to paleo observations, reconstructions, and modeling efforts. For the parts of the Earth system that influence humans and are influenced by human activities, uncertainty quantification becomes exponentially more complicated because describing the evolution of the system requires tracking the impacts of various uncertainty sources through natural and anthropogenic processes. This process is iterative in nature in the sense that these sources feed back to one another.

Our definition of Earth system science is broader than just the combination of classical disciplines that describe the components of the Earth (i.e., meteorology, oceanography, geophysics, geochemistry, geobiology, etc.). Those disciplines are well established in university structures and education, and share their own history, terminology, and definitions (for example for the term "empirical

[1] *Max Planck Institute for Meteorology, Hamburg, Germany*
[2] *Universität Hamburg, Hamburg, Germany*

Natural Hazard Uncertainty Assessment: Modeling and Decision Support, Geophysical Monograph 223, First Edition.
Edited by Karin Riley, Peter Webley, and Matthew Thompson.

orthogonal functions" in meteorology, which is methodologically indistinguishable from principal component analysis). The disciplinary definitions get less useful when people think of the Earth as a coupled system of different subsystems. But Earth system science also explicitly includes new interdisciplinary scientists who deal with questions that cannot be reduced to one or more subsystems. Interdisciplinary graduate schools around the world try to give a new institutional home to doctoral candidates who deal with questions that cannot be answered in the reference room of specific disciplines. One important aspect of this interdisciplinarity in the Earth system sciences is the mixture of deterministic and stochastic systems, that is, the fact that different components have specific scales in space and time for which this component is predictable. A lot of the uncertainty language problem arises from the fact that people mix different qualities of system predictability and associated uncertainty and, hence, the resulting uncertainty quantification gets very hard to understand.

There are different uncertainty qualities and properties [*Morgan and Henrion*, 1998], and we will not touch upon the details of philosophical interpretations or the conditions to define uncertainty consistently. Instead, we focus on the problem of communicating uncertainty between disciplines. For this purpose, uncertainty characterizes the degree to which we trust, or rather do not trust, a scientific statement, no matter if this statement is theoretical, based on observations, or a model statement. The basic categories of uncertainties that are relevant for us are explicit and implicit uncertainties and reducible and irreducible uncertainties (for a more fine-grained classification see Chapter 2 by Thompson and Warmink). In our connotation, explicit uncertainty is due to sources of imperfection in our understanding and in our description of the Earth system that we are actually aware of. Explicit uncertainty is characterized by the fact that we can describe those imperfections either qualitatively or quantitatively. Explicit uncertainties are a typical feature of observations [see for example *Matthews et al.*, 2013] and arise from instrument errors, for example due to inaccurate calibration or limited measurement resolution. And they arise when raw data are converted into higher-level information, for example when remotely sensed radiation is converted into an average value of surface temperature across several square kilometers. A simple example of explicit uncertainty from the modeling side is the spread between different model simulations that all try to simulate the same physical process. Implicit uncertainty, on the other hand, arises from processes and properties of the Earth system that we do not (yet) know. Implicit uncertainty in models comes from the difference between the model and the potentially infinite amount of processes in reality that we cannot cover with our modeling efforts. In this respect, implicit uncertainty causes differences between all model simulations and the system's true state. Implicit uncertainty in observations may, for example, be due to unrecorded changes in the observational setup, like moving a weather station a few hundred meters. This implicit uncertainty usually cannot be quantified and is colloquially known as the "unknown unknowns" while the explicit uncertainty can be interpreted as "known unknowns."

The second distinction or dimension of uncertainty, that is, reducible versus irreducible, is even more basic in nature and cannot be completely decoupled from the difference between explicit and implicit. Reducible uncertainty in a scientific statement refers to the uncertainty that could in principle be reduced if we had better observations, computers, resources, or theories. Reducible uncertainty is most often connected to the description of the system, including equations, parameters, or properties of the system from which scientific statements are derived. Irreducible uncertainty, on the other hand, refers to the uncertainty in the evolution of an Earth system component that is due to chaotic behavior or (quasi-)random events. It is an inherent property of a system and it describes the degree to which the system's evolution is not predictable or observable, even given perfect understanding of the system and its properties (the discussion that we have here can also be done with a focus on predictability and its boundaries, as is done for example in many papers by Edward Lorenz [see, e.g., *Lorenz*, 2006]). Irreducible uncertainty is therefore closely connected to the stochasticity of the involved components but can also be even more fundamental. A very extreme example would be randomly changing the orbital parameters of the Earth due to heavy meteorite impacts. Such events introduce an "unknown unknown" in the system that cannot be covered in our uncertainty description. Reducible uncertainties often represent practical scientific or technical challenges (more numerous and more accurate observations, better models, increased understanding of interaction), while irreducible uncertainties are structurally impossible to reduce; those uncertainties will remain and need to be included into decision making. Both reducible and irreducible uncertainties should be quantified. Reducible uncertainties should always be explicit (even though in reality there might be implicit uncertainties that can be reduced); irreducible uncertainties can be both explicit and implicit.

To make those qualities of uncertainty easier to grasp, we give a very simple example by posing the following question: what will be the global mean surface temperature on Earth in the year 2100? The projections done with a multitude of models of differing complexities result in a variety of warming for different emission scenarios (see Fig. 3.1). The overall spread as shown by the models is

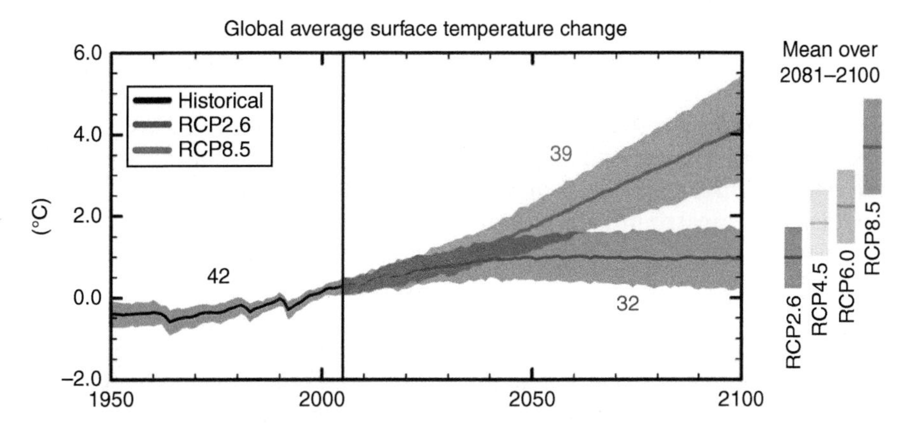

Figure 3.1 Global temperature evolution and associated uncertainties. The simulated evolution of global average surface temperature for the historical period and the projected changes are shown until the year 2100 (data from the coupled model intercomparison project CMIP5, plot from IPCC AR5, Summary for policymakers, *IPCC* [2013]). The overall spread of the projections in 2100 includes scenario (forcing) uncertainty and structural uncertainty (model differences). It is also clear from comparing the historical period 1950–2000 and the projected time period 2000–2100 that some of the variability of the real system is only implicitly accounted for in this multimodel representation.

the explicit uncertainty, not incorporating for example the implicit uncertainties that a big volcano could cool the Earth, or that an epidemic or a war could stop humans from emitting. While those uncertainties are for now implicit, they can in theory be incorporated in a full quantifiable description of the system. Some parts of the uncertainty in the model simulations are due to the fact that we do not know the response of the Earth system to a CO2 doubling in the atmosphere. This explicit and well known uncertainty is reducible up to a certain point because it can in theory be reduced if our understanding of the Earth system improves over time. At the same time, a significant part of the uncertainty of future temperature evolution is irreducible due to the random element of humanity as a whole and the corresponding behavioral effects on the climate. This highlights again that we deal with questions in which not only the influence of the Earth system on humanity matters but also vice versa.

To be able to find a global policy on climate change as a robust decision under uncertainty, policy makers have to better understand the different qualities of uncertainty for all involved disciplines. This means for the involved scientists, that we have to focus on attempts to reduce the reducible uncertainty while at the same time find ways to quantify the remaining irreducible uncertainty in a precise way. We report here on a small part of these attempts; we have organized a series of workshops and conferences on uncertainty within the Young Earth System Scientists community, specifically the Interdisciplinary Conference of Young Earth System Scientists 2013 on Understanding and Interpreting Uncertainty [ICYESS; *Rauser et al.*, 2014]. In a workshop in 2014, we have also worked on a glossary of terms related to uncertainty to define what precisely is meant by these terms. We know that this kind of work is very basic for most experts within the field, but from our experience, the exact distinction between different uncertainties is not done well enough in our field so that outsiders and newcomers in, say, climate impact studies know the difference in uncertainty quantification for precipitation or temperature sufficiently well. The difference in this example is that the largest amount of uncertainty in temperature projections is the irreducible but explicit uncertainty in anthropogenic emissions. Contrarily, the largest amount of uncertainty in precipitation projections, especially regionally, is due to structural model problems, and thus, at least in theory, reducible (structural problems are both formulation problems that cannot simply be avoided by better numerical representation or higher resolution, and representation problems). There is a longstanding discussion in the literature on which types of model problems can be reducible and which can not. More on this topic and the distinction between formulation and representation error can be found in many papers on the treatment of uncertainty in numerical models [e.g., *Oden and Prudhomme*, 2002; *Rauser et al.*, 2015].

Essential ingredients for a coherent treatment of uncertainty across disciplines are the reciprocal understanding of sources of uncertainty and the open communication of assumptions, methods, and processing steps. Only if this understanding is developed, can experiments, be they complex observations or model simulations, be independently examined and repeated to generate scientific consensus. In this manner, we can arrive at an amount of knowledge that yields dependable statements on which policy decisions can be based. Uncertainty quantification has already become common in the form of numerical error ranges, ensemble spreads, or nominal quality indicators. But this kind of information is rarely comparable across different studies, and even less across disciplines. The multi- and interdisciplinary nature of Earth system science, however, requires this comparability and eventually the inclusion and blending of discipline-specific uncertainties.

A number of steps toward a better uncertainty characterization have been taken by the Intergovernmental Panel on Climate Change (IPCC) in a guidance note on uncertainty [*Mastrandrea et al.*, 2010]. In this report, the IPCC has attempted to establish a consistent naming scheme and definition framework to discuss uncertainty for all assessment reports and special reports. The basic idea of the IPCC is to define qualitative confidence levels of scientific statements (from low to high), and only attach quantified uncertainty statements when the confidence is high. To determine if the confidence in a specific statement is low or high, the IPCC has framed the problem as two dimensional, as a question of both evidence and agreement (see Fig. 3.2). Only if there is high agreement in robust evidence, the confidence level is high. It is easy to see that this helps to create a unified language across chapters and topics around the world, but at the same time, the exact interpretation and understanding of robust evidence and high agreement are still difficult to grasp. Further, a mapping of language to quantified uncertainties, as is also done in the IPCC guidance note, is really helpful. It still falls short, though, to solve the philosophical problem of determining what a statement that is more than 95% likely to be true really means for decision makers and scientists. Still, the IPCC calibrated language is now a fundamental starting point for everybody who works on uncertainty language in the Earth system sciences and has influence on all kinds of disciplines through the multidisciplinary and interdisciplinary nature of the IPCC assessment reports, especially those of Working Groups II and III on impacts and mitigation.

In a 2014 YESS workshop, we continued the discussion that was the basis for the IPCC guidance note and we created definitions of some related terms. In doing so (see examples in Appendix at the end of this chapter), we have again been confronted with the different characteristics of uncertainty in different disciplines. We have especially seen that many scientists who work in climate-change-related fields of media reception, policy advice, and similar fields have problems understanding some terms that are of basic importance for Earth system science, such as "feedbacks" or "climate sensitivity." It has become clear that even some basic conceptual terms as "truth" or similar basic methodological terms as "trend" are not easy to define and mean different things for different scientists. While mathematicians tend to see trends as mostly linear metrics of change, many climate scientists consider trends to be long-term effects, and some researchers from the impact community directly interpret an anthropogenic component into the word, which is not necessarily true. It has also become clear that some terms from modeling and observational communities may mean different things in different contexts, especially the key concept of ensembles. In modeling the Earth system, ensembles of different types are used to either show the amount of reducible uncertainty in our ability to model processes or to cover the range of irreducible uncertainty that comes

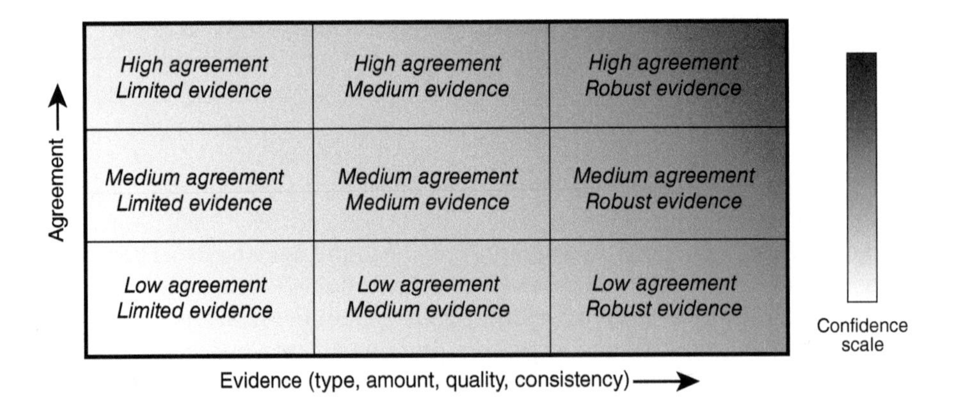

Figure 3.2 IPCC calibrated language: robustness versus agreement. This figure is taken from the IPCC guidance note on uncertainty [*Mastrandrea*, 2010] and depicts "evidence and agreement statements and their relationship to confidence. Confidence increases towards the top-right corner as suggested by the increasing strength of shading. Generally, evidence is most robust when there are multiple, consistent independent lines of high-quality evidence." This quotation and the figure show the strength and weaknesses to define a common calibrated language, as the need to use a unified approach to confidence implies common definitions for a whole set of concepts such as agreement, evidence, and robustness.

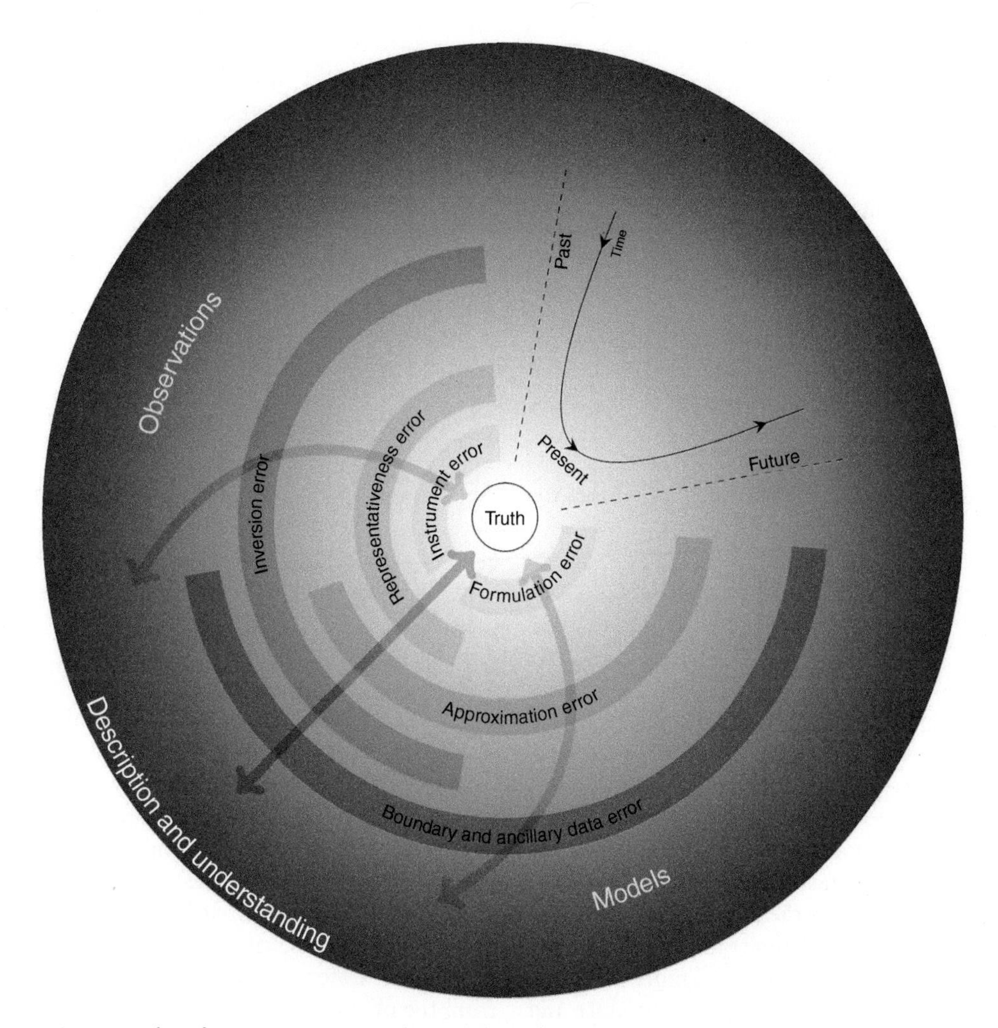

Figure 3.3 Sources of explicit uncertainty on the way from the true system to our description and understanding. Between our description and understanding of the Earth system and the true system are several layers of explicit uncertainty, shown here as known sources of error. The truth itself is unattainable because not all errors are arbitrarily reducible and because of additional implicit uncertainties. Any addition of complexity in our description, corresponding to moving outward from the center, adds more layers of uncertainty. For simple or very focused descriptions, the types of uncertainty are model- or observation-specific. But with increasing complexity, model and observation errors and thus uncertainties start to overlap and become intertwined. Moving away from the present state of the Earth system along the bent time axes also adds complexity and thus uncertainty. While the qualitative and quantitative contributions might differ for past and future, the overall uncertainties may be considered symmetric in both directions.

with the forecast of a stochastic, random system. It is therefore highly relevant to explain the type and motivation of ensembles to scientists from other fields who see identical figures in a weather-forecast ensemble prediction system (stochastic, irreducible) and in the IPCC-type multimodel ensemble (structural, reducible), while the underlying uncertainties are massively different for corresponding decision making.

The complex topography of uncertainty in the Earth system science can only be touched upon in this chapter. We have sketched two different aspects of it in Figures 3.3 and 3.4, dealing with two basic distinctions: scientific knowledge versus truth (Fig. 3.3) and the uncertainty characteristics of predictions for different timescales (Fig. 3.4). Both figures are intrinsically linked in the aspect of time, because some uncertainties change characteristics between the past and the future. Observational uncertainties, for example, are essentially fixed for the past (we unfortunately cannot add or improve instruments used, e.g., in preindustrial times), so some of that observational uncertainty is irreducible. This uncertainty translates via imperfect models, derived from those observations, in irreducible uncertainty in the future, even if in theory this uncertainty could be reduced.

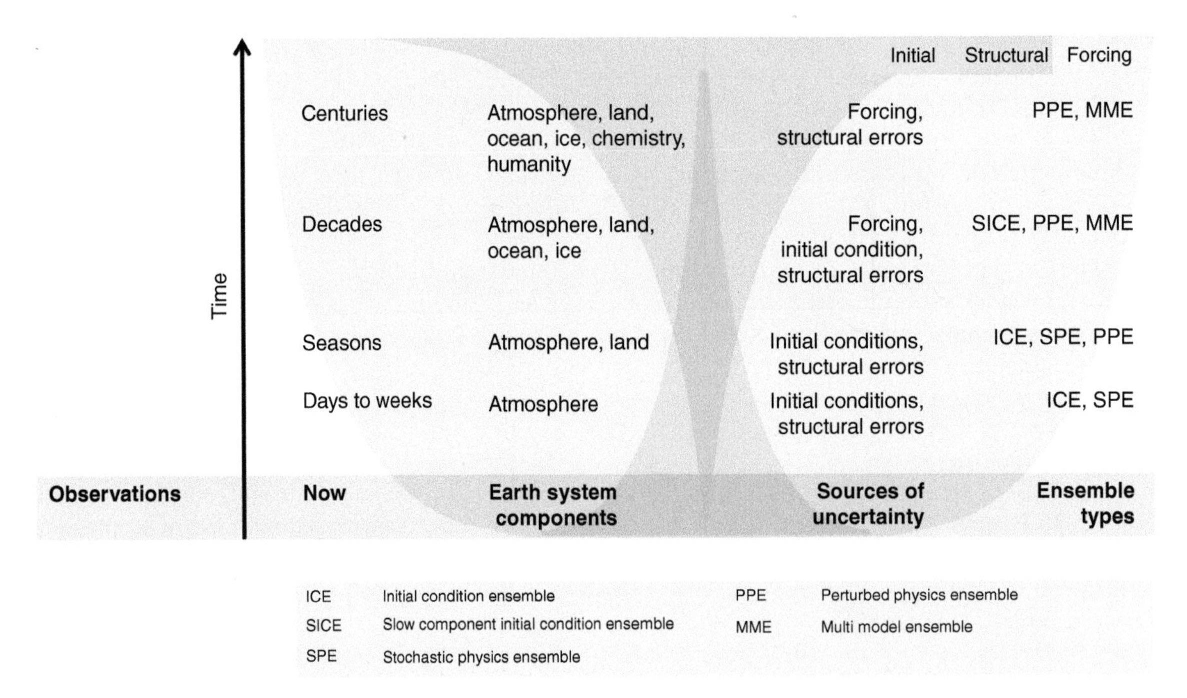

Figure 3.4 Different sources of uncertainty in the Earth system on the way to seamless environmental prediction. This schematic table shows basic relationships between the timescale of a prediction in the Earth system and involved sources of uncertainty and corresponding, technical terms for the ensemble creation in modeling.

Uncertainty can be visualized as the accumulation of possible errors when we increase the complexity of our description or of our understanding of the Earth system (see Fig. 3.3). The more comprehensive and thus the more complex the description of a system is, the more potential sources of error, or layers of uncertainty, accumulate and take us farther away from the truth. The truth itself is unattainable and can only ever be approximated with models and observations. We can choose different paths for this approximation, which will force us to deal with different types of uncertainty. Going outward from the true system, that is, increasing complexity, it is apparent that model and observation errors quickly become entangled: any higher-level, meaning further processed, observational product involves models (for example of radiative or heat transfer). Any realistic model simulation requires boundary conditions from observations. The sources of uncertainty in Figure 3.3 are named broadly. Each may be subdivided in a number of fine-scale, often technical terms. Also, the interaction of different uncertainty-induced errors is by no means as simple as just adding them up. It would be beyond the scope of this chapter to detail all possible interactions of error sources and uncertainties. Nevertheless it should become clear that simply considering one kind is rarely sufficient for a sincere uncertainty characterization.

Figure 3.4 shows a sketch of the complexities involved in the projection of the Earth system into the future. While most projections for natural and social systems look similar (line plots with time on the horizontal axis, property in question on the vertical axis, and spread to represent uncertainty), we want to reiterate here that these plots represent massively different things when applied to different timescales. When you move away from the observable present, a few things change, starting with the properties that we even can forecast (i.e., the well-known difference between deterministic weather for a few days and weather statistics for longer time periods). Corresponding to this shift in quantities (not instantaneous temperature anymore, but mean temperature) is a shift in involved and relevant components. For a regional weather forecast, no one needs a full Earth system model nor the anthropogenic emissions of well-mixed greenhouse gases, while a long-term climate projection involves the influence and interaction of most components and depends heavily on such human influences. To cover this quality difference in projections, the Earth system sciences have developed a multitude of different instruments to represent uncertainties, mostly within so-called ensembles (i.e., multiple computer simulations that vary in key properties). In Figure 3.4, we have listed them and even without a detailed discussion of the different ensembles it is fair to say that the type of ensemble mirrors the major source of uncertainty. If the biggest uncertainties for weather prediction are the initial conditions, it makes sense to create an ensemble of simulations with slightly modified initial conditions. If the reasons for the biggest uncertainties in long-term climate projections are structural errors in the models and forcing uncertainties, the ensemble construction reflects that in the use of multi-model ensembles that simulate different emission scenarios. The differentiation seems trivial, but it is key to understand the source and the quality of uncertainty that underlies a given ensemble and the respective uncertainty quantification to fully grasp the resulting communicated uncertainty. The full justification of why a specific ensemble is chosen and why it is useful for a specific question involves lots of disciplinary and interdisciplinary expert knowledge. The conclusion of Figure 3.4, as we see it, is always to keep in mind the methodology of the uncertainty quantification, both for observation and modeling, when making decisions under uncertainty.

We have seen in both figures again the focus on timescales in the Earth system and its connection to different types of uncertainty. We believe that this defines the major challenge in the natural science part of the Earth system sciences: to understand precisely for which spatial and temporal scales we currently can make scientifically sound statements, and which statements we can improve. The treatment of uncertainty in the Earth system sciences has progressed rapidly in the last years, but it is still not sufficiently rigorous: even for experts it is currently very hard to understand the full amount of explicit and implicit uncertainties in any given results, especially for complicated metrics that are derived from different components of the Earth system. We want to highlight here again that utmost care is necessary when we transfer our incomplete knowledge of the necessarily undersampled Earth system to the public. When our natural science uncertainties hit uncertainties of social, economic, and political sciences or are folded into additional normative uncertainties, it is very easy to confuse the reasoning behind error bars or any type of uncertainty quantification. To illustrate this, we return to the global warming curve in Figure 3.1 and go one step further to ask how expensive unmitigated climate change in 2100 would be. The resulting answer will necessarily be a nearly incomprehensible mixture of explicit, implicit, and normative uncertainties, as the choice of what we deem expensive will probably dwarf all remaining (explicit) uncertainties. The clearest example here is the choice of discount rate of economic impact models: how much is damage in 2100 worth, compared to 2015 or vice versa. This choice, which is purely normative since it involves only our appraisal of the next generation, will decide single-handedly if climate-change mitigation is a reasonable action. This means that all attempts to model and quantify uncertainty in the full transformation from observations, via model constructions, to forecasts can in the end be irrelevant for specific policy decisions. It is also conceivable that, given a specific normative choice of cost and damage function, the decision process will completely depend on that full system uncertainty by saying "the damage will be too high if climate sensitivity is higher than 2.5 °C." With this example, we explicitly do not want to make the point that socioeconomic and normative uncertainties dominate the full range of uncertainty in our understanding of the Earth system, but that only a consistent treatment with transparent communication of all involved sources of uncertainty can lead to robust decision processes.

Earth system science is a very ambitious endeavor, set out to describe, understand, and eventually project the evolution of the Earth including humanity. While big steps have been made with respect to scientific transparency (with the IPCC being a fundamental part of this transparency process), we believe that all Earth system scientists should be especially careful in the dissemination of their results, specifically because there is a high chance that experts from different disciplines will read and use their results. Therefore, we believe that a short, consistent, standardized description of the type and reason for uncertainties is as essential as the inclusion of error bars in the first place to make our results useful for the full community. We strongly encourage the development and usage of calibrated language attempts and suggest for the future similar developments for projection-type figures. All authors of climate change papers should be very clear and explicit as to what they think the type and characteristics of their uncertainty is, to make sure that our growing body of knowledge of the Earth system is not blemished by misconceptions of what we know, and what we do not know.

ACKNOWLEDGMENTS

We acknowledge the Young Earth System Scientists community (yess-community.org) and input of all involved Young Earth System Scientists at the ICYESS 2013 and the YESS workshop 2014, especially from the organizers Diana Süsser, Andreas Schmidt, Sebastian Sonntag, Werner Bauer, Matthias Brueck, and Iulia Polkova. We thank all members of the 2014 workshop working group on uncertainty. We especially thank Andreas Schmidt for helpful and careful comments in creating this chapter. We acknowledge funding from the Körber Foundation and support from the Cluster of Excellence "CliSAP" (EXC177) via the graduate school SICCS, Universität Hamburg, funded through the German Research Foundation (DFG), the German Climate Consortium, the Center for a Sustainable University (KNU), and Copernicus Meetings and Publications. Moreover, we are thankful for the support of various types from MARUM/GLOMAR, Future Ocean/ISOS, the University of Hamburg, and the Earth System Governance Project.

APPENDIX

Excerpt of YESS Earth System Uncertainty Language Glossary

Truth, the true state, or the true description of the Earth system is generally not attainable due to any combination of limited knowledge about processes, limited observational capacities, and limited computing resources to represent the system's full complexity. Observations and model results remain approximations or estimates of the truth, even if they can be very accurate in some cases.

Inversion errors (also called retrieval errors): The type of error that arises from the application of a retrieval algorithm that transforms immediate instrument data to higher-level, i.e., further processed, data products. Any combination of errors from statistical methods, forward models, and ancillary data may be subsumed under inversion errors. The transformation procedure from instrument to higher-level data is then considered as a black box.

Representativeness errors: The type of error that arises from differences in temporal or spatial scales of observational or model data. A data point may originate from a single point in time or space, like the temperature at the North Pole at a fixed time or it may be accumulated or averaged from a larger domain or over a larger time span, like the global mean surface temperature over 1 year. In particular when comparing remote observations, the question of whether we truly compare apples with apples becomes important. But also the comparison of model results to in situ observations is complicated by the fact that they may represent different temporal and spatial scales.

Boundary conditions describe the state of the physical system at the boundary of the simulated domain. These are usually prescribed, but can change in time. For Earth system models, the boundary conditions are commonly known as "external forcing." For regional climate change simulations, the boundary conditions are usually derived from global climate model simulations via a downscaling process, but they can also be derived from observational or reanalysis datasets.

Feedbacks in the Earth system depend on the interactions between different components of the system or different physical processes. A feedback exists if an initial change in a given physical quantity will be amplified (positive feedback) or dampened (negative feedback) due to the interactions of different processes. Feedback-related uncertainty stems from the question if all relevant physical processes and their interactions are implemented in Earth system models, and even if they are, from the question of they are appropriately simulated to an extent that allows feedbacks of comparable strength. It is very difficult to quantify feedback strength from observation. One well-known, but actively debated, feedback is the cloud feedback of global surface temperature due to a CO2-induced warming. The question is whether the initial warming will cause more or less clouds and, in either case, if this additionally warms the Earth or counteracts the initial warming.

Model formulation/structural errors: For every physical process to be simulated, a specific model formulation has to be chosen in the construction process. Additionally, a choice has to be made as to which physical processes can be included and how they are implemented. The combination of model formulation and model structure will influence the model results (see Feedbacks). Different models may have a common origin and thus share similar process descriptions and implementations. This lack of independence in model formulation and structure complicates the statistical analysis of multi-model ensembles and has to be considered when similarities in results are used to imply robust model features.

Physical processes are relations, interactions, and transformations in the Earth system, be it physical, chemical, biological, geological, social, or political. Processes in a model are represented in a form of mathematical equations as part of the model formulation; simplified process descriptions are also called parameterizations. Feedbacks in the Earth system depend on the interactions between different physical processes.

Parameterizations are simplified descriptions of physical processes occurring on temporal and spatial scales smaller than the model is able to simulate. Parameterizations can be formulated in different manners and with different complexity levels depending on scientific goals. Uncertainty in the formulation of the parameterizations arises from a variety of sources, including the representativity of the data and the lack of observational data or/and process understanding. Feedbacks in specific models can differ depending on interactions between different parameterizations and model formulations.

Model Ensembles: an ensemble of model simulations is a set of more than one simulation. Ensembles are often used because a single "true" or most likely model simulation cannot be identified. The spread of ensembles represents different uncertainties, for example in initial conditions or model formulation. Ensembles can be used to quantify the range of possible system states from the spread among the different models. There are different concepts for ensemble creation and those represent different ways to create this set of multiple simulations. Within this set of model simulations, all simulations may be equally likely or may be given different weights according to how realistic they are considered to be.

1. Multimodel ensembles comprise different Earth system models varying in their **model formulation.**

2. Perturbed physics ensemble comprise simulations with the same model but with perturbations of physical/empirical constants in the description of individual **physical processes**.

3. Stochastic physics ensemble comprise multiple simulations with the same model and same descriptions of physical processes, but changed slightly throughout the simulation runtime to include random weather effects, resulting in a set of similar but different model simulations.

REFERENCES

IPCC (2013), Summary for policymakers, in *Climate Change 2013: The Physical Science Basis*, Contribution of Working Group I to the Fifth Assessment Report of the Intergovernmental Panel on Climate Change, edited by T. F. Stocker, D. Qin, G. -K. Plattner, M. Tignor, S. K. Allen, J. Boschung, A. Nauels, Y. Xia, V. Bex, and P. M. Midgley, Cambridge University Press, Cambridge, UK, and New York.

Lorenz, E. N. (1996, reprint in 2006), Predictability, a problem partly solved, 40–58, in *Predictability of Weather and Climate*, edited by T. Palmer and R. Hagedorn, Cambridge University Press, Cambridge, UK.

Mastrandrea, M. D., C. B. Field, T. F. Stocker, O. Edenhofer, K. L. Ebi, D. J. Frame, H. Held, E. Kriegler, K. J. Mach, P. R. Matschoss, G. -K. Plattner, G. W. Yohe, and F. W. Zwiers (2010), *Guidance Note for Lead Authors of the IPCC Fifth Assessment Report on Consistent Treatment of Uncertainties*, Intergovernmental Panel on Climate Change (IPCC).

Matthews, J. L., E. Mannshardt, and P. Gremaud (2013), Uncertainty quantification for climate observations, *Bull. Amer. Meteor. Soc.*, *94*, ES21–ES25.

Morgan, M. G., and M. Henrion (1998), *Uncertainty : A Guide to Dealing with Uncertainty in Quantitative Risk and Policy Analysis*, Cambridge University Press, Cambridge, UK.

Oden, J. T., and J. T. Serge Prudhomme (2002), Estimation of modeling error in computational mechanics, *J. Computational Phys.*, *182*, 496–515.

Rauser F., J. Marotzke, and P. Korn, Ensemble-type numerical uncertainty quantification from single model integrations, *J. Computational Phys.*, in press.

Rauser, F., A. Schmidt, S. Sonntag, and D. Süsser (2014), ICYESS 2013: Uncertainty as an example of interdisciplinary language problems, *Bull. Amer. Meteor. Soc.*, *95*, ES106–ES108.

4

Uncertainty and Probability in Wildfire Management Decision Support: An Example from the United States

Matthew Thompson,[1] David Calkin,[1] Joe H. Scott,[2] and Michael Hand[1]

ABSTRACT

Wildfire risk assessment is increasingly being adopted to support federal wildfire management decisions in the United States. Existing decision support systems, specifically the Wildland Fire Decision Support System (WFDSS), provide a rich set of probabilistic and risk-based information to support the management of active wildfire incidents. WFDSS offers a wide range of decision-support components, including fire behavior modeling, fire weather information, air quality and smoke management, economics, organization assessment, and risk assessment. Here we focus on WFDSS's provision of probabilistic information and how it can facilitate strategic and tactical decision making. However, the management of active wildfire incidents can be highly complex and subject to multiple uncertainties, only some of which are addressed by WFDSS. We review remaining uncertainties, including identified issues in how fire managers interpret and apply probabilistic information, and conclude with observations and predictions for the future direction of risk-based wildfire decision support.

4.1. INTRODUCTION

Wildland fire activity around the globe is driven by complex interactions between natural and human processes [*Spies et al.*, 2014]. Wildland fire can result in significant ecological and socioeconomic loss, most notably the loss of human life. At the same time, wildland fire can be a powerful tool to achieve a wide range of purposes, including clearing vegetation for agroforestry and hunting objectives, reducing hazardous fuel loads, and restoring and maintaining habitat for fire-dependent species.

Figure 4.1 provides an overview of wildfire management illustrating the major drivers of wildfire risk as well as their respective management options, if applicable. (Note that whereas "wildland fire" is an all-encompassing term including unplanned and planned ignitions, our focus here is on unplanned ignitions, or "wildfires.") Given an ignition, weather, topography, fuel conditions, and suppression activities jointly determine the likelihood of fire reaching a given point on the landscape, as well as the intensity of wildfire. Prior to an ignition, risk mitigation options include investing in ignition prevention programs (e.g., campfire bans), reducing hazardous fuel loads (e.g., removing underbrush and reducing tree density), and investing in suppression response capacity (e.g., training and purchasing additional firefighting equipment). Factors related to the location and environment of highly valued resources and assets (HVRAs) can also be changed, by reducing HVRA exposure to fire (e.g., zoning regulations), and reducing HVRA susceptibility to fire (e.g., home construction practices).

Efficient management of wildfire activity is challenged by multiple sources of uncertainty [*Thompson and Calkin*, 2011]. First, variability surrounding weather conditions precludes deterministic prediction of fire growth and behavior, an uncertainty that is compounded by underlying knowledge gaps in fire-spread theory [*Finney et al.*, 2011a; *Finney et al.*, 2012]. Second, knowledge gaps surrounding the effects of fire preclude determination of impacts to vegetation, soil, and other ecosystem

[1] *Rocky Mountain Research Station, US Forest Service, Missoula, Montana, USA*

[2] *Pyrologix LLC, Missoula, Montana, USA*

Natural Hazard Uncertainty Assessment: Modeling and Decision Support, Geophysical Monograph 223, First Edition.
Edited by Karin Riley, Peter Webley, and Matthew Thompson.

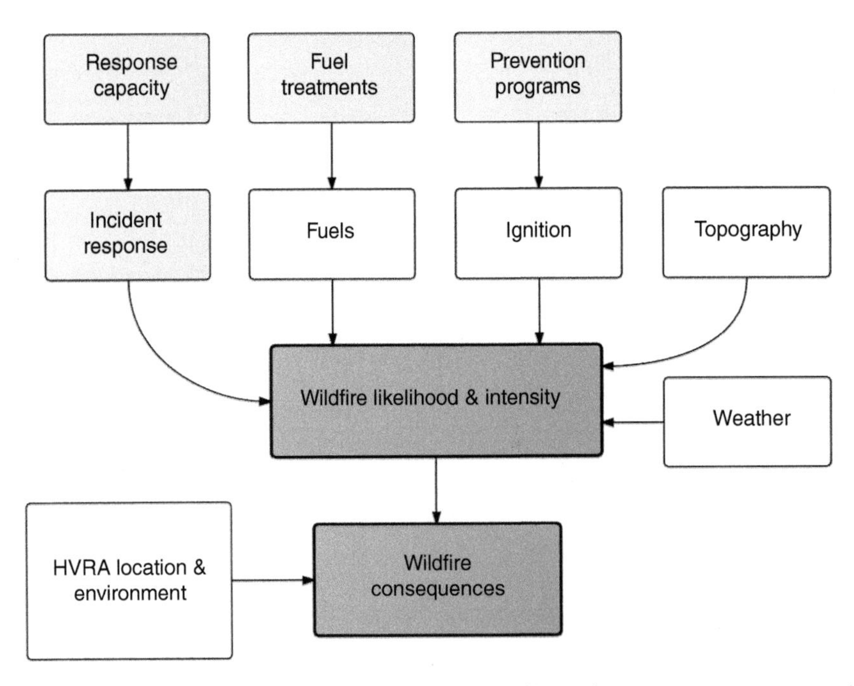

Figure 4.1 Conceptual overview of major factors influencing wildfire risk management. Boxes in light grey represent primary management options, and boxes in dark grey represent the primary components of wildfire risk [modified from *Calkin et al.*, 2011b].

components, and in turn the monetization of these impacts for cost-benefit analysis [*Venn and Calkin*, 2011; *Hyde et al.*, 2012]. Third, partial control, human error, and limited understanding of the productivity and effectiveness of firefighting resources constrain the development and implementation of optimal suppression strategies [*Holmes and Calkin*, 2013; *Thompson*, 2013]. Last, fire manager decision processes can be subject to a number of suboptimal heuristics and biases in complex, uncertain environments [*Maguire and Albright*, 2005; *Thompson*, 2014].

A wide range of models and decision support systems exist to help support risk-informed wildfire decision making, many of which specifically target one or more of the aforementioned sources of uncertainty [*Ager et al.*, 2014; *Chuvieco et al.*, 2012; *Kaloudis et al.*, 2010; *Calkin et al.*, 2011a; *Noonan-Wright et al.*, 2011; *Petrovic and Carlson*, 2012; *Rodríguez y Silva and González-Cabán*, 2010; *Salis et al.*, 2012]. In a prefire planning environment, structured decision processes can systematically and deliberatively address uncertainties with a range of techniques [*Warmink et al.*, 2010; *Marcot et al.*, 2012; *Thompson et al.*, 2013a; *Skinner et al.*, 2014]. As an illustration, *Thompson et al.* [2015] detail how stochastic simulation, expert judgment elicitation, and multicriteria decision analysis could be used to address natural variability, knowledge gaps, and preference uncertainty, respectively.

Wildfire risk assessment is increasingly being adopted across landscapes and ownerships throughout the United States for decision support [*Calkin et al.*, 2011b; *Thompson et al.*, 2013b]. Assessment of wildfire risk follows a widely adopted ecological risk assessment paradigm, the two principal components of which are exposure analysis and effects analysis [*Fairbrother and Turnley*, 2005]. A generalized framework known as the "wildfire risk triangle" (Fig. 4.2) depicts risk as a function of the likelihood of fire, the intensity at which fire burns, and the susceptibility of resources and assets to loss-benefit from fire, which can be summarized to quantify risk in terms of expected net value change [*Finney*, 2005; *Scott et al.*, 2013]. This is a value-focused approach that considers not just the possibility of wildfire occurring but also its potential ecological and socioeconomic consequences, including benefits or net gains.

Figure 4.2 Wildfire risk triangle, composed of the likelihood and intensity of wildfire along with the susceptibility of resources and assets to wildfire [*Scott et al.*, 2013].

Wildfire risk has several important features that may influence mitigation planning relative to risks presented by other natural hazards. First, wildfire risk is inherently spatial: the likelihood and intensity of fire are driven by complex spatial interactions between ignition locations, fuel conditions, topography, and weather patterns. Furthermore, the location of resources and assets determines their respective exposure to wildfire as well as their susceptibility (e.g., watersheds with steeper slopes and more erodible soils may lead to greater postfire erosion and water quality concerns). Second, wildfire can lead to substantial benefits, in terms of restoring and maintaining ecological conditions, as well as reducing future wildfire hazard. Third, in contrast to phenomena such as earthquakes and hurricanes, the likelihood and intensity of the natural hazard itself can be reduced, either preventatively or throughout the course of an event (see Fig. 4.1).

Although valuable for prioritizing mitigation needs and planning incident response in a prefire decision environment, comprehensive and systematic risk assessment is often not possible in the dynamic and time-pressed active incident decision environment. Where risk assessments have already been performed, results can inform real-time evaluations of potential consequences, but fire managers must still be responsive to changing conditions and the specific characteristics of the wildfire incident as it unfolds. With that said, existing decision-support systems can still provide a rich set of probabilistic and risk-based information to support the management of active wildfire incidents.

In this chapter, we focus on the Wildland Fire Decision Support System (WFDSS), a web-based system developed and used within the United States. Per federal policy [*Fire Executive Council*, 2009], fire managers are directed to "use a decision support process to guide and document wildfire management decisions," and WFDSS is increasingly adopted as the system of decision support, particularly for large and complex wildfires. WFDSS was designed to be a single system to replace all previous processes, to integrate fire science and information technology, and to streamline and improve fire management decision making [*Zimmerman*, 2012]. Beyond decision documentation functionality, WFDSS provides a wide range of decision support components, including fire behavior modeling, fire weather information, air quality and smoke management, economics, organization assessment, and risk assessment [*Calkin et al.*, 2011a; *Noonan-Wright et al.*, 2011]. Notably, WFDSS provides support through not only informational and analytical *content*, but also through an iterative decision *process*; both are critical for effective decision support [*Thompson et al.*, 2013a].

In the subsequent sections, we expand upon decision support functionality within WFDSS, focusing on provision of probabilistic information and how it can facilitate strategic and tactical decision making. To begin, we provide a brief overview of wildfire management in the United States. We then illustrate the role of stochastic wildfire simulation and compare and contrast modeling efforts in prefire and during-fire contexts. We next review remaining uncertainties, including identified issues in how fire managers interpret and apply probabilistic information, and conclude with observations and predictions for the future direction of risk-based wildfire decision support.

4.2. WILDFIRE MANAGEMENT

In the United States and elsewhere around the globe, the dominant management response is to aggressively suppress wildfires to keep them as small as possible. Generally speaking this approach is highly successful; in the United States, typically 95%–98% of all ignitions are rapidly contained during "initial attack" operations [*Calkin et al.*, 2005]. However, those rare fires that escape initial containment efforts account for a disproportionate share of area burned, as high as 95% depending on the geographic extent [*Short*, 2013]. Escaped large wildfires are a particularly prominent issue in the western United States, where topography is steeper, wildland areas are larger, and public acceptance of frequent prescribed burning to reduce hazard isn't as high as in other regions like the southeastern United States

Federal policy provides substantial flexibility regarding the management of large wildfires [*Fire Executive Council*, 2009], so that ecological benefits and reduced future hazard can be recognized and integrated into strategy development. However, for a variety of reasons, many of which are more sociopolitical than technical in nature, fire managers tend to be averse to implementing strategies that promote fire on the landscape [*Thompson*, 2014]. Paradoxically, the result of aggressive suppression in some ecosystems is the accumulation of fuels that would otherwise have burned by periodic fire, so that, over time, fires become increasingly intense and resistant to control [*Arno and Brown*, 1991; *Calkin et al.*, 2014a].

These larger wildfires require a more coordinated response effort that can extend over the course of multiple days to weeks. Management of active wildfire events is dynamic and entails a series of recurrent, linked decisions made by multiple actors, beginning with identification of the appropriate scale and type of response organization. On federal lands in the United States, the management of escaped wildfires follows the National Incident Management System. Under this system, local land managers have shared responsibility with Incident Management Teams (IMTs) to determine appropriate strategies and tactics to achieve land and resource objectives, subject to constraints on firefighting resource availability and firefighter safety. The complexity of the wildfire incident determines the type of IMT; more complex incidents typically require IMTs with more training, experience, and organizational structure. Factors considered in the analysis of incident complexity include potential fire behavior, threatened HVRAs, land ownership and jurisdiction, and sociopolitical concerns.

IMTs next determine the amount and type of firefighting resources to order, including hand crews, engines, bulldozers, and aerial resources. If unavailable, IMTs may request alternative firefighting resources that could act as substitutes, or may be forced to reevaluate strategies and tactics. The third level of decision making entails deploying resource mixes to achieve specific missions, which generally include restriction of fire growth or localized protection of HVRAs. Last, periodic reassessment in response to changing conditions helps ensure the appropriateness of strategies, the type of response organization, and the amount and type of firefighting resources present.

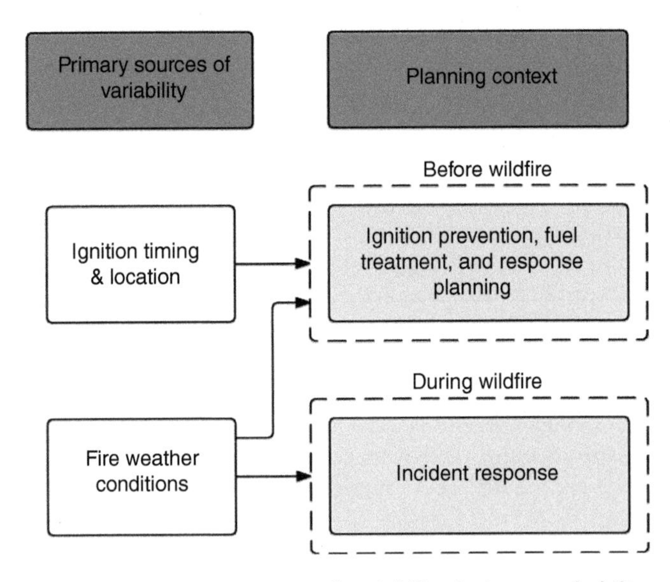

Figure 4.3 Primary sources of variability in burn probability modeling and their relation to the planning context.

4.3. PROBABILISTIC INFORMATION AND RISK-BASED WILDFIRE DECISION SUPPORT

Stochastic wildfire simulation is a foundational element of wildfire risk assessment. The state of fire modeling has significantly advanced in the past decade or so, leveraging improved fire spread algorithms with expanded computational capacity to enable spatially explicit simulation of thousands of possible realizations of fire events [*Finney,* 2002; *Finney et al.*, 2011a; *Finney et al.*, 2011b]. Further, more comprehensive fire-history databases enable improved calibration and validation of model results [*Short,* 2013].

These models rely on rasterized, or pixelated, representations of fire growth and final fire perimeters, and the aggregation of thousands of simulation runs quantifies the probability of any given pixel burning. Because most area burned comes from rare large fires [*Short,* 2013], localized burn probabilities are often influenced more by the spread of fire from remote ignitions rather than local ignitions. It is therefore critical for these models to incorporate geospatial information on topography, fuel conditions, and weather patterns to model the spread of fire across the landscape.

Burn probability modeling is now common practice in the United States, with a growing array of applications across planning contexts and geographic areas. Figure 4.3 identifies the primary sources of variability addressed with burn probability modeling, and their relation to the planning context. In both contexts, fire weather is a key source of uncertainty; temperature, humidity, and, in particular, wind speed and direction are drivers of fire behavior. Before a wildfire occurs, the exact timing and location of the ignition are unknown, although predictive models may use historical spatiotemporal patterns of human- and lightning-caused fires. The timing of ignitions is important with respect to the length of the fire season; fires that ignite earlier in the season have a longer period in which weather conditions may drive growth, whereas fires that ignite near the end of the season have a shorter window. The location of ignitions is important with respect to landscape conditions that could support fire spread as well as resources and assets that could be impacted by fire. By contrast, after an ignition has been detected, fire weather remains the primary source of uncertainty, and reliance on short-term forecasts can improve model prediction.

Within WFDSS the Fire Spread Probability (FSPro) modeling system is the main source of probabilistic information provided to fire managers [*Calkin et al.*, 2011a; *Finney et al.*, 2011a]. FSPro ingests local weather forecasts as well as historical weather data and simulates thousands of possible realizations of fire spread given the current location and size of the fire. FSPro simulation results depict burn probability contours over a defined temporal horizon (e.g., 7 days). Localized burn probabilities are calculated as the proportion of runs that a given pixel burns by the simulated fire events. Probability contours sometimes appear similar in shape to concentric circles, but variation in topography, fuels, and wind conditions influence their exact shape.

The type of advanced analysis afforded by FSPro is not used for all incidents; use is instead typically restricted to the most complex incidents with potential to be long duration and/or large in size. However, on incidents where

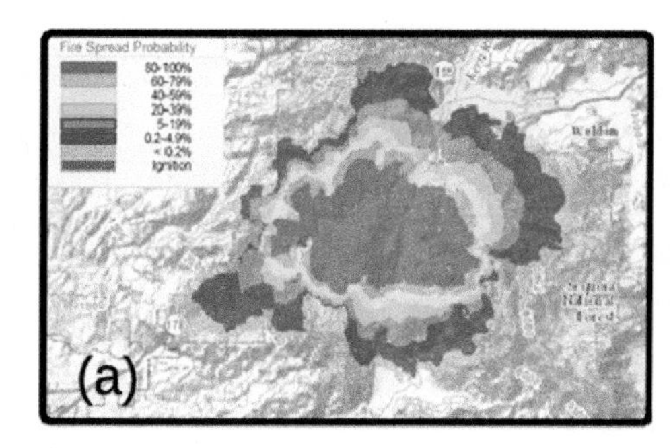

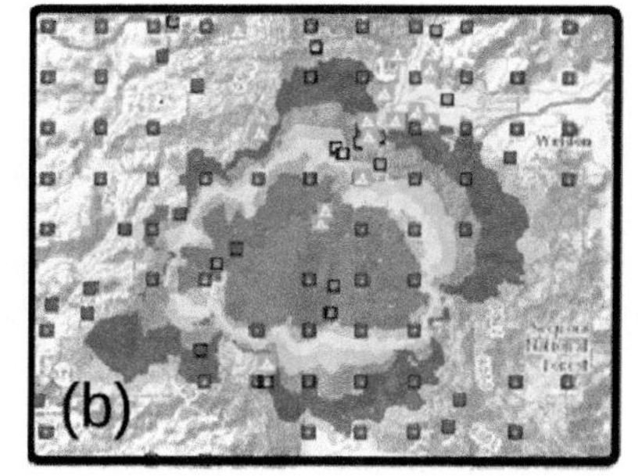

Figure 4.4 Example FSPro burn (a) probability contours and (b) exposure of a select set of resources and assets.

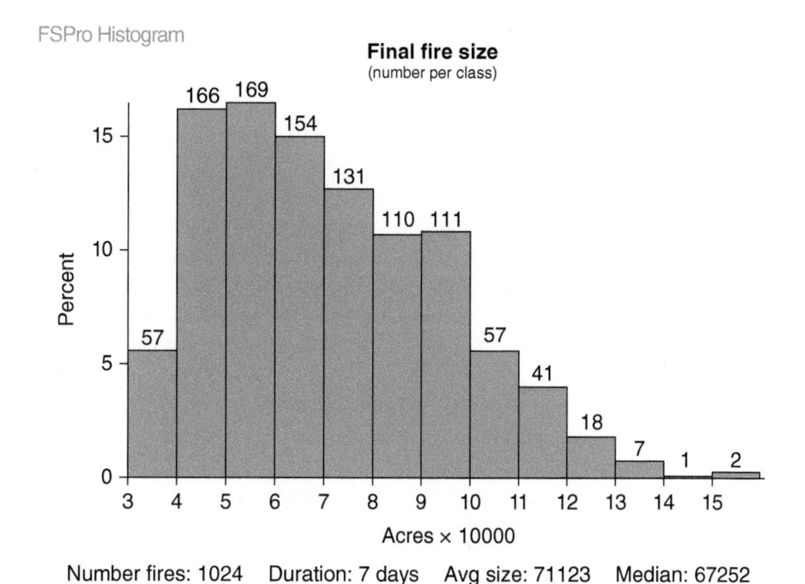

Figure 4.5 Histogram of simulated final fire sizes output from FSPro.

FSPro is used, fire managers and analysts often perform multiple FSPro runs over the course of an event in response to changing fire and weather conditions. FSPro doesn't directly simulate suppression efforts, although users can manually update landscape inputs to account for barriers to spread such as fire lines constructed by hand crews or dozers.

Figure 4.4a illustrates burn probability contours generated for the SQF Canyon Fire, which occurred in 2010 on the Kern River Ranger District of the Sequoia National Forest in California. These particular results are for a 7 day run (i.e., fire spread is modeled over the course of 7 days), for 1000 simulated growth projections. Spatial patterns in the burn probability contours can help fire managers understand fire potential in the absence of suppression and the subsequent probability of resources or assets being impacted by fire.

In addition to fire spread probability, the potential exposure of resources and assets is a key driver of risk-informed incident management. WFDSS provides this functionality as well, leveraging multiple geospatial databases compiled by different agencies to display a range of infrastructure and natural and cultural resources (i.e., HVRAs). Figure 4.4b provides an example of exposure analysis within WFDSS, specifically highlighting the locations of private building clusters (red squares), federal buildings (green and tan squares), and campgrounds (blue squares). Additional layers representing communication and energy infrastructure, roads and trails, air-quality concern areas, critical wildlife habitat, and so on, are available but not displayed here for ease of presentation. The ability to determine where fire spread may result in negative consequences can be a major driver of firefighting strategy and tactics, including division of labor between suppression and localized protection of buildings and other assets.

WFDSS provides reports with a suite of additional information from FSPro analyses to understand model results and to support decision making. Figure 4.5 presents a histogram of simulated final fire sizes, reflecting the underlying distribution of fire events from which the burn probability contours were generated. In addition to

Values list

Category	80–100%	60–79%	40–59%	20–39%	5–19%	0.2–4.9%	<0.2%	Expected value
BLM buildings	0	0	0	0	5	0	0	0.63
Building clusters: Kern	290	215	290	208	297	794	128	676
Communication towers	0	3	0	0	27	3	39	5.59
County: Kern	47,894 acres	12,029 acres	14,062 acres	15,602 acres	24,995 acres	53,989 acres	26,671 acres	67,791 acres
Des areas: Greenhorn creek IRA	9,323 acres	3,854 acres	3,003 acres	1,702 acres	2,701 acres	3,690 acres	1,974 acres	13,536 acres
Des areas: Mill creek IRA	11,596 acres	1,551 acres	1,960 acres	1,319 acres	2,236 acres	5,408 acres	2,205 acres	13,320 acres
Des areas: Piute cypress ISA WSA	2,499 acres	1,003 acres	1,183 acres	424 acres	0 acres	0 acres	0 acres	3,670 acres
Des areas: Woolstaff IRA	155 acres	506 acres	853 acres	1,095 acres	1,362 acres	10,497 acres	6,261 acres	1,698 acres
Electric power plants	0	0	0	0	2	0	0	0.25
Electric sub stations	1	0	0	0	5	0	0	1.53
Electric transmission lines	7.9 miles	0.8 miles	1.3 miles	1.1 miles	3.6 miles	6.3 miles	2.0 miles	9.27 miles
Ozone non-attainment	47,894 acres	12,029 acres	14,062 acres	15,602 acres	24,995 acres	53,989 acres	26,671 acres	67,791 acres
Particulates non-attainment	11,626 acres	3,860 acres	4,393 acres	3,336 acres	6,300 acres	15,315 acres	9,248 acres	17,558 acres
Responsible agency: BLM	8,647 acres	2,690 acres	3,750 acres	4,969 acres	9,604 acres	7,565 acres	3,247 acres	14,431 acres
Responsible agency: CDF	8,075 acres	1,429 acres	2,296 acres	3,645 acres	4,657 acres	15,179 acres	7,702 acres	11,494 acres
Responsible agency: LOCAL	0 acres	0 acres	25 acres	123 acres	144 acres	0 acres	0 acres	67.4 acres
Responsible agency: USFS	31,172 acres	7,909 acres	7,990 acres	6,865 acres	10,591 acres	31,245 acres	15,722 acres	41,798 acres
Roads	25.2 miles	2.3 miles	1.2 miles	3.8 miles	8.2 miles	13.0 miles	4.4 miles	27.4 miles
Surface Mgmt agency: BLM	8,341 acres	2,661 acres	3,689 acres	4,910 acres	9,556 acres	7,472 acres	2,794 acres	14,079 acres
Surface Mgmt agency: DOD	0 acres	0 acres	0 acres	0 acres	151 acres	54 acres	0 acres	20.3 acres
Surface Mgmt agency: USFS	30,458 acres	7,748 acres	7,435 acres	6,372 acres	9,538 acres	29,525 acres	14,365 acres	40,439 acres
USFS buildings	3	0	0	0	1	0	0	2.83

Figure 4.6 Tabular exposure analysis results summarizing FSPro results intersected with spatial value layers; results are presented across burn probability zones as well as in terms of expected values.

supporting strategic evaluation of fire potential, WFDSS also provides additional fire modeling tools focusing on short- and near-term fire behavior to facilitate tactical and operational decisions.

Figure 4.6 provides tabular exposure analysis results, which overlay FSPro burn probability contours with spatial values-at-risk inventory layers and quantify expected values. As an example, the second row in the table shows the number of building clusters in Kern County broken down by burn probability zone, resulting in an expected value of 676 building clusters impacted by fire. In addition to HVRAs that may be impacted by fire, results also indicate whether the fire might spread onto land managed by other agencies, which can be critically important to prepare for as agencies may have different mandates and different fire management objectives.

4.4. FUTURE DIRECTIONS FOR WILDFIRE DECISION SUPPORT

Returning to sources of uncertainty enumerated earlier, WFDSS and specifically FSPro provide the decision support functionality to address uncertainty over possible fire spread directions and subsequent exposure of HVRAs. However, multiple other sources of uncertainty remain that fire managers must face. Below we identify opportunities for future directions of risk modeling and economic research to directly address these remaining uncertainties, ideally to improve risk-informed and cost-effective fire management.

4.4.1. Addressing the Consequences of Wildfire

Many contemporary landscape-scale wildfire risk assessments follow the exposure and effects analysis framework first identified by *Finney* [2005] and later formalized by *Scott et al.* [2013]. That framework quantifies wildfire risk for any discrete location on the landscape as the expected net change in value (*eNVC*) of all highly valued resources and assets (HVRAs) exposed to wildfire at that location. The calculation of *eNVC* incorporates the likelihood of burning and the conditional net value change (*cNVC*) given that a fire occurs.

$$eNVC_k = BP_k * cNVC_k$$

where BP_k is the burn probability at pixel k. BP is determined through stochastic simulation of wildfire occurrence and spread. The calculation of $cNVC_k$ incorporates intensity results of a deterministic or stochastic wildfire simulation, as well as the susceptibility of HVRAs to wildfire. Specifically,

$$cNVC_k = \sum_i \sum_j \left(FLP_{ik} * RF_{ij} * RI_j \right)$$

Table 4.1 Response Functions and Relative Importance Weights for Two Stylized HVRAs

	Response function value, by flame-length class (FLC)						
HVRA	FLC 1	FLC 2	FLC 3	FLC 4	FLC 5	FLC 6	Relative importance weight
Critical infrastructure	−10	−10	−80	−80	−80	−80	100
Fire-dependent wildlife habitat	+50	+40	+30	−10	−30	−60	80

Source: See *Thompson et al.* [2015].
Note: FLCs are presented in order of increasing flame length, and users can define the number of categories and their corresponding flame lengths depending upon the application. The response function for critical infrastructure illustrates minimal loss until a flame length threshold is crossed with significant loss, whereas the response function for the wildlife habitat illustrates beneficial effects at low to moderate intensity fire. Response functions can be nonlinear, and can be multivariate with additional HVRA-specific information

where FLP_{ik} is the conditional probability burning in flame-length class i at pixel k, and RF_{ij} is a "response function" value that indicates the consequence to HVRA j of a wildfire occurring in flame-length class i, and RI is the "relative importance" per unit area of HVRA j. Response function values for physical assets like residential structures and critical infrastructure are always negative, indicating a net loss of value when exposed to wildfire. Wildfire positively affects the value of some resources, such as wildlife habitat or fire-adapted ecosystems, so RF values can also be positive. Response functions are designed by resource specialists who rely on their experience and the scientific literature. Relative importance weighting is necessary to put all coincident HVRAs into a common currency. Relative importance values are determined by the line officers ultimately responsible for managing the landscape. To illustrate, Table 4.1 presents example response functions and relative importance weights for two stylized HVRAs, where response values range from −100 to +100, and importance weights range from 0 to +100.

The quantitative framework described above is designed to support land, resource, and fuel-management planning, typically relying on more advanced fire modeling systems that simulate tens of thousands of fire seasons to generate burn probabilities and flame length distributions [*Finney et al.*, 2011b; *Thompson et al.*, 2013b]. However, its results can also support planning for the response to a wildfire before one has started and even for planning the management of a wildfire after it has escaped initial attack. For example, the $cNVC_k$ values can be summed within each simulated fire perimeter from a stochastic fire modeling system, resulting in an estimate of the overall NVC for the fire. Because the simulated fire start location is known, this new NVC attribute can be used to identify the net "risk source" associated with each ignition. Net "risk source" maps that average the consequence of all simulated fires starting across different parts of a landscape, whether positive or negative, can then succinctly summarize consequences of ignitions in various locations. Such an analysis could help to create a spatial wildfire management response plan by identifying locations on the landscape where fires tend to result in positive net effects and where they tend to cause damage. Likewise, a similar analysis can be done using simulated fire perimeters generated by FSPro for an ongoing wildfire event. That analysis would not generate a risk-source map, but would instead quantify the likelihood of exceeding threshold quantities of net value change, a small improvement over the analysis currently available in WFDSS.

To demonstrate potential applications in the real-time incident decision environment, Figure 4.7 illustrates risk modeling results generated from FSPro outputs. Fire perimeters were simulated on a real incident on a landscape in the southern Sierras in California, and paired with a preexisting $cNVC$ grid that was generated following the risk assessment framework outlined above. While the scatterplot generally indicates increasing net loss as simulated fire size grows, variation in loss stems from the shape and location of the fire with respect to potential fire intensity, HVRA location, and HVRA susceptibility. Notably, the greatest losses do not come from the largest fires but rather tend to concentrate around 15,000–20,000 acres, and this result underscores the importance of the fire's shape and location in addition to its size. Some of the most consequential fires may have burned into a community, whereas the fires that grew the largest may have done so by virtue of growing into undeveloped wildlands with few susceptible HVRAs.

Figure 4.8 further summarizes simulated fire-level net loss using an exceedance probability chart. These results display the likelihood of experiencing a given level of net loss, which can be compared against prospective suppression expenditures to inform cost-effectiveness analysis. For example, there is approximately a 20%

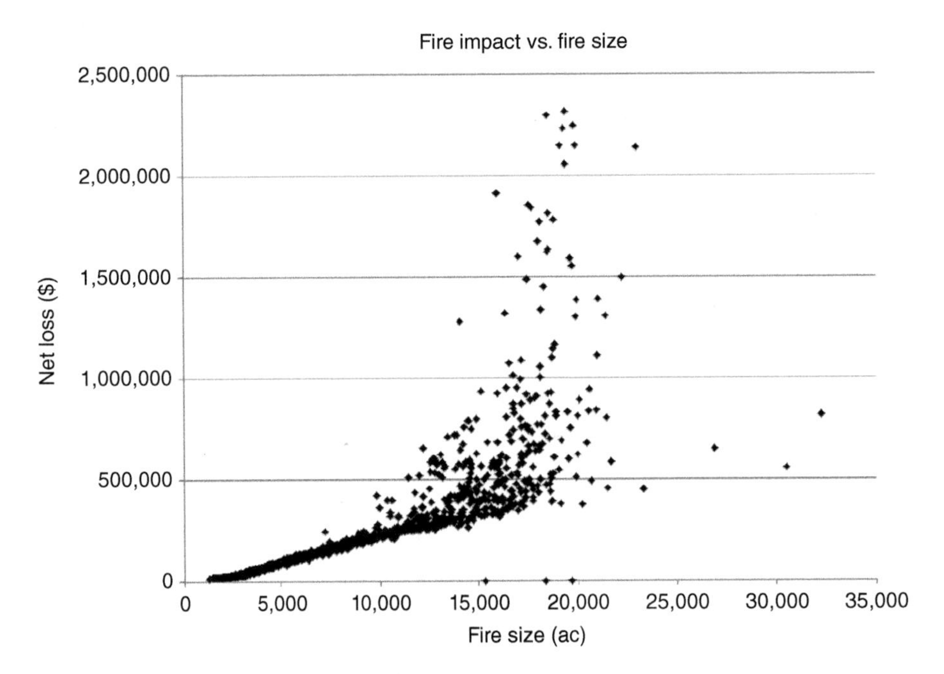

Figure 4.7 Scatterplot of fire-level *cNVC* versus fire size.

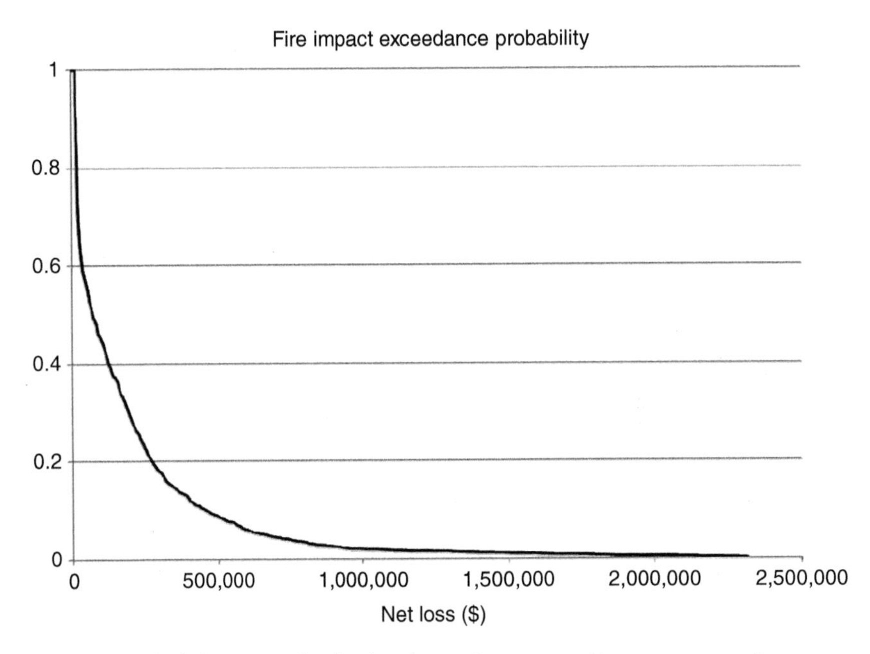

Figure 4.8 Exceedance probability curve for fire-level *cNVC*, expressed in monetary units.

chance of exceeding $250,000 in losses. For illustration, results are presented in monetary terms, although operationally the quantification of all possible market and nonmarket impacts can be challenging [*Venn and Calkin*, 2011]. Extensions include modifying these exceedance probability curves in cases where response functions indicate potential for benefit, and probabilistically analyzing prefire risk mitigation investments in terms of avoided losses.

4.4.2. Suppression Effectiveness

Despite the scale of investment in large wildfire suppression, relatively little is understood about how suppression actions influence large wildfire spread and those conditions that ultimately lead to containment [*Finney et al.*, 2009]. Wildfire containment under initial attack (IA) has typically been modeled by evaluating the elliptical rate of spread of an ignition under identified fuel and weather conditions compared with the productive capacity and arrival time of IA resources [see for example *Fried and Fried*, 1996]. However, the large fire environment presents additional complexity and it has not been demonstrated if the IA containment approach is relevant to large wildfire suppression.

There is considerable uncertainty in managing large wildfires including the quality of weather forecasts, complex environmental conditions, variation in the type and quality of suppression resources, and whether or not requested suppression resources will be assigned [*Thompson and Calkin*, 2011]. Additionally, many resources are engaged in non-line-building activities such as point protection, contingency line development, and mop up. Further, given that the wildfire escaped IA, it is likely that the characteristics of wildfire growth are such that line-building efforts may not be feasible or effective.

Data necessary to understand suppression effectiveness within the United States can be difficult to obtain. Some recent studies have relied on primary reporting systems such as the Incident Status Summary (ICS-209) Situation report. However, these data do not provide spatial characteristics of the fire environment and rely on self-reporting by the incident team responsible for managing the events. In particular, some of the most relevant data for suppression modeling (specifically percentage of the wildfire contained, growth potential, and reported values at risk) are subjective and may not be accurately reported [*Holmes and Calkin*, 2013].

Despite these challenges, several authors have examined the ICS-209 data to model suppression effectiveness. *Finney et al.* [2009] modeled the probability that on a given day a large fire would be declared fully contained by examining wildfire suppression resource assignment, daily fire growth, fuel type, and other reported data within the 209 reports. The most significant factor in achieving wildfire containment was quiescent periods during the fire. That is, containment was most directly related to the number of low growth fire days and the number of previous intervals of low growth. Containment probability was negatively related to the presence of timber fuel types. No significant relationship was found between likelihood of containment and fire size or number of personnel assigned. *Holmes and Calkin* [2013] utilized similar data from the ICS-209 to examine the relative efficiency of suppression resources by comparing published resource line-building production rates published by *Broyles* [2011] with daily line built estimated from reported fire size and percentage containment. The results indicated that the actual production rates of suppression resources on a set of large wildland fires in 2009 were relatively low; 14% for engines, 18% for dozers, and 35% for hand crews compared to the reported standard production rates. Helicopters were the one exception with actual production rates estimated to be close to published rates (93%). Limited understanding of the objectives of resource assignments and the conditions of the suppression environment limit the ability of research efforts to characterize the conditions under which suppression activities are most effective.

In-flight GPS-based systems such as the Automated Flight Following and Operational Loads Monitoring systems allow for analysis of the spatial use patterns of wildfire aviation resources such as Large Air Tankers (LATs), large planes capable of dropping greater than 1800 gal of retardant or water. Understanding the cost effectiveness of LATs has been a particular emphasis area of the US Forest Service over the last several years as the existing fleet of Korean War vintage aircraft are replaced with newer ships. *Calkin et al.* [2014b] tied individual retardant drops from LATs to wildfire outcomes during the 2010 and 2011 wildfire seasons in the United States. The authors found that approximately half of all use of LATs occurred after the fire had escaped initial attack. On those incidents where LATs were used during initial attack, 75% escaped, compared to the 2%–4% annual escape rate on all wildfire ignitions. These results suggest that LAT usage on IA is typically restricted to only those fires with a very high escape rate. Information on the effectiveness of LATs in large fire support is currently limited due to missing data on the objectives and outcomes of retardant drops in supporting large fire strategies.

The practice of wildfire management is highly complex, and, currently, there are many challenges to understanding the effectiveness of wildfire suppression actions. Improved and expanded data collection systems and continued research efforts are critical to our understanding of the conditions that lead to effective suppression, informed trade-off analyses of alternative suppression strategies and organizations, and safer and more economically efficient outcomes.

4.4.3. Fire Managers' Use and Interpretation of Probabilities and Decision Making Under Uncertainty

Decision-support tools may be able to leverage how managers respond to risk information to mitigate some cognitive biases and decision heuristics. Common biases and decision heuristics have been linked to the wildfire management environment [Table 4.2; see also *Maguire and Albright*, 2005], and wildfire managers have exhibited several of these when choosing among strategies in hypothetical wildfire scenarios [*Wilson et al.*, 2011; *Wibbenmeyer et al.*, 2013]. In some cases, the informational content or way in which information is presented can affect decisions. Framing potential wildfire outcomes in certain ways can increase the salience of certain outcomes, and trigger an analytical response from decision makers.

Table 4.2 Select Set of Identified Cognitive Biases That Influence How Fire Managers Perceive and Respond to Risk

Cognitive biases	Impacts
Discounting bias	Tendency to overweight short-term risk over long-term risk
Loss aversion	Tendency to exhibit risk-averse behavior when outcomes are framed as gains and to exhibit risk-seeking behavior for outcomes framed as losses
Overconfidence bias	Tendency to be overconfident in the state of knowledge or accuracy of beliefs
Status quo bias	Tendency to disproportionately stick with the status quo alternative
Sunk cost bias	Tendency to continue with ineffectual strategy/tactics because significant resources have already been expended

For example, presenting information about the duration of use of suppression resources as an accident or fatality rate may highlight for managers the risk to personnel of using those resources. Presenting fatality rates instead of usage rates has been shown to result in reduced exposure of personnel to risk in hypothetical fire scenarios [*Hand et al.*, 2015], although this may exacerbate other biases related to responses to outcome probabilities.

Biased responses to outcome probabilities [see *Tversky and Kahneman,* 1992; *Prelec,* 1998] may be less amenable to direct intervention through decision support tools. An option for addressing cognitive biases that may not respond well to different information framing may be to identify those managers who tend to exhibit responses to risk that are most closely aligned with agency preferences and facilitate training and knowledge transfers among managers. Heterogeneity in responses to risk (and probabilities in particular) is evident among managers, and a portion of managers appear to make decisions that minimize expected losses in a risk environment [*Hand et al.*, 2015]. Decision support could help by providing structured decision processes [*Maguire and Albright*, 2005] that mirror risk-based training efforts and knowledge transfer among managers.

4.5. CONCLUSION

Wildfire management is complex, dynamic, and uncertain, and a full investigation into wildfire risk assessment and mitigation planning is beyond the scope of a single chapter. Nevertheless, we highlighted salient uncertainties faced in the management of active large wildfire incidents, reviewed an existing decision-support system widely used in the United States (WFDSS), and illustrated how probabilistic information provided by WFDSS can inform risk-based decision making. Significant sources of uncertainty remain, which vary according to the degree of influence they may exert on decision processes as well as approaches to manage those uncertainties and improve decisions. A fundamental need for increased wildfire management efficiency is improved understanding of suppression effectiveness, requiring large-scale collection of operational data across incidents. We expect continued research into risk and decision analysis will play a role for years to come in wildfire management and risk mitigation, and that in particular advanced risk modeling techniques will be used to inform wildfire management decisions.

REFERENCES

Ager, A. A., Day, M. A., Finney, M. A., Vance-Borland, K., and N. M. Vaillant. (2014) Analyzing the transmission of wildfire exposure on a fire-prone landscape in Oregon, USA, *For. Ecol. Man.*, *334*, 377–390.

Arno, S. F., and J. K. Brown (1991), Overcoming the paradox in managing wildland fire, *West. Wildlands*, *17*(1), 40–46.

Broyles, G. (2011), Fireline production rates, USDA Forest Service, National Technology and Development Program, *Fire Man. Rep.*, 1151–1805 (San Dimas, CA).

Calkin, D., A. Ager, M. Thompson, M. Finney, D. Lee, T. Quigley, C. McHugh, and K. L. Riley (2011b), A comparative risk assessment framework for wildland fire management: the 2010 cohesive strategy science report, Gen. Tech. Rep. RMRS-GTR-262, Fort Collins, CO: U.S. Department of Agriculture, Forest Service, Rocky Mountain Research Station.

Calkin, D. E., C. S. Stonesifer, M. P. Thompson, and C. W. McHugh (2014b), Large airtanker use and outcomes in suppressing wildland fires in the United States, *Int. J. Wildland Fire*, *23*(2), 259–271.

Calkin, D. E., J. D. Cohen, M. A. Finney, and M. P. Thompson (2014a), How risk management can prevent future wildfire disasters in the wildland-urban interface, *Proc. Nat. Acad. Sci.*, *111*(2), 746–751.

Calkin, D. E., K. M. Gebert, J. G. Jones, and R. P. Neilson (2005), Forest Service large fire area burned and suppression expenditure trends, 1970–2002, *J. Forestry*, *103*(4), 179–183.

Calkin, D. E., M. P. Thompson, M. A. Finney, and K. D. Hyde (2011a), A real-time risk-assessment tool supporting wildland fire decision-making, *J. Forestry*, *109*(5), 274–280.

Chuvieco, E., I. Aguado, S. Jurdao, M. Pettinari, M. Yebra, J. Salas, S. Hantson, J. De La Riva, P. Ibarra, and M. Rodrigues (2012), Integrating geospatial information into fire risk assessment, *Int. J. Wildland Fire*, *23*(5), 606–619.

Fairbrother, A., and J. G. Turnley (2005), Predicting risks of uncharacteristic wildfires: application of the risk assessment process, *For. Ecol. Man.*, *211*(1), 28–35.

Finney, M., I. C. Grenfell, and C. W. McHugh (2009), Modeling containment of large wildfires using generalized linear mixed-model analysis, *For. Sci.*, *55*(3), 249–255.

Finney, M. A. (2002), Fire growth using minimum travel time methods, *Can. J. For. Res.*, *32*(8), 1420–1424.

Finney, M. A. (2005), The challenge of quantitative risk analysis for wildland fire, *For. Ecol. Man.*, *211*, 97–108.

Finney, M. A., C. W. McHugh, I. C. Grenfell, K. L. Riley, and K. C. Short (2011b), A simulation of probabilistic wildfire risk components for the continental United States, *Stochas. Environ. Res. Risk Assess.*, *25*(7), 973–1000.

Finney, M. A., I. C. Grenfell, C. W. McHugh, R. C. Seli, D. Tretheway, R. D. Stratton, and S. Brittain (2011a), A method for ensemble wildland fire simulation, *Environ. Mod. Assess.*, *16*(2), 153–167.

Finney, M. A., J. D. Cohen, S. S. McAllister, and W. M. Jolly (2012), On the need for a theory of wildland fire spread, *Int. J. Wildland Fire*, *22*(1), 25–36.

Fire Executive Council (2009), Guidance for Implementation of Federal Wildland Fire Management Policy, http://www.nifc.gov/policies/policies_documents/GIFWFMP.pdf; last accessed 2 February 2015.

Fried, J. S., and B. D. Fried (1996), Simulating Wildfire Containment with Realistic Tactics, *For. Sci.*, *42*(3), 267–281.

Hand, M. S., M. J. Wibbenmeyer, D. E. Calkin, and M. P. Thompson (2015), Risk preferences, probability weighting, and strategy tradeoffs in wildfire management, Submitted to *Risk Anal.*, *35*(10), 1876–1891.

Holmes, T. P., and D. E. Calkin (2013), Econometric analysis of fire suppression production functions for large wildland fires, *Int. J. Wildland Fire*, *22*(2), 246–255.

Hyde, K., M. B. Dickinson, G. Bohrer, D. Calkin, L. Evers, J. Gilbertson-Day, T. Nicolet, K. Ryan, and C. Tague (2012), Research and development supporting risk-based wildfire effects prediction for fuels and fire management: status and needs, *Int. J. Wildland Fire*, *22*(1), 37–50.

Kaloudis, S. T., C. P. Yialouris, N. A. Lorentzos, M. Karteris, and A. B. Sideridis (2010), Forest management planning expert system for wildfire damage reduction, *Comput. Electron. Agri.*, *70*, 285–291.

Maguire, L.A., and E. A. Albright (2005), Can behavioral decision theory explain risk-averse fire management decisions? *For. Ecol. Man.*, *211*(1–2), 47–58.

Marcot, B. G., M. P. Thompson, M. C. Runge, F. R. Thompson, S. McNulty, D. Cleaves, M. Tomosy, L. A. Fisher, and A. Bliss (2012), Recent advances in applying decision science to managing national forests, *For. Ecol. Man.*, *285*, 123–132.

Noonan-Wright, E. K., T. S. Opperman, M. A. Finney, G. T. Zimmerman, R. C. Seli, L. M. Elenz, D. E. Calkin, and J. R. Fiedler (2011), Developing the U.S. Wildland Fire Decision Support System (WFDSS), *J. Combust.*, Article ID 168473; doi: 10.1155/2011/168473

Petrovic, N., and J. Carlson (2012), A decision-making framework for wildfire suppression, *Int. J. Wildland Fire*, *21*(8), 927–937.

Prelec D. (1998), The probability weighting function. *Econometrica*, *66*(3): 497–527.

Rodríguez y Silva, F., and A. González-Cabán (2010), 'SINAMI': a tool for the economic evaluation of forest fire management programs in Mediterranean ecosystems, *Int. J. Wildland Fire*, *19*(7), 927–936.

Salis, M., A. A. Ager, B. Arca, M. A. Finney, V. Bacciu, P. Duce, and D. Spano (2012), Assessing exposure of human and ecological values to wildfire in Sardinia, Italy, *Int. J. Wildland Fire*, *22*(4), 549–565.

Scott, J. H., M. P. Thompson, and D. E. Calkin (2013), A wildfire risk assessment framework for land and resource management, Gen. Tech. Rep. RMRS-GTR-315, U.S. Department of Agriculture, Forest Service, Rocky Mountain Research Station.

Short, K. (2013), A spatial database of wildfires in the United States, 1992–2011, *Earth Syst. Sci. Data Disc.*, *6*(2), 297–366.

Skinner, D. J., S. A. Rocks, and S. J. Pollard (2014), A review of uncertainty in environmental risk: characterising potential natures, locations and levels, *J. Risk Res.*, *17*(2), 195–219.

Spies, T. A., E. M. White, J. D. Kline, A. P. Fischer, A. Ager, J. Bailey, J. Bolte, J. Koch, E. Platt, and C. S. Olsen (2014), Examining fire-prone forest landscapes as coupled human and natural systems, *Ecol. Soc.*, *19*(3), 9.

Thompson, M. P. (2013), Modeling wildfire incident complexity dynamics, *PloS one*, *8*(5), e63297.

Thompson, M. P. (2014), Social, institutional, and psychological factors affecting wildfire incident decision making, *Soc. Nat. Res.*, *27*(6), 636–644.

Thompson, M. P., and D. E. Calkin (2011), Uncertainty and risk in wildland fire management: A review, *J. Environ. Man.*, *92*(8), 1895–1909.

Thompson, M. P., B. G. Marcot, F. R. Thompson, S. McNulty, L. A. Fisher, M. C. Runge, D. Cleaves, and M. Tomosy (2013a), The science of decision making: Applications for sustainable forest and grassland management in the National Forest System, Gen. Tech. Rep. WO-GTR-88. Washington, DC: U.S. Department of Agriculture, Forest Service.

Thompson, M. P., J. R. Haas, J. W. Gilbertson-Day, J. H. Scott, P. Langowski, E. Bowne, and D. E. Calkin (2015), Development and application of a geospatial wildfire exposure and risk calculation tool, *Environ. Mod. Soft.*, *63*, 61–72.

Thompson, M. P., J. Scott, D. Helmbrecht, and D. E. Calkin (2013b), Integrated wildfire risk assessment: Framework development and application on the Lewis and Clark National Forest in Montana, USA, *Integrat. Environ. Assess. Man.*, *9*(2), 329–342.

Tversky, A., and D. Kahneman (1992), Advances in prospect theory: Cumulative representation of uncertainty, *J. Risk Uncert.*, *5*, 297–323.

Venn, T. J., and D. E. Calkin (2011), Accommodating nonmarket values in evaluation of wildfire management in the United States: Challenges and opportunities, *Int. J. Wildland Fire*, *20*(3), 327–339.

Warmink, J., J. Janssen, M. J. Booij, and M. S. Krol (2010), Identification and classification of uncertainties in the application of environmental models, *Environ. Mod. Soft.*, *25*(12), 1518–1527.

Wibbenmeyer, M., M. Hand, D. Calkin, T. Venn, and M. Thompson (2012), Risk preferences in strategic wildfire decision-making: A choice experiment with U.S. wildfire managers, *Risk Anal.*, *33*(6), 1021–1037.

Wilson, R. S., P. L. Winter, L. A. Maguire, and T. Ascher (2011), Managing wildfire events: Risk-based decision making among a group of federal fire managers, *Risk Anal.*, *31*(5), 805–818.

Zimmerman, T. (2012), Wildland fire management decision making, *J. Agric. Sci. Tech.*, *B*(2), 169–178.

5

Role of Uncertainty in Decision Support for Volcanic Ash Cloud Modeling

Peter Webley

ABSTRACT

As volcanoes erupt, they can produce plumes and generate dispersing ash clouds. These are a significant hazard to all forms of the aviation industry and lead to potential ground-based health hazards. Events in 2010 from the Eyjafjallajökull volcano illustrated that there is a level of uncertainty in the amount of volcanic ash emitted, the particle size range in the dispersing clouds, and the fraction of emitted volcanic ash that is then transported downwind of the volcano. Understanding and quantifying the uncertainty is the first step toward improving our knowledge of the hazard and how to then mitigate the risk and develop the best available decision-support system. In this paper, the different types of uncertainty that play a role in assessing the volcanic ash hazard are discussed, with examples of how the research community is working with the decision support/operational environment to reduce the level of uncertainty and increase confidence in the volcanic ash cloud simulations. The aim of the paper is to show how uncertainties exist in developing volcanic ash hazard assessment, illustrate how they are being mitigated to increase confidence in assessing the hazard, and document how they could be incorporated into the decision-support system.

5.1. INTRODUCTION

Volcanoes can erupt both effusively, through outpouring of lava onto the ground [see Kilauea, Hawaii; *Dzurisin et al.*, 1984; *Mattox et al.*, 1993; *HVO*, 2014] and explosively, through fragmentation of magma that often leads to significant ash plumes, such as Eyjafjallajökull in 2010 [*Gudmundsson et al.*, 2012] and Puyehue-Cordon-Caulle in 2011 [*BVGN*, 2012], see examples in Figure 5.1. Each type of eruptive process can produce its associated hazards, from lava flows that can impact local population [*HVO*, 2014] to dispersing volcanic ash clouds that can lead to aviation impacts [*Guffanti et al.*, 2009, 2010; *Harris et al.*, 2012] and be an issue to human health from ashfall, for example, Soufriere Hills, Montserrat [*Horwell et al.*, 2003].

Geophysical Institute, University of Alaska Fairbanks, Fairbanks, Alaska, and Volcanic Ash Detection, Avoidance Preparedness for Transportation (V-ADAPT), Inc., Fairbanks, Alaska, USA

The focus on the work presented in this manuscript relates to the uncertainty that exists in volcanic ash cloud modeling. First, there is a need to evaluate and assess the uncertainty, then, second, to quantify the uncertainties, and finally to reduce the uncertainties and improve confidence in all aspects of the cloud modeling. This allows those in real-time forecasting of volcanic ash clouds to provide to end users a more complete picture of the future location of the volcanic clouds and assess the potential impact that its dispersal could have on day-to-day operations and decision-support systems.

Since the mid-1990s, there have been operational groups, known as volcanic ash advisory centers (VAAC) worldwide mandated by International Civil Aviation Organization (ICAO) to produce volcanic ash advisories (typically combining observations and simulations) for an eruptive event and volcanic ash advisories (VAA; see example in Fig. 5.2). These are available to the aviation industry through local meteorological watch offices (MWO). To generate these VAAs, the VAACs will use volcanic ash transport and dispersion (VATD) models to

Natural Hazard Uncertainty Assessment: Modeling and Decision Support, Geophysical Monograph 223, First Edition.
Edited by Karin Riley, Peter Webley, and Matthew Thompson.

(a)

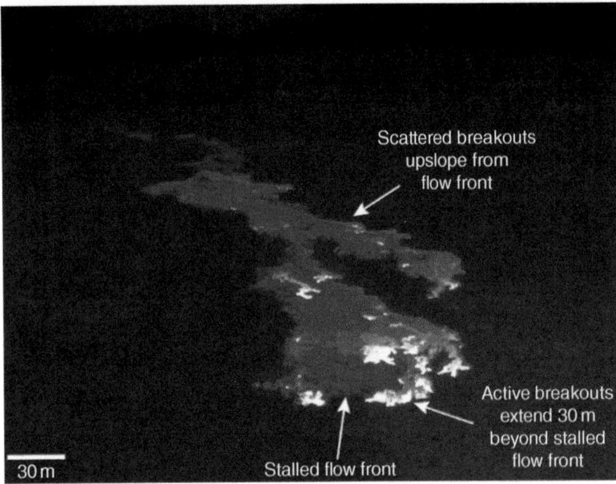

(b)

Figure 5.1 Effusive versus explosive eruptions from an active volcano. (a) Visible and thermal images of the 27 June flow from Kilauea volcano taken on 1 October 2014. The thermal image viewpoint is represented in white box in the visible imagery (adapted from *HVO* [2014]); (b) the Eyjafjallajökull eruption on 17 April 2010 showing a heavy black plume engulfing the neighboring district of Eyjafjoll (adapted from *Gudmundsson et al.* [2010]).

simulate the clouds dispersion into the atmosphere. There are several different types of dispersion modeling formulations used by the VAACs, namely, Eulerian or Lagrangian. They are applied based on the model that each VAAC uses for its volcanic ash transport. [*WMO*, 2010, 2013]. Both reports provide an overview of the different dispersion models available and used by the VAACs, along with details on their modeling formulations and operational setup (see Fig. 5.3). These different models each have their own advantages and disadvantages and the chosen model at each VAAC depends on if they have developed the original model, such as UK Meteorological Office for the NAME model at the London VAAC, or the research group that they have had a close affiliation, such as the Buenos Aires VAAC, which uses the FALL3D model. The models illustrated in Figure 5.3 are then used by the VAACs to generate the +6, +12, and +18 hr panels in Figure 5.2b in the VAA, with the first panel from any available satellite remote sensing data or the +0 hr panel.

During and since the 2010 eruptive event from Eyjafjallajökull volcano [see *Gudmundsson et al.*, 2010, 2012], there has been an interest for the VAACs to produce simulations of airborne volcanic ash concentrations. Several of these VAACs have produced such concentration maps; see Figure 5.4 for London VAAC example from Grimsvotn volcano and Toulouse VAAC example from Cordon Caulle, both in 2011. These maps document the location of the volcanic ash over defined vertical segments of the atmospheric column (e.g. surface [SFC] or flight level 0 [FL000] to flight level 200 [FL200] or approximately 20,000 ft above sea level [ASL]). Flight levels are determined from standardized pressure altitudes and thus the aircraft can determine the difference in altitude between each other, which is more important than their altitude relative to sea level. VAACs also aimed to define thresholds to the downwind ash concentrations of 0–2 mg/m^3, 2–4 mg/m^3, and > 4 mg/m^3, which were proposed during the 2010 Eyjafjallajökull eruption with the "fly"/"precautionary"/"no fly" regions, respectively.

The maps, while not standard VAAC products, are generated as supplemental information to be coupled with the ICAO mandated VAAs and VAG. Note that for the London VAAC, Figure 5.4a, the thresholds are chosen in terms of micrograms, while the Toulouse VAAC, Figure 5.4b, chose to use "low," "moderate," and "high" terms for the concentrations, but with the same color scales as for the London VAAC concentrations.

Since these quantitative concentration maps were produced, there has been a lot of discussion in the International Volcanic Ash Task Force [IVATF; see *ICAO*, 2012], and at other meetings such as the International Union of Geodesy and Geophysics (IUGG)/World Meteorological Organization (WMO) workshops in 2010

(a)

Volcanic Ash Advisory Text

FVFE01 RJTD 161800
VA ADVISORY
DTG 20141016/1800z
VAAC: TOKYO
VOLCANO: KARYMSKY 300130
PSN: N5403e15927
AREA: RUSSIA
SUMMIT ELEV: 1536M
ADVISORY NR: 2014/49
INFO SOURCE: MTSAT-2
AVIATION COLOUR CODE: NIL
ERUPTION DETAILS: VA CONTINUOUSLY OBS ON SATELLITE IMAGERY
OBS VA DTG: 16/1715Z
OBS VA CLD: SFC/FL100 N5250 E16050 - N5150 E16150 - N5120 E16130
N5235 E16015 MOV SE 30KT
FCST VA CLD +6 HR: 16/2315Z SFC/FL090 N5230 E16030 - N5235 E16150
- N5115 E16330 - N5000 E16330 - N5005 E16115 - N5115 E16030
FCST VA CLD +12 HR: 17/0515Z SFC/FL100 N5300 E16000 - N5300 E16135
- N5125 E16300 - N5030 E16300 - N5030 E16125 - N5100 E16000
FCST VA CLD +18 HR: 17/1115Z SFC/FL110 N5430 E15730 - N5430 E15855
N5310 E16100 - N5200 E16100 - N5200 E15730
RMK: NIL
NXT ADVISORY: 20141017/0000Z=

(b)

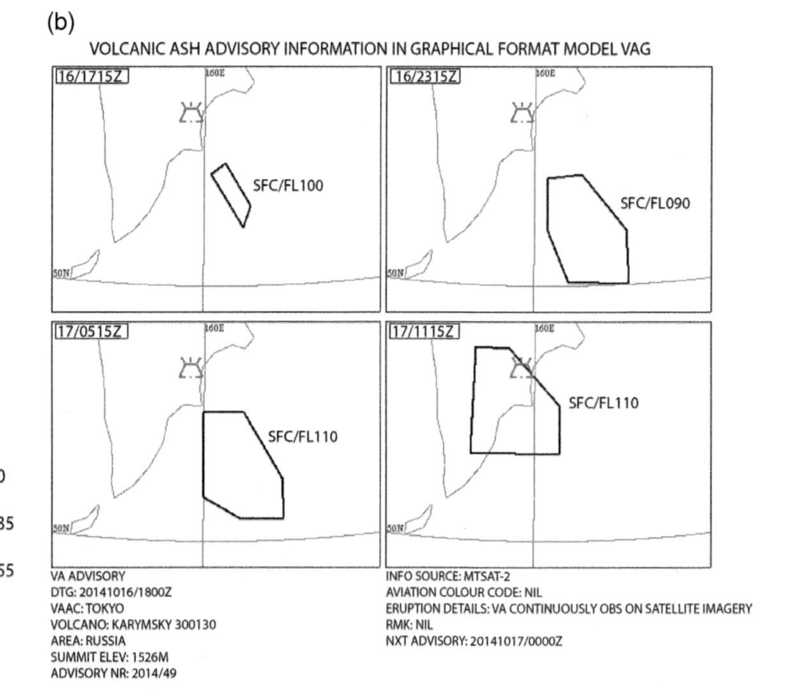

Figure 5.2 (a) An example volcanic ash advisory (VAA) from the Tokyo VAAC and (b) its accompanying volcanic ash graphic (VAG). This is for Karymsky volcano, Russia, on 16 October 2014 at 18:00 UTC or Z. Note in the VAA that satellite data were used as the source for the detection of the event. The VAG provides the satellite data at +0 hours and the forecasted locations at +6, +12, and +18 hr as reported by location points in the VAA. (The VAA is from *Tokyo VAAC* [2014a] and the VAG is from *Tokyo VAAC* [2014b].)

	ASH3D	ATHAM	FALL3D	FLEXPART	HYSPLIT	JMA-GATM JMA-RATM	MLDP0	MOCAGE	NAME	PUFF	TEPHRA2	VOL-CALPUFF
Operational												
Approach (1)	E/H	E	E	L	H	L	L	E	LH	L	E	H
Method (2)	N	N	N	N	N	N	N	N	N	N	A	S
Coverage (3)	LRG	L	LR	LRG	LRG	RG	LRG	G	LRG	LRG	L	LR
Physics												
Topography												
H wind adfection												
V wind advection												
H atm. diffusion								See (5)				
V atm. diffusion												
Particle sed.												
Other dry dep.												
Wet deposition												
Dry part. aggr.												
Wet part aggr.												
Variable part. shape												
Gas species												
Chemic. processes												
Granulometry												
Variable size class												
Variable GS distr.												
Variable size limites												
Source term												
Mass distribution (4)	PS/U/LN	O	ALL	PS/L/U/P/O	PS/L/U/P/LN	PS/L/LN	PS/L/U/P/O	PS/L	PS/L/U/BP/O	PS/L/U/P/O	L/U/LN/O	PS/BP

(1) L=Lagrangian, E=Eulerian, H=Hybrid
(2) A=Analytical, S=Semi-analytical, N=Numerical
(3) L=Local, R=Regional, G=Global
(4) PS=Point Source, L=Linear, U=Umbrella-type, P=Poisson, LN=Log-normal, BP=Buoyant Plume, O= Other
(5) Neglected. Diffusion of numerical origin appears to be sufficient, with particularly good results at 0.5°.

Figure 5.3 Suite of volcanic ash transport and dispersion models used for simulations of volcanic ash clouds. For full names of the models, see within *WMO* [2013] adapted from Table 1 in *WMO* [2013].

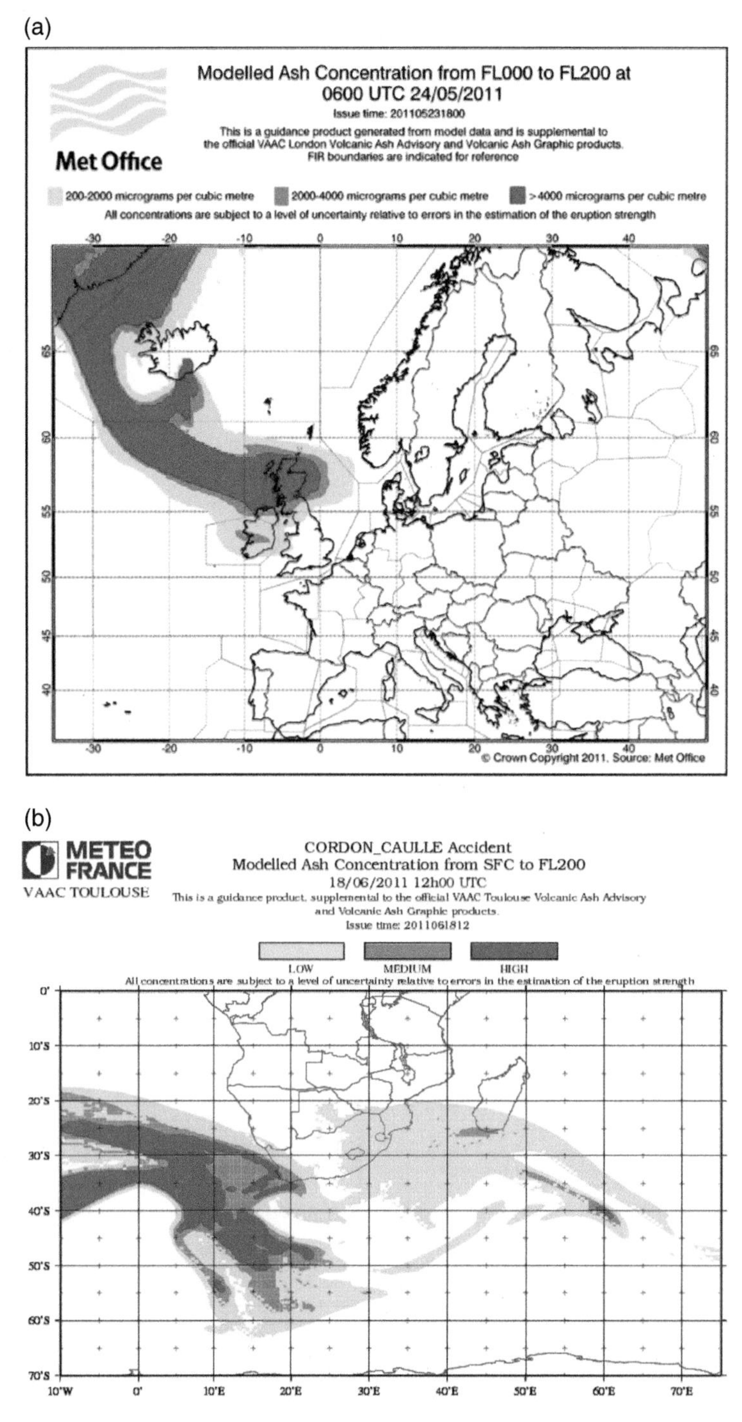

Figure 5.4 Volcanic ash concentration maps produced as additional information to the standard VAA and VAG. Examples are from the (a) London VAAC for Grímsvötn eruption and (b) from the Toulouse VAAC for the Cordon Caulle eruptive event, both from 2011 (Grimsvötn image is from GVP of Smithsonian [*GVP*, 2014]).

and 2013 [see *WMO*, 2010, 2013], into the uncertainty in the inputs to the models used for simulating the volcanic ash dispersion modeling and how to progress from deterministic modeling, one set of inputs and/or one numerical prediction model to probabilistic modeling of the dispersing ash clouds. There is now a considerable amount of research into (1) improving our understanding of the uncertainties in the model inputs, (2) how to use observational data to constrain the model simulations, and (3) how to use probabilistic simulations of modeled ash clouds with their associated uncertainty in decision making and for the VAACs to produce their VAA.

Often the first detection of a new volcanic eruption will come from a local volcano observatory, which will then inform the VAAC through the observatories reporting mechanism. One example of this mechanism is the US volcano observatories volcanic activity notice for aviation (VONA) or volcanic activity notice (VAN). During the first stages of an event, certain VATD modeling inputs may be unknown or have a large level of uncertainty and as such they will need to be approximated until any observation can be made. By assessing the potential range of the model inputs and distribution of likelihood across this range, probabilistic modeling of the downwind concentrations is possible. For example, what is the maximum and minimum mass eruption rate (MER) given the known eruption information and what is the probability density function across the range of MER (i.e., uniform likelihood of any MER or does one MER have a higher likelihood)?

By coupling all the potential variability in the modeling inputs, the VAACs and those other groups producing volcanic ash cloud models, such as those highlighted in *WMO* [2010, 2013], can then build a suite of potential ash cloud simulations, each with a weighted probability given the inputs. Once other observations of the eruption are available, such as satellite remote sensing data, ground-based radar data, or visual observations, then the full spectrum of probabilities of downwind ash concentrations and mass loadings can be constrained by those parameters measured in real time. Then for the VAACs, these probabilistic modeling results can be used in the VAA.

In this paper, there is a discussion on how to first classify the uncertainties, in terms of the parameters with the greatest impact and least knowledge, in producing the volcanic ash cloud dispersion modeling, and then quantify the uncertainties to (1) improve the confidence in the dispersion modeling, (2) reduce the overconfidence that sometimes occurs from deterministic forecasts, and (3) produce the most informative simulation of the future ash-cloud location. Finally, the paper highlights some of the methods being used to implement this uncertainty modeling and discuss how they could be applied in a real-time operational environment and be of most use in decision-support systems.

This chapter focuses on the uncertainty in ash-cloud modeling specifically for the aviation community. In terms of ashfall forecasting, additional operational services exist, such as the ashfall maps often generated by a volcano observatory or from a VATD model and provided by the VAACs, and there are uncertainties that exist both in the inputs to this ashfall modeling and in the interpretation of the results and how to compare the data to any field measurements. Adding in these uncertainties to the modeling described here would be large undertaking and would be a separate manuscript in itself. Therefore, the distinction is made here in the introduction of the clear focus on ash-cloud-dispersion modeling for aviation rather than in terms of ashfall modeling of the proximal ash hazards.

In addition to this distinction on the hazard that this work is aiming to analyze in terms of uncertainty, there is a need for further clarification that the focus is mainly on input uncertainty applied to ash-cloud modeling, with some references to research that has aimed to incorporate the uncertainties in the numerical weather prediction models used in ash-cloud modeling. This paper does not aim to analyze the uncertainties for model parameterization processes, such as sedimentation schemes, wet-dry deposition, or the approach for particle aggregation. Also, the paper does not aim to analyze the uncertainties of the structural integrity of the dispersion models and how well they perform as an approximation to reality. As with the ashfall modeling uncertainty analysis, these topics of model parameterization and structural integrity uncertainties could be discussed at great length in separate papers and as such are not discussed in this paper.

The aim of the paper is to highlight, for ash-cloud modeling for civil aviation, how groups tackle classifying or determining the uncertainty (i.e., which input parameter has the greatest impact), and then quantifying the uncertainty in the parameter of interest. The final sections of the paper discuss how one is then able to mitigate the uncertainty and apply the knowledge on uncertainty quantification to real-time decision support. Readers of the chapter will be able to increase their knowledge of uncertainty analysis as applied to dispersion modeling of ash clouds, with special attention made to the inputs of the models and their impact on the probabilistic modeling and decision-support processes.

5.2. CLASSIFYING THE UNCERTAINTY

In terms of uncertainty analysis as applied to volcanic ash-cloud modeling, the chapter aims to determine the different parameters that have an associated uncertainty in their absolute measurement that can then have an impact on the accuracy and confidence in the dispersion modeling forecasts. The paper is not defining new terms for the inputs parameters or new methods on uncertainty classification. As can be seen in Table 5.1, the different input parameters have been classified by the ability to be detected in real time or assumed from past eruptions or empirical relationships. Describing all the uncertainties as documented in the Introduction is beyond the scope of this paper and the focus of attention is on the input parameters to the dispersion models.

As a volcano erupts, there is a need to forecast the future location of any potential volcanic ash cloud. At the first stage of the eruption, there are often uncertainties in the inputs needed for the VATD models. Table 5.1 documents the input parameters that are needed by the different VATD models documented in Figure 5.3. Some of these are known at the time of a reported

Table 5.1 Inputs Needed for VATD Models to Simulate Volcanic Ash Cloud Location and Airborne Ash Concentrations After an Eruption Has Been Detected

Input parameter	Provided/detectable/assumed/derived
Plume height (m or km)	Derived/detectable
Mass eruption rate (kg/s)	Derived
Total erupted mass (kg)	Derived
Start time	Detectable
Duration (sec, min, hr)	Assumed/detectable
Vertical distr. (Uniform/exp./umbrella)	Assumed/detectable
Particle size distribution	Assumed
Numerical weather prediction data	Provided

Note: Sec = seconds, kg = kilogram, m = meters, km = kilometers, min = minutes, hr = hours, distr. = distribution.

eruption, such as the start time, and are listed as a "detectable" parameter. For others, they may be detectable (i.e., measureable) such as plume height, vertical distribution, and event duration from observations, but these may not be known for all possible eruptions. Hence, these parameters are listed as either "derived/detectable" for plume height or "assumed/detectable" for duration.

There are parameters, such as mass eruption rate (MER) and total erupted mass, that are derived from other data. *Sparks et al.* [1997] and *Mastin et al.* [2009] list just two methods, which have been developed, that link plume height above the vent and mass eruption rate. Total mass is then simply the combination of MER and total event duration. Next, there are inputs that are assumed and directly specified based on a volcano's past history of eruptive events or using a defined database, such as the *Mastin et al.* [2009] catalogue. In Table 5.1, the particle size distribution (PSD), also known as the total grain size distribution (TGSD), is one of these assumed parameters. *Mastin et al.* [2009] list a set of eruption source parameters that can be used to specify the TGSD for the volcano, while *Webley et al.* [2009] assessed the significance of different TGSD on dispersion model simulations and *Mastin et al.* [2014] adjusted the TGSD for their modeling simulations based on observed deposits.

The final input parameter type is the one termed "provided." This relates to data that is always available, irrespective of the occurrence of a volcanic eruption. For the VATD modeling, this is the numerical weather prediction (NWP) data as it is generated routinely for NWP forecasting. There are uncertainties that can exist in the NWP data from the wind fields, precipitation data, atmospheric stability, and these need to be incorporated using ensemble NWP modeling. Describing all of the different uncertainties is beyond this paper and we focus on the input uncertainties used for dispersion modeling of ash clouds.

There are global to regional data available at varying spatial resolutions dependent on the location of the volcanic event, and the chosen NWP data for each simulation depends on the spatial and temporal resolution of the most appropriate dataset. From the list of inputs in Table 5.1, the question remains where does the uncertainty in the downwind ash-cloud location and airborne concentrations originate? Which parameter has the greatest significance on the downwind concentrations? *Webley and Steensen* [2013] provide detailed discussions on the individual inputs to the model and their level of uncertainty at the time of the volcanic event. In this section of the chapter, the focus in on how to classify the uncertainty and discuss some of the more significant parameters.

Given that the volcanic plume height is recorded from other observations, such as from satellites [see *Holaske and Rose*, 1991; *Zaksek et al.*, 2013] or ground-based radar data [see *Rose et al.*, 1995; *Marzano et al.*, 2006], then the focus is on the determination of the mass eruption rate and also the defined TGSD. Knowledge of the volcanic particle shape and density would reduce the uncertainties in downwind ash concentrations, but they have a smaller impact on the uncertainties than the MER and TGSD. There are uncertainties that can exist from the method used to measure the plume height (such as pixel size of the satellite height, plume opacity, distance from ground radar to the source, and temporal frequency of the data). *Webley and Steensen* [2013] provide details on how these can impact plume height measurements and they concluded that the uncertainty in deriving the MER from the plume height is greater than the uncertainty in the plume height measurement and can have a larger impact on the total erupted mass and downwind ash concentrations.

Using the methods of *Sparks et al.* [1997] or *Mastin et al.* [2009], then past eruptions can be used to approximate the MER given the eruption plume top altitude. How representative are these methods to determine the actual MER for the current eruption? *Mastin et al.* [2009] state that the method can determine the MER with an accuracy of a factor of four, so the real question is how can we fully quantify the uncertainty in the MER given the volcanic event plume altitude? Also, is the MER constant through the duration of the event, and should the instantaneous MER for the maximum plume altitude be used through the full duration, or should a mean MER be applied with the event duration to determine the total mass erupted? Often in an operational setting, observations of plume height will be collected with a low temporal frequency and so modelers will use one or a few plume height measurements to represent the varying MER as they may not be able to obtain the data to capture the time-varying plume altitude.

The empirical relationships of *Sparks et al.* [1997] and *Mastin et al.* [2009] provide a first approximation of the MER using past events and the associated best fit relationships. They are neither volcano specific nor do they account for local environmental factors, such as local wind conditions, atmospheric stability, and structure. *Tupper et al.* [2009] assessed the sensitivity of the tropospheric instability on how "tall" clouds can be generated from small eruptions. For example, different MERs could lead to the same plume top height when comparing an eruption in a tropical versus polar region. One-dimensional models, like Plumeria [*Mastin*, 2007] and BENT [*Bursik*, 2001], can produce a second form of approximation, where local atmospheric conditions are taken into account, along with information on the volcano itself. The BENT model was adapted for local crosswinds and uses radiosonde data to help constrain how the local atmospheric environment will impact how the MER correlates to the plume top height.

There have been recent updates [*Woodhouse et al.*, 2013; *Degruyter and Bonadonna*, 2013] to improve upon these methods. The impact on the relationship between MER and plume height can be dramatic for weak eruption plumes. While the more empirical relationships are a focus here in this chapter, the work of *Woodhouse et al.* [2013] and *Degruyter and Bonadonna* [2013], along with the models of *Mastin* [2007] and *Bursik* [2001] can be used for a specific volcano and time and therefore include the impact of the local atmospheric conditions as a factor on the uncertainty of the MER. The empirical relationships provide a good first approximation of the MER from the plume height, and by using plume rise models, including those with local wind effects, the full range of uncertainty in the MER from the plume height, specific to the eruption, can be included.

For the TGSD, either analysis of the historical activity at the volcano is needed to assess the potential distribution for a given eruption or the VATD models use an eruption source parameter (ESP) distribution from work such as in *Mastin et al.* [2009]. Analysis of the archival eruptions at every volcano worldwide is a massive task, and so models assume a TGSD based on, if available from past eruptions, the experience of the model developers or local observatories or they use the ESP TGSD from *Mastin et al.* [2009] for the volcano. The *Mastin et al.* [2009] research classified each volcano by an eruption type, and so they were able to assign the TGSD for that type to the volcano of interest. Of the 60 or so eruptions used in their MER relationship, *Mastin et al.* [2009] chose the most representative eruption for each ESP type and then used this for all volcanoes worldwide with the same assigned ESP. For example, Kasatochi volcano, Alaska, was assigned the ESP type of S2, or a medium silicic eruption. This used the Mount Spurr, Alaska, USA, eruption of August 1992 [*Eichelberger et al.*, 1995] as the default event for the ESP type. So the TGSD for S2 has 40% of the erupted mass for ash particles < 63 μm. This defined size of 63 μm represented the size of particles that would disperse downwind and those coarser than 63 μm would fall out as proximal deposits. This definition did not take into account aggregation processes so that finer particles may fall out close to the volcano, proximal, and coarser particles, as aggregates, could be measured at distal locations.

The questions that then arise in classifying the uncertainty in the TGSD come from how representative is the Spurr August 1992 eruption to an eruption at Kasatochi volcano. Will the TGSD have 40% of the erupted mass for ash particles < 63 μm? How much will the actual TGSD vary from this default ESP? What is the possible range in the TGSD that could be assumed from Kasatochi? For example, the Mount Spurr eruption was for a 3 hr duration event to 11 km ASL [see *Mastin et al.*, 2009, for the defined parameters]. But the Kasatochi eruption of 2008 was the accumulation of three separate events with a maximum of 1 hr in length with altitudes of 14 km and 18 km ASL [see *Waythomas et al.*, 2010, for more details]. So given these differences in the eruption length and altitude, how accurate is the TGSD from the ESP type S2 to Kasatochi? How do those using the VATD models then characterize and quantify the uncertainty in their model forecasts needed for decision making?

The uncertainties in the model inputs together add another level of complexity to classifying and quantifying the parameters with the greatest impact on the VATD model's ability to forecast the clouds location, as there can be uncertainty propagation from one parameter to the next. The chapter focuses on the uncertainties in the model inputs rather than also including the uncertainties in the future activity of the volcano itself. Adding in this predictive uncertainty would add another dimension to the uncertainty and require considerable amount of additional work in modeling of magma dynamics and predictive analysis of seismic signals. A discussion on this uncertainty is beyond the scope of this manuscript and hence the focus is on the uncertainty of the input parameters that then impact the downwind dispersion modeling.

So at the start of an event, which parameter would have the greatest significance on the downwind ash concentrations? How can we quantify the uncertainty in the inputs and assign the probabilities of the most likely input data? The time-varying nature of the initial plume height would be one of the most significant, given how the plume top height is often used to infer MER and therefore the temporal frequency of observations can have an impact on the observed/recorded altitude. Quantification of the uncertainty that can exist in this parameter and the other inputs to the modeling will allow those in decision support to have increased confidence in the locations and concentrations of the downwind ash clouds. All the uncertainties that can exist in the inputs, documented in Table 5.1 has lead the modeling community from deterministic to probabilistic modeling. Once the uncertainties can be identified and quantified, then probabilistic maps of cloud locations and concentrations can be produced.

5.3. QUANTIFYING THE UNCERTAINTY

The previous section touched on moving from classifying uncertainties in volcanic cloud dispersion modeling into quantifying the uncertainty and build probabilistic forecasting maps. The model inputs listed in Table 5.1 each have their own level of uncertainty, with some measurable during an event and others assumed from past history or defined source parameters. Note that in the absence of an aggregation term in the dispersion modeling, the downwind concentration is directly proportional to the MER at the source. Hence, an error or uncertainty in the MER of 50% at the source implies an error/uncertainty in the modeled ash cloud at any point in the domain.

The focus in this section will be to discuss how the VATD modelers would be able to quantify the uncertainty in the mass eruption rate, or MER. This is one of the most significant parameters, when producing downwind ash concentrations, and there is a combination of multiple input parameters, each with an associated uncertainty, that can generate significant variability in the derived mass eruption rate as compared to the actual mass eruption rate. So first, there is the accuracy in the measured plume height. Table 5.2 illustrates that if the recorded height had an accuracy of 0.25 km for a plume at 5 or 10 km above the vent height, then the MER used in the VATD models could vary by 22% and 11%, respectively. This shows how significantly the derived MER can vary with only a small uncertainty in the measured plume top. *Mastin et al.* [2009] state that their equation, determined from multiple past eruptions, has an accuracy factor of four for any given eruption from the best fit equation, and this can increase for weak plumes where local atmospheric conditions can impact the plume rise. This means that for some eruptions the MER from their equation is either half of the likely MER or double that actual MER. Couple this to the potential accuracy in the plume height and then the derived MER can only be determined within a factor of 10.

The final item that can be coupled to the plume height measurement is the choice of MER to use during the eruptive event. The VATD modelers could use the maximum plume height measured, as this is often the only reported plume height of an event, and then determine the MER and apply this for the full length of the event. For example, the 18 km ASL plume height from the third event at Kasatochi 2008, as reported by *Waythomas et al.* [2010], was the maximum height reached during the 1 hr event. This height leads to a MER of $2.37*10^7$ kg/s. But this is the instantaneous MER at the time of the recorded maximum height? The mean MER across the 1 hr event is likely lower, as the maximum plume height was not sustained during the full event. *Harris et al.* [2007] show the different measureable effusion rates for volcanic eruptions and these temporally varying effusion rate methods are also applicable for the MER. There is the effusion rate or ER that is the mean rate of effusion/emission across the full cycle or full eruption. This is then the mean MER during the eruptive event length. Also, there is the "instantaneous effusion rate or IER" that is the effusion/eruption rate at one set time (or a very small time period). For the MER, this is the rate at one time given the plume height (i.e., at the time of the measurement).

Arason et al. [2011] shows the time varying plume heights from the Eyjafjallajökull eruption in 2010, where ground-based radar measurements were possible every 5 min.

Figure 5.5 shows the significance of the temporal frequency of measurements. By varying the average time

Table 5.2 Range in Possible Mass Eruption Rate (MER) Given 0.25 km Accuracy in Plume Height Measurements

Height above vent (km)	Max/min possible heights	MER (kg/s)	Variation of MER
5	4.75	$9.41*10^4$	
	5.0	$1.16*10^5$	22%
	5.25	$1.43*10^5$	
10	9.75	$1.86*10^6$	
	10.0	$2.07*10^6$	11%
	10.25	$2.29*10^6$	

Source: MER calculated from *Mastin et al.* [2009] equations.

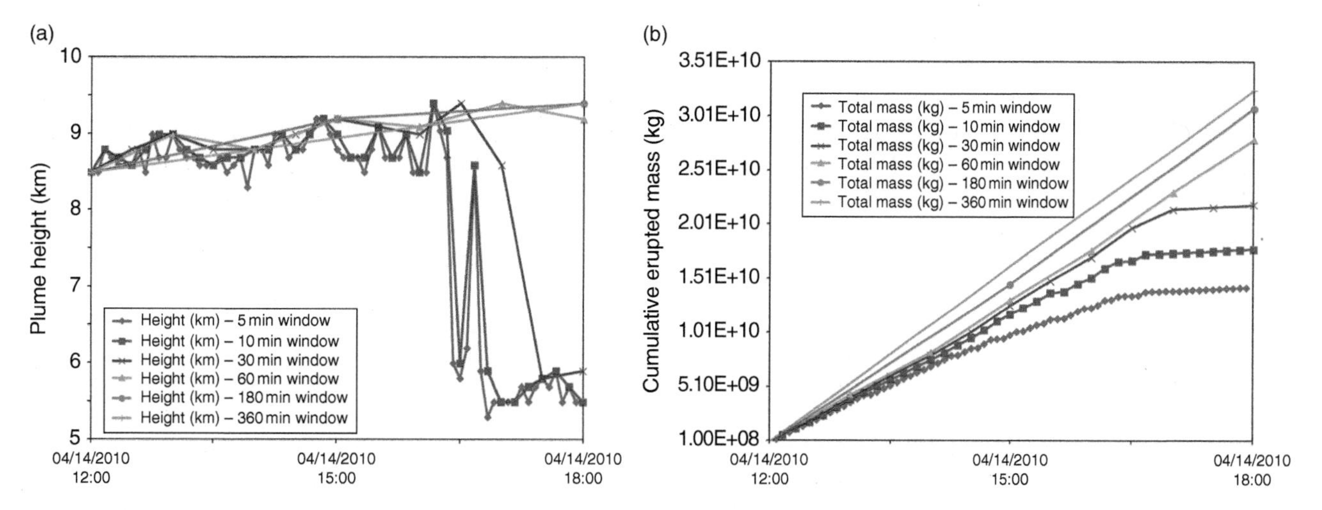

Figure 5.5 (a) Maximum plume heights from Eyjafjallajökull eruption from 14 April 2010 at 12:00—18:00 UTC over varying time windows and (b) significance of temporal repeatability of plume height measurements on mass eruption rates. (Data derived from *Arason et al.* [2011]).

period of the measured heights, and recording the maximum height in the time window, the data show that with measurements every 5 min, the total mass from 12:00–18:00 UTC on 14 April would have been $1.73*10^{10}$ kg (Fig. 5.5b). If radar data would have been available only every 30 min, then the total mass from the plume heights would have been $2.35*10^{10}$ kg or nearly 1.4 times larger (Fig. 5.5b). Varying the time interval to every 3 hr, the total mass would have been $4.05*10^{10}$ kg or 2.4 times higher. The variations in the total mass come from the varying plume heights later in the 6 hr period (Fig. 5.5a), which are not accounted for when the temporal window is increased to 3 or 6 hr. Often the temporal frequency of plume heights is not at the 5 min window in Figure 5.5 and therefore the mass eruption rate is likely overestimated by using one plume height measurement for the eruption length. This is another uncertainty in the MER, to add to the plume height accuracy and best-fit relationships used to derive the rate from the plume height above vent.

Given the quantified uncertainty in the range of VATD model inputs from Table 5.1, the questions raised are: How do we place a range on the uncertainties? What is the probability density function across this range? Is it uniform or Gaussian? How do we combine the uncertainties from the suite of inputs to build a probabilistic modeling environment? How do we sample the uncertainty space? How do we determine the skill in the resulting probabilistic modeling? What methods exist to increase our confidence both in the input parameter data (like mass eruption rate and true particle size distribution) and how we can then increase the confidence in the probabilistic modeling? With increased knowledge of the inputs, the VATD models can then produce the downwind ash concentrations with higher probabilities in both in terms of location and actual concentration. The following section will discuss some examples of how research projects are working on mitigating uncertainty and increasing confidence in the VATD modeling of volcanic ash clouds.

5.4. MITIGATING THE UNCERTAINTY AND INCREASING CONFIDENCE

The suite of VATD model inputs and their associated uncertainties has led the research and operational modeling communities to pursue probabilistic modeling of volcanic ash clouds. There is a need to both measure the uncertainty and mitigate its impact on the forecasted ash clouds location and concentrations to lead to increase confidence for decision support. The previous section focused on the uncertainties in the mass eruption rate from a volcanic eruption and how this can propagate into impacting confidence in absolute volcanic ash concentrations downwind of the volcano. While the *Mastin et al.* [2009] or *Sparks et al.* [1997] relationships are useful tools to determine range of the eruption rates, they do not take account of the actual volcano's physical properties or the surrounding atmospheric conditions. However, these two together can influence the mass eruption rate required to reach the plume top altitude.

Several one-dimensional and three-dimensional modeling efforts have developed to provide a better representation of the actual mass eruption rate. These include BENT [*Bursik*, 2001], Plumeria [*Mastin*, 2007], and the work of *Woodhouse et al.* [2013]. Here, the characteristics of the volcanic eruption and volcano itself, such as eruptive velocity, magmatic temperature, vent radii, and water content of the plume are taken into account. These models also use local atmospheric profiles of temperature, pressure, and water content to determine the vertical plume shape, particle size distribution, and mass eruption rate to reach the measured plume height from observations. This methodology can provide a closer representation to the actual eruption rate, rather than using the *Mastin et al.* [2009] and *Sparks et al.* [1997] relationships, but is still a modeling of the actual event.

To constrain the mass eruption rate, other researchers have focused on using observations, where available, of the growing volcanic plume to determine the eruption rate of the ongoing event. Some of these methods include using high temporal resolution satellite data [*Pouget et al.*, 2013], close proximity high temporal resolution radar measurements [*Schneider and Hoblitt*, 2013], and infrasound and thermal imagery [*Ripepe et al.*, 2013]. These methods measure the eruption rate from the observations of the erupting volcano, but require the specific observational data to constrain the eruption rate, which is not always available for all eruptions.

Finally, to constrain the model inputs and therefore provide an improved forecast, data assimilation or inverse modeling approaches have been applied, for example *Kristiansen et al.* [2010], *Stohl et al.* [2011], *Boichu et al.* [2014], and *Madankan et al.* [2014]. Here, observations of the dispersing volcanic cloud are used to constrain the suite of probabilistic modeling simulations to a smaller subset. There is also the ability to generate satellite-derived measurements of ash-cloud characteristics, each with its uncertainty or probability of accuracy/error estimate, (see references in *Pavolonis et al.* [2013] and application with VATD models in *Patra et al.* [2013]). These are then used to forecast the future location and ash-cloud concentrations. As more observational data are available, the simulations are again constrained to the smaller suite that matches the observations.

One important item to note here is that while probabilistic modeling can be performed, the output from this form of modeling or the ensemble forecast can still be a deterministic result (i.e., the ensemble mean), with associated uncertainties (i.e., the dispersion of the members around the mean). Unless those using the ensemble results understand how the mean and associated uncertainties are generated, the mean results can be used inappropriately as then can often be more disperse than each simulation member and need observational data to constrain the ensemble and assess the variability in the dispersion around the mean as compared to observed data.

For the uncertainties in the surrounding atmospheric conditions, dispersion models can use NWP ensemble modeling. *Stefanescu et al.* [2014] and *Vogel et al.* [2014] illustrate two examples where these data have been applied to probabilistic modeling. Here, either updated ensemble members are used to initialize the dispersion modeling simulation or each member is used to initialize a new VATD model simulation and the resultant suite of volcanic ash dispersion-modeling results is combined in a probabilistic map of airborne ash concentrations and the variability generated from the different ensemble members. The question raised here is how to weight the individual ensemble members in the downwind probabilistic modeling maps of airborne ash concentrations, and how often to reinitialize the VATD model simulation with the ensemble member.

In this monograph, *Webley et al.* [2016] and *Kristiansen et al.* [2016] document two examples where approaches to constrain the dispersion-modeling simulations are applied to improve confidence in downwind ash concentrations and the accuracy of the forecasts relative to observational data. *Webley et al.* [2016] focus on a workflow that incorporates the uncertainties in the VATD model inputs and ensemble members from NWP data to generate probability maps. These are then compared to the probabilistic satellite observations of methods in *Pavolonis et al.* [2013] to improve confidence in the modeling results. *Kristiansen et al.* [2016] document the inverse modeling approach where satellite data are used to constraint the best-fit modeling simulations using cost-benefit methods, allowing the initial conditions to be constrained to a smaller sample dataset. The simulations with the highest match results are used to forecast the clouds future location.

There has been a big push to increase confidence in the volcanic ash modeling simulations ability to produce downwind ash concentrations. *WMO* [2010; 2013] brought together the operational and research communities to better focus the research to produce operational tools and products of use in decision support and to allow the operational community to understand and quantify the capabilities that can be developed for VATD modeling. The final question on the role of uncertainty in decision support for volcanic ash is how to implement it and use the results to provide the best product to the aviation community?

5.5. APPLICATION OF UNCERTAINTY INTO THE DECISION SUPPORT SYSTEM

Figure 5.2 illustrates the current volcanic ash advisory design as used by the volcanic ash advisory centers worldwide. Here, a text product of ash-cloud location and +0 hr location from observational data is produced along with locations at +6, +12, and +18 hr.

Figure 5.4 shows how in 2010 and 2011 derived products were being produced with defined concentrations for the ash-cloud boundaries. Research since has shown the level of uncertainty that exists in modeling inputs, observational data, and weather prediction data changes the field in terms of moving from deterministic model simulations with one set of inputs to a suite of probabilistic model simulations. But how does the operational decision-support community move from graphics of ash-cloud boundaries to maps of probabilities and portray this information to the aviation community? For a VAAC to be able to change its VAA product and VAG, policies need to be changed through ICAO procedures. For this, there would need to be a standard model setup to allow one suite of model simulations to be compared to any other generated product.

Webley and Steensen [2013], in a previous American Geophysical Union monograph on Lagrangian modeling, discuss the advantages and disadvantages of how probabilistic modeling could be used in a decision-support system. The application of generating probabilities in weather forecasting has been used for decision support and weather advisories. The role of uncertainties or probabilistic modeling for volcanic ash decision support should look to learn from this community and assess how the ensemble NWP data are used in day-to-day weather forecasting.

For volcanic ash, retrievals of cloud characteristics have recently produced observations with associated probabilities. *Pavolonis et al.* [2013] document the techniques and retrieval schemes to be able to produce volcanic ash retrievals. Figure 5.6 shows an example from operational tools developed to automatically calculate these cloud characteristics, with associated uncertainty or error estimates on the measurements. Here, the results from the online tools show the eruption of San Miguel volcano in San Salvador on 29 December 2013. *Patra et al.* [2013] show how these can then be coupled to the VATD modeling to constrain the probabilistic forecast location and concentrations to improve confidence in further simulations.

It's critical to assess how these methods would be applied in real-time decision support systems and then be used for aviation and population safety. An assessment of their use and how to effectively build them in the decision tree process is beyond this chapter and ultimately it may depend on the loss model of the user and the application where it will be applied. Once the international aviation community is able to assess the generated products and update policies to use them in a real-time environment, then uncertainty/probabilistic modeling and the simulated maps of airborne ash concentrations with associated error can be used. Until then, the work will stay in the research community. Continued work by meetings such as *WMO* [2010, 2013] and the work of the International Airways volcano watch operations working group (IVAWOPSG; see *ICAO* [2014]), will focus when and how the probabilistic modeling of volcanic ash cloud is implemented into operational volcanic ash advisories.

5.6. CONCLUSIONS

Recently, there have been developments in the application of uncertainty modeling to volcanic ash cloud model simulations [see *WMO*, 2010, 2013], where groups have implemented coupled systems for the dispersion modeling with one-dimensional plume rise models [*Bursik et al.*, 2012; *Devenish*, 2013; *Woodhouse*, 2013] and also examined how to incorporate the uncertainty in the wind fields with NWP ensemble members [e.g., *Stefanescu et al.*, 2014]. The aim of all this research has been to classify the uncertainties in downwind ash concentrations and then quantify the impact of the model and NWP uncertainties on the accuracy of the ash-cloud simulations as compared to volcanic ash cloud observations. Then the decision support community and operational groups responsible for real-time advisories for aviation, the ICAO VAACs, can increase the users confidence in the forecasted ash-cloud locations and concentration data.

In this chapter, discussions have centered on how the uncertainties in volcanic ash-cloud modeling are first classified and then quantified. The focus in this chapter is on ash-cloud modeling, as this is used operationally in decision support by VAACs to produce advisories for the aviation community. Additionally, discussions provided some examples and methods being applied to reduce the uncertainty in ash-cloud-dispersion modeling and increase model confidence for downwind ash-cloud-concentration forecasts. Comparisons with the newly developed probabilities in ash-cloud observational data from groups like *Pavolonis et al.* [2013] can only lead to more application of the uncertainty estimates and model confidence probabilities in day-to-day operations.

The bigger question that arises is how to use this uncertainty information in probabilities of specific ash concentrations in a decision-support system. The uncertainty modeling within the research community can provide simulations outputs to better understand the eruption and dispersing ash cloud, but how would a 90% probability of a concentration of 10 mg/m^3 at a defined location be used in a VAA by a VAAC? How would this information be interpreted by the aviation community? Clearly, the 2010 eruption from Eyjafjallajökull volcano and other eruptions since have shown the need for a better understanding of the model uncertainties and complexities when generating a volcanic ash-cloud simulation. There are methods that can be applied from the NWP community to evaluate the probabilistic forecasts [e.g., *Anderson*, 1996; *Roulston and Smith*, 2002] and the ash-cloud modeling community could analyze how these could be applied in a real-time environment.

As the role of uncertainty is further integrated into the volcanic ash-cloud community, there are lessons that can be learned from other communities, such as the numerical weather prediction community. Here, probabilities have been used for day-to-day weather forecasts and then applied by the aviation community. Lessons learned from these groups on how they have integrated them in air traffic management can only help the transition of the probabilistic modeling of volcanic ash clouds and how model uncertainties or model simulation confidence is then applied directly in the operational environment.

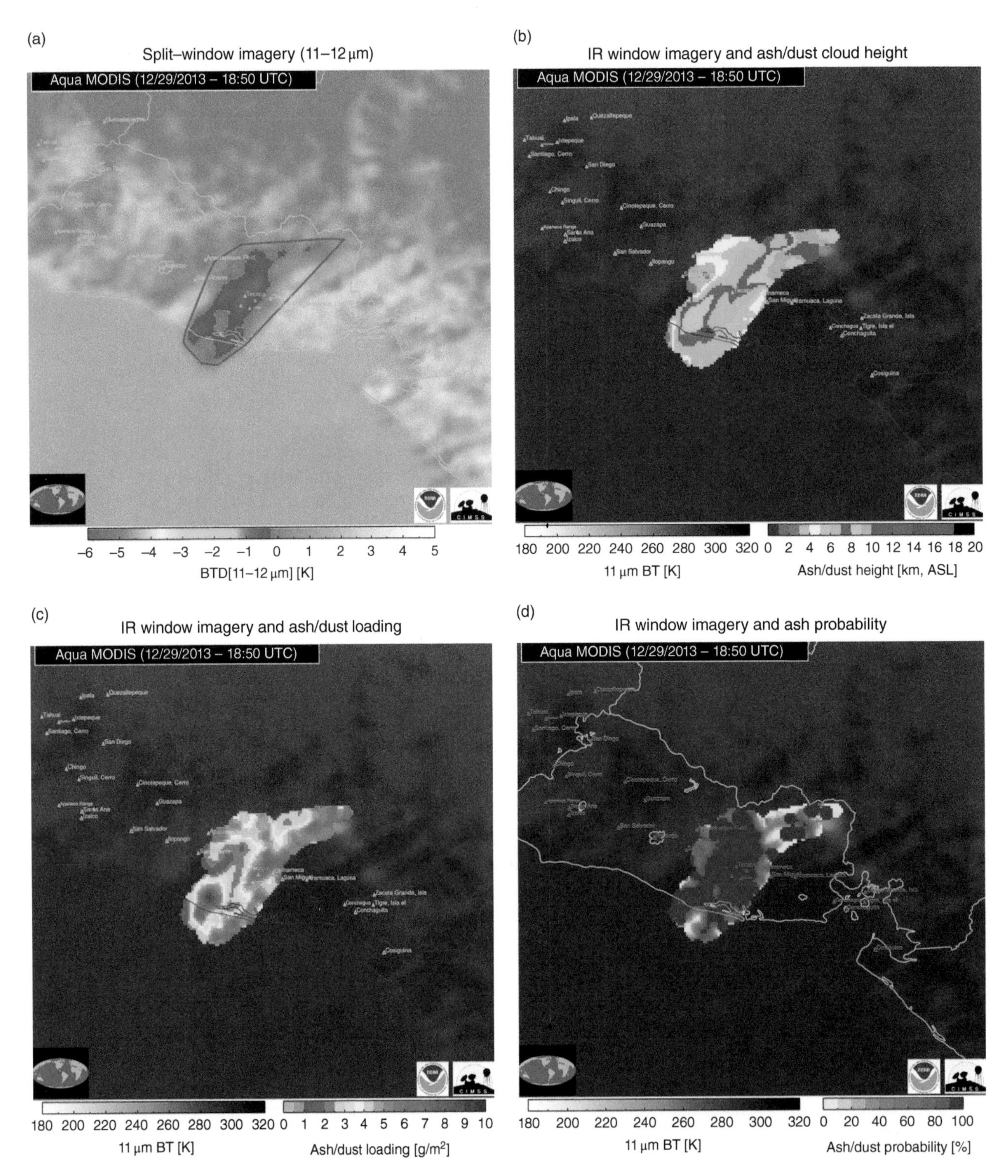

Figure 5.6 Volcanic ash-cloud products with probability of ash occurrence. Eruption from San Miguel volcano on 29 December 2013: (a) Volcanic brightness temperature split window, (b) retrieved cloud top height, (c) retrieved volcanic ash loading per pixel, and (d) ash probability of detection (Data from Aqua MODIS satellite and http://volcano.ssec.wisc.edu/tutorials).

ACKNOWLEDGMENTS

This work was funded by National Science Foundation (NSF) Interdisciplinary/Collaborative Research under grant no. CMMI-1131799. I would like to thank the two anonymous reviewers for their excellent suggestions and assistance in developing this manuscript for publication. I also thank the editors in chief and associate editors of the American Geophysical Union Monograph for the opportunity to publish this manuscript and for bringing all the different manuscripts together to develop such a unique and diverse publication.

REFERENCES

Anderson, J. L. (1996), A method for producing and evaluating probabilistic forecasts from ensemble model integrations, *J. Climate*, *9*(7), 1518–1530.

Arason, P., G. N. Petersen, and H. Bjornsson (2011), Observations of the altitude of the volcanic plume during the eruption of Eyjafjallajökull, April–May 2010, *Earth Syst. Sci. Data Discuss.*, *4*(1), 1–25.

Boichu, M., L. Clarisse, D. Khvorostyanov, and C. Clerbaux (2014), Improving volcanic sulfur dioxide cloud dispersal forecasts by progressive assimilation of satellite observations, *Geophys. Res. Lett.*, *41*(7), 2637–2643.

Bursik, M. (2001), Effect of wind on the rise height of volcanic plumes, *Geo. Res. Lett.*, *28*(18), 3621–3624.

Bursik, M., M. Jones, S. Carn, K. Dean, A. Patra, M. Pavolonis, E. B. Pitman, T. Singh, P. Singla, P. Webley, H. Bjornsson, and M. Ripepe (2012), Estimation and propagation of volcanic source parameter uncertainty in an ash transport and dispersal model: Application to the Eyjafjallajökull plume of 14–16 April 2010, *Bull. Volcanol.*, *74*(10), 2321–2338.

BVGN (2012), Monthly report on Puyehue-Cordon Caulle volcano, June 2011 eruption emits circum-global ash clouds, 37:03, *Bull. Glob. Vol. Net.*, viewed 27 October 2014, http://www.volcano.si.edu/world/volcano.cfm?vnum=1507-15=&volpage=var#bgvn_3703.

Degruyter, W., and C. Bonadonna (2013), Impact of wind on the condition for column collapse of volcanic plumes, *Earth Planet. Sci. Lett.*, *377*, 218–226.

Devenish, B. J. (2013), Using simple plume models to refine the source mass flux of volcanic eruptions according to atmospheric conditions, *J. Vol. Geotherm. Res.*, *256*, 118–127.

Dzurisin, D., R. Y. Koyanagi, and T. T. English (1984), Magma supply and storage at Kilauea Volcano, Hawaii, 1956–1983, *J. Vol. Geotherm. Res.*, *21*(3), 177–206.

Eichelberger, J. C., T. E. C. Keith, T. P. Miller, and C. J. Nye (1995), The 1992 eruptions of Crater Peak vent, Mount Spurr volcano, Alaska: Chronology and summary, in the 1992 eruptions of Crater Peak Vent, Mount Spurr volcano, Alaska, edited by T. E. C. Keith, *USGS Bull.*, *2139*, 1–18.

Gudmundsson, M. T., R. Pedersen, K. Vogfjörd, B. Thorbjarnardóttir, S. Jakobsdóttir, and M. J. Roberts (2010), Eruptions of Eyjafjallajökull volcano, Iceland, *Eos, Trans. AGU*, *91*(21), 190–191.

Gudmundsson, M. T., T. Thordarson, Á Höskuldsson, G. Larsen, H. Björnsson, F. J. Prata, B. Oddsson, E. Magnussin, T. Hognadotir, G. N. Petersen, C. L. Hayward, J. A. Stevenson, and I. Jónsdóttir (2012), Ash generation and distribution from the April-May 2010 eruption of Eyjafjallajökull, Iceland, *Nature Sci. Rep.*, *2*.

Guffanti, M., D. J. Schneider, K. L. Wallace, T. Hall, D. R. Bensimon, and L. J. Salinas (2010), Aviation response to a widely dispersed volcanic ash and gas cloud from the August 2008 eruption of Kasatochi, Alaska, USA, *J. Geophys. Res. Atmos.*, *115*(D2).

Guffanti, M., G. C. Mayberry, T. J. Casadevall, and R. Wunderman (2009), Volcanic hazards to airports, *Nat. Hazards*, *51*(2), 287–302.

Global Volcanism Program (GVP) (2014), Modeled ash concentrations from Grimsvotn volcano: FL000 to FL200 at 06:00 UTC May 24, 2011, From the UK Met. Office for the London VAAC, http://www.volcano.si.edu/volcanoes/region17/ice_ne/grimsvot/3606gri6.jpg, viewed on 26 October 2014.

Harris, A. J., J. Dehn, and S. Calvari (2007), Lava effusion rate definition and measurement: a review, *Bull. Volcanol.*, *70*(1), 1–22.

Harris, A. J., L. Gurioli, E. E. Hughes, and S. Lagreulet (2012), Impact of the Eyjafjallajökull ash cloud: A newspaper perspective, *J. Geophys. Res. Sol. Earth*, *117*(B3).

Hawaii Volcano Observatory (HVO) (2014), Photographs and videos of the Kilauea lava flow, June 27th flow, Photo taken on 1 October 2014 and viewed on 26 October 2014, http://hvo.wr.usgs.gov/multimedia/uploads/multimediaFile-823.jpg

Holaske, R. E., and W. I. Rose (1991), Anatomy of 1986 Augustine volcano eruptions as recorded by multispectral image processing of digital AVHRR weather satellite data, *Bull. Volcanol.*, *53*(6), 420–435.

Horwell, C. J., R. S. J. Sparks, T. S. Brewer, E. W. Llewellin, and B. J. Williamson (2003), Characterization of respirable volcanic ash from the Soufrière Hills volcano, Montserrat, with implications for human health hazards, *Bull. Volcanol.*, *65*(5), 346–362.

International Civil Aviation Agency (ICAO) (2012), International Volcanic Ash Task Force, http://www.icao.int/safety/meteorology/ivatf/Pages/default.aspx, last viewed 29 October 2014.

International Civil Aviation Agency (ICAO) (2014), International Airways Volcano Watch Operations Group (IAVWOPSG) website, last viewed 7 November 2014; http://www.icao.int/safety/meteorology/iavwopsg/Pages/default.aspx.

Kristiansen, N., et al. (2016), Inverse modeling for ash dispersion: Using obs. data to constrain the uncertainty in VATD inputs, *AGU Monograph: Natural Hazard Assessment: Uncertainty Modeling and Decision Support, this issue.*

Kristiansen, N. I., A. Stohl, A. J. Prata, A. Richter, S. Eckhardt, P. Seibert, A. Hoffmann, C. Ritter, L. Bitar, T. J. Duck, and K. Stebel (2010), Remote sensing and inverse transport modeling of the Kasatochi eruption sulfur dioxide cloud, *J. Geophys. Res.Atmos.* (1984–2012), *115*(D2).

Madankan, R., S. Pouget, P. Singla, M. Bursik, J. Dehn, M. Jones, A. Patra, M. Pavolonis, E. B. Pitman, T. Singh, and P. Webley (2014), Computation of probabilistic hazard maps and source parameter estimation for volcanic ash transport and dispersion. *J. Computational Phys.*, *271*, 39–59.

Marzano, F. S., S. Barbieri, G. Vulpiani, and W. I. Rose. (2006), Volcanic ash cloud retrieval by ground-based microwave weather radar, *Geoscience and Remote Sensing, IEEE Trans.*, *44*(11), 3235–3246.

Mastin, L. G. (2007), A user-friendly one-dimensional model for wet volcanic plumes, *Geochem. Geophys. Geosyst.*, *8*(3).

Mastin, L. G., A. R.. Van Eaton, and J. B. Lowenstern (2014), Modeling ash fall distribution from a Yellowstone Supereruption, *Geochem. Geophys. Geosyst.*, *15*(8), 3459–3475.

Mastin, L. G., M. Guffanti, R. Servranckx, P. Webley, S. Barsotti, K. Dean, A. Durant, J. W. Ewert, A. Neri, W. I. Rose, D. Schneider, L. Siebert, B. Stunder, G. Swanson, A. Tupper, A. Volentik, and C. F. Waythomas (2009), A multi-disciplinary effort to assign realistic source parameters to models of volcanic ash-cloud transport and dispersion during eruptions, *J. Volcanol. Geotherm. Res.*, *186*(1), 10–21.

Mattox, T. N., C. Heliker, J. Kauahikaua, and K. Hon (1993), Development of the 1990 Kalapana flow field, Kilauea volcano, Hawaii, *Bull. Volcanol.*, *55*(6), 407–413.

Patra, A. K., M. Bursik, J. Dehn, M. Jones, R. Madankan, D. Morton, M. Pavolonis, E. B. Pitman, S. Pouget, T. Singh, P. Singla, E. R. Stefanescu, and P. Webley (2013), Challenges in developing DDDAS based methodology for volcanic ash hazard analysis–effect of numerical weather prediction variability and parameter estimation, *Procedia Computer Sci.*, *18*, 1871–1880.

Pavolonis, M. J., A. K. Heidinger, and J. Sieglaff (2013), Automated retrievals of volcanic ash and dust cloud properties from upwelling infrared measurements, *J. Geophys. Res. Atmos.*, *118*(3), 1436–1458.

Pouget, S., M. Bursik, P. Webley, J. Dehn, and M. Pavolonis (2013), Estimation of eruption source parameters from umbrella cloud or downwind plume growth rate, *J. Volcanol. Geotherm. Res.*, *258*, 100–112.

Ripepe, M., C. Bonadonna, A. Folch, D. Delle Donne, G. Lacanna, E. Marchetti, and A. Höskuldsson (2013), Ash-plume dynamics and eruption source parameters by infrasound and thermal imagery: The 2010 Eyjafjallajökull eruption, *Earth Planet. Sci. Lett.*, *366*, 112–121.

Rose, W. I., A. B. Kostinski, and L. Kelley (1995), Real-time C-band radar observations of 1992 eruption clouds from Crater Peak, Mount Spurr volcano, Alaska, *USGS Bull.*, *2139*, 19–26.

Roulston, M. S., and L. A. Smith (2002), Evaluating probabilistic forecasts using information theory, *Monthly Weather Rev.*, *130*(6), 1653–1660.

Schneider, D. J., and R. P. Hoblitt (2013), Doppler weather radar observations of the 2009 eruption of Redoubt Volcano, Alaska, *J. Volcanol. Geotherm. Res.*, *259*, 133–144.

Sparks, R. S. J., M. I. Bursik, S. N. Carey, J. Gilbert, L. S. Glaze, H. Sigurdsson, and A. W. Woods, A. W. (1997), *Volcanic Plumes*, John Wiley, Hoboken, NJ.

Stefanescu, E. R., A. K. Patra, M. I. Bursik, R. Madankan, S. Pouget, M. Jones, P. Singla, T. Singh, E. B. Pitman, M. Pavolonis, D. Morton, P. Webley, and J. Dehn (2014), Temporal, probabilistic mapping of ash clouds using wind-field stochastic variability and uncertain eruption source parameters: Example of the 14 April 2010 Eyjafjallajökull eruption, *J. Adv. Model. Earth Syst.*, *6*(4), 1173–1184.

Stohl, A., A. J. Prata, S. Eckhardt, L. Clarisse, A. Durant, S. Henne, N. Kristiansen, A. Minikin, U. Schumann, P. Seibert, K. Stebel, H. E. Thomas, T. Torsteinsson, K. Torseth, and B. Weinzierl (2011), Determination of time-and height-resolved volcanic ash emissions and their use for quantitative ash dispersion modeling: the 2010 Eyjafjallajökull eruption, *Atmos. Chem. Phys.*, *11*(9), 4333–4351.

Tokyo VAAC (2014a), Volcanic ash advisory text on October 16, 2014 at 18:00 UTC or Z, http://ds.data.jma.go.jp/svd/vaac/data/TextData/20141016_KARY_0049_Text.html, last viewed on 26 October 2014.

Tokyo VAAC (2014b), Volcanic ash advisory graphic on October 16, 2014 at 18:00 UTC on Z, http://ds.data.jma.go.jp/svd/vaac/data/Figure/20141016_KARY_0049_PF15.html, last viewed on 26 October 2014.

Tupper, A., C. Textor, M. Herzog, H. F. Graf, and M. S. Richards (2009), Tall clouds from small eruptions: The sensitivity of eruption height and fine ash content to tropospheric instability, *Nat. Hazards*, *51*(2), 375–401.

Vogel, H., J. Förstner, B. Vogel, T. Hanisch, B. Mühr, U. Schättler, and T. Schad (2014), Time-lagged ensemble simulations of the dispersion of the Eyjafjallajökull plume over Europe with COSMO-ART, *Atmos. Chem. Phys.*, *14*, 7837–7845; doi:10.5194/acp-14-7837-2014.

Waythomas, C. F., W. E. Scott, S. G. Prejean, D. J. Schneider, P. Izbekov, and C. J. Nye, (2010), The 7–8 August 2008 eruption of Kasatochi Volcano, central Aleutian Islands, Alaska, *J. Geophys. Res.Sol. Earth (1978–2012)*, *115*(B12).

Webley, P. W., and T. Steensen (2013), Operational volcanic ash cloud modeling: Discussion on model inputs, products, and the application of real-time probabilistic on forecasting, 271–298, in *Lagrangian Modeling of the Atmosphere*, edited by John Lin, Geophysical Monograph Series, vol. 200, American Geophysical Union, Washington, DC.

Webley, P. W., B. J. B. Stunder, and K. G. Dean (2009), Preliminary sensitivity study of eruption source parameters for operational volcanic ash cloud transport and dispersion models: A case study of the August 1992 eruption of the Crater Peak vent, Mount Spurr, Alaska, *J. Volcanol. Geotherm. Res.*, *186*(1), 108–119.

Woodhouse, M. J., A. J. Hogg, J. C. Phillips, and R. S. J. Sparks (2013), Interaction between volcanic plumes and wind during the 2010 Eyjafjallajökull eruption, Iceland, *J. Geophys. Res. Sol. Earth*, *118*(1), 92–109.

World Meteorological Organization (WMO) (2010), Workshop on ash dispersal forecast and civil aviation: Results from the 1st Workshop: Model Definition Document, viewed 29 October 2014; http://www.unige.ch/sciences/terre/mineral/CERG/Workshop/results/Model-Document-Geneva10.pdf.

World Meteorological Organization (WMO) (2013), Workshop on ash dispersal forecast and civil aviation, Results from the 2nd Workshop: Updated Model Definition Document, viewed 29 October 2014; http://www.unige.ch/sciences/terre/mineral/CERG/Workshop2/results-2/Model-Definition-Document-2013.pdf.

Zakšek, K., M. Hort, J. Zaletelj, and B. Langmann (2013), Monitoring volcanic ash cloud top height through simultaneous retrieval of optical data from polar orbiting and geostationary satellites, *Atmos. Chem. Phys.*, *13*(5), 2589–2606.

Part II
Geological Hazards

Peter Webley

Editor-in-Chief

6

Building an Uncertainty Modeling Framework for Real-Time VATD

Peter Webley,[1] Abani Patra,[2] Marcus Bursik,[3] E. Bruce Pitman,[4] Jonathan Dehn,[1] Tarung Singh,[2] Puneet Singla,[2] Matthew D. Jones,[5] Reza Madankan,[2] E. Ramona Stefanescu,[2] and Solene Pouget[3]

ABSTRACT

When forecasting the future location of volcanic-ash clouds, uncertainties exist in the input parameters used in dispersion modeling and in the weather prediction data used for modeling the advection terms. Recent developments have shown that probabilistic modeling provides the tools to assess the variability in downwind ash concentrations. We show a probabilistic modeling approach where ensembles of forecasts are generated from a suite of simulations using a coupled one-dimensional plume model and a Lagrangian dispersion model. This approach produces charts of the probability of ash-cloud concentrations and mass loadings exceeding user-defined thresholds. We focus on the initial plume uncertainties and discuss how uncertainties in numerical weather prediction data could also be applied within our approach. Our results show how, by assigning the initial likelihoods of input parameters, the probabilistic approach can produce mean ash concentrations and mass loadings as well as probabilities of breaching a defined threshold. We show how, given the variability in the inputs, the probabilistic modeling can be used to assess the confidence in the ash-mass loadings. This is critical for real-time volcanic-hazard assessment and our approach illustrates how a new tool could be developed for those in decision support.

6.1. INTRODUCTION AND BACKGROUND

Volcanic-ash plumes and dispersing clouds can be a hazard to both the aviation community and population centers downwind of the volcano [*Horwell and Baxter*, 2006; *Prata and Tupper*, 2009]. Operational organizations, such as Volcanic Ash Advisory Centers (VAACs), simulate the clouds' future location for use in their decision-support systems. Then volcanic ash advisories [VAA; see within *ICAO*, 2012] can be generated for the aviation industry. Also, fallout advisories often provided by local volcano observatories can provide advice on the potential impact to human health (see *Horwell and Baxter* [2006] for more on impact of volcanic ash on human health). To forecast the ash plumes' and clouds' future position and concentration levels, volcanic-ash transport and dispersion (VATD) models have been used. These VATD models are being used in either an operational setting to produce the cloud forecasts required for the VAACs' VAA [*Met Office*, 2012; *JMA*, 2014] (Fig. 6.1a), or in a research mode [see *Witham et al.*, 2007; *Webley et al.*, 2009a,b, 2010; *Folch et al.*, 2012] to

[1] Geophysical Institute, University of Alaska Fairbanks, Fairbanks, Alaska, and Volcanic Ash Detection, Avoidance and Preparedness for Transportation (V-ADAPT), Inc., Fairbanks, Alaska, USA

[2] Department of Mechanical and Aerospace Engineering, University at Buffalo, SUNY, Buffalo, New York, USA

[3] Department of Geology, University at Buffalo, SUNY, Buffalo, New York, USA

[4] Department of Mathematics, University at Buffalo, SUNY, Buffalo, New York, USA

[5] Center for Computational Research, University at Buffalo, SUNY, Buffalo, New York, USA

Natural Hazard Uncertainty Assessment: Modeling and Decision Support, Geophysical Monograph 223, First Edition.
Edited by Karin Riley, Peter Webley, and Matthew Thompson.

(a)

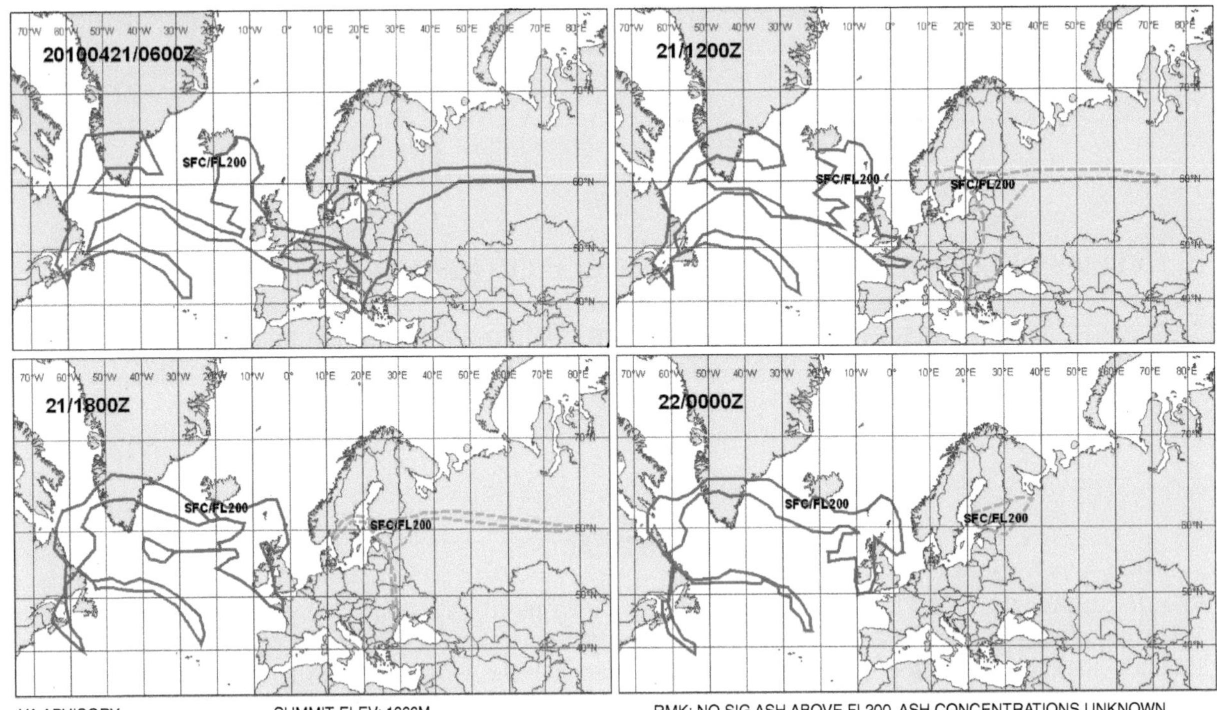

VA ADVISORY
DTG: 20100421/0600Z
VAAC: LONDON
VOLCANO:
EYJAFJALLAJOKUL 1702-02
PSN: N6338 W01937
AREA: ICELAND

SUMMIT ELEV: 1666M
ADVISORY NR: 2010/029
INFO SOURCE: ICELAND MET OFFICE
AVIATION COLOUR CODE: RED
ERUPTION DETAILS: ERUPTION CONTINUING TO AROUND FL120 TO FL160.

RMK: NO SIG ASH ABOVE FL200. ASH CONCENTRATIONS UNKNOWN
THE TWO PLUMES ON 21/1200Z, 21/1800Z AND 22/0000Z CHARTS ARE BOTH AT SFC/FL200.
NXT ADVISORY: 20100421/1200Z

(b)

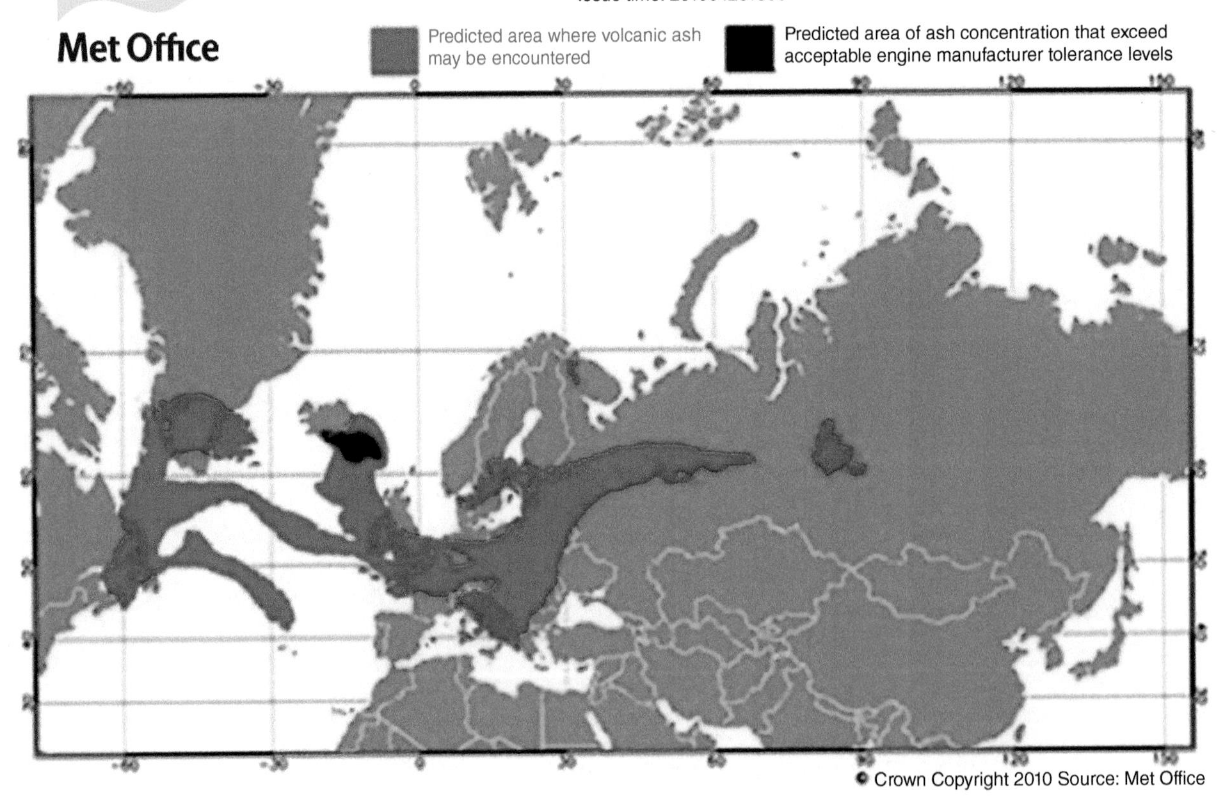

Figure 6.1 (a) London VAAC's VAA produced during Eyjafjallajökull eruption on 14 April 2010; (b) additional concentration product from the same date at 06:00 UTC; and (c) the progression to concentration thresholds during the Grimsvotn eruption on 25 May 2011.

(c)

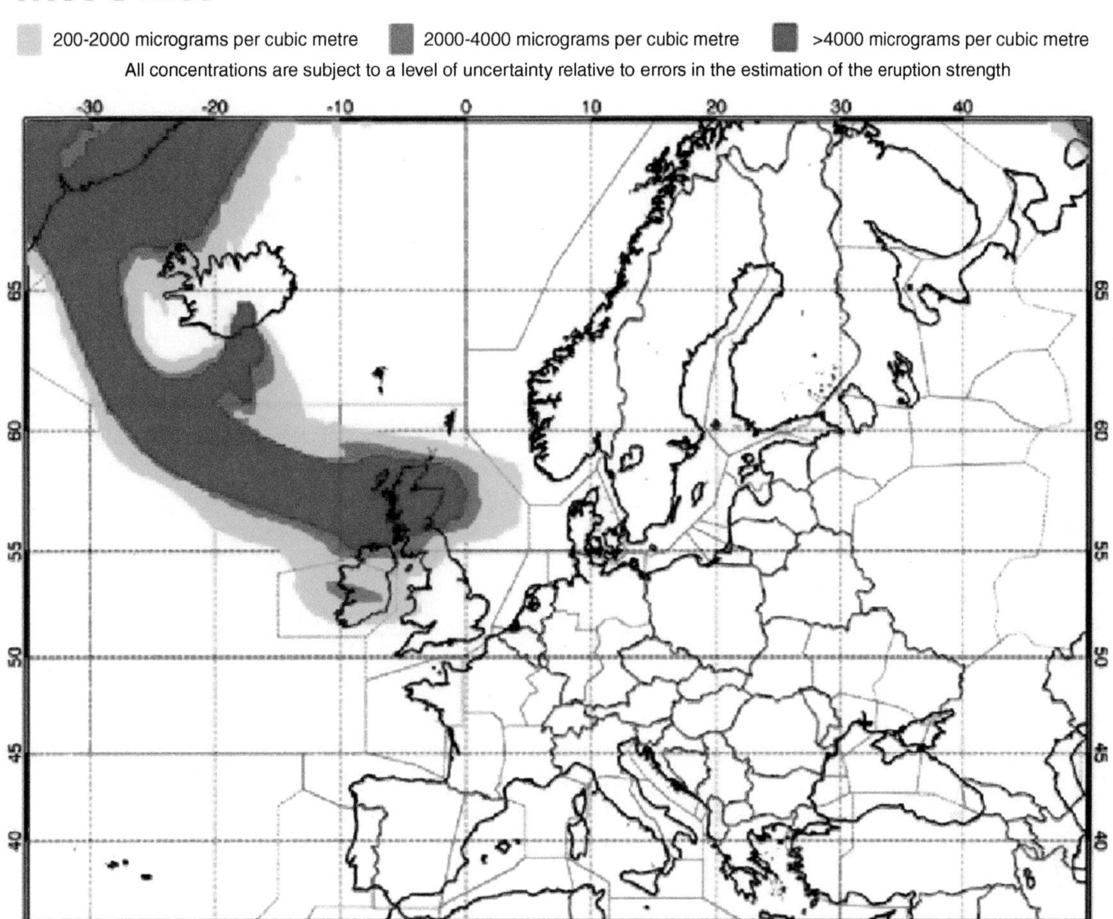

Figure 6.1 (Continued)

better understand the volcanic event and develop new processing algorithms and analysis tools for future volcanic-hazard assessment.

The eruption of Eyjafjallajökull volcano in 2010 [*Gudmundsson et al.*, 2010, 2012] illustrated the uncertainties that exist in performing volcanic-ash plume and cloud modeling in a real-time environment. The operational VAAC for the region produced, in addition to its standard VAA (Fig. 6.1a), a nonoperational product (Fig. 6.1b) that displayed the ash-cloud concentrations that exceeded aviation engine tolerance levels. These products were the result of a deterministic model simulation from one set of input parameters and used one deterministic numerical weather prediction (NWP) model. During the eruption of Grimsvotn in 2011 [*Tesche et al.*, 2012], the concentration forecast changed to display thresholds of 0–2 mg/m^3, 2–4 mg/m^3, and > 4 mg/m^3 (Fig. 6.1c). These were at the time deemed as acceptable by aircraft-engine manufacturers so there would be no or minimal risk of immediate damage to any aircraft [*Guffanti and Tupper*, 2014]. Uncertainties in plume-height estimation, vertical plume shape, initial particle- or grain-size distribution (PSD or GSD), event length, and mass eruption rate indicate that the potential range of the input parameters can vary on scales greater than the sensitivity of the concentration thresholds. Thus, the research and operational communities [*WMO*, 2010a, 2010b] held discussions on the need to progress toward adding probabilistic ash-cloud modeling to the deterministic forecasting that would place exceedance threshold estimates in proper context and allow better decision making.

Since the end of the International Civil Aviation Organization (ICAO) lead international volcanic-ash task force (IVATF) in 2012 [*ICAO,* 2012], there has been a progression toward probabilistic VATD modeling environments. These incorporate both the uncertainties in the model inputs [e.g., *Bursik et al.*, 2012] and the potential variability in the NWP [e.g., *Stefanescu et al.*, 2014; *Vogel et al.*, 2014]. Observational data from ground measurements [*Schneider and Hoblitt,* 2013], airborne campaigns [*Weber et al.*, 2012], and satellite remote sensing data [*Ellrod et al.*, 2003; *Pavolonis et al.*, 2013] can then be used to constrain these approaches through inverse modeling [e.g., *Madankan et al.*, 2014].

In this chapter, we present a new modeling approach that incorporates the uncertainties in the volcanic eruption initial conditions and the stochastic nature of the NWP data to generate a volcanic-ash-cloud forecast with associated probabilistic estimates in the location and four-dimesional concentrations (x, y, z, and t). We focus only on the input variability in this paper. We couple the Puff VATD model [*Searcy et al.*, 1998], to a one-dimensional model for plume rise called BENT [*Bursik*, 2001], which we refer to as Puffin. This approach provides the uncertainty estimates in the initial conditions of the eruption volcano. We have developed a sophisticated workflow that builds probabilistic Puff model simulations for a range of inputs from the Puffin tool. We will provide, in this chapter, an overview of the developed workflow focusing on one NWP dataset and illustrate some of the output products available that can then be used to compare to any available observational data.

6.2. METHODOLOGY

6.2.1. Probabilistic Modeling Workflow

Our approach incorporates eruption and NWP variability and uncertainty together to provide a probabilistic estimate of the ash-cloud location and concentration downwind of the volcano. Just as the Puff VATD model is able to analyze past volcanic eruptions using reanalysis [*Webley et al.*, 2012] and hindcast [*Steensen et al.*, 2013] data, our tool is applicable for past eruptive event analysis as well as for real-time model simulations for use in operational decision making. The workflow (Fig. 6.2) demonstrates how the source parameter uncertainty is applied to the BENT-Puff (Puffin) tool to build a set of dispersion model simulations/ensemble members. The mean and covariance of this set of simulations are updated by assimilating any available observational data (i.e., a satellite data) to then produce a posterior mean and covariance of the uncertain parameters (see *Madankan et al.* [2014] for more details).With each available satellite dataset, a new source parameter input distribution would be generated. With subsequent iterations, the workflow reduces the uncertainties in the inputs and hence produces a simulated ash-cloud product with higher confidence levels for the location and downwind concentrations. If no satellite data are available, then simulations with the prior input parameters are used for the full model simulation.

In the initial phase of the workflow, any input parameter for the BENT model can be defined with its associated variability. If observational data are available, such as eruption height from ground or space-borne observations, then the initial weightings can be edited to reflect the recorded plume height. We currently chose four parameters: vent diameter, vent velocity, mean particle size (log scale), and standard deviation of the size distribution (using a Gaussian shape centered on the mean particle size). For these four parameters, a set number of simulations are defined that target the potential range of each parameter with each simulation given an associated weight based on a minimization of the moments in the probabilistic analysis. More details on the definition of the methodology can be found in *Patra et al.* [2013], *Madankan et al.* [2014], and *Stefanescu et al.* [2014].

6.2.2. Near-Real-Time Processing Routines

In addition to the overall workflow design, we have built a set of processing routines to complete the probabilistic modeling in near real-time (NRT; Fig. 6.3). Given the start time and date, the routines download data from the closest radiosonde to the volcano's location extracting vertical profiles of temperature, wind speed, and relative humidity, at either 00 or 12 UTC depending on the start time for the simulations. These will be updated to use NWP to determine the atmospheric conditions at the volcano. Next, the routine builds the template for the Puffin tool. For our current setup in Fig. 6.3, we use four parameters to represent the

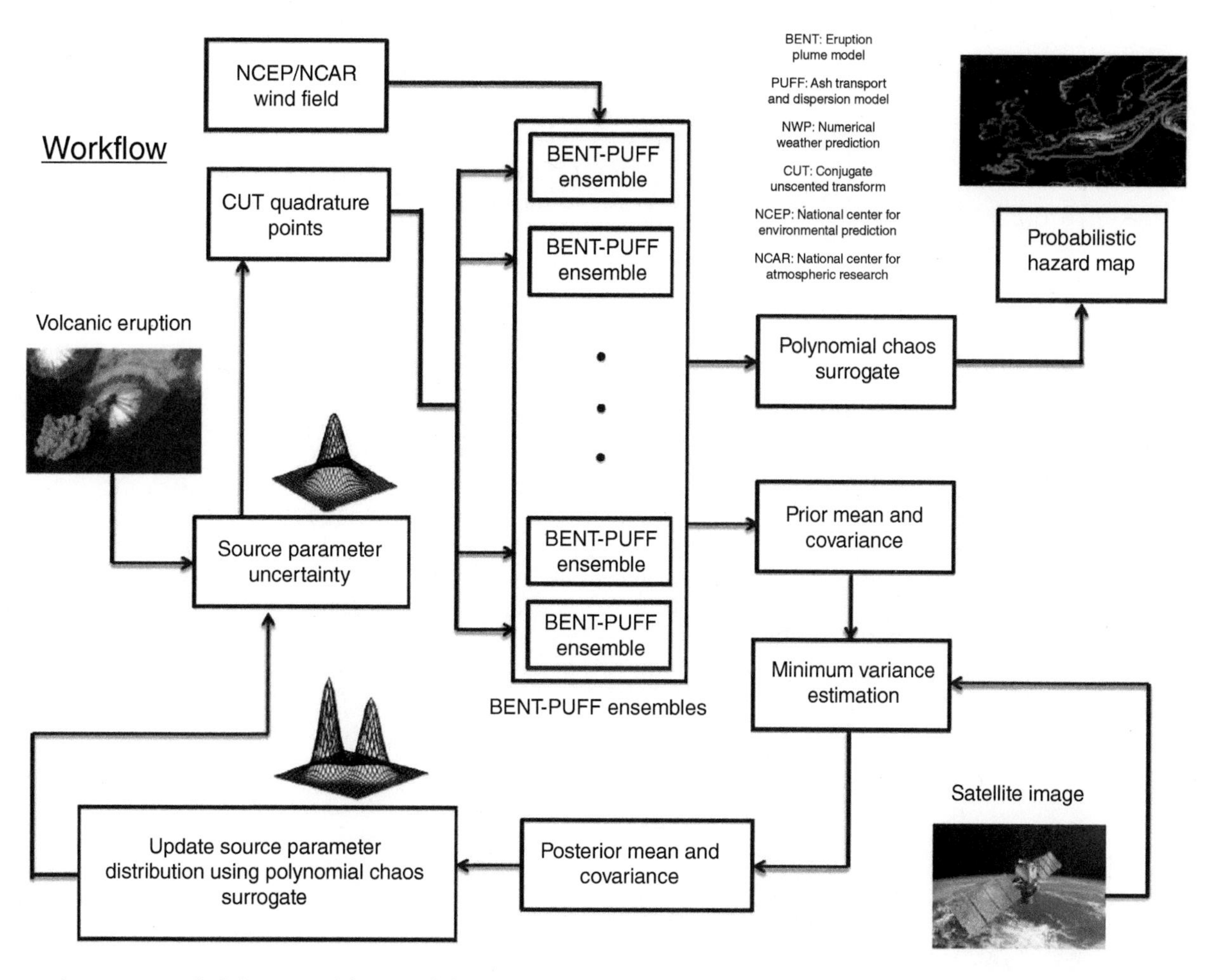

Figure 6.2 Probabilistic modeling workflow, adapted from *Madankan et al.* [2014], using the Puff VATD model and coupled one-dimensional plume rise model, BENT.

uncertainty in the eruption source input. We have chosen the vent radius (m), vent velocity (m/s), log mean grain size (μm), and standard deviation of log grain size. Given the local radiosonde and the range of these four values along with the BENT model results, we end up with 161 different sets of source parameter inputs for the Puff VATD model. Each has a specific likelihood of occurrence based on the volcano's physical characteristics (vent size and velocity) and previous eruption style (mean ash grain size and standard deviation of distribution). This can be adapted to any volcano of interest. The BENT model within Puffin provides the maximum plume altitude (km), vertical plume shape, initial grain-size distribution for dispersal, and mass eruption rate (MER, kg/s). We specify the event length to convert the MER to total erupted mass (kg). Several Puffin simulations can run in parallel to reduce the time to complete the full suite of simulations. For example, with two Puff simulations running in parallel, each with a dedicated processing node and 1e5 (100,000) ash particles, the wall-clock time for the 161 simulations was 2.4 hr (approx. 1.8 min per simulation pair) on a 23 CPU-node server and using approximately 100 MB of allocated memory.

The processing routines will run the Puffin tool for each of the 161 simulations to generate the Puff input file. As each parallel run completes, the Puff particle location and ash-concentration output files are generated and the routines moves down the list of the 161 simulation members. In the final part of the NRT processing routines, the outputs from the simulations are generated using the initial weightings defined in for the source parameters to generate a mean ash concentration. These results are sent to the second postprocessing routine to produce the probabilistic maps as GEOTIFF data, JPEG imagery, and Google Earth KML and KMZ files.

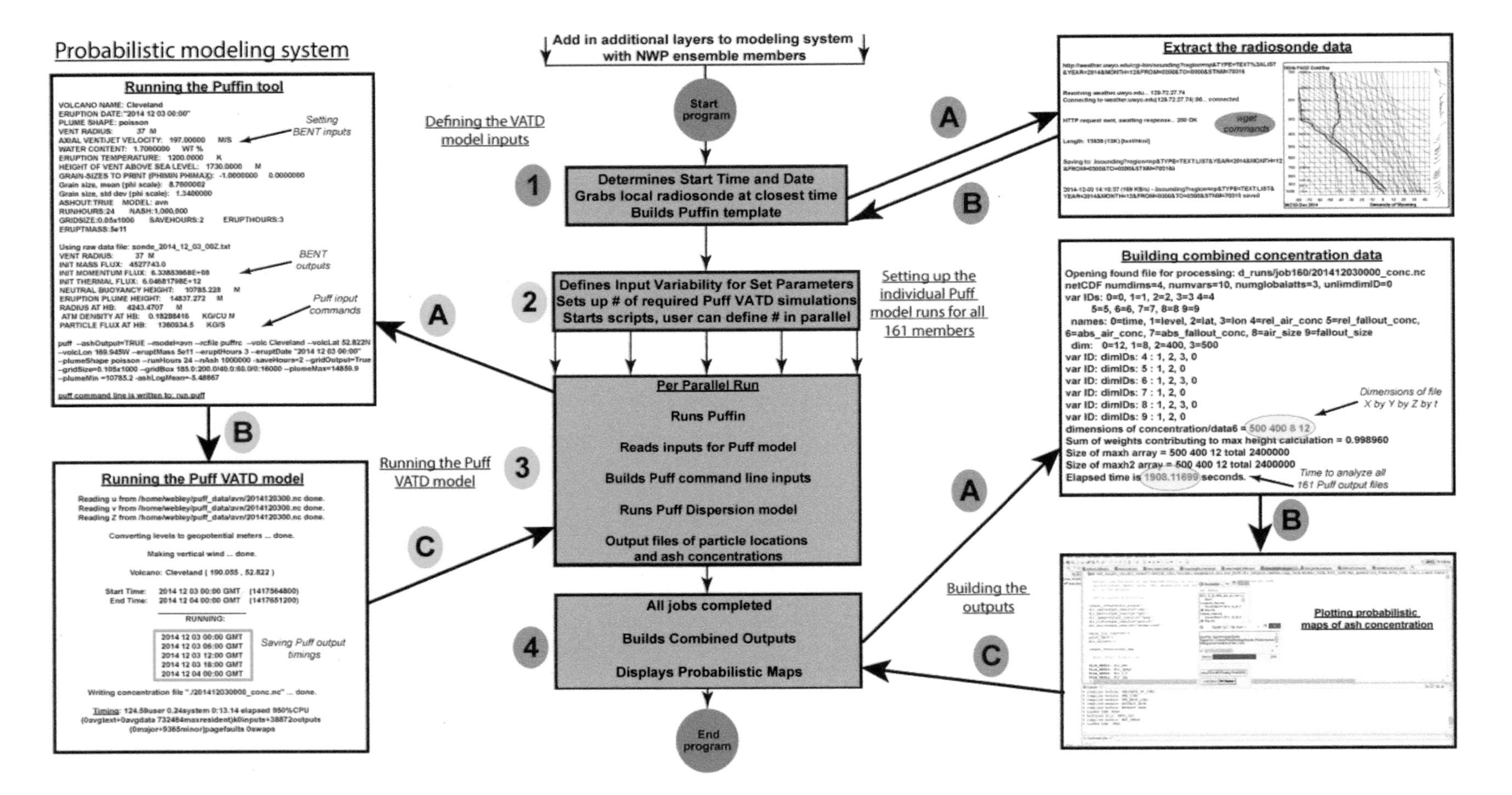

Figure 6.3 Real-time processing routines from the probabilistic modeling of volcanic ash clouds. Results from these routines include mean ash mass loadings and ash concentrations at defined altitudes from all 161 ensemble members.

6.3. PROBABILISTIC MODELING RESULTS

6.3.1. Cleveland Volcano, Alaska, USA: 3 December 2014

Cleveland volcano was at elevated alert during early December 2014 [*AVO*, 2014] so we set up probabilistic model simulations to assess the capability of the system to develop timely results and to assess if improvements were needed in our workflow for future eruptions. The volcano did not erupt but our example illustrates how the system could provide pre-event warnings to assist in VAA and VAG generation for aviation safety. The Puffin tool determined plume top heights to range from 12.2 km to 18.95 km above sea level (ASL) and mean particle size to range from 2.3 to 3.7 μm. Figure 6.4 shows output from one Puff model simulation (i.e., Simulation Number 1). For this simulation, Puffin determined the plume top height to be 14.6 km ASL, the mean particle size to be 3 μm, and the total mass to be 5×10^{11} kg. For this and all the Puff model simulations, we used the North American model (NAM) 216 grid at 45 km spatial resolution to provide the atmospheric data for the simulations.

Figure 6.4 displays the two outputs available from the Puff VATD model. Being Lagrangian in form, the particle locations in four dimensions (x, y, z, and t), along with the ash concentrations (g/m^3) can be generated per time step. The volcanic-ash mass loadings (g/m^2) are derived as total atmospheric column loadings from the ash-concentration gridded data. Figure 6.4 shows the ash locations and mass loadings at + 12 hr into the simulation, or 12:00 UTC on 3 December 2014. There is evidence of fallout close to the volcano with a highest mass loading of 40.5 mg/m^2.

The next step in our routine is to combine the 161 model simulations together. Figures 6.5, 6.6, and 6.7 document the averaged mass loadings and concentrations from all the simulation members along with the probabilities associated with exceeding specific mass and concentration thresholds. For the averaged results, the location of the cloud closely matches that from Simulation Number 1 (Fig. 6.4), while the concentrations at 2 km, 10 km, and 16 km ASL illustrate the different footprints predicted by the Puff model as a result of the variations in wind patterns at these altitudes. The highest altitude portion, 16 km ASL, is centered in the westerly section of the cloud, and the lower altitude portions, below 10 km ASL, are focused in the easterly and southeasterly sections. Figure 6.5a documents the ash-mass loading from the averaged results of the 161 simulations. Figure 6.5b shows that there is little variability in the 161 members and that the uncertainties in the model inputs led to a (spatially) well-constrained set of simulations. Figure 6.5a and Figure 6.5b together illustrate that mass loadings greater than 0.1 mg/m^2 correlate to the location of the higher probabilities of ash presence.

Figure 6.6a shows the mean concentrations from all 161 members at 2 km ASL, while Figure 6.7a shows the corresponding probabilities of the concentration $> 1 \times 10^{-6}$ g/m^3 (=1μg/m^3) our chosen ash versus no-ash boundary. Figure 6.7a produces a very conservative representation of ash-cloud location. The 2010 eruption of Eyjafjallajökull volcano [*Gudmundsson et al.*, 2010] showed the impact of how the ash-cloud "edge" is defined in model data where the modeled cloud extent was much greater than seen in the satellite data. Our conservative no-ash versus ash boundary at 1μg/m^3 is a factor of 1000 lower than the 1–4 mg/m^3 proposed during the Eyja events. Satellite observations would be needed to compare to our probability of occurrence to generate posterior model inputs for an improved simulation and asses our chosen no-ash versus ash boundary threshold.

Figures 6.6b and 6.7b show the mean ash concentrations and probabilities of ash occurrence $\geq 1 \times 10^{-6}$ g/m^3 at 10 km ASL, while Figures 6.6c and 6.7c show the same parameters at 16 km ASL. The results at 16 km ASL show that some of the simulations have a smaller spatial footprint. Figure 6.7c shows that the region of 100% probability of concentrations exceeding 1×10^{-6} g/m^3 is confined to the northwest segment of the cloud footprint matching the higher concentrations from the 161 members (Fig. 6.6c). Significantly reducing the concentration threshold to 1×10^{-12} g/m^3 would result in the cloud footprints at 2 km, 10 km, and 16 km ASL being almost identical.

6.3.2. Zhupanovsky Volcano, Kamchatka, Russia: 29 December 2014

GVP [2015] reported that Zhupanovsky volcano had continuing activity leading to an eruption with a plume top height of 6–9 km ASL on 29 December 2014. For our probabilistic modeling, we set a start time of 00:00 UTC. The Puffin tool determined the plume heights ranged from 10.7 to 13.3 km ASL and the mean size from 2.3 to 3.7 μm across our 161 simulation members. Here we used the NCEP Global Forecast Systems (GFS) 1.25° spatial resolution NWP data with the Puff VATD model. Figure 6.8 illustrates the results for Simulation Number 1 from the Zhupanovsky event with the particle locations and ash-mass loadings presented at + 12 hr after the event start. There is evidence of ashfall close to the volcano in the particle location output while the highest mass loadings occur in the eastern extent of the dispersing cloud. For this simulation, the plume top height was 11.2 km ASL with a mean particle size of 3.2 μm and at +12 hr into the simulation the maximum mass loadings was 5.6 mg/m^2.

+ 12 hours

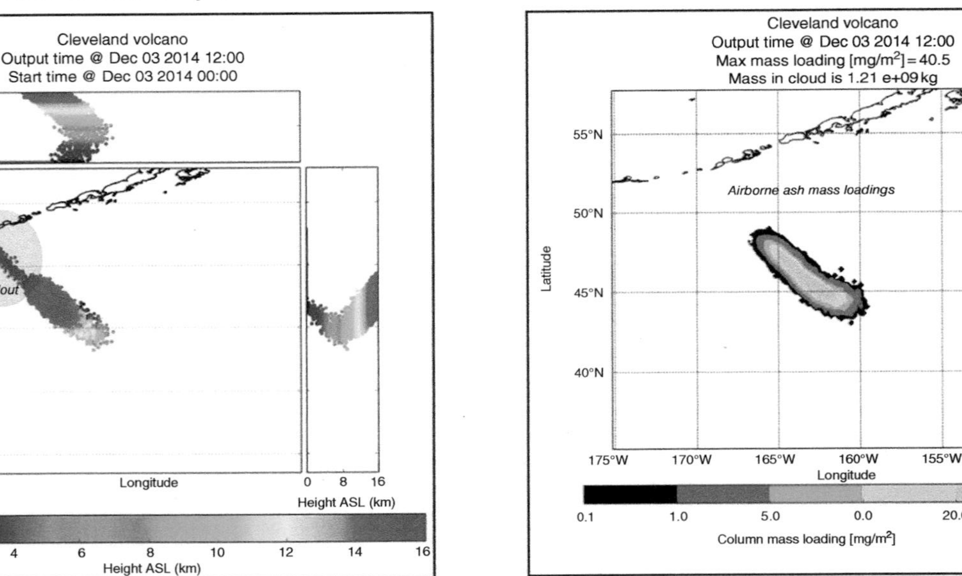

Cleveland volcano

Simulation number 1

Cleveland volcano
Start time: Dec 3, 2014 at 00:00 UTC
3 hour eruptive event
Plume height = 14.6 km ASL

Mean particle size = 3 μm
PSD has 63% from 0.3 μm to 30 μm

Total mass = 5×10^{11} kg

Very fine PSD

Figure 6.4 Cleveland Volcano Puff VATD model simulation for Probabilistic Simulation number 1. This is for start time on 3 December 2014 at 00:00 UTC with the particle locations and ash mass loading (mg/m^2) at + 12 hr after the eruption start, or 12:00 UTC.

(a)

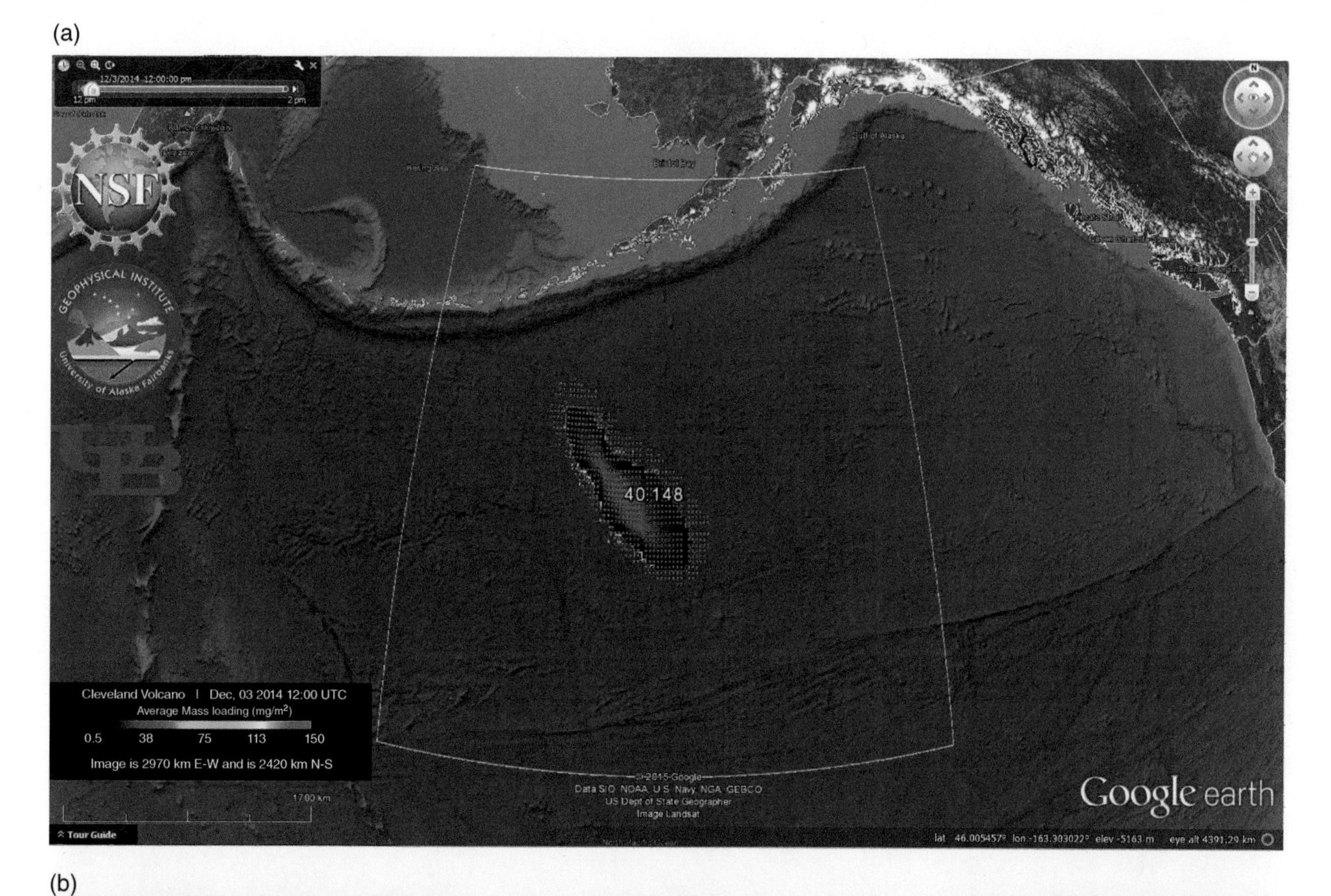

(b)

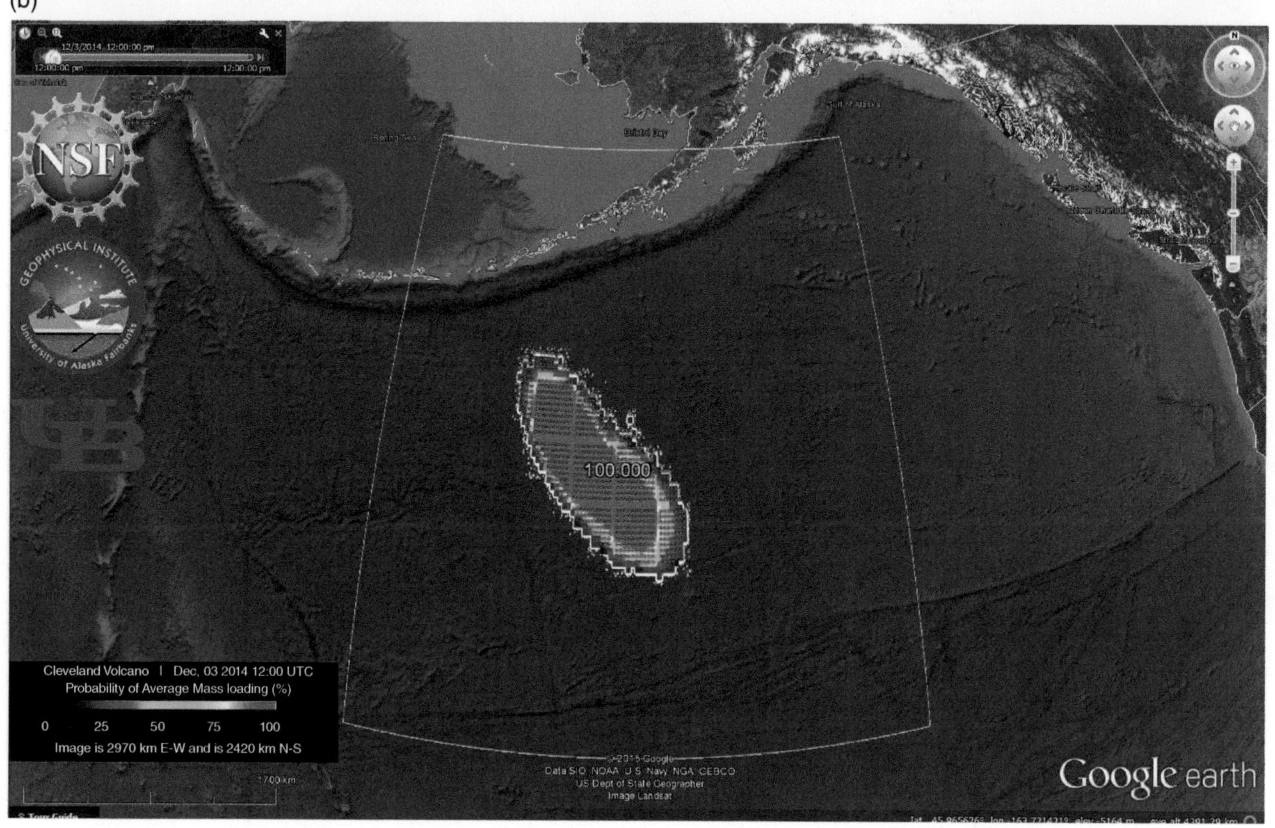

Figure 6.5 Probabilistic modeling outputs at 12:00 UTC, + 12 hr after eruption, for Cleveland 3 December 2014 model simulation. (a) Mean of the 161 simulation members showing ash mass loadings (mg/m^2) and (b) probabilities (%) of ash mass loading exceeding predefined threshold.

(a)

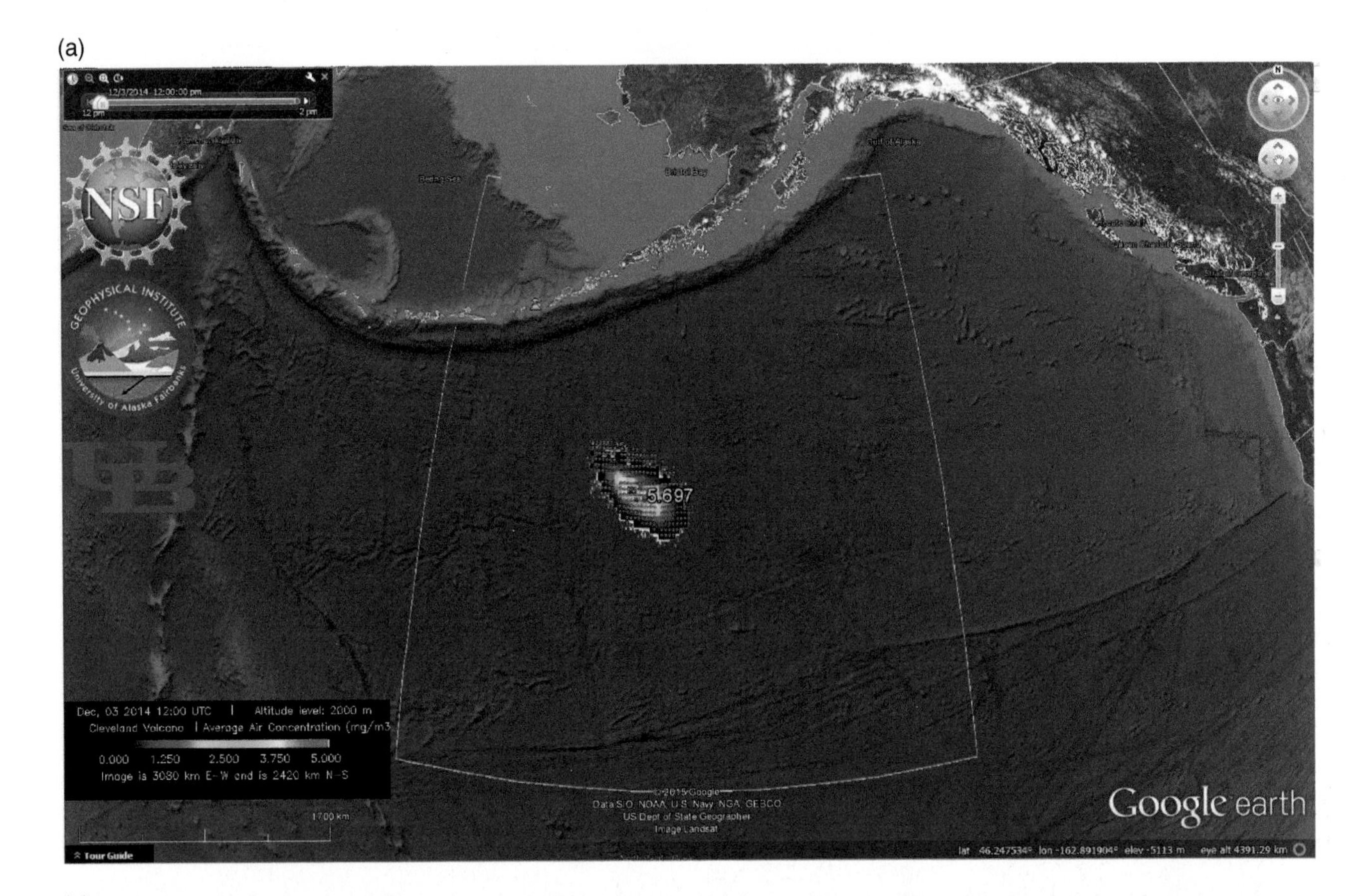

(b)

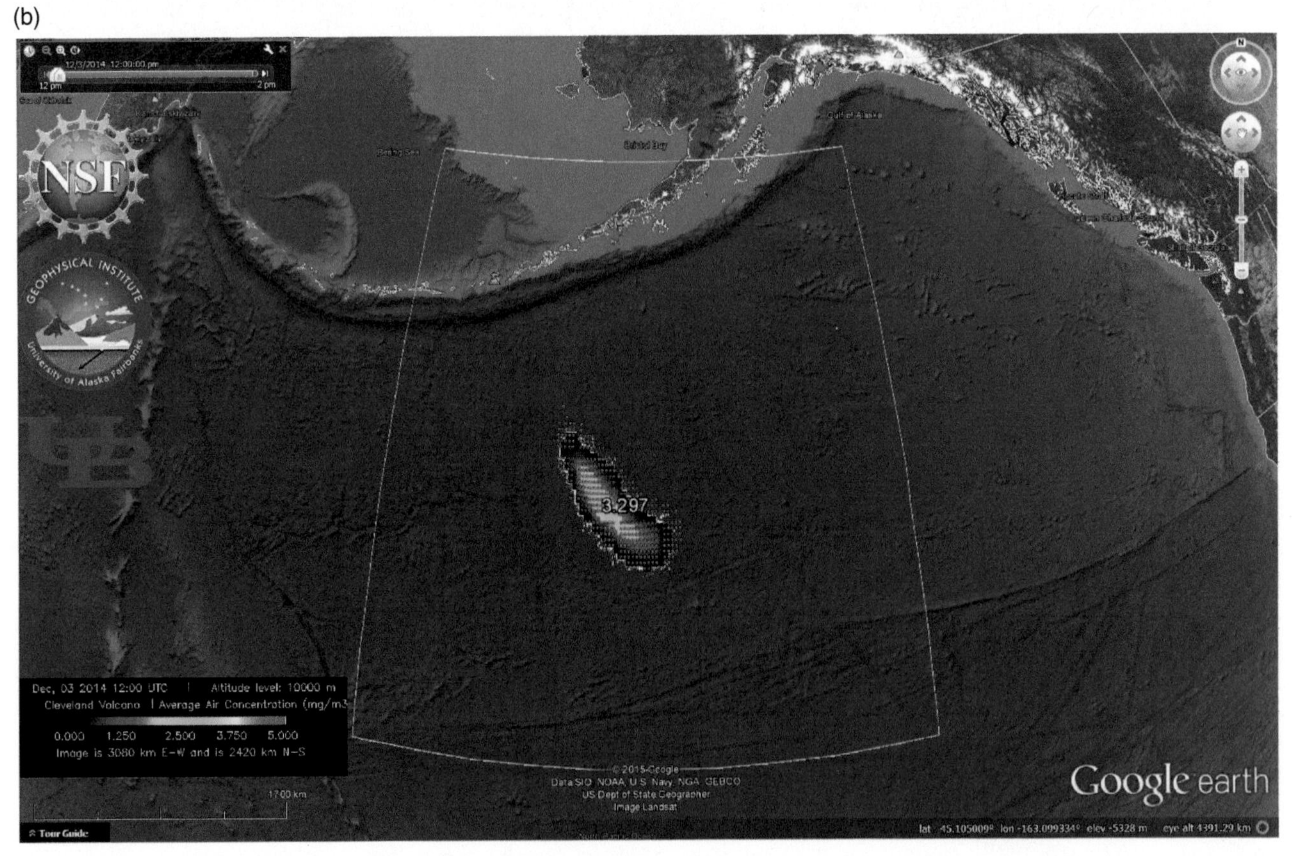

Figure 6.6 Probabilistic modeling outputs at 12:00 UTC, + 12 hr after eruption, for Cleveland 3 December 2014 model simulation. Mean results from the 161 simulation members showing ash concentrations (mg/m³) at 2 (a), 10 (b), and 16 (c) km ASL.

(c)

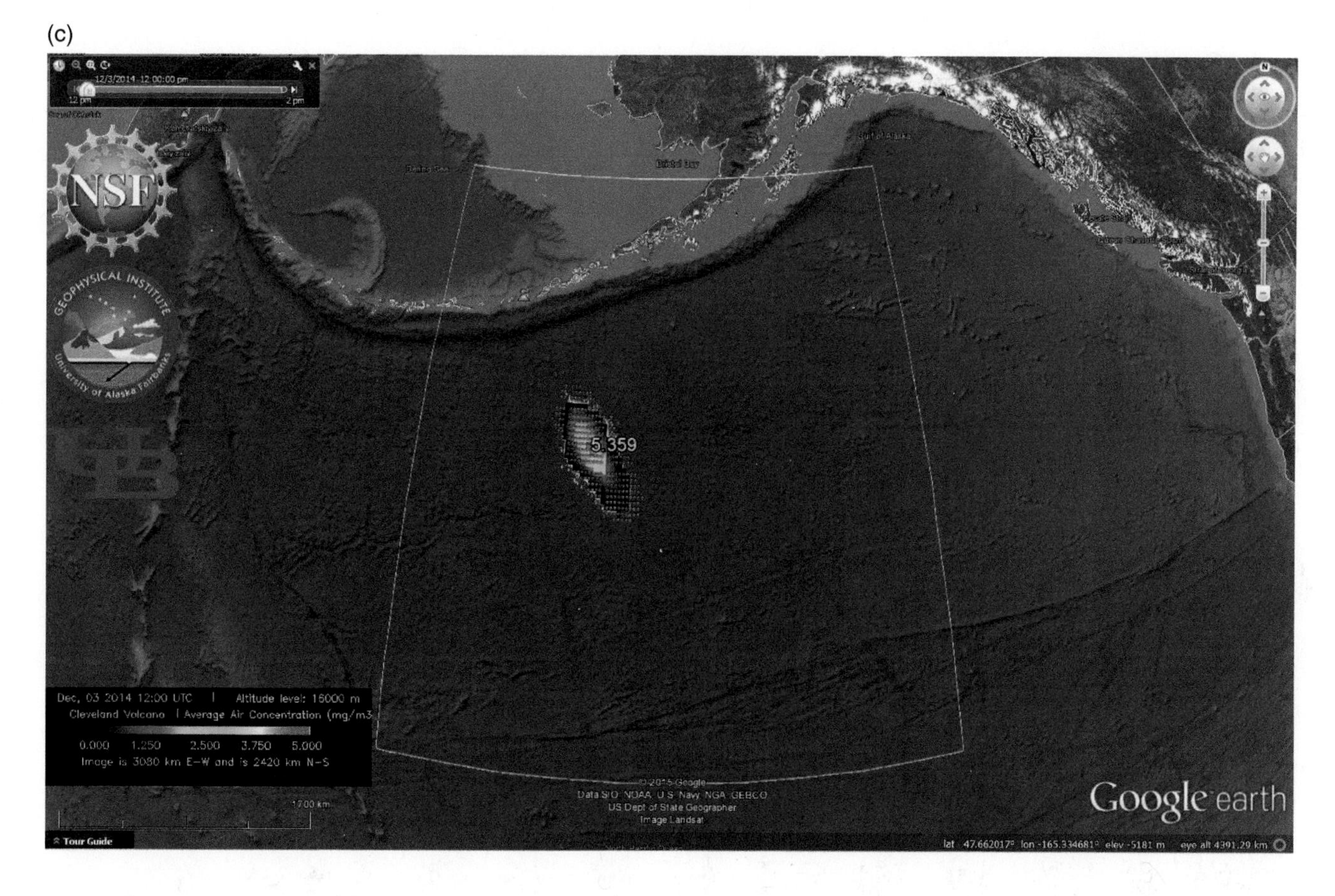

Figure 6.6 (Continued)

6.3.2.1. Comparison of the Mean and Probabilities of All Simulations

Figure 6.8 illustrates a problem in using predefined grids for the ash-concentration data. The particle location map is defined by the maximum extent of the ash particles in the 24 hr period while the concentration grid is set prior to the simulation. Therefore, the cloud could disperse beyond the extensions of the concentration grid. Further developments are needed to our approach to define the concentration grid domain at the end of the simulation rather than as a predefined domain. Puff simulates the ash-particle dispersal and generates a final concentration grid. We can predefine this grid with a fine spatial resolution and a large domain to cover the maximum possible extent of the cloud dispersal in 24 hr. This would generate large (> 4 GB) gridded datasets that are not optimal in terms of file size and spatial resolution for operational data analysis. Running the model simulations with a small number of ash particles would allow us to evaluate the maximum extent of the ash cloud to then optimally design the concentration grid to capture the full cloud dispersal and set the finest possible spatial resolution. However, to implement this for our model simulations is beyond the scope of the research shown in this chapter. Figure 6.9 shows the results of the probabilistic modeling for the 29 December simulations from Zhupanovsky volcano at + 12 hr after the start of the event. Figure 6.9b shows the ash-loading probabilities where concentrations exceeded 10^{-6} g/m^3 or 1 μg/m^3. Further examples can be seen in Figures 6.10 and 6.11 for ash concentrations and their probabilities at 2 km, 6 km, and 10 km ASL.

For Zhupanovsky volcano, we also illustrate the significance of the minimum threshold chosen for the probabilistic analysis and how it could impact the spatial footprint applied in developing a VAA. Figure 6.12 shows the probabilities of measureable ash-mass loadings for six different minimum exceedance thresholds for the concentration data at +12 hr into the simulations. None of the 161 members forecasted a concentration ≥ 100 mg/m^3 (Fig. 6.12a). Moving from exceedance thresholds ≥ 10 mg/m^3 (Fig. 6.12b) to ≥ 0.1 mg/m^3 (Fig. 6.12e) there is evidence of a growing cloud footprint and in the region of 100% probability of exceeding the threshold. As we

(a)

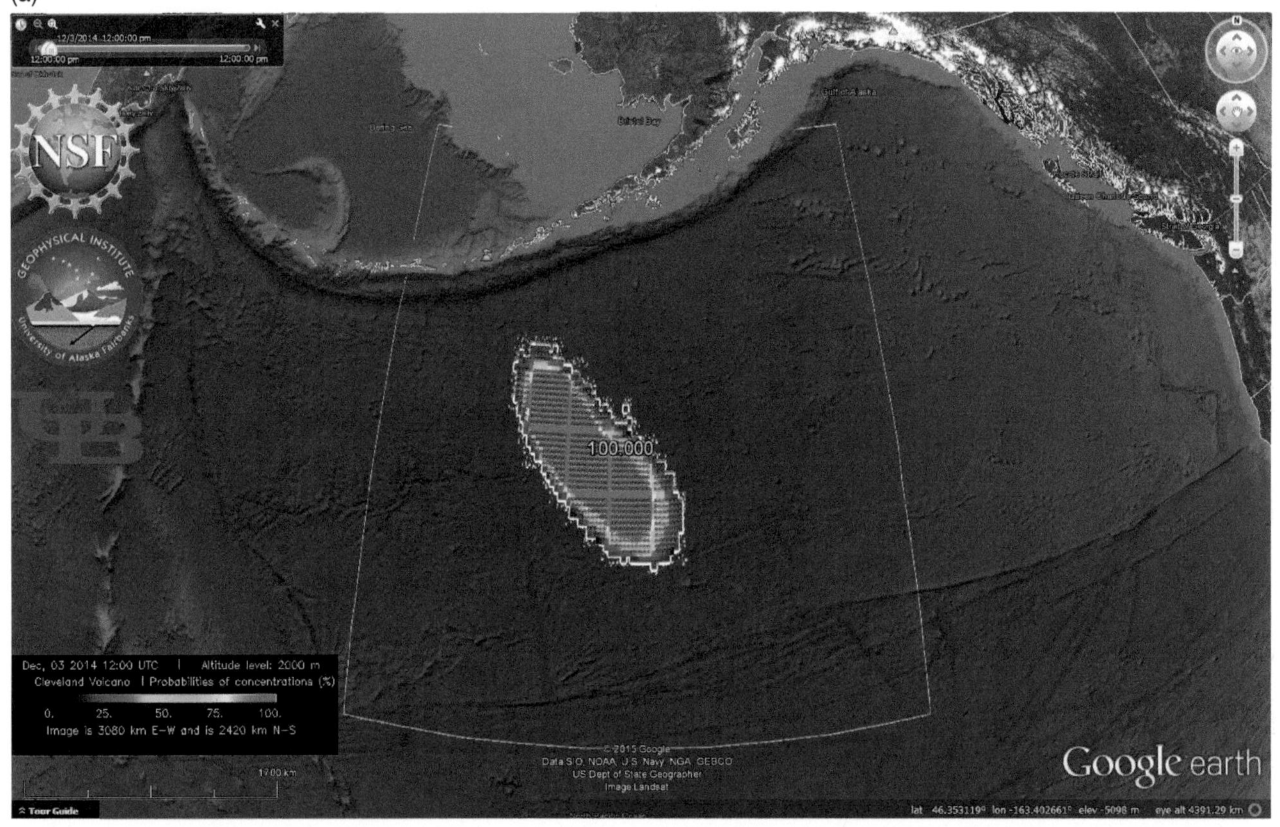

(b)

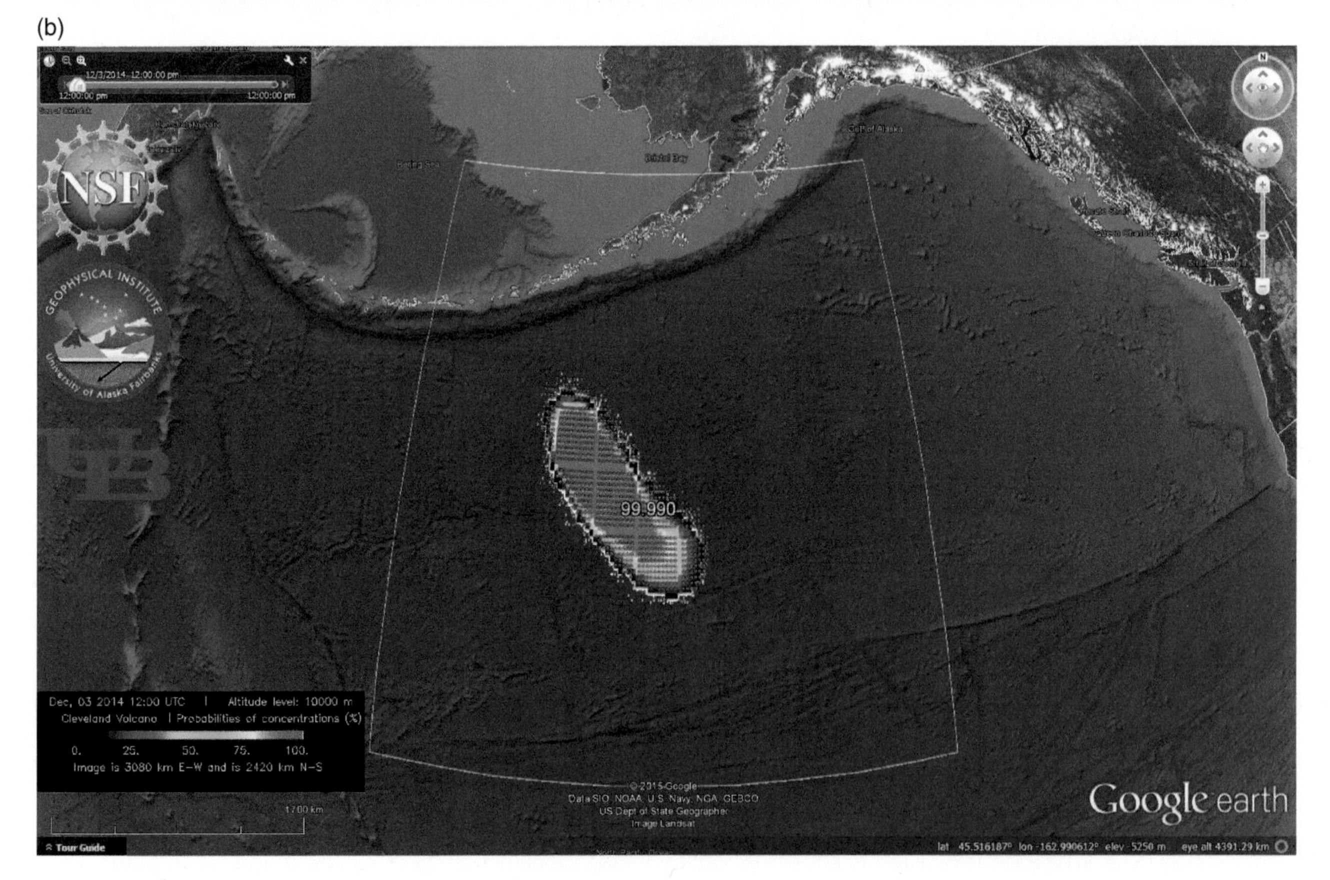

Figure 6.7 Probabilistic modeling outputs at 12:00 UTC, + 12 hr after eruption, for the Cleveland 3 December 2014 model simulation. Probabilities (%) of ash concentration (mg/m^3) exceedances at 2 (a), 10 (b), and 16 (c) km ASL.

(c)

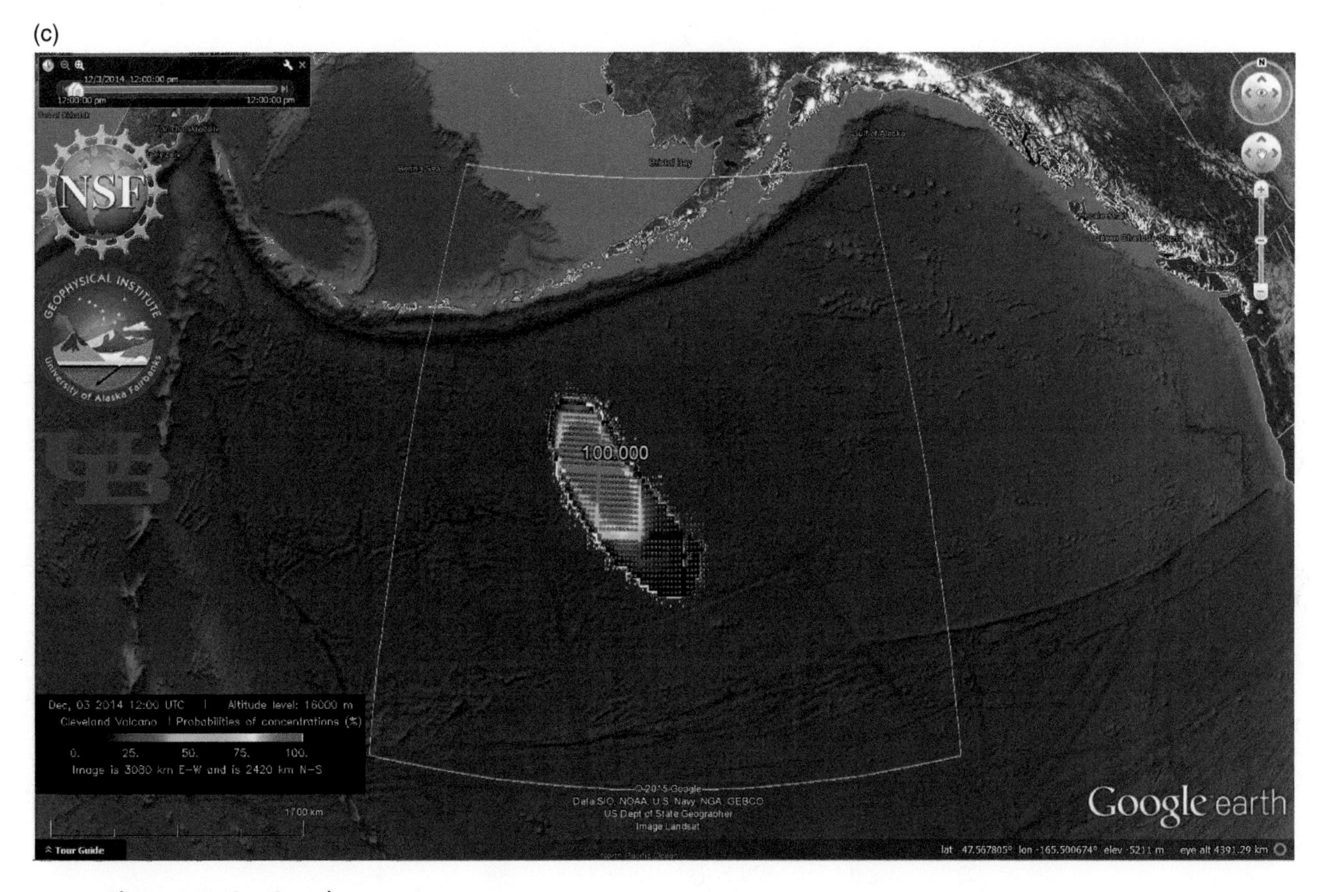

Figure 6.7 (Continued)

reduce the constraints on the minimum concentration threshold, the probability of a measureable ash-mass loading increases.

By examining the probabilities of exceeding specific concentrations, there is evidence of significant differences at varying altitudes in the atmosphere. For the probabilities at 2 km ASL, the spatial footprint is similar to that from the mass loadings (Fig. 6.12). Comparing the probabilities at 2 km ASL (Fig. 6.13) to 10 km ASL (Fig. 6.14), significant differences occur. For a concentration threshold of 100 mg/m^3, there was no probability of exceedance in the gridded concentration at either 2 km ASL (Fig. 6.13a) or at 10 km ASL (Fig. 6.14a). By relaxing the concentration threshold by a factor of 10, Figure 6.13b–f for 2 km ASL and Figure 6.14b–f for 10 km ASL illustrate an increasing spatial extent to the simulated cloud occurs.

Our results show that the probability of exceeding the same specific concentration threshold varies significantly by altitude. Awareness of this vertical variability is critical for those in real-time hazard assessment where there is a need to produce maps at critical altitudes or flight levels for the aviation industry. Figures 6.15 and 6.16 show how the spatial extent of the probabilities varies with time for differing concentration thresholds. Here we fix the altitude to only compare concentration probabilities at 2 km ASL or the lowest vertical level of the Puff VATD model outputs. As the cloud grows in size, the spatial extent of the probability of the defined thresholds being exceeded also grows directly correlated to the level of cloud dispersal as simulated by Puff.

6.3.2.2. Comparing Individual Members and the Mean of All Simulations

There is a need to compare individual simulation members to the mean of all simulations as well as compare the probabilistic results to any observational data. Then the probabilistic modeling approach can be elevated to determine if it provides more information on the potential variability in the ash-cloud dispersion and is useful for operational hazard assessment. For the Zhupanovsky simulation, rather than examine all 161 members and compare them one by one to each other and the mean results, we compared simulation members numbered 51 and 160. These two members represent the Puff model runs with the maximum and minimum initial

+ 12 hours

Particle locations by altitude : Atmos. ash mass loadings

<u>Zhupanovsky volcano</u>

<u>Simulation number 1</u>

Zhupanovsky volcano
Start time: Dec 29, 2014 at 00:00 UTC
3 hour eruptive event
Plume height = 11.2 km ASL

Mean particle size = 3.2 μm
PSD has 63% from 0.32 μm to 32 μm

Total mass = 5×10^{11} kg

Very fine PSD

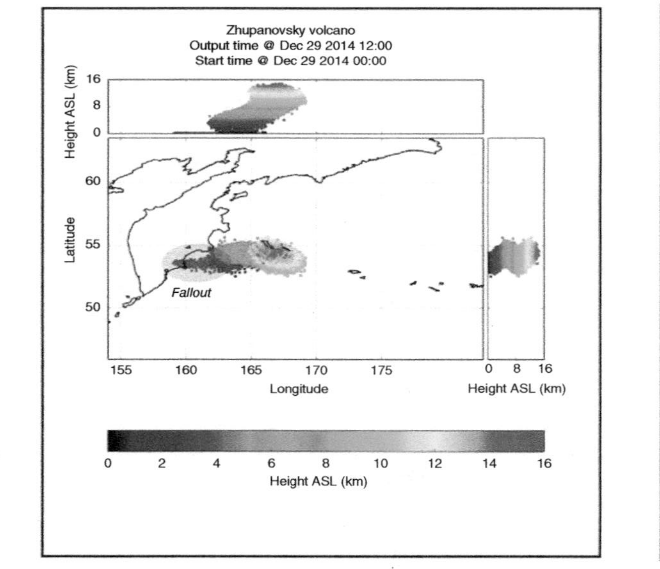

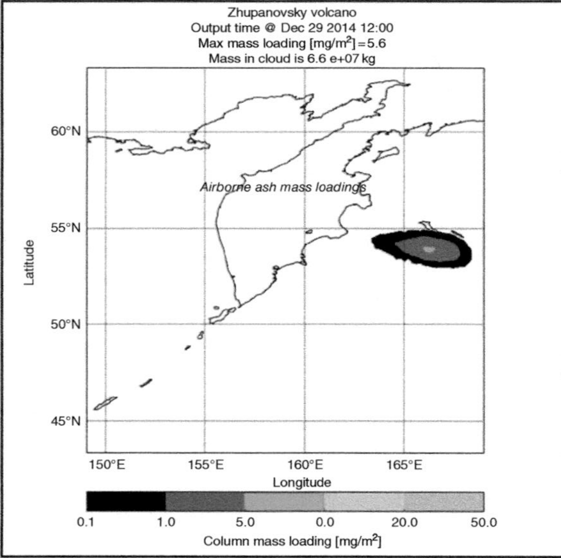

Figure 6.8 Zhupanovsky Volcano Puff VATD model simulation for probabilistic run number 1. This is for start time on 29 December 2014 at 00:00 UTC with the particle locations and ash mass loading (mg/m^2) at + 12 hr after the eruption start, or 12:00 UTC.

(a)

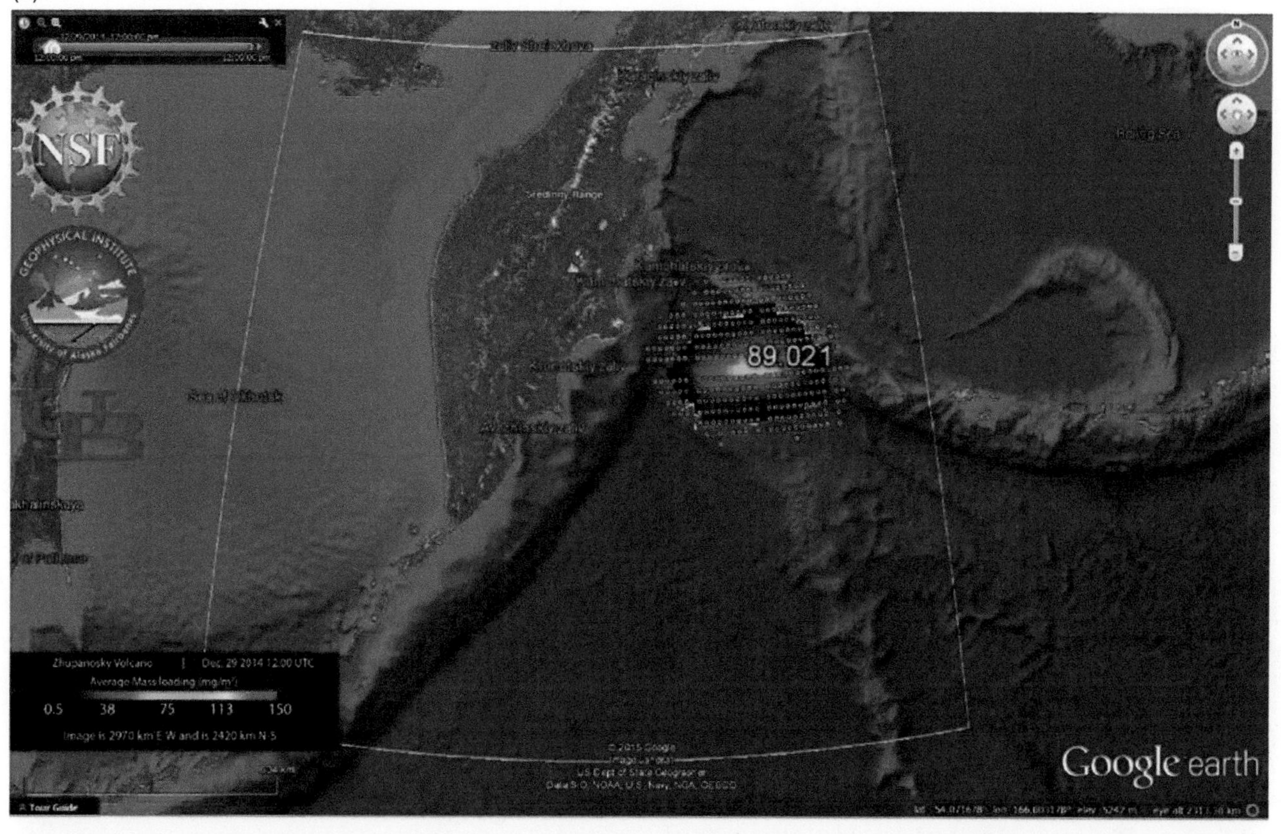

(b)

Figure 6.9 Probabilistic modeling outputs at 12:00 UTC, + 12 hr after eruption, for Zhupanovsky 29 December 2014 model simulation. (a) Mean of the 161 simulation members showing ash mass loadings (mg/m^2) and (b) probabilities (%) of ash mass loading exceeding predefined threshold.

(a)

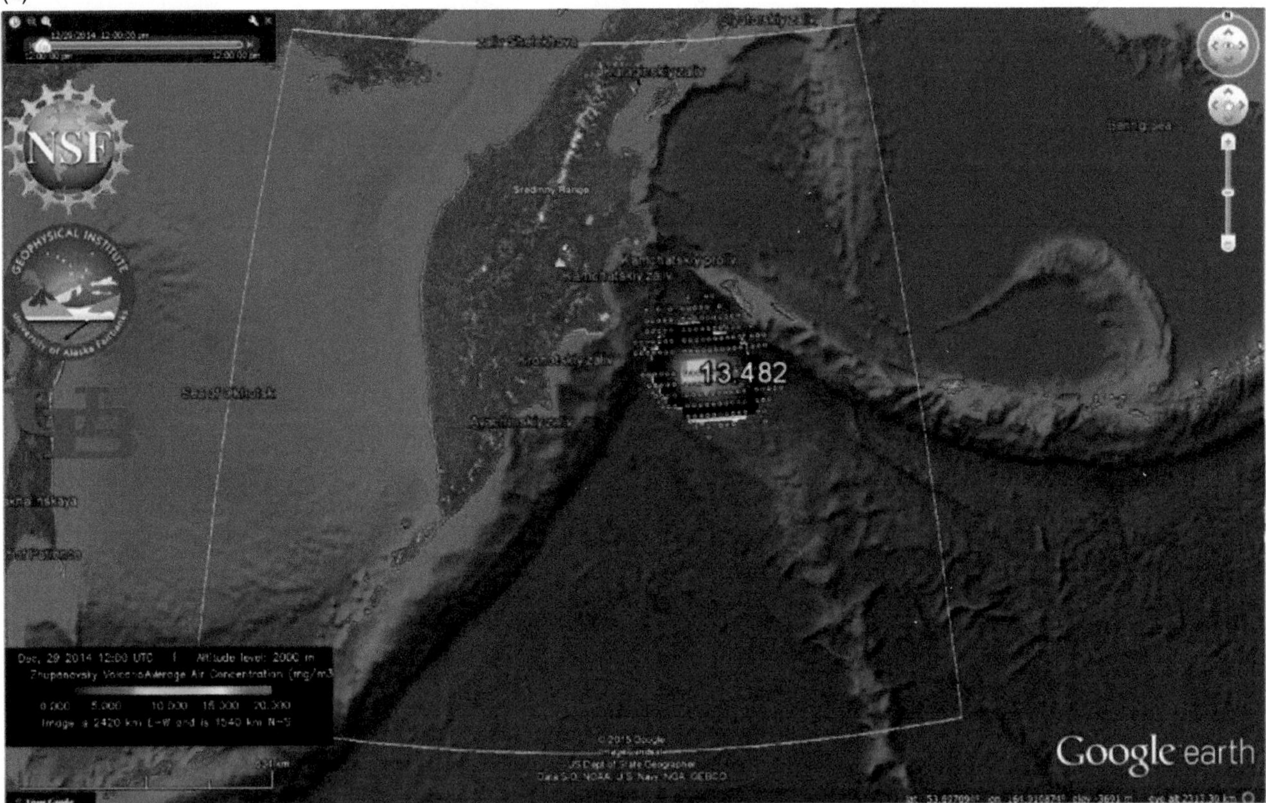

(b)

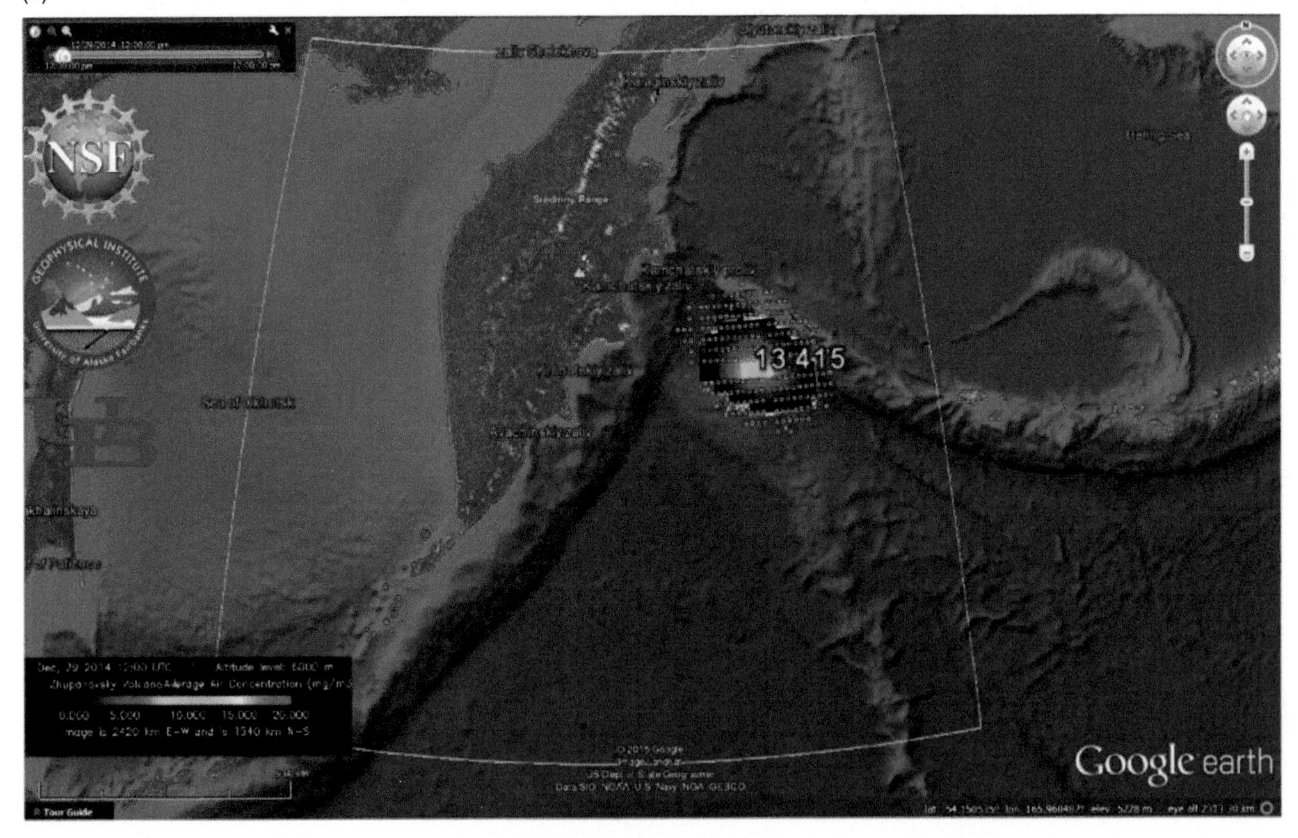

Figure 6.10 Probabilistic modeling outputs at 12:00 UTC, + 12 hr after eruption, for Zhupanovsky 29 December 2014 model simulation. Mean results from the 161 simulation members showing ash concentrations (mg/m^3) at 2 (a), 10 (b), and 16 (c) km ASL.

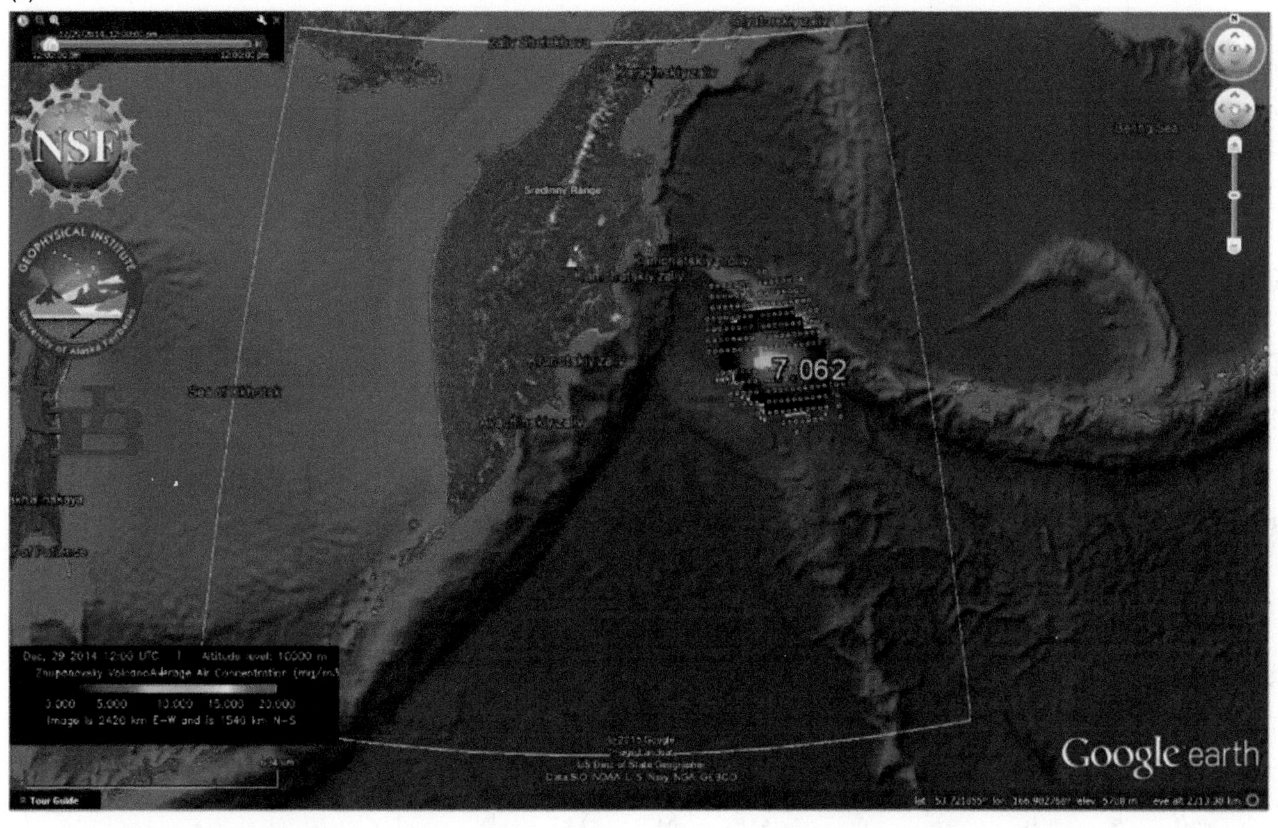

Figure 6.10 (Continued)

plume heights, 10.98 km ASL for Run Number 51 and 13.63 km ASL for Run Number 160. Each model simulations used the same initial vertical shape (Poisson distribution to represent an umbrella-shaped cloud) while the PSD was defined directly from the Puffin model. For Simulation Number 51, Figure 6.17a shows the Puff particle locations in plan view along with longitudinal and latitudinal cross sections. Figure 6.17b shows the mass loadings with a polygon defining the spatial extent for all locations where mass loadings exceed 1 mg/m^2. Figure 6.17c shows the ash concentrations at 10–12 km ASL with its associated polygon for concentrations exceeding 1 mg/m^3.

These can be directly compared to the corresponding results from Simulation Number 160 : Figure 6.17d for particle locations, Figure 6.17e for mass loadings, and Figure 6.17f for ash concentrations. The higher altitude initial plume for Simulation Number 160 has an impact on the footprint of the mass loadings and ash concentrations. This is highlighted in Figure 6.18, which compares the polygons for mass loadings and 10–12 km ASL concentrations to the mean of the 161 simulations. Figure 6.18a, for the mass loadings, shows small differences in the total footprints (Number 51 at 44,500 km^2, Number 160 at 47,500 km^2, and the mean of all the runs at 54,000 km^2). The impact of the higher initial plume height in Run Number 160 is seen to greater effect in the area of the ash concentrations at 10–12 km ASL (Number 51 at 16,600 km^2, Number 160 at 26,300 km^2, and the mean of all the runs at 32,700 km^2). Here, the Simulation Number 160 (Fig. 6.18b) extends farther to the northwest as compared with Simulation Number 51. As the cloud disperses, this displacement between the two runs grows with time.

6.4. DISCUSSION

Our results show how probabilistic modeling can be used to assess the probability of exceeding a ash concentration and/or mass loading threshold in both space and time. The higher the probability, the more likely this threshold would be exceeded. This can provide a higher degree of confidence in the modeling results and be used to build a map of the area most at risk to concentrations/ mass loadings greater than the specific threshold. Figures 6.5 to 6.7 for Cleveland volcano and Figures 6.9

(a)

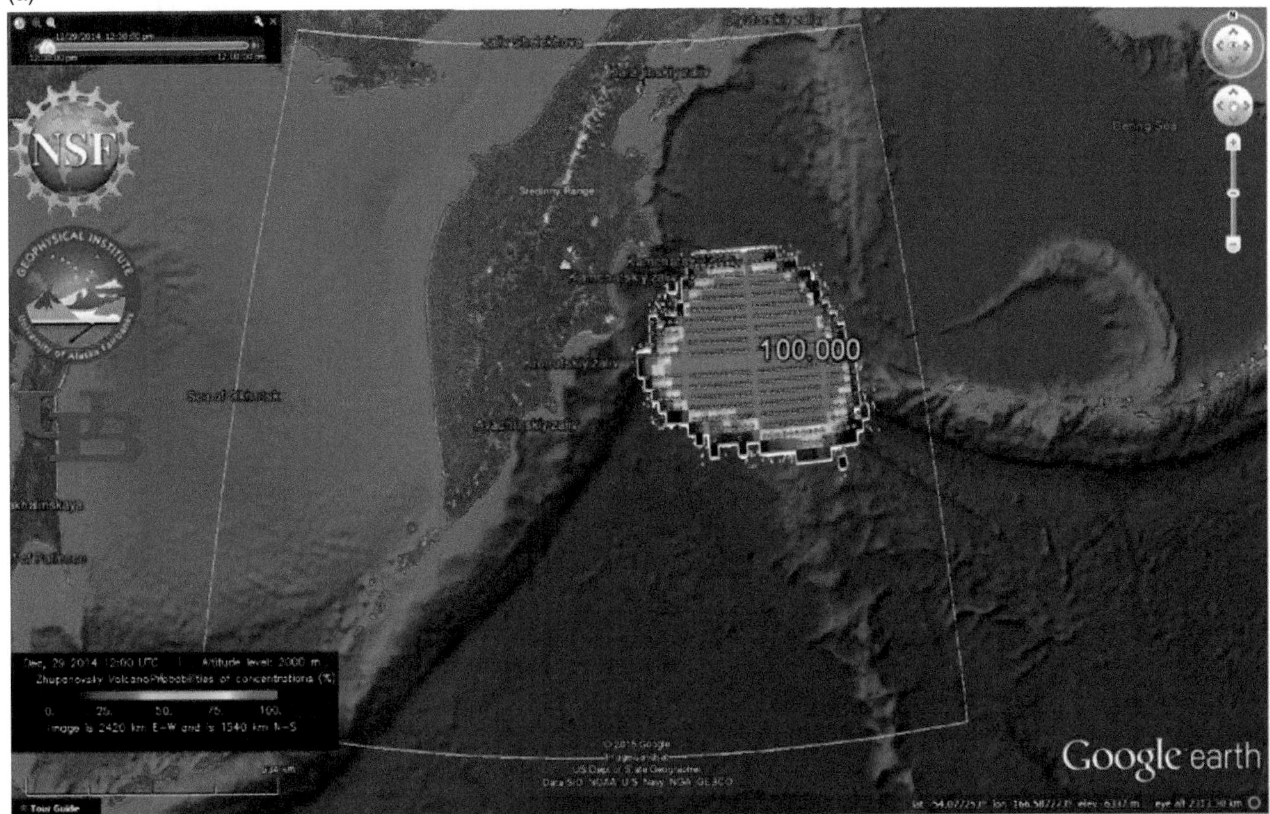

(b)

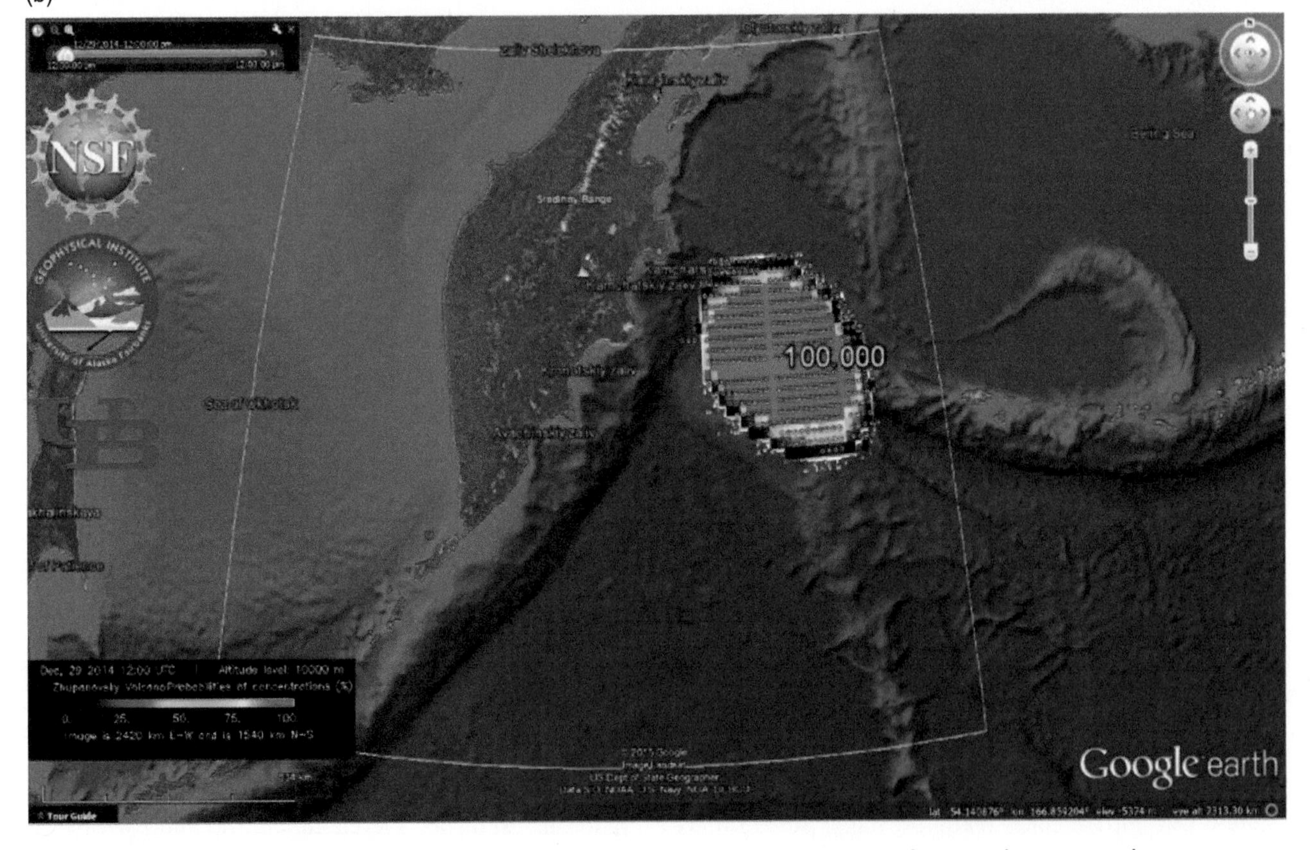

Figure 6.11 Probabilistic modeling outputs at 12:00 UTC, + 12 hr after eruption, for Zhupanovsky 29 December 2014 model simulation. Probabilities (%) of ash concentration (mg/m^3) exceedances at 2 (a), 10 (b), and 16 (c) km ASL.

(c)

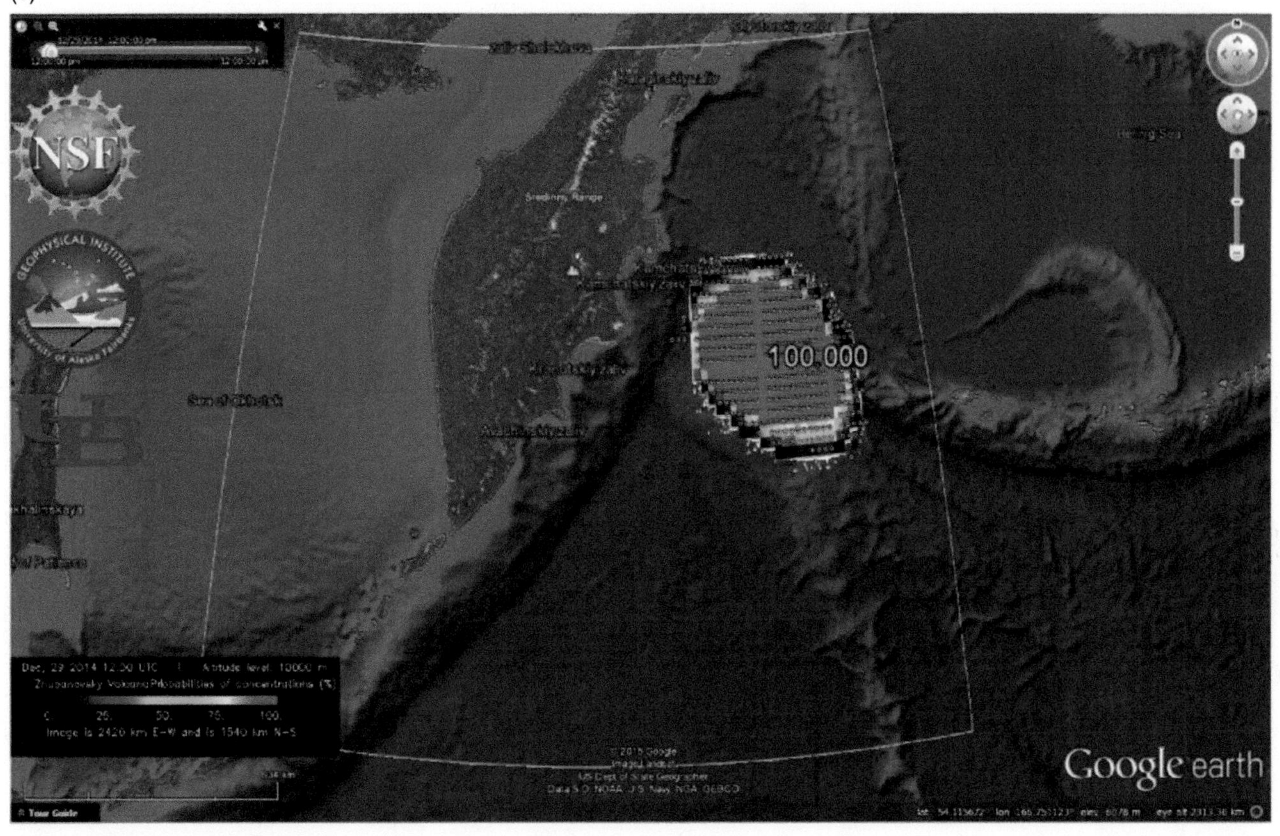

Figure 6.11 (Continued)

to 6.16 for Zhupanovsky volcano show outputs from our probabilistic modeling workflow. For real-time assessments of the ash-cloud impact and likely location and concentration, the communication of our and any available probabilistic results to the end user becomes critical. Displaying the results in a common interface and ensuring they are compatible with tools currently used to generate VAA and VAG should be a focus of researchers and operational organizations developing the probabilistic modeling capabilities.

For our contribution to the monograph, we focused on four eruption source inputs to the VATD modeling of the cloud. Observational data are needed to constrain the results from any dispersion modeling and update prior knowledge of the input parameters and associated uncertainties into posterior input data for updated VATD model simulations. Prior to an eruption, the likely maximum, minimum, and mean/median value for the modeling input parameters can be chosen. Several questions need to be evaluated as the workflow is developed. Does one choose a Gaussian distribution to sample the parameter values? How many sample points are required to fully represent the uncertainties and produce useful probabilistic modeling results without increasing the number of required simulations for real-time applications?

We plan to build upon the approach shown here by adding in the NWP ensembles, following on from the work in *Stefanescu et al.* [2014], to the real-time programming environment highlighted in Figure 6.3. Additionally, we will develop the real-time routines to integrate with different VATD models in a plug-and-play approach as well as with a time-varying version of the 1-D BENT model.

6.5. CONCLUSIONS

Eruptions like Eyjafjallajökull in 2010 [*Gudmundsson et al.*, 2010] can change the landscape for both the scientific and operational hazard-assessment communities. At the time of the 2010 events, there were a request and a need for a better understanding of the uncertainties in the ash-modeling simulations. Meetings such as those reported on in *WMO* [2010a, 2010b, 2013] brought

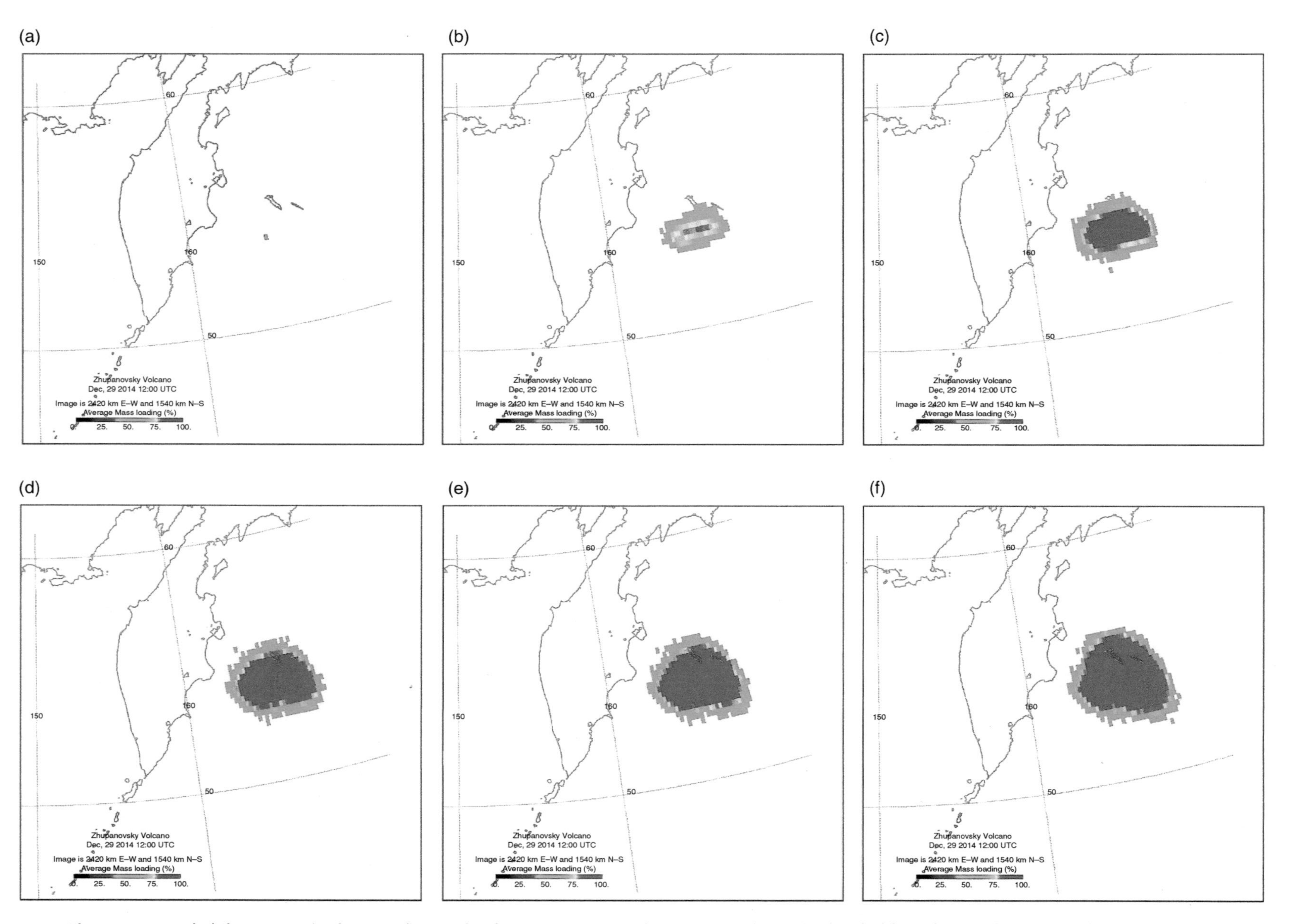

Figure 6.12 Probabilities (%) of volcanic ash mass loading (mg/m^2) exceeding a range of specific thresholds in the simulations for Zhupanovsky volcano, 12:00 UTC 29 December 2014. (a) 100 mg/m^3, (b) 10 mg/m^3, (c) 1 mg/m^3, (d) 0.1 mg/m^3, (e) 0.01 mg/m^3, and (f) 0.001 mg/m^3 or 1 μ mg/m^3.

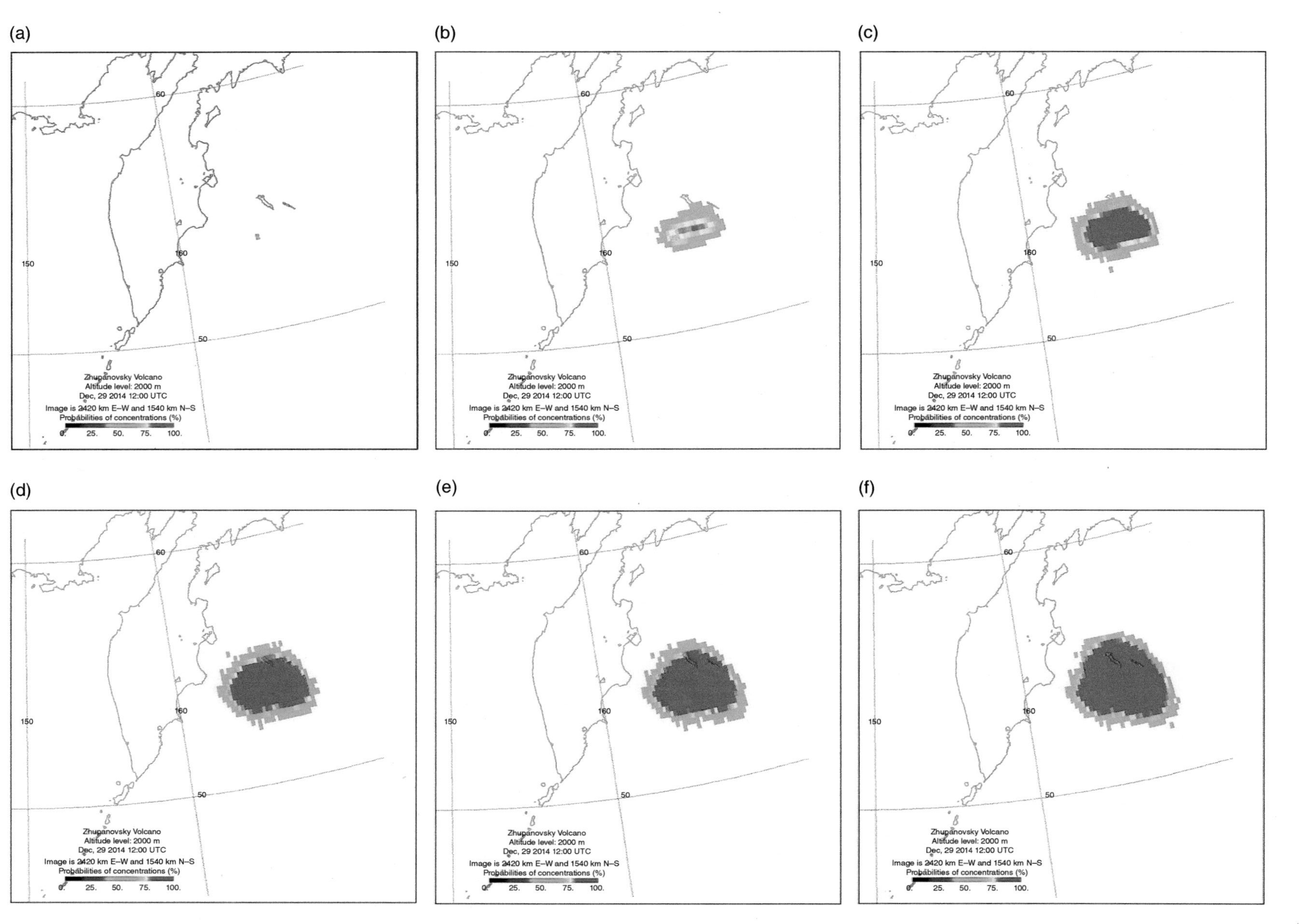

Figure 6.13 Probabilities (%) of volcanic ash concentration (mg/m^3) occurrence at 2 km ASL exceeding a range of specific thresholds in the simulations for Zhupanovsky volcano, 12:00 UTC 29 December 2014, when concentration threshold set at (a) 100 mg/m^3, (b) 10 mg/m^3, (c) 1 mg/m^3, (d) 0.1 mg/m^3, (e) 0.01 mg/m^3, and (f) 0.001 mg/m^3 or 1 μ mg/m^3.

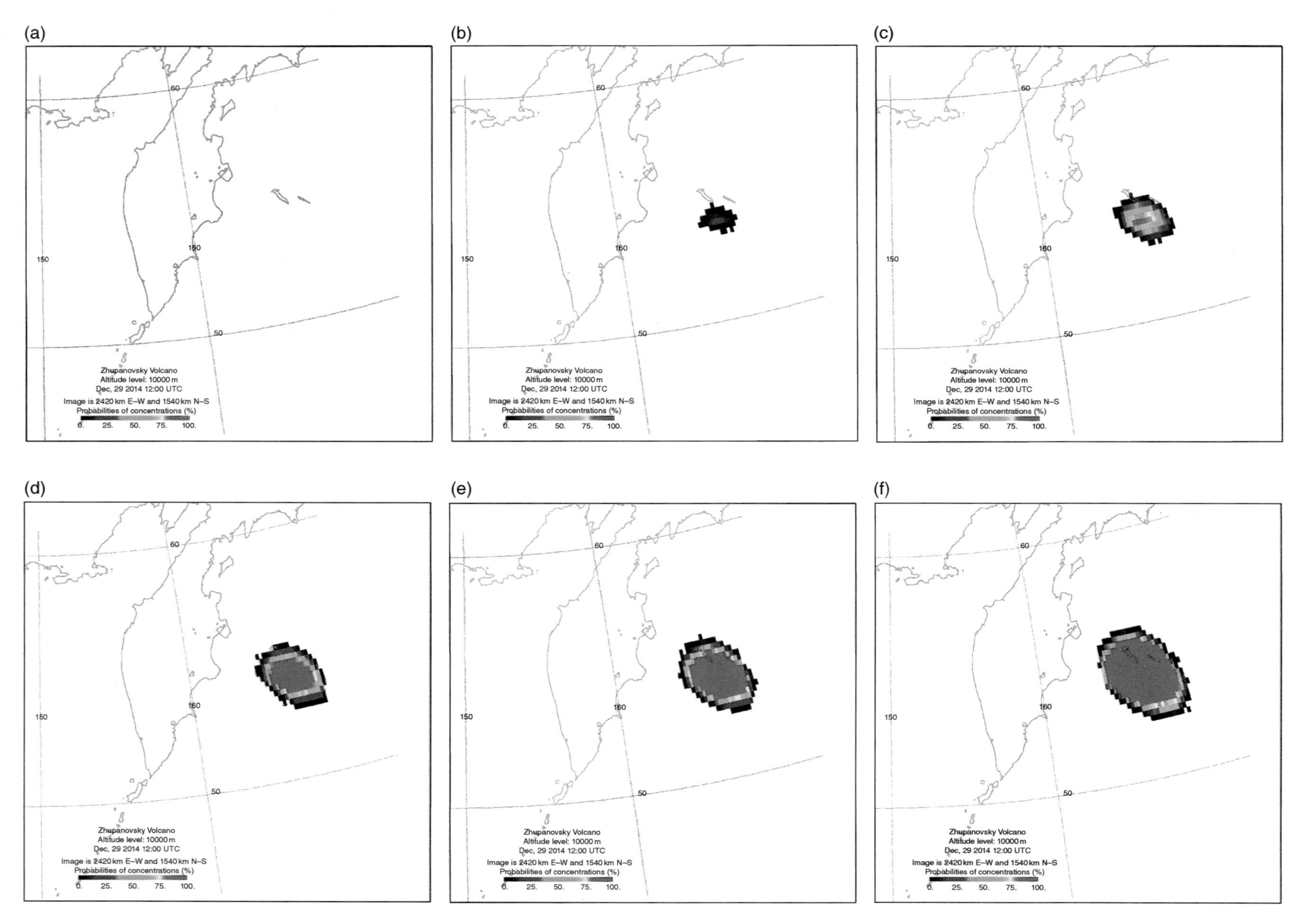

Figure 6.14 Probabilities (%) of volcanic ash concentration (mg/m^3) occurrence at 10 km ASL exceeding a range of specific thresholds in the simulations for Zhupanovsky volcano, 12:00 UTC 29 December 2014, when concentration threshold set at (a) 100 mg/m^3, (b) 10 mg/m^3, (c) 1 mg/m^3, (d) 0.1 mg/m^3, (e) 0.01 mg/m^3, and (f) 0.001 mg/m^3 or 1 μ mg/m^3.

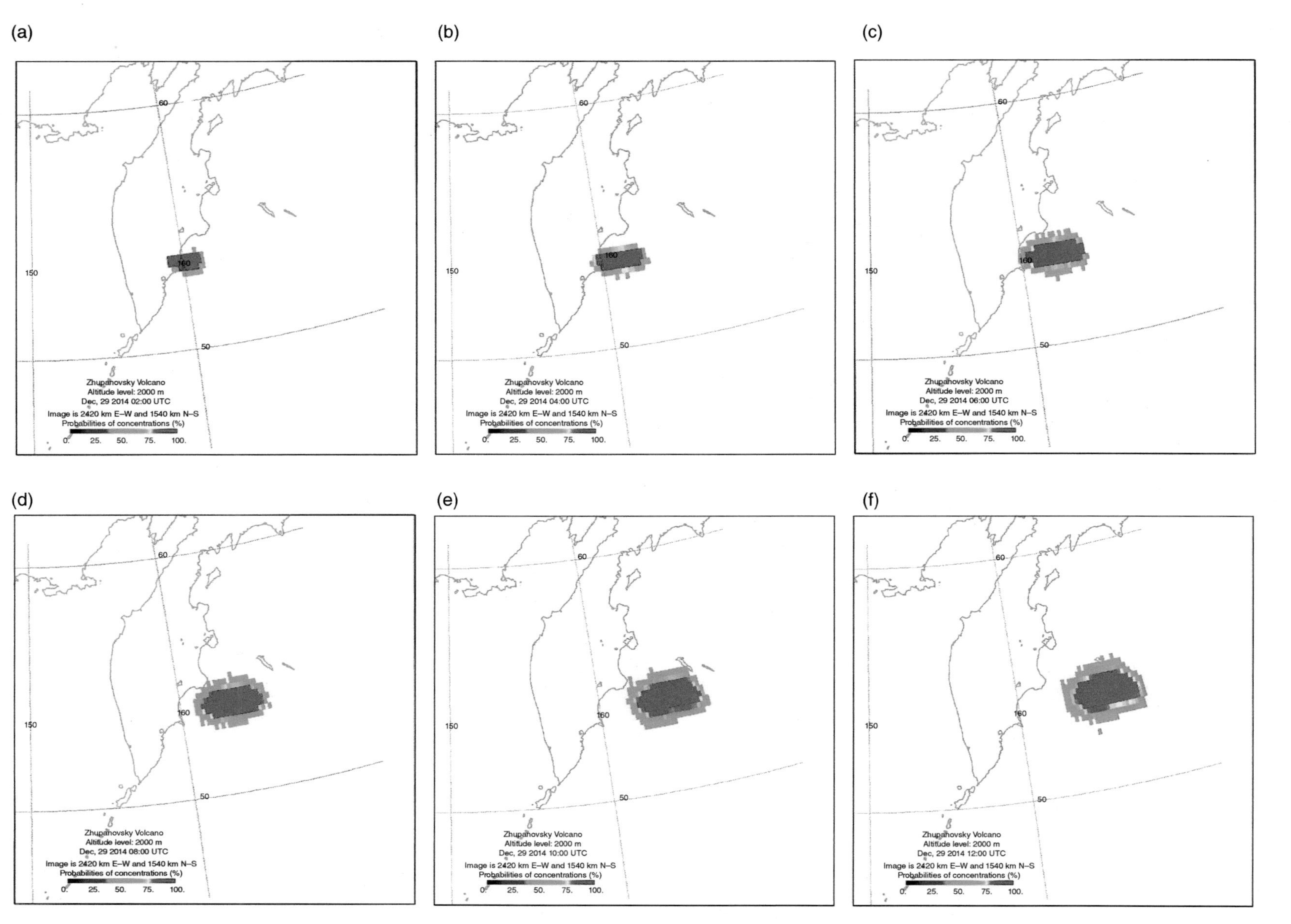

Figure 6.15 Probabilities (%) at 2 km ASL from the simulations for Zhupanovsky volcano, 29 December 2014, when concentrations exceeding at 1 mg/m^3, (a) 02:00, (b) 04:00, (c) 06:00, (d) 08:00, (e) 10:00, and (f) 12:00.

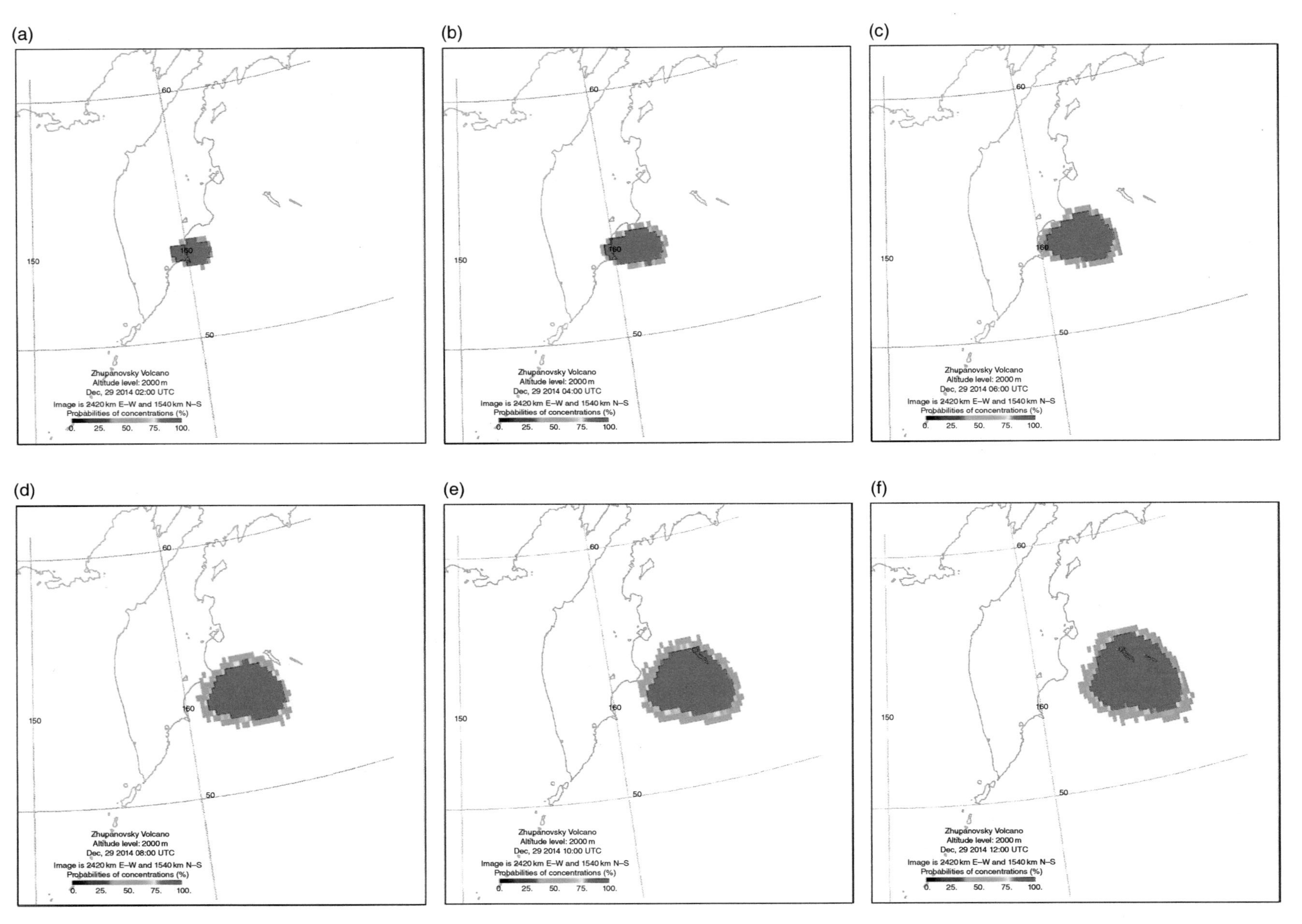

Figure 6.16 Probabilities (%) at 2 km ASL from the simulations for Zhupanovsky volcano, 29 December 2014, when concentrations exceeding at 0.001 mg/m^3, (a) 02:00, (b) 04:00, (c) 06:00, (d) 08:00, (e) 10:00, and (f) 12:00.

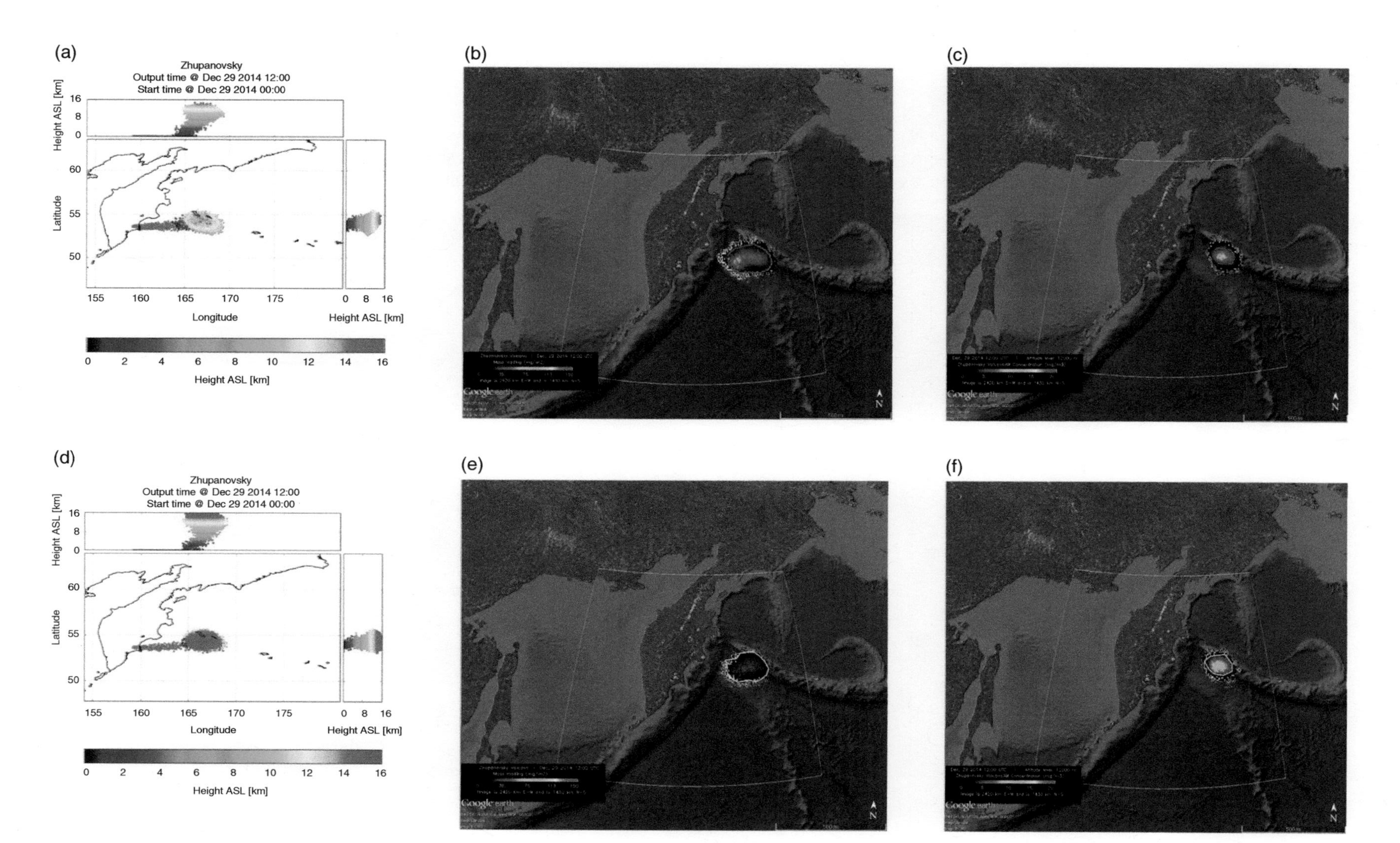

Figure 6.17 Puff volcanic ash cloud simulations #51 and #160, for Zhupanovsky volcano on 29 December 2014 at 12:00 UTC showing Puff particle locations ([a] for #51; [d] for #160), mass loadings ([b] for #51; [e] for #160), and ash concentrations from 10 to 12 km ASL ([c] for #51; [f] for #160).

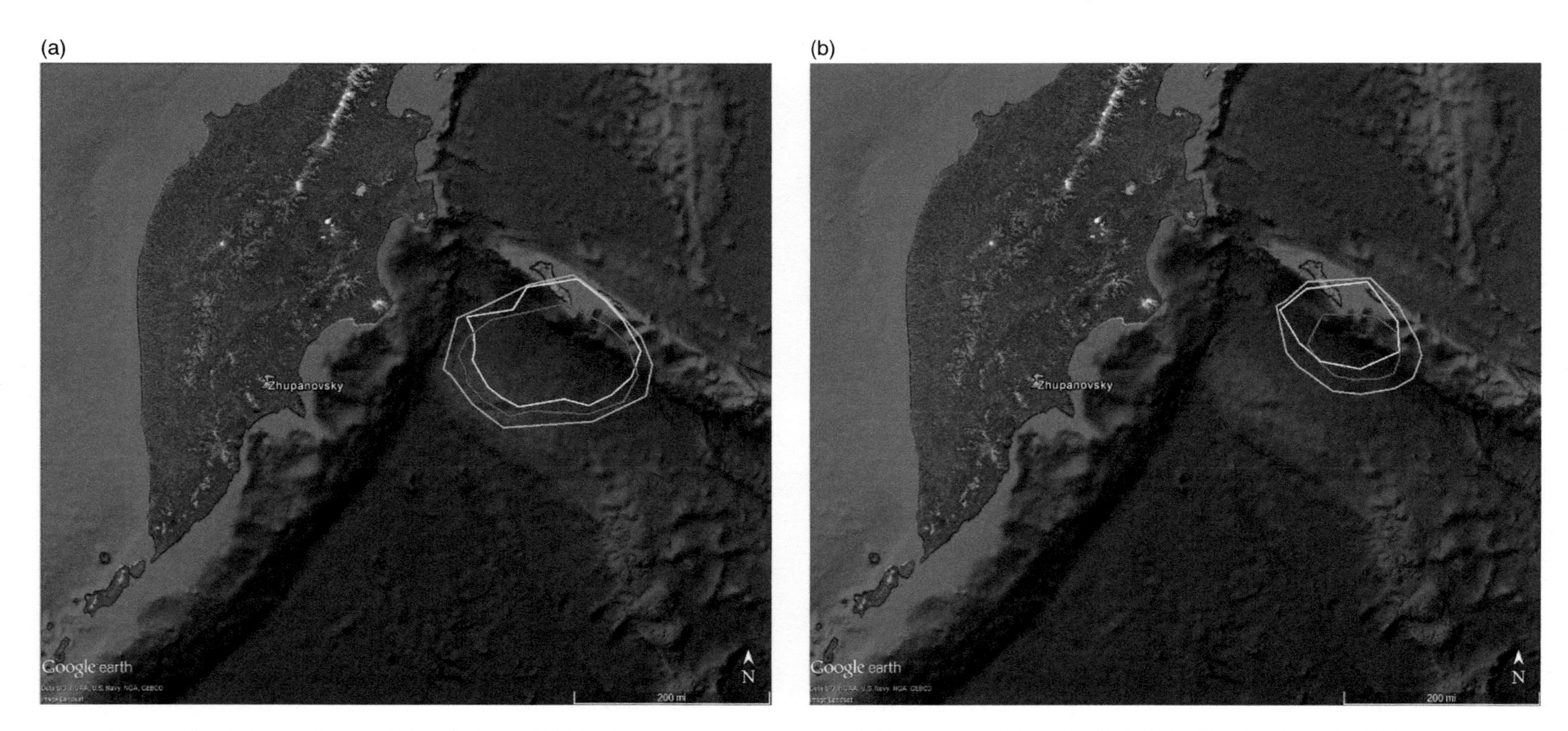

Figure 6.18 Polygons for simulation #51 and #160 for Zhupanovsky volcano on 29 December 2014 at 12:00 UTC. (a) Mass loadings for #51, yellow polygon, and #161, red polygon, as well as the mean from all simulation members, green polygon. (b) Ash concentrations at 10–12 km ASL for #51, yellow polygon, and #161, red polygon as well as the mean from all simulation members, green polygon.

together these two communities where a common theme emerged on the need to move from deterministic modeling to combined deterministic and probabilistic approach. As *WMO* [2013] states, certain eruption source parameters, such as plume height (km ASL) and eruption length (s), can be measured during the event while others, such as eruption rate (m^3/s) and particle size distribution, can either be estimated from past eruptions or derived directly from the measured eruption data.

We presented a workflow for probabilistic modeling where a 1-D plume rise model, BENT [*Bursik,* 2001] has been coupled to a four-dimensional volcanic-ash transport and dispersion (VATD) model. We focused on four BENT model parameters to build our probabilistic modeling approach. We built a complete workflow that coupled the input uncertainties from the 161 simulation members with the Puffin tool to develop downwind atmospheric ash concentrations and mass loadings with the associated probabilities of exceeding specific thresholds. We have shown how, in using our system, maps of the mean ash-mass loadings with time from our 161 simulation members can be produced along with probabilities of exceeding defined ash-mass loadings and atmospheric ash-concentration thresholds. Being able to quantify the likelihood of exceeding a specific concentration or mass loading threshold delivers a new tool for those in real-time ash-cloud hazard assessment to add to their advisories needed for aviation safety and human health impact.

However, the more critical question is how to communicate these probabilities to the end user and how to transition the research to operations? What does it mean to say that half of the 161 members breached the threshold? How would a 95% probability of exceeding a mass loading of 1 mg/m^2 in a model simulation be used in a VAA by a VAAC forecaster? How would this information be interpreted by the aviation community? Our modeling results could be compared and evaluated with available remote sensing data to provide additional tools as the VAAC produces its VAA and VAG. To fully develop these probabilistic tools so they can move directly from research to operations requires the research community to demonstrate how the probabilistic modeling provides additional and useful information on the dispersing cloud that can then assist in the advisories produced by each VAAC.

Learning from the NWP community in how they use and communicate probabilities from their ensemble member NWP model simulations, such as *Roebber et al.* [2004], will be critical in how the probabilistic modeling approach is used in real-time volcanic ash-cloud hazard assessment. To develop a probabilistic approach into a real-time system that can produce results in a timely manner is as important as the research itself into the probabilistic analysis techniques and sensitivity of dispersion results. If the computation takes too long or has not been developed to integrate with operational hazard assessment tools in the VAAC, then it will be very difficult for the operators to use the tool. Working directly with those in operations and supporting them on the integration of the tool as the probabilistic modeling system is developed means that the final product can be used from day one of operations.

ACKNOWLEDGMENTS

This work was funded by National Science Foundation (NSF) Interdisciplinary/Collaborative Research under grant no. CMMI-1131799. We would like to thank Dr. Donald Morton, Arctic Region Supercomputing Centre, Geophysical Institute, UAF, for his assistance on the NSF IDR project and Dr. Rorik Peterson, College of Engineering and Mines, UAF, for his time and expertise in building the MPI version of the Puff VATD model. We thank the Vhub team for housing the Puffin interface for users of the tool and for providing the computing support for the NSF IDR project discussions and meeting notes. Finally, we would like to thank the editors in chief and associate editors of the American Geophysical Union monograph series for the opportunity to publish this manuscript and for bringing all the different manuscripts together to develop such a unique and diverse publication.

REFERENCES

Alaska Volcano Observatory (AVO) (2014), Cleveland Volcano: Current Volcanic Activity, http://avo.alaska.edu/activity/Cleveland.php, last viewed 9 January 2015.

Bursik, M. (2001), Effect of wind on the rise height of volcanic plumes, *Geophys. Res. Lett.*, *28*(18), 3621–3624.

Bursik, M., M. Jones, S. Carn, K. Dean, A. Patra, M. Pavolonis, E. B. Pitman, T. Singh, P. Singla, P. Webley, H. Bjornsson, and M. Ripepe (2012), Estimation and propagation of volcanic source parameter uncertainty in an ash transport and dispersal model: application to the Eyjafjallajökull plume of 14–16 April 2010, *Bull. Volcanol.*, *74*(10), 2321–2338.

Ellrod, G. P., B. H. Connell, and D. W. Hillger (2003), Improved detection of airborne volcanic ash using multispectral infrared satellite data, *J. Geophys. Res. Atmos. (1984–2012)*, *108* (D12); doi: 10.1029/2002JD002802.

Folch, A., A. Costa, and S. Basart, (2012), Validation of the FALL3D ash dispersion model using observations of the 2010 Eyjafjallajökull volcanic ash clouds, *Atmos. Environ.*, *48*, 165–183.

Gudmundsson, M. T., R. Pedersen, K. Vogfjörd, B. Thorbjarnardóttir, S. Jakobsdóttir, and M. J. Roberts, (2010), Eruptions of Eyjafjallajökull Volcano, Iceland, *Eos, Trans. AGU*, *91*(21), 190–191.

Gudmundsson, M. T., T. Thordarson, Á. Höskuldsson, G. Larsen, H. Björnsson, F. J. Prata, B. Oddsson, E. Magnússon, T. Högnadóttir, G. N. Petersen, C. L. Hayward, J. A. Stevenson, and I. Jónsdóttir (2012), Ash generation and distribution from the April–May 2010 eruption of Eyjafjallajökull, Iceland, *Sci. Rep.*, *2*.

Guffanti, M., and A. Tupper (2014), Chapter 4, Volcanic ash hazards and aviation risk, *Volcanic Hazards, Risks and Disasters,* edited by P. Papale and J. F. Shroder, 87–105.

Global Volcanism Program (GVP) (2015), Weekly activity report, 31 December 2014–6 January 2015; http://www.volcano.si.edu/reports_weekly.cfm#vn_211060, last viewed 9 January 2015.

Horwell, C. J., and P. J. Baxter (2006), The respiratory health hazards of volcanic ash: a review for volcanic risk mitigation, *Bull. Volcanol.*, *69*(1), 1–24.

International Civil Aviation Agency (ICAO) (2012), International Volcanic Ash Task Force, http://www.icao.int/safety/meteorology/ivatf/Pages/default.aspx, last viewed 8 January 2015.

Japanese Meteorology Agency (JMA) (2014), Tokyo Volcanic Ash Advisory Center, http://ds.data.jma.go.jp/svd/vaac/data/index.html, last viewed 8 January 2015.

Madankan, R., S. Pouget, P. Singla, M. Bursik, J. Dehn, M. Jones, A. Patra, M. Pavolonis, E. B. Pitman, T. Singh, and P. Webley (2014), Computation of probabilistic hazard maps and source parameter estimation for volcanic ash transport and dispersion, *J. Computational Phys.*, *271*, 39–59.

Meteorological Office (2012), The volcanic ash modelling setup: London VAAC, viewed 8 January 2015; http://www.metoffice.gov.uk/media/pdf/p/7/London_VAAC_Current_Modelling_SetUp_v01-1_05042012.pdf.

Patra, A. K., M. Bursik, J. Dehn, M. Jones, R. Madankan, D. Morton, M. Pavolonis, E. B. Pitman, S. Pouget, T. Singh, P. Singla, E. R. Stefanescu, and P. Webley (2013), Challenges in developing DDDAS based methodology for volcanic ash hazard analysis–effect of numerical weather prediction variability and parameter estimation, *Procedia Computer Sci.*, *18*, 1871–1880.

Pavolonis, M. J., A. K. Heidinger, and J. Sieglaff (2013), Automated retrievals of volcanic ash and dust cloud properties from upwelling infrared measurements, *J. Geophys. Res. Atmos.*, *118*(3), 1436–1458.

Prata, A. J., and A. Tupper (2009), Aviation hazards from volcanoes: The state of the science, *Nat. Hazards*, *51*(2), 239–244.

Roebber, P. J., D. M. Schultz, B. A. Colle, and D. J. Stensrud (2004), Toward improved prediction: High-resolution and ensemble modeling systems in operations, *Weath. Forecast.*, *19*(5), 936–949.

Schneider, D. J., and R. P. Hoblitt (2013), Doppler weather radar observations of the 2009 eruption of Redoubt Volcano, Alaska, *J. Volcanol. Geotherm. Res.*, *259*, 133–144.

Searcy, C., K. Dean, and W. Stringer (1998), PUFF: A high-resolution volcanic ash tracking model, *J. Volcanol. Geotherm. Res.*, *80*(1), 1–16.

Steensen, T., M. Stuefer, P. Webley, G. Grell, and S. Freitas (2013), Qualitative comparison of Mount Redoubt 2009 volcanic clouds using the PUFF and WRF-Chem dispersion models and satellite remote sensing data, *J. Volcanol. Geotherm. Res.*, *259*, 235–247.

Stefanescu, E. R., A. K. Patra, M. I. Bursik, R. Madankan, S. Pouget, M. Jones, P. Singla, T. Singh, E. B. Pitman, M. Pavolonis, D. Morton, P. Webley, and J. Dehn (2014), Temporal, probabilistic mapping of ash clouds using windfield stochastic variability and uncertain eruption source parameters: Example of the 14 April 2010 Eyjafjallajökull eruption, *J. Adv. Model. Earth Syst.*, *6*(4), 1173–1184.

Tesche, M., P. Glantz, C. Johansson, M. Norman, A. Hiebsch, A. Ansmann, D. Althausen, R. Engelmann, and P. Seifert (2012), Volcanic ash over Scandinavia originating from the Grímsvötn eruptions in May 2011, *J. Geophys. Res. Atmos. (1984–2012)*, *117*(D9).

Vogel, H., J. Förstner, B. Vogel, T. Hanisch, B. Mühr, U. Schättler, and T. Schad (2014), Time-lagged ensemble simulations of the dispersion of the Eyjafjallajökull plume over Europe with COSMO-ART, *Atmos. Chem. Phys.*, *14*, 7837–7845; doi:10.5194/acp-14-7837-2014.

Weber, K., J. Eliasson, A. Vogel, C. Fischer, T. Pohl, G. van Haren, M. Meier, B. Grobety, and D. Dahmann (2012), Airborne in-situ investigations of the Eyjafjallajökull volcanic ash plume on Iceland and over north-western Germany with light aircrafts and optical particle counters, *Atmos. Environ.*, *48*, 9–21.

Webley, P. W., B. J. B. Stunder, and K. G. Dean (2009a), Preliminary sensitivity study of eruption source parameters for operational volcanic ash cloud transport and dispersion models: A case study of the August 1992 eruption of the Crater Peak vent, Mount Spurr, Alaska, *J. Volcanol. Geotherm. Res.*, *186*(1), 108–119.

Webley, P. W., K. Dean, J. E. Bailey, J. Dehn, and R. Peterson (2009b), Automated forecasting of volcanic ash dispersion utilizing Virtual Globes, *Nat. Hazards*, *51*(2), 345–361.

Webley, P. W., K. Dean, R. Peterson, A. Steffke, M. Harrild, and J. Groves (2012), Dispersion modeling of volcanic ash clouds: North Pacific eruptions, the past 40 years: 1970–2010, *Nat. Hazards*, *61*(2), 661–671.

Webley, P. W., K. G., Dean, J. Dehn, J. E. Bailey, and R. Peterson (2010), Volcanic-ash dispersion modeling of the 2006 eruption of Augustine volcano using the Puff model, chapter 21 of J. A. Power, M. L. Coombs, and J. T. Freymueller, eds., The 2006 eruption of Augustine Volcano, Alaska, U.S. Geological Survey Professional Paper 1769, 482–501.

Witham, C. S., M. C. Hort, R. Potts, R. Servranckx, P. Husson, and F. Bonnardot (2007), Comparison of VAAC atmospheric dispersion models using the 1 November 2004 Grímsvötn eruption, *Meteor. App.*, *14*(1), 27–38.

World Meteorological Organization (WMO) (2010a), Workshop on ash dispersal forecast and civil aviation:

Results from the 1st Workshop, Model Definition Document, last viewed 8 January 2015; http://www.unige.ch/sciences/terre/mineral/CERG/Workshop/results/Model-Document-Geneva10.pdf.

World Meteorological Organization (WMO) (2010b), Workshop on Ash Dispersal Forecast and Civil Aviation: Results from the 1st Workshop: Benchmark Document, last viewed 8 January 2015; http://www.unige.ch/sciences/terre/mineral/CERG/Workshop/results/ADCAW2010-benchmark-doc.pdf.

World Meteorological Organization (WMO) (2013), Workshop on Ash Dispersal Forecast and Civil Aviation: Results from the 2nd Workshop: Updated model definition document, viewed 8 January 2015; http://www.unige.ch/sciences/terre/mineral/CERG/Workshop2/results-2/Model-Definition-Document-2013.pdf.

7

Uncertainties in Estimating Magma Source Parameters from InSAR Observation

Wenyu Gong,[1] Zhong Lu,[2] and Franz Meyer[1]

ABSTRACT

Satellite radar interferometry (InSAR) has been proven to be an essential technique for measuring volcano-wide surface deformations at a spatial resolution of tens of meters and centimeter-level accuracy. For many volcanic systems, InSAR observations have been the main data source for studying geophysical processes at active volcanoes and are used frequently to estimate volcano source parameters using inverse modeling techniques. This chapter evaluates how the accuracy of estimated source model parameters is affected by typical errors in radar phase measurement such as baseline-induced phase signals, atmospheric distortions, and decorrelations noise. To mathematically study the influence of these errors, we use the Mogi source model as example and discuss how the different error components in InSAR deformation measurement affect the uncertainties of estimated source model parameters. To understand this inherent processes, nonlinear least squares and Monte Carlo simulations are used to generate the posterior probability distribution of source parameters. In addition, the impact of approximations in the Mogi source model are also discussed in this paper, including the flat-surface assumption, the disregard of magma compressibility, and the Poisson's ratio.

7.1. INTRODUCTION

There are approximately 1500 volcanoes on Earth that have erupted in the past 10,000 years. Today volcanic activity affects the lives and livelihoods of a rapidly growing number of people around the globe. About 20 volcanoes are erupting on Earth at any given time; 50–70 erupt each year, and about 160 erupt each decade [*GVP*, 2015]. Volcano monitoring is key to mitigating the adverse impacts of volcanic activity [*VHP*, 2015]. In general, the "eruption cycle" of a volcano can be conceptualized as a series of events from deep magma generation to surface eruption, including such stages as partial melting, initial ascent through the upper mantle and lower crust, crustal assimilation, magma mixing, degassing, shallow storage, and finally ascent to the surface [*Dzurisin*, 2003; 2007]. This process is complex, varying from one eruption to the next and from volcano to volcano. In many cases, volcanic eruptions are preceded by pronounced ground deformation in response to increasing pressure from magma chambers or due to the upward intrusion of magma. Therefore, monitoring and modeling volcanic deformation constitutes a critical element in providing warning of impending eruptions, reducing loss of life, and mitigating impact on property [*Dzurisin*, 2007].Together with seismology, continuous ground deformation measurements (like GPS), geology/petrology, and gas geochemistry observations, the spatially dense deformation field derived satellite interferometric synthetic aperture radar (InSAR) can play a pivotal role in constraining the most important parameters needed to assess short-term volcanic hazards and better understand volcanic processes.

InSAR imaging of ground surface deformation is contaminated by multiple error sources, (e.g., atmospheric artifacts), contributions from inaccurate satellite orbit and auxiliary topography model, reducing the sensitivity

[1] *Geophysical Institute, University of Alaska Fairbanks, Fairbanks, Alaska, USA*
[2] *Southern Methodist University, Dallas, Texas, USA*

Natural Hazard Uncertainty Assessment: Modeling and Decision Support, Geophysical Monograph 223, First Edition.
Edited by Karin Riley, Peter Webley, and Matthew Thompson.

and accuracy of this technique in geodetic applications. In the studies of earthquake source models by using InSAR-derived observations, previous efforts have been made on assessing the uncertainties and trade-offs of estimated earthquake source parameters, where the simulated interferograms with realistic noise levels have been generated and inverted [e.g., *Dawson and Tregoning*, 2007; *Lohman and Simons*, 2005]. Even though InSAR has been widely used to estimate the volcanic source parameters (geometry and volume change), there are limited reports on how InSAR observations could reduce the quality of estimated source parameters. For example, *Lu et al.* [2003] reported that the atmospheric contamination could lead to biased estimates of source depth in studying Westdahl volcano, Alaska, USA.

In this study, we present a systematic study on estimating uncertainties associated with volcano source parameters from InSAR observations, through the simulated interferometric data and the inversion of Mogi source model [*Mogi*, 1958]. The Mogi model is selected because of its capability in approximating volcano sources, simplicity (hence computational efficiency), and wide use. In the rest of this paper, we will present a section on uncertainties related to InSAR-measured volcano deformation, and a brief introduction to the analytical Mogi source model. The inherent impacts from InSAR measurement uncertainties on the estimated volcano source model parameters will be investigated quantitatively. Finally, a brief discussion is provided regarding the impacts of the assumptions in Mogi source model on the estimated source parameters.

7.2. VOLCANO DEFORMATION FROM INSAR AND THE ASSOCIATED UNCERTAINTIES

7.2.1. Uncertainties in InSAR Observations

Interferometric SAR (InSAR) involves the use of two or more synthetic aperture radar (SAR) phase images of the same area to generate interferograms [*Burgmann et al.*, 2000; *Lu and Dzurisin*, 2014; *Massonnet and Feigl*, 1998; *Rosen et al.*, 2000; *Simons and Rosen*, 2007]. The phase component of a SAR image is controlled mainly by the round-trip travel time from SAR to ground. Hence, the interferogram (i.e., the phase difference of SAR images) along with a digital elevation model (DEM) can be used to extract any surface deformation that might have occurred during the interval between image acquisitions and resulted in a change of distance between satellite and ground. InSAR has been proven an important observational tool for measuring volcano-wide deformation at a spatial resolution of tens of meters with centimeter-level accuracy [e.g., *Amelung et al.*, 2000; *Biggs et al.*, 2010; *Dzurisin et al.*, 2006; *Lu et al.*, 2000, 2005, 2010; *Lu and Dzurisin*, 2014; *Massonnet and Feigl*, 1998; *Poland et al.*, 2006; *Pritchard and Simons*, 2002; *Wicks et al.*, 1998, 2002, 2006].

Uncertainties in our knowledge of the satellite position can cause baseline-induced phase ramps in an interferogram. In addition, the radar phase measurement is also affected by atmospheric conditions (water vapor in the troposphere and electron density in the ionosphere) and by the interaction of the radar signal with the ground surface; therefore, an InSAR image can contain spurious phase anomalies due to atmospheric artifacts and decorrelation noise. These errors should be modeled and removed as much as possible to minimize uncertainties in deformation source parameters.

To mitigate phase ramps as well as topography-dependent phase artifacts related to uncertainties in the SAR antenna positions at the times of image acquisitions (ϕ_{orb}) [e.g., *Lu and Dzurisin*, 2014], baseline refinement steps are often employed. A commonly used method is to determine the baseline vector based on an existing DEM via a least-squares approach [*Rosen et al.*, 1996]. For this method, areas of the interferogram that are used to refine the baseline should have negligible deformation or deformation that is well characterized by an independent data source. Because baseline-induced anomalies are characterized by primarily long-wavelength ramping fringes, they are alternatively often modeled and removed using two-dimensional polynomial functions [*Biggs et al.*, 2007].

Inhomogeneity in the atmosphere that results in path anomalies (ϕ_{aps}) in InSAR phase images is the most significant error source in repeat-pass InSAR deformation measurements. Differences in tropospheric water-vapor content as well as ionospheric variations at two observation times can cause differing path delays and consequent anomalies in an InSAR deformation image. The tropospheric artifact can be classified into two types [*Hanssen*, 2011]: (1) stratified artifact, which is caused by changes in vertical refractivity profile at two image acquisition times, is correlated to the local topography; and (2) turbulent effect, which is caused by three-dimensional heterogeneity in the tropospheric refractivity at image acquisition times, impacts both flat and mountainous areas. Atmospheric delay anomalies can reduce the accuracy of InSAR-derived deformation measurements from several millimeters under ideal conditions to a few centimeters under more typical conditions, thus obscuring subtle changes that could hold clues to the cause of the deformation [e.g., *Lu and Dzurisin*, 2014]. Generally speaking, atmospheric artifacts have shorter spatial wavelength than those due to baselines. The difficulty with estimating water-vapor or ionosphere conditions with the needed accuracy and spatial density is an important limiting factor in deformation monitoring with InSAR. Several methods have been proposed to estimate the water-vapor content and remove its effect from deformation interferograms:

1. The first method is to estimate water-vapor concentrations in the target area at the times of SAR image acquisitions using short-term predictions from weather models [*Foster et al.*, 2006; *Gong et al.*, 2014; *Jolivet et al.*, 2014]. While weather models would be sufficient to correct stratified atmospheric artifacts over a volcano [*Cong and Eineder*, 2012], the problem with this approach is that current weather models have much coarser resolution (a few kilometers) than InSAR measurements (tens of meters). This deficiency can be remedied to some extent by integrating weather models with high-resolution atmospheric measurements, but this approach requires intensive computation.

2. The second method is to estimate water-vapor concentration from continuous global positioning system (CGPS) observations in the target area [*Li et al.*, 2005]. The spatial resolution (i.e., station spacing) of local or regional CGPS networks at volcanoes is typically a few kilometers to tens of kilometers, which renders this method ineffective in most cases.

3. The third approach to correcting atmospheric delay anomalies in InSAR observations is to utilize water-vapor measurements from optical satellite sensors such as the Moderate Resolution Imaging Spectroradiometer (MODIS), Advanced Spaceborne Thermal Emission and Reflection Radiometer (ASTER), and European Medium Resolution Imaging Spectrometer (MERIS) [*Li et al.*, 2003]. The disadvantage of this method is the requirement of nearly simultaneous acquisitions of SAR and cloud-free optical images.

4. The fourth method extracts the atmospheric contributions from interferometric phases themselves, including many approaches for mitigating ionospheric artifacts, for example, split-spectrum, Faraday rotation related method [*Meyer and Nicoll*, 2008], azimuth offset method, and multiple-aperture interferometry [*Jung et al.*, 2013]. Additionally, methods that account for the correlation between atmospheric delays and the local topography [*Bejar-Pizarro et al.*, 2013], as well as the spatial variability [*Bekaert et al.*, 2015; *Lin et al.*, 2010] can be used to correct stratified atmospheric delays. However, those solutions have limited impacts on reducing turbulent mixing tropospheric component.

5. The fifth technique is to correct atmospheric delay anomalies using a multi-temporal InSAR technique [e.g., *Berardino et al.*, 2002; *Ferretti et al.*, 2001; *Hooper et al.*, 2007; *Lu and Dzurisin*, 2014]. When more than several SAR images are available for a given study area, multi-interferogram InSAR processing can be employed to improve the accuracy of deformation maps. If there is sufficient knowledge on the deformation behavior, a temporal deformation model can be built and used to constrain the atmospheric phase mitigation. Otherwise, atmospheric artifacts can be removed via spatial-temporal filtering processes, as they are characterized as spatially correlated but temporally uncorrelated fringes while deformation signals are correlated in both space and time domains. This technique is most promising when a large SAR stack is available, because more SAR observations allow better characterization of the spatial-temporal properties of different interferometric phase components.

The accuracy of interferometric phase is also affected by decorrelation noise (ϕ_{decor}), including the temporal, geometric, and volumetric decorrelation as well as other radar receiver noise [*Hanssen*, 2001; *Lu and Dzurisin*, 2014]. In volcano studies, decorrelation is aggravated by surface changes related to volcanic eruptions (lava and lahar flows, ashfall), seasonal or perennial changes of snow and ice in high-latitude regions, dense vegetation in middle to lower regions, and layover or shadowing effects under complex terrain conditions. In a typical scenario, the region around a volcano's summit loses coherence. Several examples from C-band data are given in Figure 7.1. These examples show than the volcanic summits often lose coherence even if only summer images separated by one satellite revisiting cycle are used. Decorrelation over the volcano peaks, where volcanic deformation is expected to be largest, can compromise the accuracy of source parameters estimated from InSAR data.

An InSAR image might also contain errors due to inaccurate DEMs (ϕ_{topo}). Despite the increasing availability of remote-sensing-derived DEM data, this is a continuous issue at active volcanoes, as their topography is constantly modified by eruptive activity. The availability of TanDEM DEM [*Krieger et al.*, 2007], newly released enhanced 1-arc-second SRTM DEM [*SRTM*, 2015], and tandem ERS-1/ERS-2 images [e.g., *Lu et al.*, 2011] can reduce this error source. Additionally, if multiple SAR images are available over a study area, the existing DEM can be updated or refined using interferograms with large spatial baselines but short temporal separations [e.g., *Lu et al.*, 2013; *Lu and Dzurisin*, 2014]. This scheme of baseline setting is used to reduce impacts from the deformation signal and other phase artifacts on the DEM as large spatial-baseline interferograms are more sensitive to the ϕ_{topo} and the small temporal-baseline reduces the contamination by the deformation signal. The non-topography-related signals can be further mitigated based on temporal-spatial statistical properties of different phase components (e.g., reduce atmospheric signal

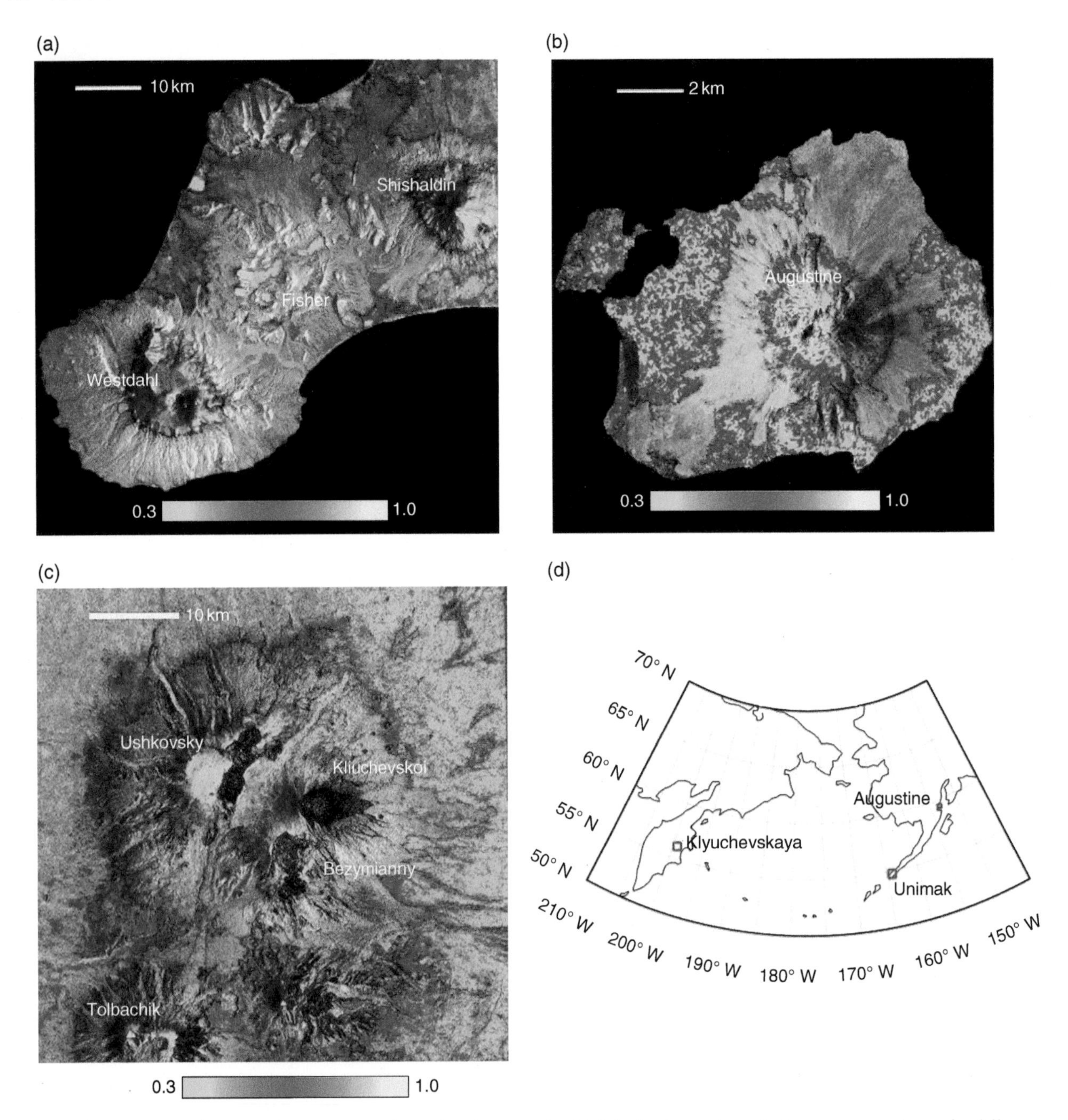

Figure 7.1 Coherence image examples of volcanoes in high-latitude region during summer to early fall. Coherence images are overlaying on the corresponding radar intensity images and major volcanoes in the example sites are annotated: (a) Envisat data pair of Unimak Island (Alaska, USA) with a 35-day time interval, (b) Radarsat-1 data pair of Augustine volcano (Alaska, USA) with a 24-day time interval, (c) Envisat data pair of Klyuchevskaya group of volcanos (Kamchatka, Russia) with a 35-day time interval, (d) geographic locations of the coherence images shown in (a), (b), and (c).

through stacking). Errors of the inaccurate DEM, which is a function of ϕ_{topo} and perpendicular baseline, is then estimated.

Finally, for volcano studies, regional tectonic signals or other geophysical signals might introduce long wavelength "artifacts" (ϕ_{artf}) on volcano-wide deformation, for example, ocean tidal loading that may however have only small influence on volcano studies (e.g., 3 cm over 100 km [*DiCaprio and Simons*, 2008]). Often, this type of long wavelength signal can be confused with baseline errors [*Zhang et al.*, 2014] and is likely removed along with the phase ramp due to baseline error.

Both ϕ_{topo} and ϕ_{orb} are considered as deterministic errors, which are error terms that can be modeled explicitly [*Hanssen*, 2001]. They can be removed in the deformation phase reconstruction step or jointly be estimated in the geophysical model inversion. For example, ϕ_{orb} is modeled as ramps and integrated in volcano and earthquake model inversions [*Gong et al.*, 2014; *Lu et al.*, 2010; *Sudhaus and Jonsson*, 2009]. The remaining error terms, mainly ϕ_{aps} and ϕ_{decor}, are typically considered in the stochastic model that contributes to the variance-covariance matrix (or noise model) of the measurements [*Hanssen*, 2001]. Previous efforts have been made to explicitly integrate these error sources in the InSAR noise model [e.g., *Hanssen*, 2001; *Knospe and Jonsson*, 2010; *Lohman and Simons*, 2005], such that each pixel can be properly weighted in the geophysical parameter inversion and uncertainties of estimated model parameters can be predicted.

7.2.2. Satellite Line-of-Sight Measurements

The error characteristics of geophysical parameters derived from InSAR are additionally impacted by the oblique imaging geometry of SAR sensors, which limits the measurement sensitivity of InSAR to the projection of the three-dimensional deformation signal is in the sensor's line-of-sight (LOS) direction. Using the notation by *Wright et al.* [2004] and *Lu and Dzurisin* [2014], the deformation signal for ground pixel i in east, north, and up directions ([$d_{i,e}$, $d_{i,n}$, $d_{i,u}$]) is projected into the sensor's LOS direction by multiplying by the unit vector $\Lambda = [-cos\phi \cdot sin\theta,\ \sin\phi \cdot sin\theta,\ cos\theta]$, where θ is the incidence angle and ϕ is the satellite track angle clockwise from north.

$$d_{i,los} = \Lambda \cdot \left[d_{i,e},\ d_{i,n},\ d_{i,u} \right]^T \qquad (7.1)$$

In equation (7.1), $d_{i,los}$, $d_{i,e}$, $d_{i,n}$, and $d_{i,u}$ are deformation measurements in the LOS, east, north, and up directions in the ith interferogram, respectively. Thus, an interferogram provides a scalar one-dimensional deformation measurement at every coherent pixel. When interferograms in multiple geometries are available, one can solve equation (7.1) to derive the 3-D deformations. However, it is difficult to resolve the displacements in N-S direction with the near-polar orbiting satellite data in general [*Wright et al.*, 2004]. The offset tracking method or multiple aperture InSAR (MAI) can be used to extract the along-track displacement under favorable conditions [*Bechor and Zebker*, 2006; *Jung et al.*, 2011].

7.3. RETRIEVAL OF MAGMA SOURCE PARAMETERS FROM INSAR AND ASSOCIATED MEASUREMENT UNCERTAINTIES

7.3.1. Mogi Source

To better understand volcanic processes, mathematical models are often employed to relate the physical parameters of a magma source to the deformation measurements made at the surface (e.g., via InSAR or GPS). The high spatial resolution and large spatial extent of surface deformation data provided by InSAR make it possible to constrain and compare models with various source geometries and gain insight into the most likely shape, location, and volume change within a volcanic source. Typical source geometries include the point pressure source [*Mogi*, 1958], finite spherical source [*Mctigue*, 1987], dislocation sources (sill or dike source) [*Okada*, 1985], ellipsoidal sources [*Davis*, 1986; *Yang et al.*, 1988], and penny-crack sources [*Fialko et al.*, 2001]. Among the physical parameters, the location and volume change of the source are usually of most interest.

The most widely used source in volcano deformation modeling is the point pressure source (widely referred to as the Mogi source) embedded in an elastic homogeneous half-space [*Mogi*, 1958]. In a Cartesian coordinate system, the predicted displacement u at the free surface due to a change in volume ΔV or pressure ΔP of an embedded Mogi sphere is:

$$\begin{aligned} u_i\left(x_1 - x_1', x_2 - x_2', -x_3\right) &= \Delta P\left(1-v\right)\frac{r_s^3}{G}\frac{x_i - x_i'}{\left|R^3\right|} \\ &= \Delta V\frac{\left(1-v\right)}{\pi}\frac{x_i - x_i'}{\left|R^3\right|} \end{aligned} \qquad (7.2)$$

where x_1', x_2', and x_3' are the horizontal coordinates and depth of the center of the sphere, R is the distance between the center of the sphere and the observation point (x_1, x_2, and 0), ΔP and ΔV are the pressure and volume changes in the sphere, and ν is Poisson's ratio of the host rock (typical value is 0.25). Furthermore, r_s is the radius of the sphere, and G is the shear modulus of the host rock [*Delaney and Mctigue*, 1994; *Johnson*, 1987].

7.3.2. Nonlinear Least-Squares Estimation of Mogi Source Parameters

The determination of the Mogi source geometry and volume change from InSAR deformation measurements is a typical nonlinear inverse problem. The problem can be solved by a nonlinear optimization that searches for the optimal source model parameters by minimizing a misfit function. The misfit (S) describes the level of disagreement between the observed deformation ($d_{i,obs}$) and the source model prediction [*Cervelli et al.*, 2001; *Lu and Dzurisin*, 2014] that accounts for unmolded residuals, including InSAR measurement errors and impacts from oversimplifications in the source model (e.g., elastic half-space), among others. Equation (7.3) shows the minimum misfit constraint in the residual variance:

$$S_{\min} = \arg\min \frac{\Sigma\left(\Lambda \cdot u_i\left(X'\right) - d_{i,obs}\right)^2}{N - p} \tag{7.3}$$

in which N is the number of coherent pixels in an InSAR image and p is the number of unknowns in the analytical volcano source model; X' is the vector of the source geometry (x_1', x_2', and x_3').

Various strategies are available to solve the inverse problem by finding global minima (e.g., simulated annealing) and/or local minima (e.g., nonlinear least-squares schemes) [*Cervelli et al.*, 2001; *Feigl et al.*, 1995; *Lu and Dzurisin*, 2014; *Press et al.*, 2007; *Sudhaus and Jonsson*, 2009; *Wright et al.*, 1999]. The benefit of global minima algorithms is the ability to discriminate among multiple local minima in the solution space [*Mosegaard and Tarantola*, 1995]. Otherwise, with a reasonable initial guess of model parameters, the nonlinear least-squares algorithms can be used [*Cervelli et al.*, 2001; *Feigl et al.*, 1995], which is computationally more efficient. In the rest of this paper, the Levenberg-Marquardt algorithm, a nonlinear least-squares algorithm that optimizes the solution between iterative linearization and gradient methods [*Marquardt*, 1963; *Press et al.*, 2007], is used to solve for the inversion of Mogi source parameters with InSAR data.

7.3.3. Deriving Uncertainty Measures for Estimated Mogi Model Parameters

After the source geometry and associated volume change are determined, the next goal is to assess the quality of estimated parameters. Thus, we need to find the associated uncertainties (or confidence level, posterior probability distribution) of model parameters. In this section, we will discuss impacts on uncertainties (posterior probability estimates) due to measurement errors. Synthetic deformation interferograms are generated from a predefined Mogi source and then converted to radar's LOS geometry and a coherence map of Westdahl volcano (Fig. 7.1a) is used to approximate realistic coherence conditions for complex volcanic terrains.

There are various methods that can be used to propagate observational errors to the estimate parameters, for example, linearization, likelihood, Monte Carlo simulation, and bootstrap [*Donaldson and Schnabel*, 1987; *Draper and Smith*, 1966; *Press et al.*, 2007]. A linearization method calculates the uncertainties of estimated parameters with the estimated residual variance $S' = \frac{\Sigma(\Lambda \cdot u_i(X') - d_{i,obs})^2}{N' - p}$ [*Donaldson and Schnabel*, 1987]. Notice that N' is the number of independent deformation measurements. For instance, the parameter variance-covariance matrix can be computed with Jacobian matrix (J) as $S'(J^T \cdot J)^{-1}$. However, due to the dependency of InSAR deformation measurements, N' is generally less than N. Without a properly determined N', the linearization method will underestimate the uncertainties of estimated parameters. The advantage of the linearization method is that it is less computationally intensive.

The Monte Carlo simulation and the bootstrap methods, which can be performed with less computational efficiency, have been suggested to be more reliable in constructing the confidence intervals of the estimated parameters [*Arnadottir and Segall*, 1994; *Donaldson and Schnabel*, 1987]. The bootstrap method can be applied without sufficient knowledge of the measurement error, while special attention is needed to draw random samples to satisfy the statistical preassumption [*Lisowski*, 2007; *Press et al.*, 2007]. Additionally, there has been overestimation reported on confidence intervals when using the bootstrap method [*Cervelli et al.*, 2001]. When prior information or valid assumptions about the statistical properties of the measurement errors are available, the Monte Carlo method is supposed to correctly characterize the error properties of the estimated parameters [*Donaldson and Schnabel*, 1987; *Press et al.*, 2007]. The Monte Carlo method has been implemented in many geodetic studies to determine the posterior probability distribution of volcanic or earthquake model parameters [*Feigl et al.*, 1995; *Hooper et al.*, 2007; *Wright et al.*, 2003]. In order to provide accurate results on source model parameter uncertainties, we generate simulated error-contaminated unwrapped interferograms so that statistical properties of InSAR residuals are known and have been set up to approximate the reality conditions. Taking the advantage of the known InSAR error properties and stable performance of Monte Carlo method in providing reliable results, we choose the Monte Carlo method in our study to derive uncertainties of the estimated source model parameters.

7.3.3.1. Summary of Main InSAR Measurement Errors and Error Propagation Approach

Summing up the errors discussed in Section 7.2.1, the main contribution of InSAR deformation phase residuals ϵ is shown in equation (7.4).

$$\epsilon = \phi - \phi_{defo} = \phi_{aps} + \phi_{orb} + \phi_{topo} + \phi_{decor} \quad (7.4)$$

According to the spatial properties of above error components, ϵ can be categorized into three groups: (1) spatial uncorrelated error (ϵ^u) that contributes to diagonal elements (variance) of measurements' variance-covariance matrix. The variance characterizes the quality of the single point and can be predicted or estimated from coherence condition (ϕ_{decor}) [*Hanssen*, 2001]; (2) locally correlated errors (ϵ^a) that are mainly caused by localized atmospheric structures (ϕ^s_{aps}), which are correlated within a few kilometers [*Colesanti et al.*, 2003]; and (3) long-wavelength errors ϵ^l, which include long-wavelength atmospheric signal components ϕ^l_{aps} and ϕ_{orb} [*Biggs et al.*, 2007]. ϕ_{topo} is related to the accuracy of external terrain model; this term will not be considered here as this deterministic term can be removed when interferograms of different baselines are available. Additionally, topography-related atmospheric signal will also not be discussed in the following part, given it can be modeled and reduced via many previous discussed methods (see Section 7.2.1). Overall, both ϵ^a and ϵ^l create spatial correlations among InSAR pixels that leads to nonzero covariance and affect model parameter confidence intervals [*Hanssen*, 2001; *Lohman and Simons*, 2005].

To conduct our error analysis, we start with a simulated surface inflation data set corresponding to a Mogi-type volcanic source with source geometry of $x_1' = 21\text{km}$, $x_2' = 16$, $x_3' = 6\text{km}$ in local coordinates and $\Delta V = 10 \times 10^6\ \text{m}^3$. The lower-left corner is defined as the origin of the local coordinate used in the inversion. For every error type, we use Monte Carlo simulation techniques to create a sufficient number of simulated phase screens that are representative for the individual error type. We add these phase screens to the simulated Mogi-type deformation map to create a random sample of distorted InSAR data. For every simulated phase map, we apply the previously mentioned nonlinear least-squares inversion with the same initial bounds to estimate Mogi source parameters. The posterior probability distributions for these parameters can then be generated and analyzed according to input InSAR error types and mean values of distributions are expected to approximate the true Mogi source parameters.

7.3.3.2. Effects of ϵ^u, ϵ^a, and ϵ^l on the Accuracy of Source Model Parameters

The approaches to generate random realizations of the three different error types together with the resulting posterior distributions for the four Mogi model parameters are summarized in the following:

1. The spatially uncorrelated errors (ϵ^u) are simulated as normally distributed random signals and added to true deformation maps. The standard deviation (σ) of ϵ^u (one way error contribution) in the individual interferogram ranges from about 2 to 10 mm with median value of about 6 mm. An example of InSAR deformation map with ϵ^u is given in Figure 7.2b where the noise σ is about 6 mm. Assuming the analyzed interferometric targets are single-looked distributed scatterers, their phase variances can be predicted from the magnitude of complex coherence (γ) and Euler's dilogarithm [*Bamler and Hartl*, 1998]. Hence, a phase noise σ of 6 mm corresponds to an unfiltered coherence level of above 0.5 in case of C-band data. The posterior probability distribution of estimated parameters due to ϵ^u error conditions are approximately normally distributed for all four Mogi source parameters (Fig. 7.3). The confidence intervals at 95% confidence level (2σ) are shown as red lines in Figure 7.3 and the standard deviations and mean values of the posterior distribution are listed in Table 7.1. Figure 7.3 shows that the addition of random noise does not lead to significant biases of the estimated parameters.

To further analyze the influence of random errors, we increase the noise level of ϵ^u to $6mm < \sigma < 14mm$ with median value of about 10 mm that might correspond to severe decorrelation of distributed targets in C-band interferograms. For this setup, the standard deviations of the posterior distributions increase (see Table 7.1) but still maintain normal distribution with mean values that are statistically identical to their true values (at the 95% confidence level).

2. The locally correlated errors (ϵ^a), which are dominated by local scale atmospheric distortions [*Emardson et al.*, 2003; *Hanssen*, 2001; *Lohman and Simons*, 2005], are simulated as a set of atmospheric phase screens. The simulated ϵ^a were generated using a fractal surface with a dimension factor of 2.67, which is an typical value for the dimension factor as suggested for the atmospheric signal simulation [*Hanssen*, 2001; *Kampes*, 2006]. The family of ϵ^a used in this experiment has standard deviations of less than 12 mm with a median value of 3.9 mm. This value range has been set to resemble the realistic situation: for example, atmosphere error standard deviation has been reported to be less than 15 mm in southern San Andreas [*Lyons and Sandwell*, 2003], and less than 12 mm in multiple test sites (Mexico City, Netherlands, and South Australia; *Liu* [2012]). An example is given in Figure 7.2c with a noise standard deviation of 3.3 mm. Note that the simulated ϵ^a mainly accounts for the atmosphere turbulence mixing; the standard deviation of the total atmospheric artifact in an interferogram is supposed to be larger, given that the stratification is also included. Similar to the last experiment, through Monte Carlo simulations, we can draw the posterior probably distributions for all estimated parameters (Fig. 7.4). The source parameter's standard deviations are shown in Table 7.2 and range from 1 km to 1.6 km for the positions of the Mogi source.

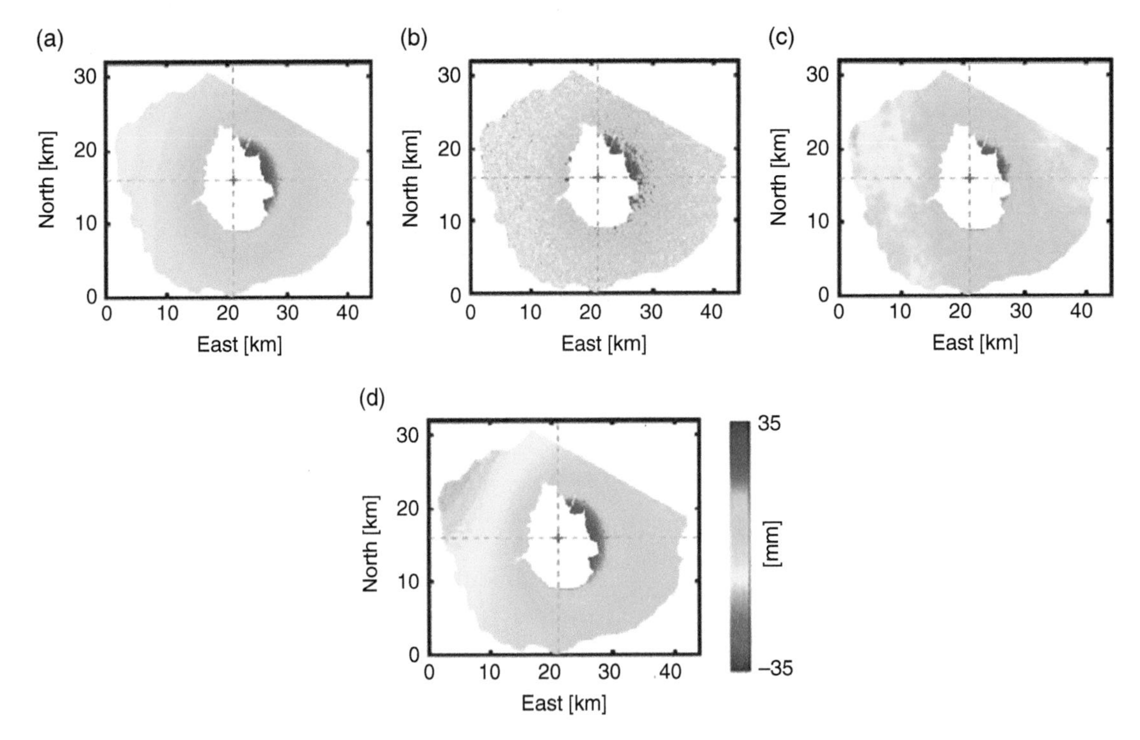

Figure 7.2 Example of one way LOS deformation images in a descending imaging geometry without/with measurement errors. The red marker is the location of Mogi source center. (a) True LOS inflation of Mogi source; (b) deformation contaminated by ϵ^u; (c) deformation contaminated by ϵ^a; (d) deformation contaminated by ϵ^l.

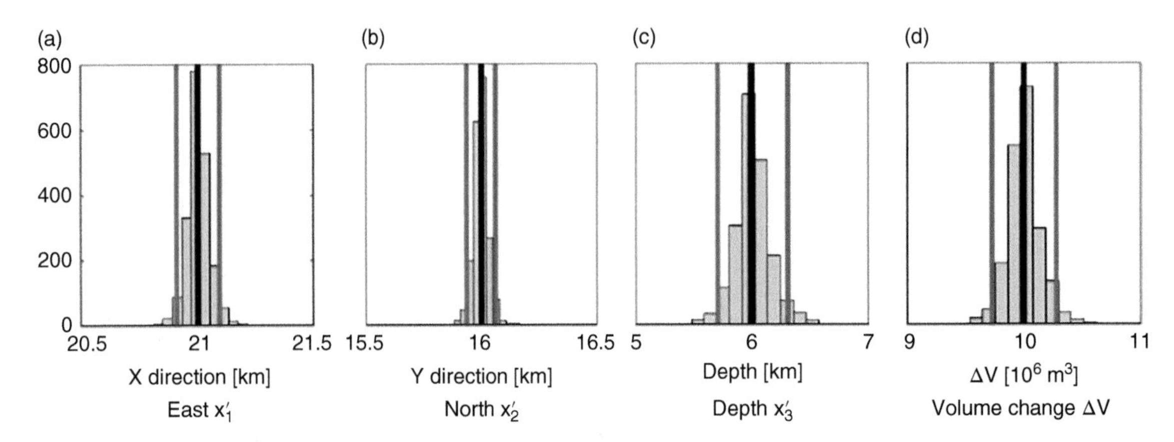

Figure 7.3 Example of probability distribution of estimated parameters where measurements have spatially uncorrelated errors ϵ^u. Red bars denote 2σ confidence bounds and black bars denote the true value of the Mogi source parameters (the same holds for Figures 7.4 and 7.5).

For each of the four parameters, the probability distribution is approximately normal. In Figure 7.4c, the estimated probability distribution of the depth is distorted on the high end by our choice for the initial bounds (2–8 km) of the nonlinear least-squares inversion. While the mean values of x_3' and ΔV differ from the true values, they are still statistically identical to their theoretical values at the 95% confidence level. Given the trade-off between these two parameters [*Mann and Freymueller*, 2003], it was suggested that the ϵ^a errors could cause difficulties in distinguishing these two parameters.

3. The third dataset was generated by simulating long-wavelength errors (ϵ^l) whose shape is approximated by planar phase screens as shown in equation (7.5). These phase screens are used to model both ϕ_{orb} and long-wavelength atmospheric signal components [*Biggs et al.*, 2007; *Lu and Dzurisin*, 2014]. An example is given in Figure 7.2d with noise standard deviation of about 3.3 mm and N-S slope coefficient of 0.26 mm/km and W-E slope coefficient of −0.23 mm/km.

$$\epsilon^l = a \cdot x_1 + b \cdot x_2 + c \tag{7.5}$$

Table 7.1 Mean and Standard Deviation (σ) of Estimated Mogi Source Parameters With ϵ^u Errors in Deformation Measurements

		x_1' [km]	x_2' [km]	x_3' [km]	$\Delta V \times 10^6$ m^3
Lower	Mean	21.00	16.00	6.01	10.00
ϵ^u	σ	0.05	0 .03	0.15	0.14
Higher	Mean	21.00	16.00	6.02	10.00
ϵ^u	σ	0.07	0.05	0.23	0.22

Note: The true values of source parameters are $x_1' = 21$ km, $x_2' = 16$ km, $x_3' = 6$ km, and $\Delta V = 10 \times 10^6$ m^3.

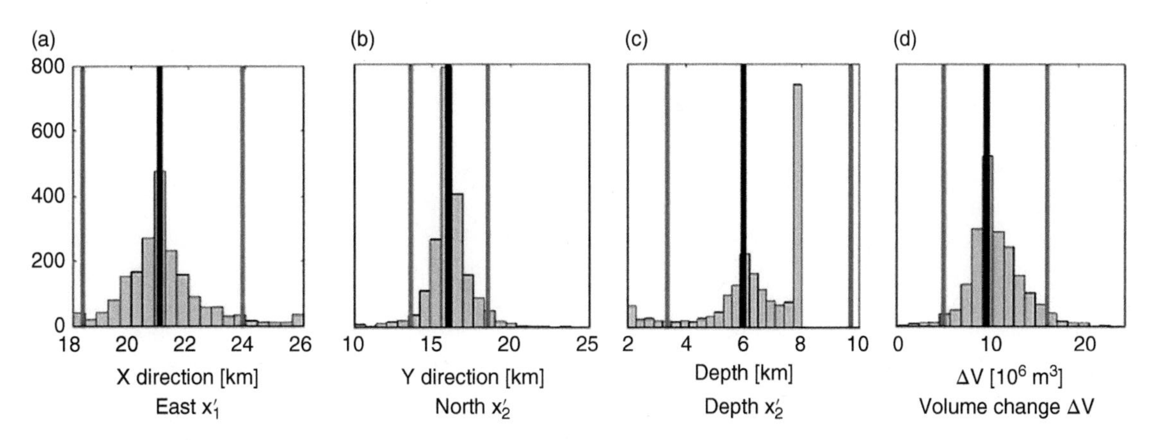

Figure 7.4 Probability distribution of estimated parameters when measurements have spatially correlated errors ϵ^a.

Table 7.2 Mean and Standard Deviation (σ) of Estimated Mogi Source Parameters When the Deformation Measurement Contains ϵ^a and ϵ^l, Respectively

		x_1' [km]	x_2' [km]	x_3' [km]	$\Delta V \times 10^6$ m^3
ϵ^a	Mean	21.12	16.06	6.53	10.97
	σ	1.39	1.23	1.59	2.82
ϵ_1^l	Mean	20.99	15.91	7.81	12.58
	σ	2.09	2.69	0.58	3.60
ϵ_2^l	Mean	21.09	16.00	7.39	11.59
	σ	1.27	1.40	0.90	2.41

Note: The true values of source parameters are $x_1' = 21$ km, $x_2' = 16$ km, $x_3' = 6$ km, and $\Delta V = 10 \times 10^6$ m^3.

The parameters $[x_1, x_2]$ in equation (7.5) are the local coordinates of a pixel in the synthetic datasets, while a and b are the simulated east-west and north-south slope coefficients, and c is the intercept. *Biggs et al.* [2007] has found that for typical ERS-type SAR interferograms, long-wavelength phase ramps have slope coefficients that are roughly normal distributed, centered to zero, and with absolute values of smaller than 2 mm/km. Based on these empirical results, we use Monte Carlo simulation to create a family of phase screens ϵ_1^l with slope coefficients whose statistical properties resemble the observations in the study by *Biggs et al.* [2007]. The phase screens were added to the deformation data and a nonlinear least-squares inversion was used to estimate the Mogi model parameters from the distorted data. The inversion results are shown in Table 7.2, indicating that the performance on x_3' and ΔV estimation is poor. With this level (ϵ_1^l) of long-wavelength errors, the trade-off between depth and volume change intensively hampers the parameter estimation.

Thus, we reduce long wavelength noise level (ϵ_2^l) to a set of smaller slope coefficients (within $[-1, 1]$ mm/km) in both directions, by assuming that most of the ramp signals can be compensated using preprocessing procedures. The posterior probability distributions obtained based on these reduced long-wavelength noise levels (ϵ_2^l) are shown in Figure 7.5 and their statistics are listed in Table 7.2. The

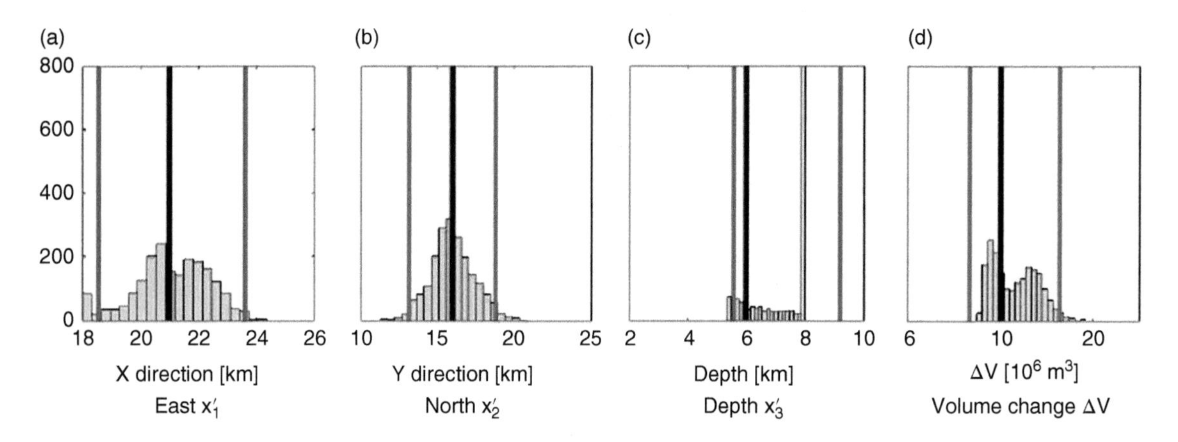

Figure 7.5 Probability distribution of estimated parameters when measurements have spatially local correlated errors ϵ^l.

source depth and volume change still appear to be poorly constrained in this case and the histogram of ΔV is bimodal. The spatial standard deviation of ϵ_2^l and ϵ^a in the last case study are within the same value range (less than 12 mm) yielding similar uncertainties of the estimated Mogi parameters (e.g., $1-\sigma$ uncertainties of source horizontal location is about 1–1.5 km). However, the posterior probability distributions, especially for x_3' and ΔV, are less normal for data affected by long-wavelength errors. For x_3' and ΔV, the mean values of the estimated source parameters differ from their theoretical values but this difference remains statistically insignificant (at the 95% confidence level).

Overall, from our experiments, we can draw the following conclusions for the dependence of the posterior probabilities of the estimated Mogi source parameters on different phase noise types:

1. Spatially correlated noise (ϵ^l and ϵ^a) can significantly impact the accuracy with which the parameters of a volcanic source model can be determined. Strongest impacts are observed for the accuracy of source depth and volume change. In comparison, the influence of ϵ^u on the estimated model parameters is small.

2. InSAR deformation maps containing ϵ^l and ϵ^a errors reduce our ability to determine source depth and volume change parameters. Here, especially long wavelength ϵ^l errors are harmful.

3. Reducing ϵ^l and ϵ^a errors prior to source model inversion is highly recommended; alternatively, ϵ^l could also be jointly modeled in the inversion.

7.3.3.3. Estimation Errors Related to the Flat-Surface Assumption

Many volcanic inflation/deflation analytical source models (including the Mogi model) apply a flat-surface assumption, that is, they ignore the local topography of the area of interest and assume that all measurement points are located at the same level. Some researchers have suggested that ignoring the local surface topography may lead to biases in the estimated model parameters [*Cayol and Cornet*, 1998; *Williams and Wadge*, 1998]. Correction methods have been proposed that use simple first-order approximations of the topography variation relative to the reference frame or include a higher order correction scheme [*Williams and Wadge*, 1998; 2000]. Simply speaking, neglecting local surface topography will lead to increased misfits in the parameter inversion.

To analyze the impact of ignoring surface topography, we first compute and compare surface deformation signals caused by an inflating Mogi source for (1) a model that uses the flat-surface assumption (u_f) and (2) a situation where the local topography is considered when evaluating the Mogi forward model in equation (7.2) (u_T).The topography of Westdahl volcano is used as an example in this study and full coherence is assumed (Fig. 7.6). These results indicate that ignoring local topography can lead to significant overestimation of displacement, especially in areas near the center of the volcano where effects caused by ignoring topography in modeling are the largest. The overestimation is caused by the fact that the distance between the source location and the measurement points is underestimated if surface topography is ignored. Errors are generally smaller on the flank farther away from source center. Generally, the effect of the flat-surface assumption depends on the ratio of topographic height to source depth. The deeper the source and, hence, the smaller this ratio, the smaller the error introduced by ignoring topography. Overall, comparing Figure 7.6(a) and (b) suggests that residuals from ($u_T - u_f$) are a function of height and distance from the volcano source center. For example, the highest peak that locates north of the source center contains relatively small residual. It also explains the widespread residual values with increasing height as shown in Figure 7.6(c). Note that pixels with larger errors are located around the volcano peak where InSAR coherence is generally lost in real data application, as discussed earlier.

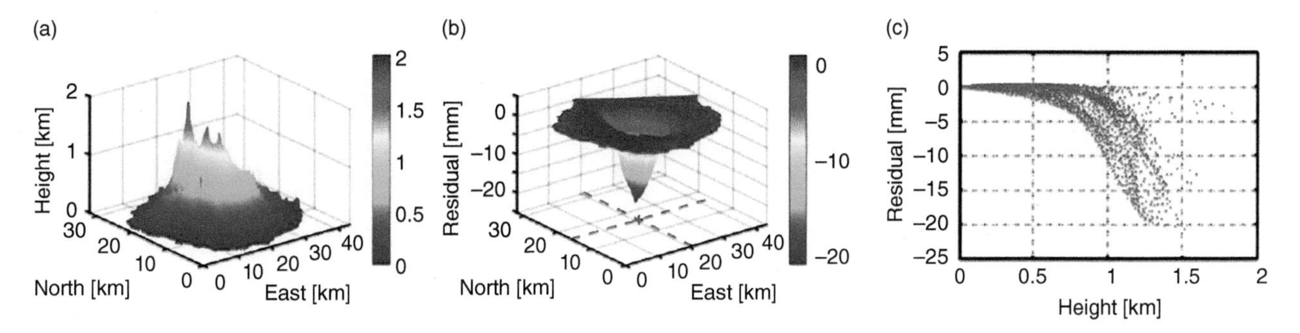

Figure 7.6 Predicted LOS displacements for flat-surface Mogi model (u_f) and elevation-varying Mogi model (u_T): (a) 3D view of the terrain condition of test site, and (b) the spatial distribution of residual displacements of $u_T - u_f$ in millimeters. The red cross indicates the horizontal location of the source center. (c) The relation between the residual of $u_T - u_f$ and the local topography.

Table 7.3 Biased Estimation From Flat-Surface Assumption in Mogi Model

	x_1' [km]	x_2' [km]	x_3' [km]	$\Delta V \times 10^6$ m^3
Coherence everywhere	21.48	15.88	7.16	10.51
Decorrelation	21.14	15.92	6.45	10.10

Note: The true values of source parameters are x_1' =21 km, x_2' =16 km, x_3' =6 km, and ΔV=10 $\times 10^6$ m^3.

To further investigate how the flat-surface assumption affects the source parameter estimation process, we generate two LOS displacement maps for two different coherence conditions, one where full coherence is considered and one where we assume decorrelation near the summit area (see previous examples for the shape and size of the decorrelated area). These LOS deformation maps are prepared while fully considering surface topography resulting in unbiased deformation maps u_T. Finally, to solve for the volcano source parameters, we make the typically used flat-surface assumption in the inversion process. Shown in Table 7.3 are the estimated source parameters compared to their true value (shown in brackets). It can be seen that while biases are introduced to all four model parameters, these biases are most severe for the source depth (x_3') and volume change (ΔV) parameters. It can be seen that biases are larger if coherence near the summit is preserved. Note that this finding is realistic if the source is located roughly underneath the center of the volcano.

7.3.3.4. Joint Inversion with Descending and Ascending Orbit Data

Previous studies have suggested that inversions for an analytical deformation model can be greatly improved if InSAR-based displacement measurements from different geometries are combined in a joint inversion [*Biggs et al.*, 2007; *Lu and Dzurisin*, 2014; *Wright et al.*, 2004]. With sufficient independent measurements, the 3-D deformation map can be reconstructed and then used to constrain the inversion. Alternatively, the vertical and east-west deformations can be derived from interferograms from descending and ascending orbits [*Yun et al.*, 2006].

To discuss how joint inversion would impact the Mogi source parameter estimation, we first perform the nonlinear inversion and Monte Carlo method to compute the posterior probability distribution of source parameters based on synthetic datasets for descending orbit-only (Fig. 7.2a) and ascending orbit-only (Fig. 7.7a) configurations. Phase residuals ϵ, the summation of all InSAR error terms, are included with standard deviation less than 9.5 mm and median values of 4.1 mm. From these inversions, we find that descending-only and ascending-only measurements produce source-model estimates with similar error characteristics (see Table 7.2 for descending-type and Table 7.4 for ascending-type results).

Finally, we add another model run where ascending and descending data are used jointly in the Mogi source parameter inversion. The results show a reduction in the parameter's uncertainties if data from both orbit directions are combined. This performance improvement is particularly large for the horizontal location of the Mogi source. However, when the coherence condition is poor, the inversion results from single tracks and joint tracks still have large uncertainties in the depth and volume change parameters, and also the biases in their mean values remain ($x_3' = 6$ km and $\Delta V = 10 \times 10^6 \text{m}^3$). Our results furthermore show that the benefit of joint ascending and descending data improves if coherence over the source

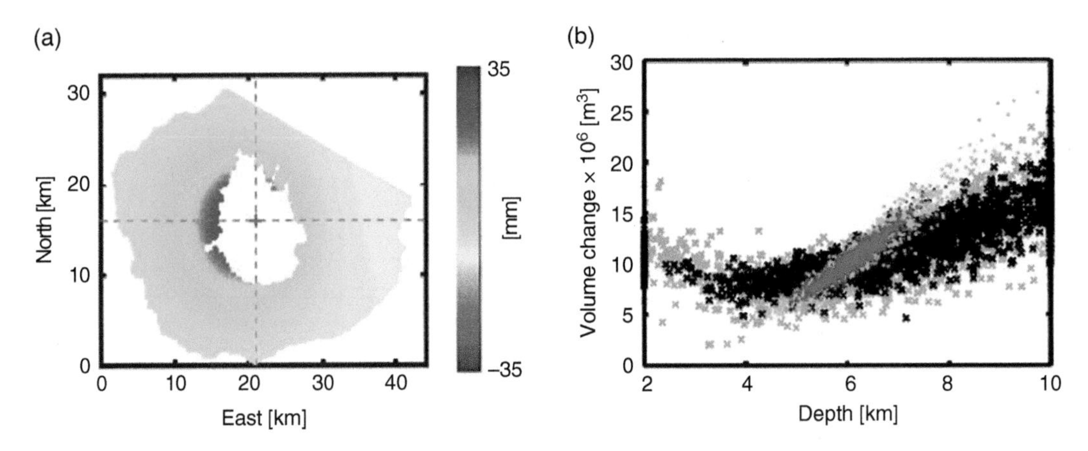

Figure 7.7 An example of Mogi model inversion with descending and ascending track. (a) Synthetic inflation map in ascending geometry; (b) tradeoffs between source depth and volume change. Grey crosses denote the result with ascending orbit data only under poor coherence condition; black crosses denote the result with two orbits under poor coherence condition; green dots denote the result with ascending orbit data only at full coherence; red dots denote the result with two orbits at full coherence.

Table 7.4 Standard Deviation (σ) of Estimated Mogi Source Parameters With Single Track and Multitrack Deformation Maps

			x_1' [km]	x_2' [km]	x_3' [km]	$\Delta V \times 10^6$ m^3
Joint	Decorrelation	mean	21.00	15.99	7.31	12.12
		σ	0.96	0.90	1.96	3.22
	Coherence everywhere	mean	21.00	16.00	6.08	10.30
		σ	0.33	0.31	0.41	1.58
Ascending	Decorrelation	mean	20.86	16.00	7.12	12.13
		σ	1.24	1.18	2.10	4.00
	Coherence everywhere	mean	20.99	16.01	6.16	10.68
		σ	0.44	0.44	0.68	2.60

Note: The true values of source parameters are $x_1' = 21$ km, $x_2' = 16$ km, $x_3' = 6$ km and $\Delta V = 10 \times 10^6$ m^3.

can be maintained. In this case, biases in source depth and volume change estimates can be reduced.

To further investigate the impact of observation geometry on the separability of x_3' and ΔV we plot the correlation between x_3' and ΔV in Figure 7.7b for different combination of input measurements and coherence conditions. It shows that, as a characteristic of the Mogi model, the trade-off between x_3' and ΔV exists even if observations with full coherence from multiple geometries are available. The results suggest that (1) adding measurements from different look directions will reduce the uncertainties of estimated parameters; (2) if observations are available near the summit (and, hence, above the source location), the uncertainties of source parameters can be reduced. Although we compare only the ascending-only data with the joint-orbit data in this example, the result should be similar if descending-only data were used as long as they possess similar signal-to-noise level and coherence coverage.

7.4. DISCUSSION ON IMPACTS OF OTHER GEOPHYSICAL ASSUMPTIONS ON THE MOGI SOURCE MODEL

Ideally, with accurate ground deformation measurements and adequate inversion strategies, the Mogi source parameters can be well constrained. However, to interpret the estimated volcano source parameters under realistic conditions, the assumptions of the Mogi source model need to be considered (i.e., impacts from the local topography as discussed in the previous section), because they can induce additional uncertainties in the estimated source parameters.

First, the Mogi model has inability to separate pressure change (ΔP) and volume change (ΔV). For instance, the deformation pattern caused by smaller ΔP and larger chamber volume (Vc) cannot be distinguished from that by a combination of larger ΔP and smaller V_c. This is because a point source can be formed by either volume change (ΔV) as $\frac{\Delta V(1-\upsilon)}{\pi}$ or by the pressure change (ΔP) as $\frac{\Delta \mathrm{P} r_S^3}{G(1-\upsilon)}$ as shown in equation (7.2), thus there is $\Delta V = \frac{\Delta P}{G}\pi r_S^3$.

Second, the magma compressibility needs to be considered in understanding the computed volume change from a Mogi model. Notice that the derived ΔV accounts for only the change in the cavity size, which is actually the superposition of injected/withdrawal magma volume (ΔV_m), magma compression (ΔV_c), and volume change of the volatiles [*Johnson*, 1992]. Potentially additional gravitational data and rock physics information would be preferred in order to assist inversion or the interpretation.

Another parameter contributing to the inferred volume change is Poisson's ratio ν. The Mogi model uses an average value of 0.25 for ν, however, the real value of ν depends on the petrological nature of the host-rock and the average ν determined from seismology is considered to be between 0.265 and 0.30 for continental and oceanic crusts [*Christensen*, 1996]. Varying ν has significant impact on the estimated ΔV: the estimated ΔV will increase by 20% if $\nu\nu$ is increased from 0.25 to 0.3 [e.g., *Lu et al.*, 2003]. Thus, it is necessary to consider the uncertainty of volume change due to the usage of an averaged $\nu = 0.25$ in the inversion.

There are also many other assumptions in analytical volcano source models that we will not discuss here (e.g., anisotropic and non-homogeneous expansion environment, inelastic condition of rock [*Masterlark*, 2007]). Hence, we should be aware of the uncertainties from the assumptions used in Mogi model and try not to over-interpret the inversion result.

7.5. CONCLUSION

This chapter presents a study on InSAR measurement uncertainties and inherent impacts on estimated source-model parameters for volcano deformation. The Mogi model is used in the discussion given its simplicity and wide use. Nonlinear least squares and Monte Carlo simulations are used to generate the posterior probability distribution of source parameters. Through experiments with simulated datasets, it suggests that spatially correlated noise is critical to estimate the actual Mogi source parameters and the associated parameter uncertainties. The nonideal coherence condition and existence of spatially correlated InSAR measurement errors emphasize the difficulty in determining the optimal volume change and source depth parameters and their associated uncertainties. The surface topography can impact the estimate of the Mogi source model, thus we suggest a Mogi model with the topography correction incorporated. Adding deformation measurements from different viewing geometry will better constrain the model inversion by reducing uncertainties of estimated parameters.

To interpret the Mogi source parameters for volcano studies in real volcanic conditions, it is important to keep in mind that the computed volume change from geodetic inversion can account for net change of the magma reservoir conceptualized by the Mogi source. Therefore, source model assumptions, for example, the magma compressibility, choice of Poisson's ratio, and volatile concentration, also impact uncertainties of the estimated source volume change. Potentially, together with geological information or/and gravitational data, one can better understand the magma changes in volcano applications.

In this study, we have conducted the tests based on nonlinear inversion with equal weight. However, in the real data analysis, if an adequate InSAR noise model is available, the input deformation maps can be weighted differently so that measurements with large errors will have less impact in the source-model inversion. The InSAR measurements can be weighted using (1) a diagonal variance matrix [e.g., *Wright et al.*, 2004] that only considers the spatially uncorrelated errors, or (2) a more sophisticated full variance-covariance matrix that is formed by also considering the spatial correlation of errors (e.g., building covariance structure by assuming errors are stationary and isotropic [*Lohman and Simons*, 2005]). With a proper weighting scheme, we expect the Mogi source parameter inversion can be better constrained.

ACKNOWLEDGMENTS

Z. Lu acknowledges support from the NASA Earth Surface and Interior Program (NNX14AQ95G) and the Shuler-Foscue Endowment at Southern Methodist University. Constructive comments and reviews from Michael Poland (USGS), Editor-in-Chief Peter Webley, and an anonymous reviewer improved the manuscript.

REFERENCES

Amelung, F., S. Jonsson, H. Zebker, and P. Segall (2000), Widespread uplift and "trapdoor" faulting on Galapagos volcanoes observed with radar interferometry, *Nature*, *407*, 993–996.

Arnadottir, T., and P. Segall (1994), The 1989 Loma-Prieta earthquake imaged from inversion of geodetic data, *J. Geophys. Res. Sol. Earth*, *99*(B11), 21835–21855; doi: 10.1029/94jb01256.

Bamler, R., and P. Hartl (1998), Synthetic aperture radar interferometry, *Inv. Prob.*, *14*, 54; doi:10.1088/0266-5611/14/4/001.

Bechor, N. B. D., and H. A. Zebker (2006), Measuring two-dimensional movements using a single InSAR pair, *Geophys. Res. Lett.*, *33*(16); doi: Artn L1631110.1029/2006gl026883.

Bejar-Pizarro, M., A. Socquet, R. Armijo, D. Carrizo, J. Genrich, and M. Simons (2013), Andean structural control on interseismic coupling in the North Chile subduction zone, *Nat. Geosci.*, *6*(6), 462–467; doi: 10.1038/Ngeo1802.

Bekaert, D. P. S., A. Hooper, and T. J. Wright (2015), A spatially-variable power-law tropospheric correction technique for InSAR data, *J. Geophys. Res. Sol. Earth*, 2014JB011558; doi:10.1002/2014jb011558.

Berardino, P., G. Fornaro, R. Lanari, and E. Sansosti (2002), A new algorithm for surface deformation monitoring based on small baseline differential SAR interferograms, *IEEE Trans. Geosci. Remote Sens.*, *40*(11), 2375–2383; doi: 10.1109/Tgrs.2002.803792.

Biggs, J., T. Wright, Z. Lu, and B. Parsons (2007), Multi-interferogram method for measuring interseismic deformation: Denali fault, Alaska, *Geophys. J. Int.*, *170*(3), 1165–1179; doi: 10.1111/j.1365-246X.2007.03415.x.

Burgmann, R., P. A. Rosen, and E. J. Fielding (2000), Synthetic aperture radar interferometry to measure Earth's surface topography and its deformation, *Ann. Rev. Earth Planet. Sci.*, *28*, 169–209; doi: 10.1146/annurev.earth.28.1.169.

Cayol, V., and F. H. Cornet (1998), Effects of topography on the interpretation of the deformation field of prominent volcanoes: Application to Etna, *Geophys. Res. Lett.*, *25*(11), 1979–1982; doi: 10.1029/98gl51512.

Cervelli, P., M. H. Murray, P. Segall, Y. Aoki, and T. Kato (2001), Estimating source parameters from deformation data, with an application to the March 1997 earthquake swarm off the Izu Peninsula, Japan, *J. Geophys. Res. Sol. Earth*, *106*(B6), 11217–11237; doi: 10.1029/2000jb900399.

Christensen, N. I. (1996), Poisson's ratio and crustal seismology, *J. Geophys. Res. Sol. Earth*, *101*(B2), 3139–3156; doi: 10.1029/95jb03446.

Colesanti, C., A. Ferretti, R. Locatelli, and G. Savio (2003), Multi-platform permanent scatterers analysis: first results, paper presented at Remote Sensing and Data Fusion over Urban Areas, 2003, Second GRSS/ISPRS Joint Workshop on, 22–23 May *2003*.

Cong, X., and M. Eineder (2012), Volcano deformation measurement using persistent scatterer interferometry with atmospheric delay corrections, paper presented at Synthetic Aperture Radar, 2012, EUSAR, Ninth European Conference on 23–26 April 2012.

Davis, P. M. (1986), Surface deformation due to inflation of an arbitrarily oriented triaxial ellipsoidal cavity in an elastic half-space, with reference to Kilauea volcano, Hawaii, *J. Geophys. Res. Solid*, *91*(B7), 7429–7438; doi: 10.1029/Jb091ib07p07429.

Dawson, J., and P. Tregoning (2007), Uncertainty analysis of earthquake source parameters determined from InSAR: A simulation study, *J. Geophys. Res. Sol. Earth*, *112*(B9); doi: 10.1029/2007jb005209.

Delaney, P. T., and D. F. Mctigue (1994), Volume of magma accumulation or withdrawal estimated from surface uplift or subsidence, with application to the 1960 collapse of Kilauea volcano, *B. Volcanol.*, *56*(6–7), 417–424; doi: 10.1007/Bf00302823.

DiCaprio, C. J., and M. Simons (2008), Importance of ocean tidal load corrections for differential InSAR, *Geophys. Res. Lett.*, *35*(22), L22309; doi:10.1029/2008gl035806.

Donaldson, J. R., and R. B. Schnabel (1987), Computational experience with confidence-regions and confidence-intervals for nonlinear least-squares, *Technometrics*, *29*(1), 67–82; doi: 10.2307/1269884.

Draper, N. R., and H. Smith (1966), *Applied regression analysis*, *ix*, Wiley, New York.

Dzurisin, D. (2003), A comprehensive approach to monitoring volcano deformation as a window on the eruption cycle (41), 1001, 2003), *Rev. Geophys.*, *41*(2); doi: 10.1029/2003rg000134.

Dzurisin, D. (2007), *Volcano deformation: Geodetic monitoring techniques*, *xxxv*, Springer-Praxis Berlin; New York; Chichester, UK.

Emardson, T. R., M. Simons, and F. H. Webb (2003), Neutral atmospheric delay in interferometric synthetic aperture radar applications: Statistical description and mitigation, *J. Geophys. Res. Sol. Earth*, *108*(B5); doi: 10.1029/2002jb001781.

Feigl, K. L., A. Sergent, and D. Jacq (1995), Estimation of an earthquake focal mechanism from a satellite radar interferogram - Application to the December 4, 1992 Landers aftershock, *Geophys. Res. Lett.*, *22*(9), 1037–1040; doi: 10.1029/94gl03212.

Ferretti, A., C. Prati, and F. Rocca (2001), Permanent scatterers in SAR interferometry, *IEEE Trans. Geosci. Remote Sens.*, *39*(1), 8–20.

Fialko, Y., Y. Khazan, and M. Simons (2001), Deformation due to a pressurized horizontal circular crack in an elastic half-space, with applications to volcano geodesy, *Geophys. J. Int.*, *146*(1), 181–190; doi: 10.1046/j.1365-246X.2001.00452.x.

Foster, J., B. Brooks, T. Cherubini, C. Shacat, S. Businger, and C. L. Werner (2006), Mitigating atmospheric noise for InSAR using a high resolution weather model, *Geophys. Res. Lett.*, *33*(16); doi: 10.1029/2006gl026781.

Gong, W., F. J. Meyer, C. W. Lee, Z. Lu, and J. Freymueller (2014), Measurement and interpretation of subtle deformation signals at Unimak Island from 2003 to 2010 using Weather Model-Assisted Time Series InSAR, *J. Geophys. Res. Sol. Earth*, 2014JB011384; doi:10.1002/2014jb011384.

GVP (2015), Global Volcanism Program, Smithsonian Institution, http://www.volcano.si.edu/faq.cfm#q3.

Hanssen, R. (2001), *Radar Interferometry: Data Interpretation and Error Analysis*, Kluwer Academic Publishers, Netherlands.

Hooper, A., P. Segall, and H. Zebker (2007), Persistent scatterer interferometric synthetic aperture radar for crustal deformation analysis, with application to Volcán Alcedo, Galápagos, *J. Geophys. Res.*, *112*(B7), B07407; doi:10.1029/2006jb004763.

Johnson, D. J. (1987), Elastic and inelastic magma storage at Kīlauea volcano, in *Volcanism in Hawaii,* edited by Robert W. Decker, Thomas L. Wright, and Peter H. Stauffer, U.S. Geological Survey Professional Paper 1350.

Johnson, D. J. (1992), Dynamics of magma storage in the summit reservoir of Kilauea volcano, Hawaii, *J. Geophys. Res. Sol. Earth*, *97*(B2), 1807–1820; doi: 10.1029/91jb02839.

Jolivet, R., P. S. Agram, N. N. Y. Lin, M. Simons, M. P. Doin, G. Peltzer, and Z. H. Li (2014), Improving InSAR geodesy using Global Atmospheric Models, *J. Geophy.s Res. Sol. Earth*, *119*(3), 2324–2341; doi: 10.1002/2013jb010588.

Jung, H. S., D. T. Lee, Z. Lu, and J. S. Won (2013), Ionospheric correction of SAR interferograms by multiple-aperture interferometry, *IEEE Trans. Geosci. Remote Sens.*, *51*(5), 3191–3199; doi: 10.1109/Tgrs.2012.2218660.

Jung, H. S., Z. Lu, J. Won, M. Poland, and A. Miklius (2011), Mapping three-dimensional surface deformation by combining multiple aperture interferometry and conventional interferometry: application to the June 2007 eruption of Kīlauea volcano, Hawaii, *IEEE Geosci. Remote Sens. Lett.*, *8*(1), 34–38.

Kampes, B. (2006), *Radar Interferometry: Persistent Scatterer Technique*, Springer, Dordrecht, Netherlands.

Knospe, S. H. G., and S. Jonsson (2010), Covariance estimation for InSAR surface deformation measurements in the presence of anisotropic atmospheric noise, *IEEE Trans. Geosci. Remote Sens.*, *48*(4), 2057–2065; doi: 10.1109/Tgrs.2009.2033937.

Krieger, G., A. Moreira, H. Fiedler, I. Hajnsek, M. Werner, M. Younis, and M. Zink (2007), TanDEM-X: A satellite formation for high-resolution SAR interferometry, *IEEE Trans. Geosci. Remote Sens.*, *45*(11), 3317–3341; doi: 10.1109/Tgrs.2007.900693.

Li, Z. H., J. P. Muller, and P. Cross (2003), Comparison of precipitable water vapor derived from radiosonde, GPS, and Moderate-Resolution Imaging Spectroradiometer measurements, *J. Geophys. Res. Atmos.*, *108*(D20); doi: 10.1029/2003jd003372.

Li, Z. H., J. P. Muller, P. Cross, and E. J. Fielding (2005), Interferometric synthetic aperture radar (InSAR) atmospheric correction: GPS, moderate resolution imaging spectroradiometer (MODIS), and InSAR integration, *J. Geophys. Res. Sol. Earth*, *110*(B3); doi: 10.1029/2004jb003446.

Lin, Y. N. N., M. Simons, E. A. Hetland, P. Muse, and C. DiCaprio (2010), A multiscale approach to estimating topographically correlated propagation delays in radar interferograms, *Geochem. Geophys. Geosysy.*, *11*; doi: 10.1029/2010gc003228.

Lisowski, M. (2007), Analytical volcano deformation source models, in *Volcano Deformation*, 279–304, Springer Praxis, Chichester, UK.

Liu, S. (2012), Satellite Radar Interferometry: Estimation of Atmospheric Delay, Doctoral Dissertation, Delft University of Technology.

Lohman, R. B., and M. Simons (2005), Some thoughts on the use of InSAR data to constrain models of surface deformation: Noise structure and data downsampling, *Geochem. Geophys. Geosyst.*, *6*; doi: 10.1029/2004gc000841.

Lu, Z., and D. Dzurisin (2014), *InSAR Imaging of Aleutian Volcanoes: Monitoring a Volcanic Arc from Space*, Springer; doi:10.1007/978-3-642-00348-6.

Lu, Z., D. Dzurisin, J. Biggs, C. Wicks, and S. McNutt (2010), Ground surface deformation patterns, magma supply, and magma storage at Okmok volcano, Alaska, from InSAR analysis: 1. Interruption deformation, 1997–2008, *J. Geophys. Res. Sol. Earth*, *115*(B5), B00B02; doi:10.1029/ 2009jb006969.

Lu, Z., D. Mann, J.,Freymueller, and D. Meyer (2000), Synthetic aperture radar interferometry of Okmok volcano, Alaska: Radar observations, *J. Geophys. Res.*, *105*, 10791–10806.

Lu, Z., H. S. Jung, L. Zhang, W. J. Lee, C. W. Lee, and D. Dzurisin (2013), Advances in mapping from aerospace imagery: Techniques and applications, in *DEM generation from satellite InSAR*, edited by X. Yang and J. Li, 119–144, CRC Press, New York.

Lu, Z., T. Masterlark, and D. Dzurisin (2005), Interferometric Synthetic Aperture Radar (InSAR) study of Okmok volcano, Alaska, 1992–2003: Magma supply dynamics and post-emplacement lava flow deformation, *J. Geophys. Res.*, *110*, B02403; doi:10.1029/2004JB003148.

Lu, Z., T. Masterlark, D. Dzurisin, R. Rykhus, and C. Wicks, Jr. (2003), Magma supply dynamics at Westdahl volcano, Alaska, modeled from satellite radar interferometry, *J. Geophys. Res.*, *108*(B7), 2354; doi:10.1029/2002jb002311.

Lyons, S., and D. Sandwell (2003), Fault creep along the southern San Andreas from interferometric synthetic aperture radar, permanent scatterers, and stacking, *J. Geophys. Res. Sol. Earth 108*(B1); doi: 10.1029/2002jb001831.

Mann, D., and J. Freymueller (2003), Volcanic and tectonic deformation on Unimak Island in the Aleutian Arc, Alaska, *J. Geophys. Res.*, *108*(B2), 2108; doi:10.1029/2002jb001925.

Marquardt, D. W. (1963), An algorithm for least-squares estimation of nonlinear parameters, *J. Soc. Ind. App. Math.*, *11*(2), 431–441; doi:10.2307/2098941.

Massonnet, D., and K. L. Feigl (1998), Radar interferometry and its application to changes in the Earth's surface, *Rev. Geophys.*, *36*(4), 441–500.

Masterlark, T. (2007), Magma intrusion and deformation predictions: Sensitivities to the Mogi assumptions, *J. Geophys. Res.Sol. Earth*, *112*(B6); doi: 10.1029/2006jb004860.

Mctigue, D. F. (1987), Elastic stress and deformation near a finite spherical magma body: resolution of the point-source paradox, *J. Geophys. Res. Sol.*, *92*(B12), 12931–12940.

Meyer, F., and J. Nicoll (2008), Mapping ionospheric TEC using Faraday rotation in full- polarimetric L-band SAR data, *Synthetic Aperture Radar (EUSAR), 2008 7th European Conference on*, 1–4.

Mogi, K. (1958), Relations between the eruptions of various volcanoes and the deformations of the ground surfaces around them, paper presented at Bulletin of the Earthquake Research Institute, Earthquake Research Institute, University of Tokyo.

Mosegaard, K., and A. Tarantola (1995), Monte-Carlo sampling of solutions to inverse problems, *J. Geophys. Res. Sol. Earth*. *100*(B7), 12431–12447; doi: 10.1029/94jb03097.

Okada, Y. (1985), Surface deformation due to shear and tensile faults in a half-space, *Bull. Seism. Soc. Amer.*, *75*(4), 1135–1154.

Press, W. H., S. A. Teukolsky, W. T. Vetterling, and B. P. Flannery (2007), *Numerical Recipes in C : The Art of Scientific Computing*, 3rd ed., Cambridge University Press, Cambridge, New York.

Rosen, P. A., S. Hensley, H. A. Zebker, F. H. Webb, and E. J. Fielding (1996), Surface deformation and coherence measurements of Kilauea volcano, Hawaii, from SIR-C radar interferometry, *J. Geophys. Res. Planet*, *101*(E10), 23109–23125; doi: 10.1029/96je01459.

Rosen, P. A., S. Hensley, I. R. Joughin, F. K. Li, S. N. Madsen, E. Rodriguez, and R. M. Goldstein (2000), Synthetic aperture radar interferometry: Invited paper, *P. IEEE*, *88*(3), 333–382; doi: 10.1109/5.838084.

Simons, M., and P. Rosen (2007), Interferometric Synthetic Aperture Radar Geodesy, in *Treatise on Geophysics*, edited by G. Schubert, 391–446, Elsevier Press, Amsterdam; Boston.

SRTM (2015), Enhanced Shuttle Land Elevation Data (http://www2.jpl.nasa.gov/srtm/).

Sudhaus, H., and S. Jonsson (2009), Improved source modelling through combined use of InSAR and GPS under consideration of correlated data errors: application to the June 2000 Kleifarvatn earthquake, Iceland, *Geophys. J. Int.*, *176*(2), 389–404; doi: 10.1111/j.1365-246X.2008.03989.x.

VHP (2015), USGS Volcano Hazards Program (http://volcanoes.usgs.gov/).

Williams, C. A., and G. Wadge (1998), The effects of topography on magma chamber deformation models: Application to Mt Etna and radar interferometry, *Geophys. Res. Lett.*, *25*(10), 1549–1552; doi: 10.1029/98gl01136.

Williams, C. A., and G. Wadge (2000), An accurate and efficient method for including the effects of topography in three-dimensional elastic models of ground deformation with applications to radar interferometry, *J. Geophys. Res. Sol. Earth*, *105*(B4), 8103–8120; doi: 10.1029/1999jb900307.

Wright, T. J., B. E. Parsons, J. A. Jackson, M. Haynes, E. J. Fielding, P. C. England, and P. J. Clarke (1999), Source parameters of the 1 October 1995 Dinar (Turkey) earthquake from SAR interferometry and seismic bodywave modelling, *Earth Planet. Sci. Lett.*, *172*(1–2), 23–37; doi: 10.1016/S0012-821x(99)00186-7.

Wright, T. J., Z. Lu, and C. Wicks (2003), Source model for the M w 6.7, 23 October 2002, Nenana Mountain earthquake (Alaska) from InSAR, *Geophys. Res. Lett.*, *30*(18), 1974; doi:10.1029/2003gl018014.

Wright, T. J., B. E. Parsons, and Z. Lu (2004), Toward mapping surface deformation in three dimensions using InSAR, *Geophys. Res. Lett.*, *31*(1); doi: 10.1029/2003gl018827.

Yang, X. M., P. M. Davis, and J. H. Dieterich (1988), Deformation from inflation of a dipping finite prolate spheroid in an elastic half-space as a model for volcanic stressing, *J. Geophys. Res. Sol.*, *93*(B5), 4249–4257; doi: 10.1029/Jb093ib05p04249.

Yun, S., P. Segall, and H. Zebker (2006), Constraints on magma chamber geometry at Sierra Negra Volcano, Galápagos Islands, based on InSAR observations, *J. Volcanol. Geotherm. Res.*, *150*(1–3), 232–243; doi:10.1016/j.jvolgeores.2005.07.009.

Zhang, L., X. Ding, Z. Lu, H.-S. Jung, J. Hu, and G. Feng (2014), A novel multitemporal InSAR model for joint estimation of deformation rates and orbital errors, *IEEE Trans. Geosci. Remote Sens.*, *52*(6), 3529–3540; doi:10.1109/tgrs.2013.2273374.

8

Improving Model Simulations of Volcanic Emission Clouds and Assessing Model Uncertainties

Nina Iren Kristiansen,[1] Delia Arnold,[2] Christian Maurer,[2] Julius Vira,[3] Razvan Rădulescu,[4] Damien Martin,[4] Andreas Stohl,[1] Kerstin Stebel,[1] Mikhail Sofiev,[3] Colin O'Dowd,[4] and Gerhard Wotawa[2]

ABSTRACT

Volcanic emissions of ash and SO_2 can be harmful to both human health, infrastructure, and aviation. In assessing these hazards, satellite remote sensing and transport modeling are useful tools. Satellite observations can track the volcanic clouds in the atmosphere, their main limitation being little vertical information. Transport models are used to forecast the three-dimensional atmospheric dispersion of the volcanic emissions, but large uncertainties are involved. The main sources of uncertainties are lacking knowledge of the emissions (source term), errors in the meteorological data driving the dispersion model, and the model's description of physical processes such as loss mechanisms. One of the keys for achieving more reliable simulations is incorporating observation data into the models. In this chapter, we demonstrate three techniques useful for enhancing model simulations and for evaluating model uncertainties. First, the source term can be constrained by satellite observations via inverse modeling. Second, errors in the transport model and input data can be corrected for with data assimilation by updating the modeled fields with up-to-date observations. Third, uncertainties can be inferred from ensemble modeling. The Grímsvötn 2011 eruption is used to demonstrate how these three techniques together can improve the model simulations of volcanic clouds and determine model uncertainties.

8.1. INTRODUCTION

Material injected into the atmosphere during volcanic eruptions does not only pose a hazard for aviation [*Prata and Tupper*, 2009] and human health [*Horwell and Baxter*, 2006] but can also affect climate [*Robock*, 2000]. Quantifying these impacts requires accurate knowledge of the composition of the emissions (e.g., toxic gases, sulphur dioxide [SO_2], volcanic ash), their amounts and how they are distributed in the atmosphere. For example, climate impacts are caused by sulphate formed in the stratosphere from high-altitude injections of SO_2, aviation problems are mainly due to volcanic ash at flight altitudes (typically, near the tropopause in midlatitudes) and health hazards are determined by low-altitude outgassing or aerosol emissions. While large volcanic eruptions typically constitute a multihazard, minor eruptions can be damaging as well. For instance, the large ash production of the phreatomagmatic eruption of Eyjafjallajökull in 2010 stalled air traffic in Europe, and the low-altitude outgassing of SO_2 from the Bárdarbunga (Holuhraun) eruption in 2014 was a health hazard to Icelanders [*Icelandic Directorate of Health*, 2014]. Notably, these effects were totally different and demonstrate the highly variable and unpredictable nature of volcanic eruptions, which further highlights some of the difficulties faced by atmospheric scientists tasked with assessing (or predicting) the severity and/or spatial extension of the hazard.

[1] *Norwegian Institute for Air Research (NILU), Kjeller, Norway*
[2] *Central Institute for Meteorology and Geodynamics (ZAMG), Vienna, Austria*
[3] *Finnish Meteorological Institute (FMI), Helsinki, Finland*
[4] *School of Physics and Centre for Climate and Air Pollution Studies, National University of Ireland Galway (NUIG), Galway, Ireland*

Natural Hazard Uncertainty Assessment: Modeling and Decision Support, Geophysical Monograph 223, First Edition.
Edited by Karin Riley, Peter Webley, and Matthew Thompson.

For analyzing or predicting the hazards caused by volcanic material in the atmosphere, transport and dispersion models are used. For quantifying climate impacts, climate models are needed; however, they typically assume that the distribution of sulphate and/or volcanic ash is already known and obtaining this information may also involve dispersion modeling in a first step (e.g. *Flanner et al.* [2014]). In this paper, we briefly review the current state of art in dispersion modeling for volcanic hazards and emphasize some recent developments to improve the simulations and quantify uncertainties.

Most of currently operational dispersion models have not been developed exclusively for simulating volcanic ash (or SO_2) dispersion but have also been used for other purposes, e.g., simulating the dispersion of radionuclides after a nuclear accident. Examples of such models are HYSPLIT [*Draxler and Hess*, 1998], NAME [*Ryall et al.*, 2002], FLEXPART [*Stohl et al.*, 1998], or SILAM [*Sofiev et al.*, 2008; *Vira and Sofiev*, 2012], which are all used for research as well as in operational emergency response centers. These models run off line and require meteorological input data from numerical weather prediction models. In addition, specialized volcanic ash transport and dispersion models have been developed (e.g. *Folch et al.* [2009]), and mesoscale meteorological forecast models enabled with modules for atmospheric chemistry simulation have been adapted for volcanic hazard predictions (e.g., WRF-Chem [*Stuefer et al.*, 2013]). Recent progress has also been possible through the introduction of techniques that allow constraining the model simulations with observation data.

Uncertainties in simulating volcanic emission clouds arise mainly from three components:

1. The magnitude of the emissions (for a given substance) and their variations with time and height. Using classical methods to estimate this so-called source term, even the total emissions can be uncertain by orders of magnitude [e.g., *Sparks et al.*, 1997; *Mastin et al.*, 2009] and this translates directly into errors of the same magnitude in dispersion model predictions. Promising new methods to reduce these uncertainties make use of satellite observations to indirectly quantify the source term [*Stohl et al.*, 2011; *Theys et al.*, 2013; *Flemming and Inness*, 2013].

2. Meteorological forecast or analysis errors. Errors in the wind fields or other meteorological data used to drive a dispersion model (or produced directly by an online model) can lead to erroneous transport of volcanic emission clouds and can also affect simulated removal processes. Progress in quantifying meteorological forecast errors has come mainly from the use of ensemble techniques [*Palmer et al.*, 2005], and these techniques can be applied to the simulation of volcanic emission clouds as well. Further, such errors can be corrected for with data assimilation techniques.

3. The model's formulations (parametrizations) of physical processes. For instance, loss processes such as gravitational settling, dry deposition, and wet scavenging remove mass from the atmosphere; turbulence reduces peak concentrations by dilution; and chemical transformation removes SO_2 but can form sulphate. These uncertainties can also cause substantial errors in the simulated concentrations, but it is likely that they have a smaller effect on the prediction of the location of the emission clouds than for uncertainties related to emissions and wind fields.

A recent overview of modeling techniques for volcanic ash dispersion modeling and their uncertainties was provided by *Webley and Steensen* [2013]. In this chapter, we demonstrate and discuss the modeling capacities and shortcomings of current models at the example of the Grímsvötn 2011 eruption, using model simulations of SILAM, WRF-Chem, and FLEXPART, which are of very different design (Eulerian vs. Lagrangian; off-line vs. on-line). We apply different concepts for incorporating observation data into the models (inverse modeling vs. data assimilation), and we explore the capacities of multi-input ensemble modeling techniques to quantify uncertainties. Last, we evaluate and compare the three models in a model intercomparison exercise and also compare the models to independent measurements.

8.2. METHODS

We demonstrate three techniques useful for improving and assessing uncertainties in modeling of volcanic emission clouds: inverse modeling, data assimilation, and ensemble modeling. The two first methods make use of satellite observations to constrain the volcanic emission clouds. This chapter starts with a description of such satellite observations (Section 8.2.1), followed by introductions to the three methods (Sections 8.2.2–8.2.4). Similar inversion or data assimilation techniques are also being implemented by other researchers [*Boichu et al.*, 2013; *Flemming and Inness*, 2013; *Pelley et al.*, 2014; *Wilkins et al.*, 2014] with the aim of improving ash and/or SO_2 forecasts.

8.2.1. Satellite Observations

Satellites provide a unique way of detecting atmospheric composition from space, yet no sensor has been exclusively designed for detecting volcanic emission clouds. However, satellite measurements in the visible (VIS), ultraviolet (UV), and infrared (IR) spectral regions, performed with sensors originally developed for climate and weather prediction applications, can also be used to detect and track volcanic substances. Of all volcanic debris, the most easily detected by remote sensing methods are SO_2, ash, and ice due to their unique spectral signatures. VIS measurements are very useful for qualitative assessments of ash clouds and possibly distinguishing them visually from meteorological clouds. Quantitative retrievals of ash or SO_2 are carried out using measurements in the IR or UV channels. IR measurements give good global coverage because they can sample the Earth during the day and night, whereas UV measurements are restricted to measurements during the sunlight hours of the day.

An overview of the current and future perspectives of observing volcanic emissions from space is given by *Thomas and Watson* [2010]. No single satellite instrument fulfils all requirements for a volcanic cloud observing system such as high temporal resolution (minutes), high spatial resolution (~110 km^2) and vertical resolution (~100-300 m) and a spectral range that permits detection and quantification of volcanic ash and SO_2 [*Prata et al.*, 2014].

SO_2 possesses distinctive absorption signatures in both the UV and IR regions of the electromagnetic spectrum and it is possible to retrieve the total atmospheric columns (vertically integrated; mass per square meter) of SO_2 from measurements in specific UV and IR channels. UV instruments have a better sensitivity to SO_2 in the middle and lower troposphere, whereas for IR the sensor sensitivity peaks near the tropopause (~16 km altitude), where the thermal contrast between the volcanic cloud and the background radiation is highest. Frequently used satellite instruments for observations of volcanic SO_2 are the Ozone Monitoring Instrument (OMI) on the Aura satellite [*Krotkov et al.*, 2006; *Carn et al.*, 2013] and the Global Ozone Monitoring Experiment (GOME-2) on MetOp [*Rix et al.*, 2009, 2012], both measuring in the UV range. In the thermal IR spectral ranges, the Infrared Atmospheric Sounding Interferometer (IASI) on the MetOp satellite [*Clarisse et al.*, 2008, 2012; *Walker et al.*,2012; *Carboni et al.*, 2012] and the Atmospheric InfraRed Sounder (AIRS) on Aqua [*Prata and Bernardo*, 2007] are commonly used. The SO_2 retrievals from various instruments can differ due to, for example, the impact of the calculated or assumed SO_2 cloud height on the retrievals or to different sensitivities of the UV and IR bands to SO_2, aerosols, and clouds. For example, *Kristiansen et al.* [2010] reviewed estimates of the Kasatochi emissions, which vary from 1.2 to 2.5 Tg based on retrievals from OMI, GOME-2, IASI, and AIRS. Recently, efforts have been made to provide SO_2 cloud heights from OMI [*Yang et al.*, 2009], GOME-2 [*Rix et al.*, 2012], and IASI [*Carboni et al.*, 2012].

The discrimination of meteorological clouds and volcanic-ash clouds is not always clear from VIS channels. *Prata* [1989] found that a combination of certain IR channels could be used to separate ash clouds from meteorological clouds, and the optical depth, effective radius of the ash particles, and volcanic-ash mass loadings (vertically integrated; mass per square meter) can be retrieved. An "effective height" of the ash clouds can also be estimated [e.g., *Wen and Rose,* 1994]. Geostationary satellites with high temporal resolution (about 15 min) like the Meteosat Second Generation (MSG) Spin-stabilised Enhanced Visible and Infrared Imager (SEVIRI) are well suited for near real-time ash detection [*Prata*, 2013] and quantification of the total column airborne ash loadings [*Prata and Kerkmann*, 2007]. Ash detection is a major challenge in the IR ash retrieval, that is, the separation between ash, clear sky, water clouds, ice, and ice-coated ash. Polar-orbiting satellites like IASI and AIRS can provide more detailed information on ash microphysics [*Gangale et al.*, 2010; *Clarisse et al.*, 2010]. Mass loadings between about 0.2 g m^{-2} to 20 g m^{-2} can be retrieved, with uncertainties of ~40% [*Wen and Rose*, 1994; *Gu et al.*, 2003], whereas optically thick or very faint ash clouds may cause problems for the retrieval [e.g., *Kristiansen et al.*, 2015].

The vertical profile of volcanic-ash clouds can be gained from satellite-based multiview, like the Along Track Scanning Radiometer (ATSR) and the Advanced Along Track Scanning Radiometer (AATSR) on the ERS satellites [e.g., *Prata and Turner*, 1997; *Grainger et al.*, 2013], or active instruments like CALIOP (Cloud-Aerosol Lidar with Orthogonal Polarization) on CALIPSO [*Vernier et al.*, 2013]. Altitude information from CALIOP has been used for validation of modeled cloud altitudes [e.g., *Kristiansen et al.*, 2010], and combinations of retrieved ash mass loadings with altitude information from CALIOP, can allow ash concentration retrieval [*Prata and Prata*, 2012].

In general, uncertainty characterization and validation of ash and SO_2 satellite retrievals is still a major task. Additional improvements on sensitivity and repeated coverage are expected from recent and upcoming missions.

8.2.2. Inverse Modeling

Inverse modeling tackles the source-term uncertainty explained in Section 8.1. The basis of inverse modeling is to refine the input data for the atmospheric transport model to give the best possible fit to observations. Statistical methods are used to determine the most probable source flux given atmospheric observations and independent a priori knowledge of the emissions. With the estimated source flux from the inversion, the transport models can be run with the aim of predicting the position and the concentrations of the substance in the atmosphere with an increased accuracy.

An inversion scheme was specifically developed for volcanic emissions by *Eckhardt et al.* [2008]. The satellite observations used to derive the volcanic source term are column measurements (i.e., measurements that provide no vertical information). The basic idea of the algorithm is to extract information on the emission heights and times from the horizontal dispersion patterns as observed by satellite, which vary with time in response to changing emissions and changing meteorological conditions and depend on altitude because of the vertical shear of the horizontal wind. All possible emission scenarios are established from forward dispersion modeling. The algorithm finds the linear combination of the emissions as a function of height and time, which brings the model in best agreement with the satellite measurement data. The source-term determination is most robust if strong wind shear exists, which transports emissions in different directions depending on emission altitude.

Eckhardt et al. [2008] successfully retrieved the vertical profile of the SO_2 emissions from the 2007-eruption of Jebel-at-Tair, Yemen, and *Kristiansen et al.* [2010] used the same method to accurately simulate the dispersion of the SO_2 emission cloud from the 2008 Kasatochi eruption, Alaska. The method was further developed by *Stohl et al.* [2011] to estimate the time-varying mass eruption rate (MER) of fine ash from the 2010 Eyjafjallajökull eruption, Iceland, and *Kristiansen et al.* [2012] performed sensitivity studies of the same eruption to better understand observational and modeling uncertainties. *Moxnes et al.* [2014] used the technique to study the separation of ash and SO_2 emissions during the Grímsvötn 2011 eruption. Most recently, *Kristiansen et al.* [2015] estimated the emissions of ash from the 2014 Kelut eruption, which had a direct impact on aviation.

The source terms derived with this method are "effective" emissions of SO_2 and fine ash, which represent the emissions that survive near-source removal processes and thus are available to long-range transport, and that are observed by the satellite instruments.

8.2.3. Data Assimilation for Determining the Volcanic Cloud Distribution

The concept of data assimilation originates in the field of numerical weather prediction, where it refers to the process of obtaining a statistically and physically consistent estimate (referred as the analysis) of the atmospheric state. The state estimate, which typically initializes the subsequent model simulation, is based on updating the previous model simulation (referred as the background field) with up-to-date observation data. Following a similar approach, predictions of volcanic clouds can be updated by assimilating observations, regardless of the source term used by the dispersion model.

The main assimilation methods in current numerical weather prediction (NWP) systems are either variational [*Lorenc*, 1986; *Le Dimet and Talagrand*, 1986] or sequential. The sequential assimilation systems are typically based on the Ensemble Kalman Filter [*Evensen*, 1994, 2003]. In air pollution modeling, data assimilation schemes have been extended to cover source-term estimation [*Elbern et al.*, 2000; *Barbu et al.*, 2009; *Miyazaki et al.*, 2012], which has been found to provide additional forecast skill for short-lived pollutants [*Elbern et al.*, 2007; *Vira and Sofiev*, 2012]. However, in this chapter, we will mainly consider data assimilation for initializing the volcanic tracer fields and distinguish it from source-term inversions. Compared to model simulations driven with the source-term inversion only, reinitializing the model with data assimilation provides a mechanism to correct errors due to inaccuracies in the transport model and/or the driving meteorological data.

While the data assimilation methods developed for weather prediction have been applied widely in air-quality forecasting (see, e.g., *Carmichael et al.* [2008]), their applications in volcanic ash and SO_2 dispersion are relatively scarce. *Flemming and Inness* [2013] used the four-dimensional variational (4D-Var) data assimilation method to assimilate satellite observations of SO_2 in hindcast experiments for the Eyjafjallajökull 2010 and Grímsvötn 2011 eruptions, and found the combination of data assimilation and inverse modeling to provide the best-performing forecasts. Satellite data were also used for model initialization by *Wang et al.* [2013], who employed a statistical interpolation method for modeling the SO_2 transport in the Kasatochi (2008) eruption. These studies are both based on global-scale, Eulerian chemistry-transport models.

8.2.4. Ensemble Dispersion Modeling

The sources of uncertainties in volcanic emission cloud modeling arise from several components, as explained in Section 8.1. Ideally one would aim at propagating the uncertainties coming from the different sources, starting with the critical uncertainties in the source term, to finalize with a quantified total uncertainty along with the simulation. Although attempts in this direction exist [e.g., *Patra et al.*, 2013] the origin of the uncertainties and the complexities associated to them have initiated pragmatic and probabilistic approaches such as ensemble modeling.

Ensemble dispersion prediction systems are since two decades well established and increasingly used in atmospheric transport modeling studies [e.g., *Galmarini et al.*,

2004, 2010; *Galmarini and Rao*, 2011; *Solazzo and Galmarini*, 2015], including more recently for volcanic applications [*Kristiansen et al.*, 2012; *Stefanescu et al.*, 2014; *Vogel et al.*, 2014]. In such systems, multiple forecast realizations should account for the errors introduced by the use of imperfect initial meteorological conditions (that are amplified by the chaotic nature of the dynamical system), as well as errors introduced because of imperfections in the model formulation (approximations in the mathematical methods to solve the model equations, e.g., parameterizations) or lack of knowledge about the source term. The "true" forecast should ideally fall within the predicted ensemble spread, and the amount of spread should be related to the uncertainty (error) of the forecast. This approach is nonetheless limited by the fact that ensemble forecasts do not truly account for all the potential sources of uncertainties and that often different models used in the ensembles include similar formulations/parameterizations and therefore similar/nonindependent error sources. To date, and for atmospheric transport ensemble modeling, there is still ongoing work on assessing the quality of an ensemble system, how the actual spread of the real ensemble members is related to the uncertainty of the ensemble forecast,and also how ensemble modeling should best be performed and evaluated [e.g., *Kioutsioukis and Galmarini*, 2014].

There are three main approaches in ensemble modeling:

1. **Multi-input single model (MI)** ensembles are based on slightly different but plausible input data to start the model simulations. These data may be the input forcing (source term) as well as the driving meteorological fields, or variable internal model parameters.

2. **Single-input multimodel (MM)** ensemble uses several different atmospheric transport models (with different parameterizations of physical processes) all driven, as much as possible, by the same meteorological input data (e.g., MACC air-quality ensemble system). This approach addresses the uncertainties inherent in the atmospheric transport models, which is not considered in the MI approach.

3. **Multi-input and multimodel (MI-MM)** ensemble combines the two aforementioned approaches. The ensemble members may then include different atmospheric transport models, different meteorological data for driving the models, and different source terms (e.g., JRC ENSEMBLE platform, CTBTO Ensemble backtracking system).

In this chapter, we demonstrate the use of a MI ensemble dispersion modeling with variable meteorological input data but a fixed source term, and we perform a model intercomparison of three atmospheric transport models (off-line and on-line) using an identical source term and meteorological drivers.

8.3. EXAMPLE CASE STUDY: THE GRÍMSVÖTN 2011 ERUPTION

The eruption of Grímsvötn, Iceland, on 21 May 2011 released both volcanic ash and SO_2 into the atmosphere, which were transported and tracked by satellite observations across the Northern Hemisphere [e.g., *Moxnes et al.*, 2014; *Flemming and Inness,* 2013]. We use this eruption as an example case for the demonstration of the techniques outlined above with the aim of improving volcanic ash and SO_2 modeling and assessing the associated uncertainties. There is no "typical" volcanic eruption, but this case was chosen due to the readily available and evaluated source terms, strong wind shear making the source-term determination possible and robust, and a relatively short eruption period limiting computational demands. We retrieve the ash and SO_2 source terms with inverse methods, correct for errors in source-term information and meteorological data using data assimilation, and quantify the uncertainties through ensemble modeling. Both ash (inverse and ensemble modeling) and SO_2 (inversion and data assimilation) results are shown.

8.3.1. Ash Source-Term Estimation with Inverse Modeling

The separation of ash and SO_2 during the Grímsvötn eruption was studied in detail by *Moxnes et al.* [2014] who performed source-term estimates based on inverse modeling, for both SO_2 and ash. They used satellite retrieval data from IASI (Section 8.2.1) to constrain the source terms. As a priori source information they assumed a constant emission flux released uniformly from the volcano vent up to maximum plume heights observed from radar. The inversion optimizes the source term such that the best agreement between modeled and IASI-observed volcanic clouds was found. Notice that the derived ash emissions is an "effective" source term for the fraction of ash that includes small particles (fine ash) available to long-range transport, and not the larger ash particles that quickly fall out of the ash cloud in the vicinity of the volcano and are not observed by satellites.

Figure 8.1 demonstrates the improvement in simulating volcanic ash clouds using the source term constrained by IASI satellite data (Fig. 8.1a). The a priori (Fig. 8.1b) and satellite-constrained a posteriori (Fig. 8.1c) source terms differ significantly with more ash being released to lower altitudes and in a later time period in the latter source term. The Lagrangian particle dispersion model FLEXPART is used to perform transport simulations initialized with the two different source terms, and run on ECMWF operational analyses and forecast data with $0.18^\circ \times 0.18^\circ$ resolution and 91 model levels. The transport

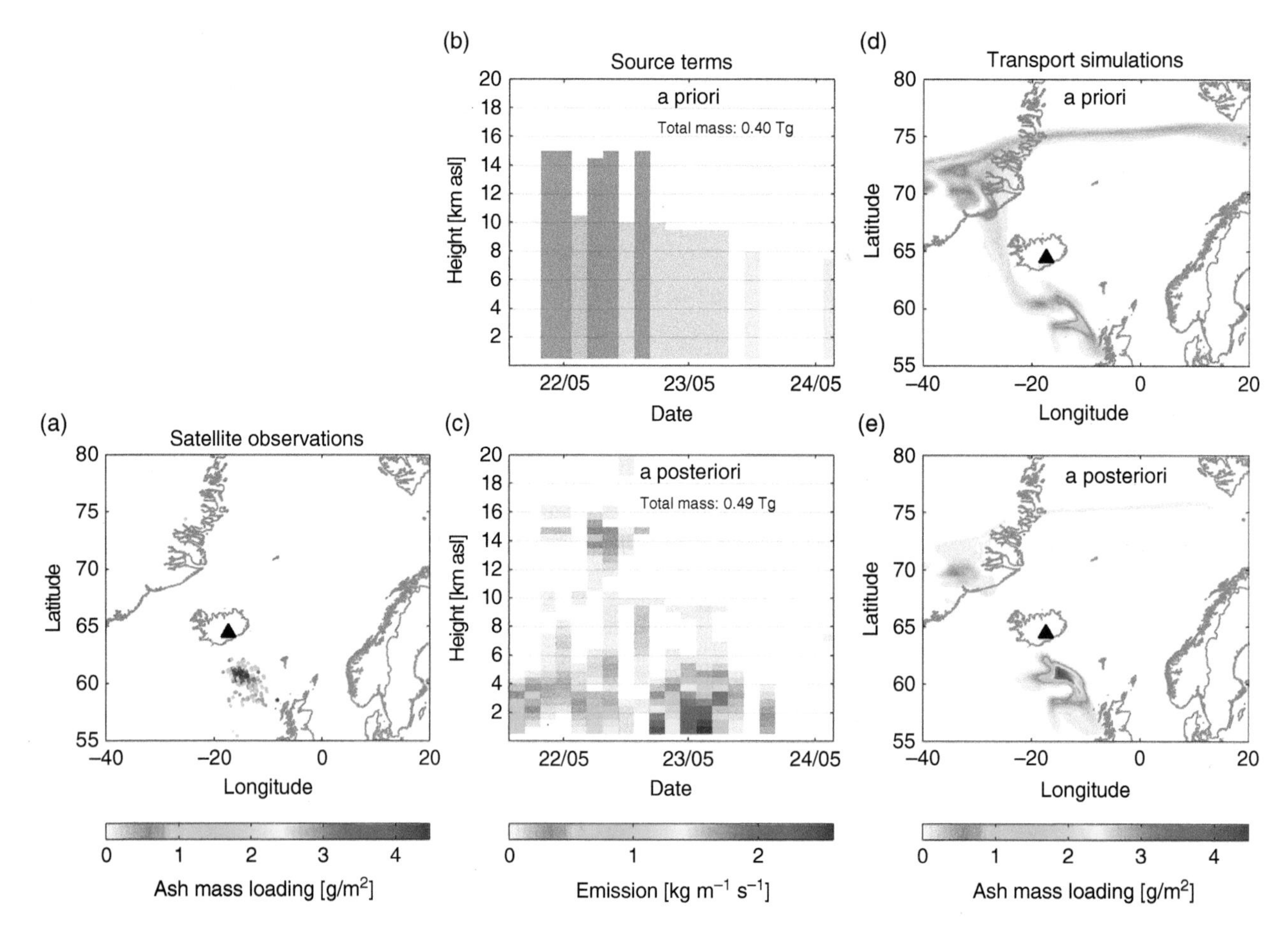

Figure 8.1 Ash source term inversion for the Grímsvötn 2011 eruption: (a) IASI satellite retrievals of ash for 23 May 2011 at 21–22 UTC; (b) a priori source term for fine ash derived from radar observations; (c) a posteriori source term constrained by IASI observations; (d) FLEXPART a priori transport simulation using the source term in (b); (e) the FLEXPART a posteriori simulation using the satellite-constrained source term in (c) showing better agreement with the satellite observations than the a priori simulation. Both FLEXPART simulations are using ECMWF operational forecast and analysis data. The Grímsvötn volcano is marked by a black triangle. All data and results are from *Moxnes et al.* [2014].

simulation using the a priori source term (Fig. 8.1d) shows that, compared to the satellite observations, this simulation has too much ash transported northeastward and too little ash transported southeastward. Also, the densest part of the ash cloud is located farther south than the observations. Using the a posteriori source term in the transport model (Fig. 8.1e) results in better agreement with the satellite observed ash cloud and the simulation nicely captures the position and magnitude of the maximum mass loading values. Notice that the figure only demonstrates the use of the technique and is not an independent validation as the IASI-data are used to determine the a posteriori source term. The results are from *Moxnes et al.* [2014] where detailed validation of the results, including comparisons to independent measurement data, can be found.

In summary, the inverse method provides an estimate for the ash emissions from the Grímsvötn eruption, which when used in the transport model improves the simulation and gives a better agreement with satellite observations.

8.3.2. SO_2 Forecasting with Data Assimilation and Inverse Modeling

The source-term inversions implicitly assume that the uncertainty of the volcanic cloud modeling is dominated by the uncertainty of the source term. As the volcanic cloud gets transported farther from the source, this assumption could be violated due to errors in meteorology and/or model parametrizations. In such cases, it would be useful to update the modeled concentration field using

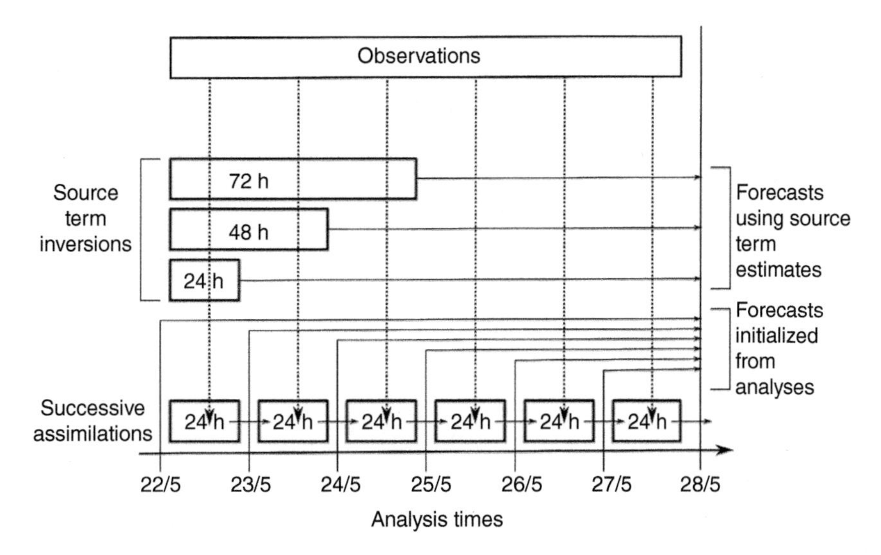

Figure 8.2 Schematic showing the procedures for the SO_2 source term inversion and data assimilation (4D-Var) for the Grímsvötn 2011 eruption.

up-to-date observations in addition to or instead of the source-term inversion.

This section compares different approaches of using satellite observations in forecasting the Grímsvötn SO_2 cloud. Satellite observations by OMI (Section 8.2.1) are used either for source-term determination (inversion, Section 8.2.2) or for updating the modeled field (data assimilation, Section 8.2.3) in a series of hindcasts. The lower stratosphere (STL) version of the OMI/Aura Sulfur Dioxide (SO_2) Total Column product [*Yang et al.*, 2007] is used. This product is intended for use with explosive volcanic eruptions where the SO_2 cloud is placed in the upper troposphere or in the stratosphere, as was the case for the Grímsvötn SO_2 cloud [*Moxnes et al.*, 2014].

The experiment covers the period from 22 May 00 UTC to 28 May 00 UTC, and aims to simulate the real-time conditions only based on the observations that are available up until the start of the forecast, that is, observations are initially not available for the entire simulation period. A forecast is launched every day at 00 UTC based on the observations available until its initial time.

In the data assimilation mode, this is realized by performing the assimilation in windows of 24 hours (Fig. 8.2). The simulated SO_2 cloud at the end of each assimilation window is the initial condition for the subsequent forecast, and also provides the background field for the subsequent assimilation. The source-term inversion, in contrast, is rerun from the start of the eruption period each time new observations are included (Fig. 8.2). The inversion is performed three times using the observations for the first 24, 48, and 72 hr. This allows us to analyze how additional observations affect the source term and subsequent forecast.

The 4D-Var data assimilation subsystem [*Vira and Sofiev*, 2012] of the SILAM dispersion model is used for both inversion and assimilation. Similar to the inversion method of Section 8.3.1, the 4D-Var inversion seeks for the source term that minimizes a quadratic cost function measuring the misfit between the model and observations. In the inversion, the SO_2 emission flux is estimated for each model level in steps of 24 hr resulting in total release of 0.22 Tg. This is slightly less than what was found by *Moxnes et al.* [2014] (0.61 Tg) based on a different dispersion model and satellite observations. The assimilation runs rely on the initialization and have no source term.

The SILAM dispersion model is Eulerian, driven by the ECMWF ERA-Interim re-analyses, and configured with 17 vertical levels and 0.5° horizontal resolution. The technical setup of the 4D-Var assimilation includes background errors for initial state, which are prescribed with a constant standard deviation of $4.5 \cdot 10^{-5}$ mol m^{-3}. This choice results in a standard deviation of approximately 100 Dobson Units (DU) for a uniform, 1 km thick layer. The horizontal correlation radius is set to 50 km; no vertical correlation is assumed for the background error. The observation errors are fixed to 10 DU.

Figure 8.3 shows the satellite observed SO_2 columns for 25 May 2011 along with three simulations. In the simulation labeled as analysis + 0 h (bottom left), the observations shown (bottom right) have been included in assimilation, and the model fields therefore match the observations closely. The analysis +24 hr (top right) simulation has only assimilated the observations until 24 hr earlier and therefore corresponds to a 24 hr forecast. The source-term inversion (top left) shown is based on the same observations as the analysis +24 hr simulation. Both

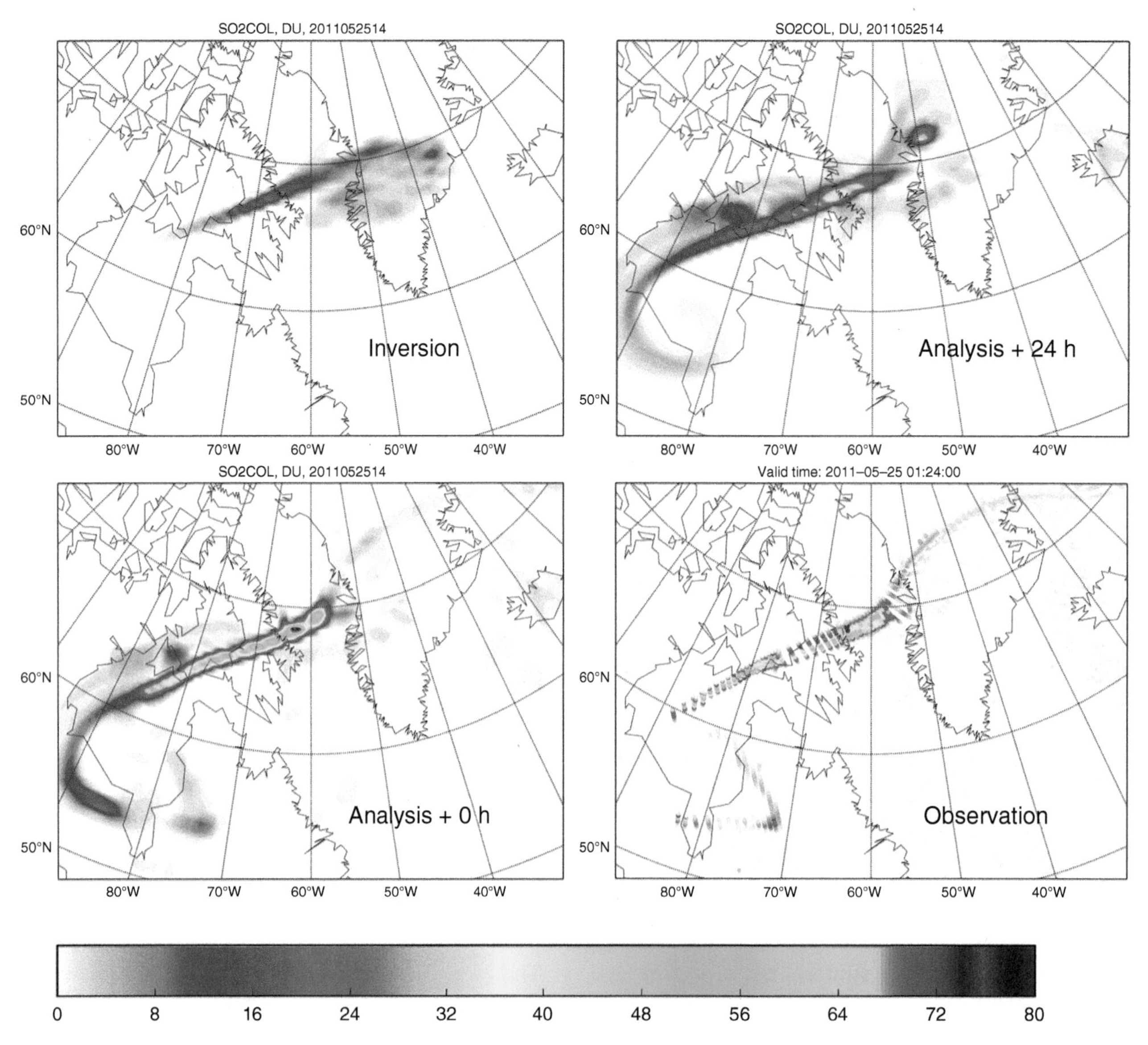

Figure 8.3 SO_2 source term inversion and data assimilation (4D-Var) for the Grímsvötn 2011 eruption: Modeled and retrieved SO_2 column densities (Dobson Unit: DU) for 25 May 2011. Top row: 24 hr forecasts using the source term inversion (left) and 4D-Var initialization from 24 hr earlier (based on the observations available until initialization time) (right). Bottom row: 4D-Var analysis field (left) updated with the up-to-date OMI satellite observations (right). The model fields are valid for 14 UTC, while the observations are combined from multiple overpasses occurring between 12 and 16 UTC.

inversion and analysis +24h fields appear less concentrated than the observed SO_2 cloud.

The spatial correlation coefficient between the modeled and observed SO_2 column loads is shown in Figure 8.4 as a function of forecast length for the period of 22–27 May. The forecasts initialized by data assimilation (red lines) outperform those based on source-term inversion (blue lines) for most 24hr forecasts. However, for 48hr and longer forecasts, the inversion leads to slightly better agreement with the data. The first two assimilation-based 24hr forecasts might especially suffer from the emitted SO_2 being incorrectly attributed to the initial condition.

The advantage of the source-term inversion in longer forecasts can be explained by it giving a better three-dimensional (3D) description of the cloud. As discussed in Section 8.2.2, the source-term inversion is in favorable conditions able to recover the vertical distribution of the emissions. However, the 4D-Var assimilation method, as employed here for forecast initialization, is unable to attribute the column observations vertically unless

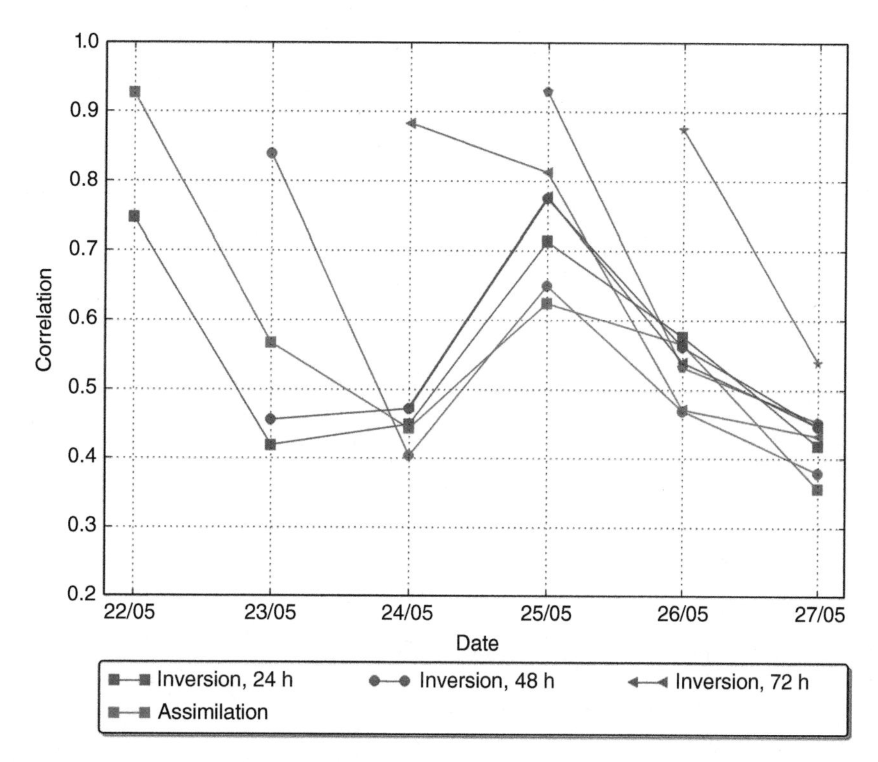

Figure 8.4 Inversion and data assimilation (4D-Var) for the Grímsvötn 2011 eruption: The spatial correlation coefficient between modeled and observed SO_2 columns as a function of forecast length. The lines and markers correspond to forecasts initialized at different analysis times (red lines). The source term inversions using 24, 48, or 72 hr of observations are shown with blue lines.

additional assumptions are made (see *Flemming and Inness*, 2013). Consequently, the analysis fields in the current study are essentially two-dimensional (2D) despite the 3D dispersion model.

The inversion using observations only for the first 24 hr (Fig. 8.4, blue lines) results in lower forecast scores than the inversion using 48 hr of observations. However, increasing the coverage to 72 hr had almost no impact on the subsequent forecasts. This indicates that the inversion becomes insensitive to additional observations as the transport distance increases.

In summary, the forecasts initialized using 4D-Var data assimilation showed better skill on short-term forecasts for the Grímsvötn SO_2 cloud. However, in the current setup, the emission inversion provided more robust predictions for longer forecast lengths due to a better representation of the 3D volcanic cloud.

8.3.3. Multi-input Ensemble Dispersion Modeling

An important source of uncertainty in dispersion model forecasts are the uncertainties associated with the meteorological driving data. In order to account for this uncertainty, a multi-input ensemble modeling approach can be taken whereby several atmospheric transport model calculations are performed using different, but equally probable, meteorological data. The ECMWF Ensemble Prediction System (ENS) is a useful data source for ensemble studies [*Galmarini et al.*, 2010] dealing with the uncertainty of the medium-range weather forecast. It consists of 50 equally probable forecasts originating from a perturbed initial state [*Molteni et al.*, 1996].

In this example case, we perform MI ensemble modeling of the Grímsvötn May 2011 eruption using the FLEXPART dispersion model driven by the 50 ENS members starting at forecast hour 12. The ENS data have three hourly time resolution, 1.0° horizontal resolution, and 62 vertical levels. In addition, these ensemble simulations are compared with the FLEXPART "operational run" based on the 3-hourly operational ECMWF forecast data (1.0° horizontal resolution and 91 model levels), and a "reference run" driven by ERA-Interim reanalysis data (1° horizontal resolution and 60 model levels). All simulations used the ash source term derived by *Moxnes et al.* [2014; Figure 8.1c].

The 50 FLEXPART dispersion calculations driven by the ENS members will generate different transport patterns and thus different areas affected by volcanic ash. This variation associated to the meteorological data can be easily identified when the pixel-wise maximum concentrations of all the 50 ensemble members, for a single model level, are compared to the concentrations obtained by the operational run (Fig. 8.5, upper panels). The operational run (at the 2–3 km a.g.l model layer) transports most of the ash in a quite narrow plume toward the British Isles reaching Scotland by the end of 23 May, while the combined ensemble runs produce a much wider spread of the simulated ash influence leading to a larger affected area over all of Scandinavia. The ensemble output also includes transport of the ash toward the northwest even reaching Greenland. The maximum concentrations of all the ensemble members show a larger area between Iceland and the British Isles with concentrations above 200 $\mu g\ m^{-3}$ and also above 2000 $\mu g\ m^{-3}$ than in the operational run.

This type of ensemble approach can provide probabilistic forecasts of the volcanic cloud evolution such as the probability of airborne ash and/or SO_2 occurring at a specific height or exceeding a specific concentration. The latter is defined by *Galmarini et al.* [2004] as the Agreement in Threshold Level, ATL, by which the number of ensemble members exceeding a specified threshold is determined. As an example, such an approach may provide the ATL of the currently recommended ash concentration thresholds for aviation in Europe [*European Commission,* 2010] including the enhanced procedure zones (200–2000 $\mu g\ m^{-3}$ and 2000–4000 $\mu g\ m^{-3}$, Figure 8.5, lower panels) and the no-fly zone (above 4000 $\mu g\ m^{-3}$, not reached in this case study). This probabilistic approach can be particularly relevant for aviation, where decisions taken solely on one model realization, with limited observations to compare with, could lead to unintentional ash encounters. In the same way, probability of exceedance of air quality health standards may be given by such approaches.

The ensemble can be further evaluated using typical statistical representations of ensembles, like the ensemble mean and median [e.g., *Galmarini et al.*, 2010]. These statistics are used as measures of verification and are ideally compared to independent measurement data to evaluate how well the ensemble reflects a real situation. Here we compare the ensemble statistics to the IASI satellite observations, the operational FLEXPART simulation, and the reference FLEXPART simulation driven by ERA-Interim reanalysis data. The IASI satellite observations are not entirely independent since those data were used to derive the source term used in all simulations. However, the comparison is still thought to provide some useful insights. The reference run is expected to represent a situation close to reality as the reanalysis data include constraints toward real meteorological observations. Comparison to other measurement data (as in the following section) was not possible due to the short length of the MI ensemble model runs dictated by computational constraints.

Figure 8.6 shows the simulated and observed total columns of ash on 23 May 13 UTC. The ensemble mean (average), median, and maximum total columns (upper panel) are compared with the FLEXPART operational simulation (upper left), the FLEXPART ERA-Interim reference simulation (bottom left), and the IASI satellite measurements (bottom right). The mean and median total column values are calculated for each grid cell without any threshold applied. The median of the ensemble simulations clearly overestimates the ash-cloud size and underestimates the ash-cloud maximum concentrations compared with the measurements, as it smooths 50 different realizations of the concentration field that do not fully overlap. Equally expected, also the mean of the ensemble simulations shows this behavior. The maximum of all ensemble members, on the other hand, shows much higher values spread over a comparably large area.

Figure 8.7 (upper panel) shows a time-series of the total amount of ash in the computational domain for the different simulations. All simulations show the expected increase in the total amount of ash due to ongoing emissions until 23 May, and thereafter a decrease in the total amount of ash due to removal processes and exiting of ash from the computational domain. Over the first few hours of the simulations, both the operational run and the mean of the ensemble runs behave close to the reference ERA-Interim-driven simulation. For later times, the differences between the simulations increase and less ash is present in the operational and ensemble simulations than in the reference run. A similar time series of the simulated horizontal extent of the ash clouds (Fig. 8.7, lower panel) demonstrate that the mean of the ensemble runs generally shows a larger extent than the ERA-Interim reference run whereas the operational run remains closer to the reference simulation, as also seen previously on Figure 8.6. In all cases, the median of the ensemble simulations clearly underestimates the total domain ash and ash extent.

In summary, the variability in the dispersion of the volcanic clouds as simulated by the different ensemble members is a useful measure of the uncertainty caused by using alternative but equally probable weather forecasts. This analysis is useful to assess the quality of a forecast of ash dispersion and transport, as provided by various centers (e.g., the Volcanic Ash Advisory Centers). Since forecast errors are typically considerably larger compared with analysis errors, neither the ensemble mean nor the median can be expected to outperform any analysis in comparison with monitoring data. Regarding the interpretation of results in practical applications, one needs to consider the smoothing effects of any averaging procedure, meaning that volcanic clouds will get significantly larger and cloud center concentrations will decrease.

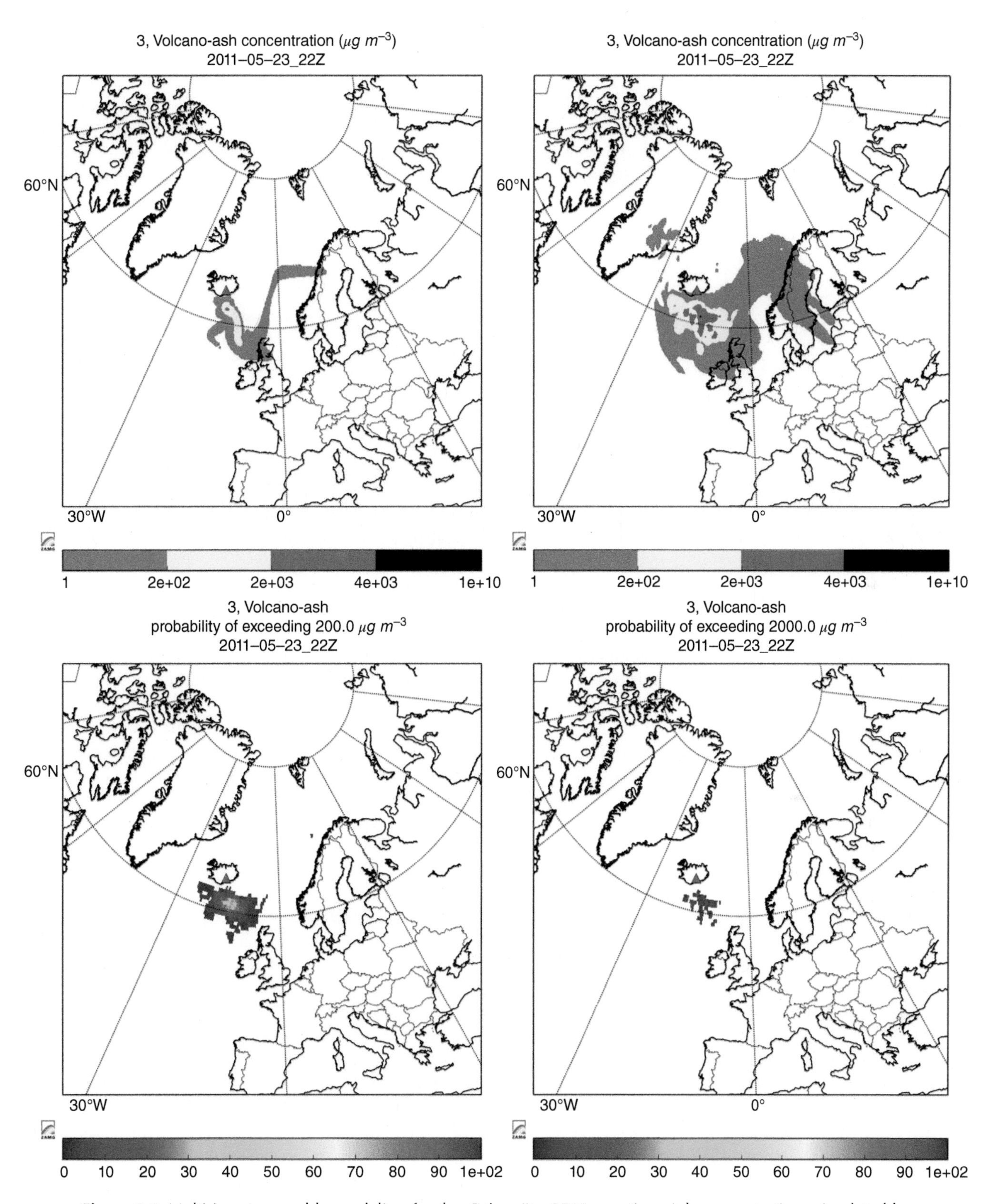

Figure 8.5 Multi-input ensemble modeling for the Grímsvötn 2011 eruption: Ash concentrations simulated by FLEXPART run on the ECMWF operational forecast data (upper left), maximum ash concentration from the simulations run on the 50-member ENS ensemble (upper right), probability of ash concentration above 200 $\mu g\, m^{-3}$ (lower left; mean [max] is 9[36] %) and 2000 $\mu g\, m^{-3}$ (lower right; mean [max] is 3[6]%). All plots are valid for 23 May 2011 at 22 UTC for the 2–3 km a.g.l model layer and the simulations are using the source term from Figure 8.1c.

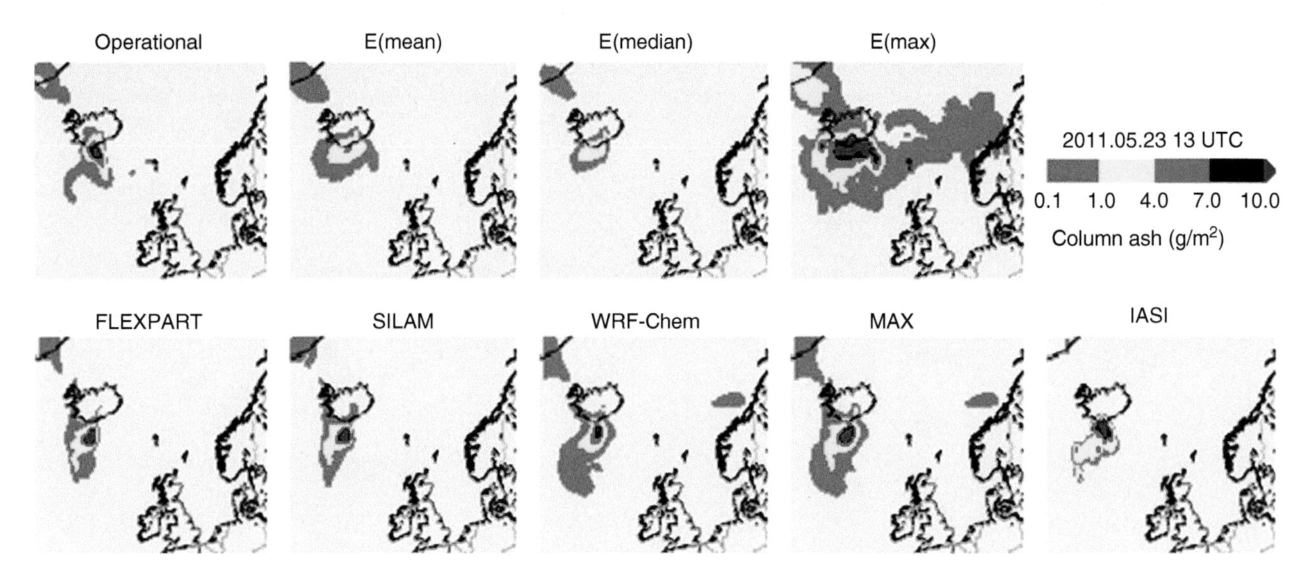

Figure 8.6 Multi-input ensemble modeling (top) and model intercomparison (bottom) for the Grímsvötn 2011 eruption. Top row: Total column ash simulated by FLEXPART run on ECMWF operational meteorological input fields and 50 ENS member ensemble. For the ensemble, the mean, median, and maximum of the members are shown. Bottom row: Simulations run on ECMWF ERA-Interim meteorological input data by the three models FLEXPART, SILAM, and WRF-Chem. The maximum values obtained from the three models are also shown. All model simulations are using the a posteriori source term from Figure 8.1c. The last plot (bottom right) shows the IASI satellite observations. All panels are valid for 23 May 13 UTC.

Therefore, for impact assessments, other parameters are usually of higher interest, for example the probability of exceedance of certain threshold concentrations.

8.3.4. Model Intercomparison

To evaluate and quantify the similarities and differences between model simulations associated with on-line and off-line meteorology and the models' formulations of physical process such as loss mechanisms (wet deposition, dry deposition, sedimentation), a model intercomparison comprising the three models FLEXPART, SILAM, and WRF-Chem was generated. All three models were initiated with the same a posteriori ash source term (Fig. 8.1c) and ash particle size distribution as in *Moxnes et al.* [2014], as well as the same meteorological fields from ECMWF ERA-Interim reanalysis data. Note that the FLEXPART simulation shown in this section differs from the one in Section 8.3.1, which was based on ECMWF operational data. Differences in the simulations here arise solely from differences in the formulation of the dispersion model used; Eulerian (WRF-Chem and SILAM) versus Lagrangian (FLEXPART), meteorological coupling; on-line (WRF-Chem) versus off-line (SILAM and FLEXPART), and the treatment of physical processes that differ in all three models.

The WRF-Chem model [*Stuefer et al.*, 2013] is the Weather Research and Forecasting (WRF) model coupled with chemistry, and is capable of simulating the emission, transport, mixing, and chemical transformation of trace gases and aerosols simultaneously with the meteorology (on-line). WRF-Chem was set to use the ARW (Advanced Research WRF) [*Skamarock et al.*, 2008] dynamics solver for computing the necessary meteorological parameters at each time step. This is a crucial difference between the on-line versus off-line modeling, which may result in differences in meteorological parameters (wind speed, wind direction, precipitation, etc.), which may have a direct impact on the ash-plume dispersion and transport. The WRF-Chem simulation was initialized at the first time step using ECMWF ERA-Interim reanalyses data and simulations commenced 12 hr before the first release of volcanic ash to allow for meteorological spin up. In order to do the analysis, FLEXPART and SILAM outputs were constrained to WRF-Chem's smaller domain (30°W to 30°E and 40°N to 70°N) at the same 0.25° horizontal resolution.

Figure 8.6 shows the individual simulations by FLEXPART, SILAM, and WRF-Chem (bottom row), and the three-model maximum (MAX) for 23 May 2011 at 13 UTC. All models simulate large total columns of ash in a concentrated region south of Iceland, while WRF-Chem also simulates some small total column values over Norway, which is not present in FLEXPART or SILAM. There is very good agreement between the two off-line models (FLEXPART and SILAM) in terms of spatial dispersion with the differences being more significant for the on-line model (WRF-Chem).

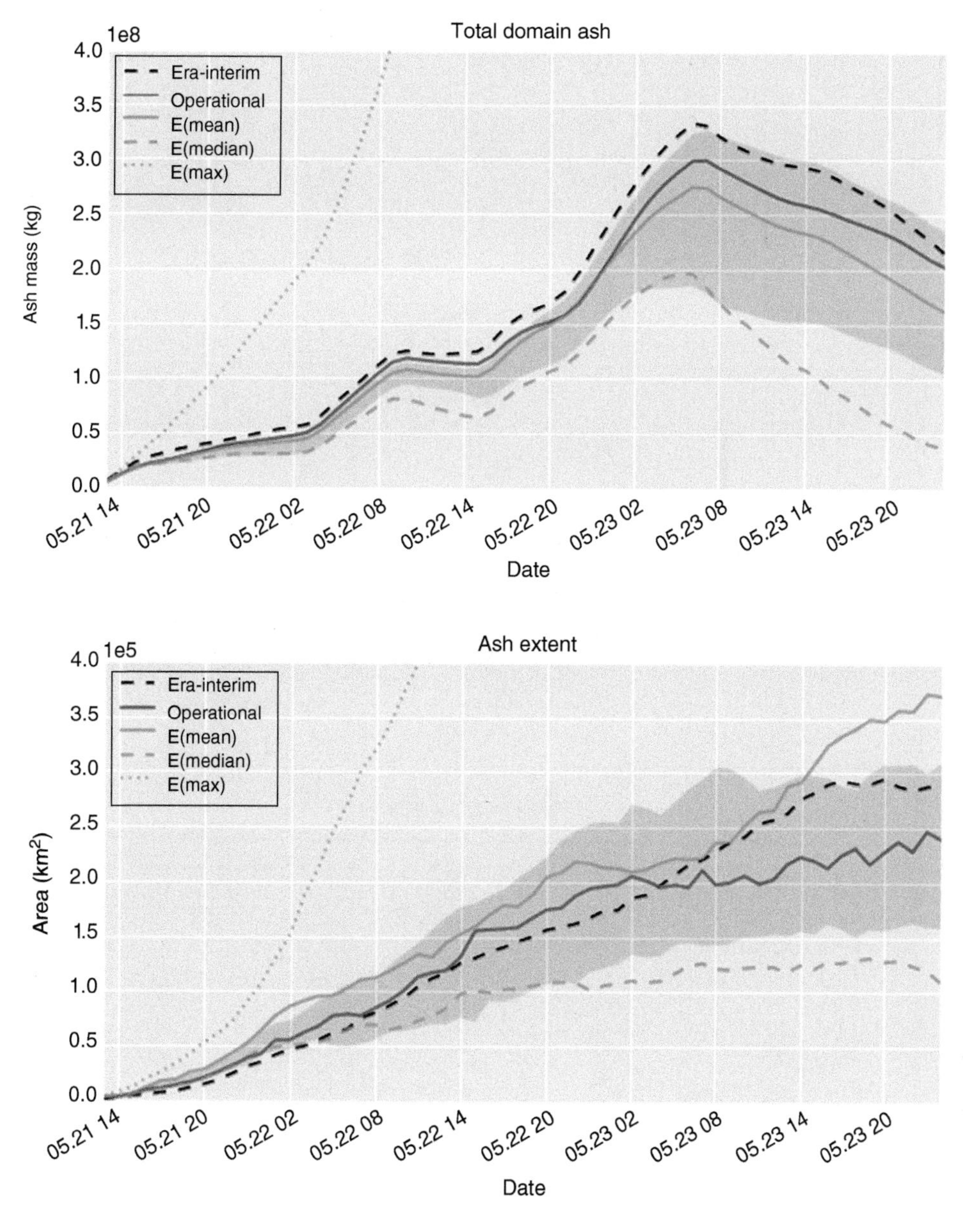

Figure 8.7 Multi-input ensemble modeling for the Grímsvötn 2011 eruption: Time series of the total amount of ash in the model domain (upper panel) and the total horizontal ash extent (lower panel) for the reference FLEXPART run driven by ECMWF ERA-Interim reanalyses data (dashed black lines), the FLEXPART simulation run on operational ECMWF data (red lines), and the FLEXPART simulations run on the 50-member ENS data (green lines/area). The mean, median, and max of the ensemble members are shown (green lines) and the full range of the member runs is highlighted within the envelope (shaded green area).

The time series of the total amounts of ash within the models' domain during and after the eruption, are shown in Figure 8.8a. There is very good agreement between the off-line models FLEXPART and SILAM over the entire period, while the on-line model WRF-Chem exhibits higher values after 23 May. The total amount of ash in the combination of models (MAX) is always larger or equal to the three single models. This is due to the fact that the maximum is calculated for each grid cell before the total domain ash loading is determined (see also Fig. 8.6). During the eruptive phase (21–23 May), when the ash emissions dominate, the deviation between the on-line model and the off-line models is negligible. As the eruptive phase ceases (24–25 May), the total amount of ash within the domain becomes more dominated by transport and loss mechanisms. The deviation between

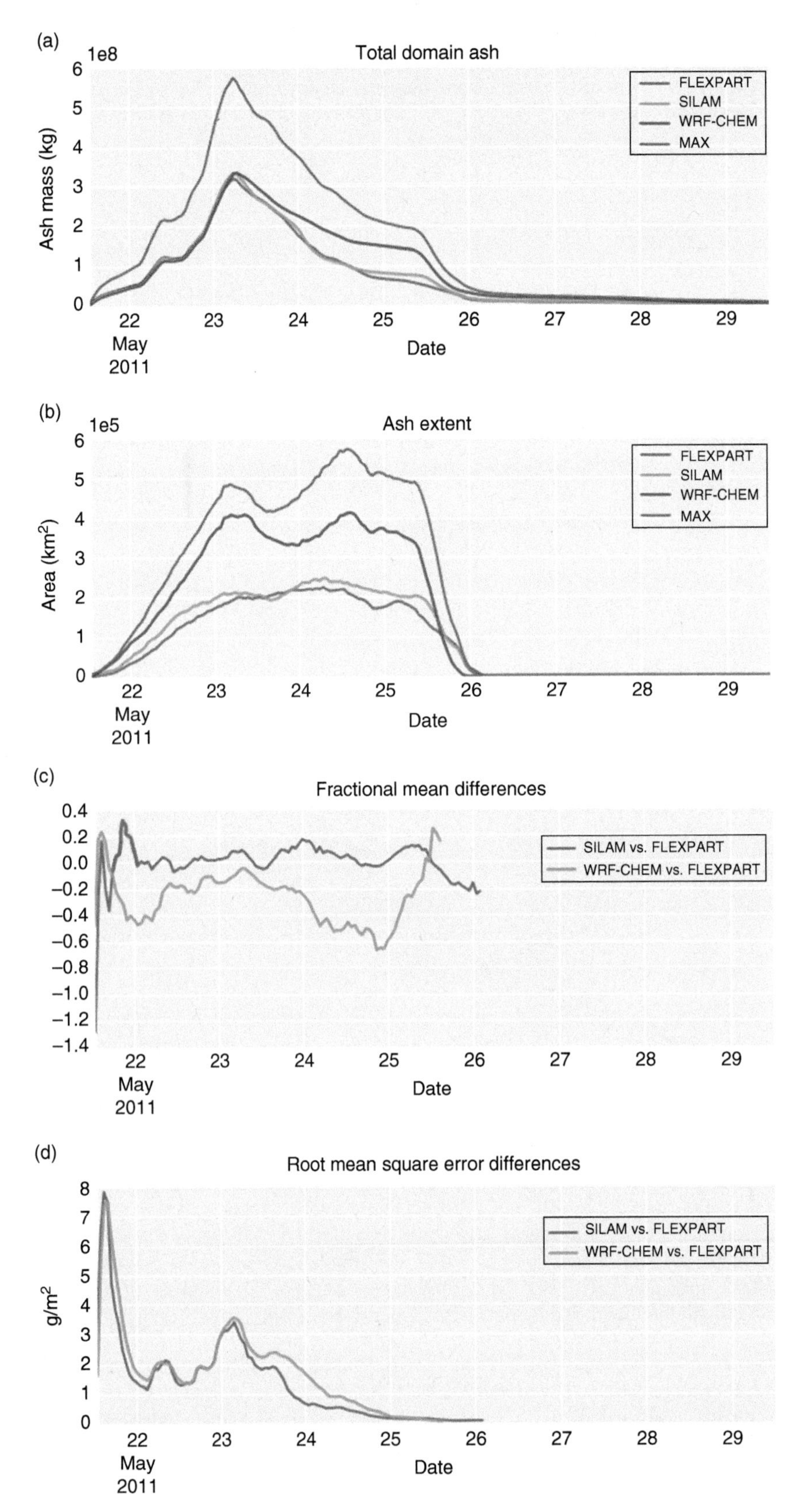

Figure 8.8 Model intercomparison for the Grímsvötn 2011 eruption: (a) Time series of the total amount of ash in the model domain and (b) the horizontal ash extent for the three models WRF-Chem, FLEXPART, and SILAM, and for the three-model maximum product. For the ash extent, a threshold value of $0.1\,\text{g}\,\text{m}^{-2}$ was used. All model simulations are using the a posteriori source term from Figure 8.1c and ERA-Interim meteorological fields. Statistical measures (two bottom panels): (c) Fractional mean (FM) differences, and (d) root mean square error (RMSE) between the ash mass loadings from the three models, calculated over the eruptive phase + 48 hr for colocated ash grid cells and using the same threshold value as for the ash extent.

WRF-Chem and the other two models increases due to differences in the simulated transport and ash removal rates between the models.

The total horizontal ash cloud extent over time (Fig. 8.8b) for the different model simulations is based on a threshold of 0.1 g m^{-2}. Again the agreement between FLEXPART and SILAM is very good, while the ash cloud extent in WRF-Chem is consistently higher which suggests that the on-line model disperses the ash plume over a larger area compared to the off-line models. This is clearly seen by the difference in dispersion patterns shown in Figure 8.6, where additional cloud advection over Norway seen in WRF-Chem results in differences in the ash extent compared with the off-line models.

Statistical measures (fractional mean [FM] and root mean square error [RMSE] differences) calculated on the ash-mass loadings between FLEXPART and both SILAM and WRF-Chem models are shown in Figure 8.8c and d, respectively. These statistical parameters are calculated for colocated ash grid cells and above the threshold value of 0.1g m^{-2} (colocated refers to grid cells that have values above the threshold value in both models entering the calculation). The first period of the eruption (up to 22 May) reflects the buildup of ash in the models' domain, where large relative differences are to be expected because of the small number of grid cells containing ash as well as the relatively small ash concentration values in those grid cells. Consequently, this period is disregarded in the analysis. Overall, the FM and the RMSE differences confirm that the FLEXPART and SILAM models have very similar ash-mass loadings for the colocated grid cells, and that WRF-Chem shows larger differences. On average, WRF-Chem presents an underestimation relative to FLEXPART. This means that for the grid cells where both FLEXPART and WRF-Chem have ash, lower ash values are found in WRF-Chem, and that the overestimations of the total domain ash and the ash extent in WRF-Chem mainly come from the grid cells where only WRF-Chem has ash. The second high peak in the RMSE coincides with the maximum total amount of ash in the atmosphere on 23 May.

To evaluate the accuracy of the predictions, comparisons against independent PM10 ground measurements at a number of locations in Scandinavian were done (Fig. 8.9). Modelled PM10 values are calculated by summing up ash concentrations for the ash particle size bins up to 10 μm diameter. To compensate for the background PM10 value before and after the ash overpass, a value of 20 μg m^{-3} was added to the modeled values as in *Moxnes et al.* [2014]. In general, all the models show an increase in PM10 coincidental within 6 hr of a rise in the ambient PM10 levels associated with the passage of the ash cloud. WRF-Chem generally simulates an earlier arrival time of the ash-cloud passage than the other two models. With respect to the magnitude, WRF-Chem compares well at most sites and overpredicting only at Oslo by about four times the measurement value. On the contrary, FLEXPART underpredicts at three sites by a factor between 1 and 2. For two of the sites (Oslo and Stavanger), SILAM considerably underestimates the magnitude of PM10. This is mainly because SILAM transports the ash cloud more to the south across Denmark and southern Sweden, while in the FLEXPART simulation, the ash cloud passes closer to Stavanger and Oslo. For the other three sites. there are small differences in SILAM compared with FLEXPART with respect to the arrival time (<3 hr), although the magnitude is uniformly the smallest of the three models.

The total amount of ash, ash-cloud extent, the statistical measures, and the PM-evaluation together suggest that the FLEXPART and SILAM models have a similar treatment of the transport and removal of the ash, while WRF-Chem dilutes and advects the ash more rapidly than the two off-line models.

The results here indicate that in the case of the Grímsvötn eruption different methods of coupling to the meteorological data (on-line versus off-line) and different formulations of physical processes such as loss mechanisms in the transport models investigated here can contribute significantly to differences in the simulated spread of the ash clouds. This result may be case dependent and further investigation of other volcanic eruptions may increase the understanding of these model differences.

8.4. SUMMARY AND CONCLUSIONS

Numerous and complex uncertainties are involved in quantifying the atmospheric impacts of volcanic emissions on local air pollution, aviation, and climate. Simulations of the volcanic emission clouds using dispersion models are useful but must carefully address all sources of uncertainty. This includes the definition of the source term (the emissions and their variation in the vertical and throughout the eruption period), the model's description of physical processes such as deposition, and possible errors in the wind fields and other meteorological data driving the dispersion model. In this chapter, we demonstrate three techniques useful for improving model simulations of volcanic emission clouds and for addressing these uncertainties. Embedding observations into the modeling has been proved to be the key to achieving more reliable simulations as long as the observations themselves are accurate, reliable, and available in due time. Methods for combining results from different model simulations can further provide useful uncertainty estimates.

1. Inversion methods can be used to determine indirectly the source term of the eruption, which is perhaps the dominating uncertainty in volcanic cloud modeling. With the

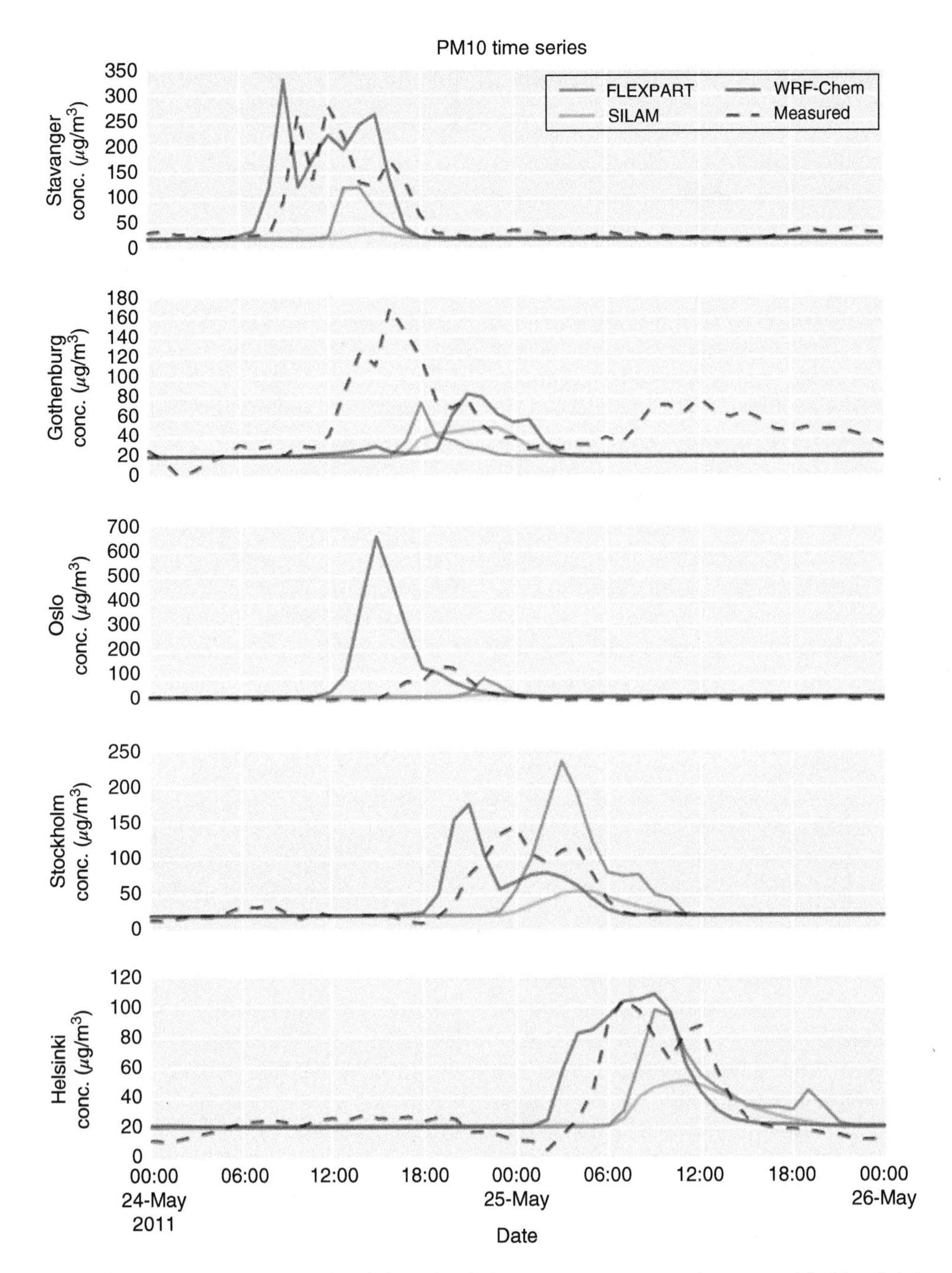

Figure 8.9 Model intercomparison for the Grímsvötn 2011 eruption: Measured and modeled PM10 time series at a number of locations in Scandinavia. A background value of 20 $\mu g m^{-3}$ was added to the model output to compensate for the measured background values at these locations.

estimated source term, the model simulations are brought in best agreement with the satellite observations. The main limitation with this method is that for longer forecast times, the errors in the wind fields and other meteorological variables, and/or model parametrizations, might become dominating, which can lead to inaccuracies.

2. Data assimilation techniques correct the modeled concentration fields using up-to-date observations. The corrections are independent of the source term, and therefore become useful especially as the transport distance increases. However, the current assimilation methods are restricted by the need of 3D initialization with only 2D column observations available, and thus the need of additional assumptions about the vertical attribution. Data assimilation can be combined with inverse modeling, which provides a better 3D description, and the combination would likely have better forecast performance than either approach alone.

3. Ensemble modeling is a way to combine several plausible model simulations, either based on different input data (multi-input) and/or different models (multi-model). The combination of simulations can provide uncertainty ranges in the forecasts and can provide the probability of emission clouds occurring at specific locations, or exceeding specific concentration thresholds. A limitation of this method is its requirement of considerable computational resources. In addition, increasing the number of ensemble members does not guarantee an increase in the quality of the ensemble results. Careful evaluation of the conclusions derived from ensemble results is therefore needed.

For the Grímsvötn May 2011 eruption, we demonstrated that the aforementioned techniques are of value to produce improved volcanic ash and SO_2 cloud simulations and to account, to a certain extent, for the uncertainties associated with the problem. Initializing the transport models with a source term constrained by satellite data led to improvements in both the simulated position and absolute values of the emission clouds. Updating the simulated concentration fields with up-to-date satellite data via data assimilation showed better skill on short-term forecasts. A 50 member multi-input ensemble from the ECMWF ENS system using variations of the initial conditions of meteorological input data allowed the quantification of uncertainties associated with the model simulations in forecast mode. From these ensembles, probabilities of ash concentrations exceeding specific thresholds could also be estimated. A model intercomparison, comprising three independent models that were run with the same source term and meteorological driver, illustrated that significant differences in the amount and location/extent of the volcanic clouds can occur due to differences in the description of loss processes (e.g., wet and dry deposition, sedimentation) in the models, as well as the method of coupling to the driving meteorological data (off-line versus on-line). These differences may be case specific and further test cases can help understand the variation in these uncertainties.

Further improvements in the incorporation of observations into the modeling may involve the use of additional data (lidar, aircraft, and ground measurements) in the applications. Also, improvements can be sought from using an iterative satellite-model approach where the cloud altitude information inferred from the source-term inversion and subsequent model simulations is used as input to the satellite retrieval with the aim of achieving more accurate retrievals. Last, further work on ensemble forecasting of volcanic-eruption clouds, both multi-input and multimodel and their combination, is needed. Due to the possible case dependency, an evaluation of a larger number of cases is important.

ACKNOWLEDGMENTS

This work was partially funded by the European Space Agency under the VAST project. We acknowledge ECMWF for providing access to the ECMWF data and archives. L. Clarisse at Université Libre de Bruxelles, Brussels, has provided the IASI ash retrieval data. The Science Foundation Ireland project INFORM is acknowledged. We also thank the editors of the AGU monograph for the ability to publish this work, and the reviewers for comments that helped improve the manuscript.

REFERENCES

Barbu, A. L., A. J. Segers, M. Schaap, A. Heemink, and P. Builtjes (2009), A multi-component data assimilation experiment directed to sulphur dioxide and sulphate over Europe, *Atmos. Environ.*, *43*(9), 1622–1631; doi:10.1016/j.atmosenv.2008.12.005.

Boichu, M., L. Menut, D. Khvorostyanov, L. Clarisse, C. Clerbaux, S. Turquety, and P.-F. Coheur (2013), Inverting for volcanic SO_2 flux at high temporal resolution using spaceborne plume imagery and chemistry-transport modelling: the 2010 Eyjafjallajökull eruption case study, *Atmos. Chem. Phys.*, *13*(17), 8569–8584; doi:10.5194/acp-13-8569-2013.

Carboni, E., R. Grainger, J. Walker, A. Dudhia, and R. Siddans (2012), A new scheme for sulphur dioxide retrieval from IASI measurements: application to the Eyjafjallajökull eruption of April and May 2010, *Atmos. Chem. Phys.*, *12*(23), 11,417–11,434; doi:10.5194/acp-12-11417-2012.

Carmichael, G., A. Sandu, T. Chai, D. Daescu, E. Constantinescu, and Y. Tang (2008), Predicting air quality: Improvements through advanced methods to integrate models and measurements, *J. Computational Phys.*, *227*(7), 3540–3571; doi:10.1016/j.jcp.2007.02.024.

Carn, S., N. Krotkov, K. Yang, and A. Krueger (2013), Measuring global volcanic degassing with the Ozone Monitoring Instrument (OMI), *Geological Society, London, Special Publications*, *380*(1), 229–257; doi:10.1144/SP380.12.

Clarisse, L., D. Hurtmans, A. J. Prata, F. Karagulian, C. Clerbaux, M. De Maziere, and P.-F. Coheur (2010), Retrieving radius, concentration, optical depth, and mass of different types of aerosols from high-resolution infrared nadir spectra, *Appl. Optics*, *49*(19), 3713–3722.

Clarisse, L., D. Hurtmans, C. Clerbaux, J. Hadji-Lazaro, Y. Ngadi, and P.-F. Coheur (2012), Retrieval of sulphur dioxide from the infrared atmospheric sounding interferometer (IASI), *Atmos. Meas. Tech.*, *5*(3), 581–594; doi:10.5194/amt-5-581-2012.

Clarisse, L., P. F. Coheur, A. J. Prata, D. Hurtmans, A. Razavi, T. Phulpin, J. Hadji-Lazaro, and C. Clerbaux (2008), Tracking and quantifying volcanic SO_2 with IASI, the September 2007 eruption at Jebel at Tair, *Atmos. Chem. Phys.*, *8*(24), 7723–7734; doi:10.5194/acp-8-7723-2008.

Draxler, R., and G. D. Hess (1998), An overview of the HYSPLIT_4 modelling system for trajectories, dispersion and deposition, *Austral. Meteor. Mag.*, *47*, 295–308.

Eckhardt, S., A. J. Prata, P. Seibert, K. Stebel, and A. Stohl (2008), Estimation of the vertical profile of sulfur dioxide injection into the atmosphere by a volcanic eruption using satellite column measurements and inverse transport modeling, *Atmos. Chem. Phys.*, *8*(14), 3881–3897.

Elbern, H., A. Strunk, H. Schmidt, and O. Talagrand (2007), Emission rate and chemical state estimation by 4-dimensional variational inversion, *Atmos. Chem. Phys. Disc.*, *7*(1), 1725–1783; doi:10.5194/acpd-7-1725-2007.

Elbern, H., H. Schmidt, O. Talagrand, and A. Ebel (2000), 4D-variational data assimilation with an adjoint air quality model for emission analysis, *Environ. Mod. Soft.*, *15*(6–7), 539–548; doi:10.1016/S1364-8152(00)00049-9.

European Commission (2010), Report on the actions undertaken in the context of the impact of the volcanic ash cloud crisis on the air transport industry, http://ec.europa.eu/transport/doc/ash-cloud-crisis/2010_06_30_volcano-crisis-report.pdf, online; accessed 27.11.2014.

Evensen, G. (1994), Sequential data assimilation with a nonlinear quasi-geostrophic model using Monte Carlo methods to forecast error statistics, *J. Geophys. Res.*, *99*(C5), 10,143–10,162.

Evensen, G. (2003), The Ensemble Kalman Filter: theoretical formulation and practical implementation, *Ocean Dynam.*, *53*(4), 343–367; doi:10.1007/s10236-003-0036-9.

Flanner, M. G., A. S. Gardner, S. Eckhardt, A. Stohl, and J. Perket (2014), Aerosol radiative forcing from the 2010 Eyjafjallajökull volcanic eruptions, *J. Geophys. Res. Atmos.*, *119*(15), 9481–9491; doi:10.1002/2014JD021977.

Flemming, J., and A. Inness (2013), Volcanic sulfur dioxide plume forecasts based on UV satellite retrievals for the 2011 Grmsvtn and the 2010 Eyjafjallajökull eruption, *J. Geophys. Res. Atmos*, *118*(17), 10,172–10,189; doi:10.1002/jgrd.50753.

Folch, A., A. Costa, and G. Macedonio (2009), FALL3D: A computational model for transport and deposition of volcanic ash, *Comput. Geosci.*; doi:10.1016/j.cageo.2008.08.008.

Galmarini, S., and S. T. Rao (2011), The AQMEII two-continent regional air quality model evaluation study: Fueling ideas with unprecedented data, *Atmos. Environ.*, *45*(14), 2464; doi: http://dx.doi.org/10.1016/j.atmosenv.2011.03.025.

Galmarini, S., F. Bonnardot, A. Jones, S. Potempski, L. Robertson, and M. Martet (2010), Multi-model vs. eps-based ensemble atmospheric dispersion simulations: A quantitative assessment on the etex-1 tracer experiment case, *Atmos. Environ.*, *44*(29), 3558–3567;doi:http://dx.doi.org/10.1016/j.atmosenv.2010.06.003.

Galmarini, S., R. Bianconi, W. Klug, T. Mikkelsen, R. Addis, S. Andronopoulos, P. Astrup, A. Baklanov, J. Bartniki, J. Bartzis, R. Bellasio, F. Bompay, R. Buckley, M. Bouzom, H. Champion, R. DAmours, E. Davakis, H. Eleveld, G. Geertsema, H. Glaab, M. Kollax, M. Ilvonen, A. Manning, U. Pechinger, C. Persson, E. Polreich, S. Potemski, M. Prodanova, J. Saltbones, H. Slaper, M. Sofiev, D. Syrakov, J. Srensen, L. der Auwera, I. Valkama, and R. Zelazny (2004), Ensemble dispersion forecastingpart i: concept, approach and indicators, *Atmos. Environ.*, *38*(28), 4607–4617; doi: http://dx.doi.org/10.1016/j.atmosenv.2004.05.030.

Gangale, G., A. J. Prata, and L. Clarisse (2010), The infrared spectral signature of volcanic ash determined from high-spectral resolution satellite measurements, *Remote Sens. Environ.*, *114*(2), 414–425; doi:10.1016/j.rse.2009.09.007.

Grainger, R., D. Peters, G. Thomas, A. Smith, R. Siddans, E. Carboni, and A. Dudhia (2013), Measuring volcanic plume and ash properties from space, *Remote Sensing of Volcanoes and Volcanic Processes: Integrating Observation and Modelling*, edited by D. M. Pyle, T. A. Mather, and J. Biggs, The Geological Society Special Publication, 380.

Gu, Y., W. I. Rose, and G. J. S. Bluth (2003), Retrieval of mass and sizes of particles in sandstorms using two modis ir bands: A case study of april 7, 2001 sandstorm in China, *Geophys. Res. Lett.*, *30*(15), n/a–n/a; doi:10.1029/2003GL017405.

Horwell, C., and P. Baxter (2006), The respiratory health hazards of volcanic ash: a review for volcanic risk mitigation, *Bull. Volcanol.*, *69*(1), 1–24; doi:10.1007/s00445-006-0052-y.

Icelandic Directorate of Health (2014), Volcanic eruption in holuhraun - human health effect: Pollution from the eruption in holuhraun, http://www.landlaeknir.is/english/volcanic-eruption-in-holuhraun-human-health-effext/, online; accessed 27.11.2014.

Kioutsioukis, I., and S. Galmarini (2014), De praeceptis ferendis: good practice in multi-model ensembles, *Atmos. Chem. Phys.*, *14*(21), 11,791–11,815; doi:10.5194/acp-14-11791- 2014.

Kristiansen, N. I., A. J. Prata, A. Stohl, and S. A. Carn (2015), Stratospheric volcanic ash emissions from the 13 february 2014 kelut eruption, *Geophys. Res. Lett.*, *42*(2), 588–596; doi:10.1002/2014GL062307.

Kristiansen, N. I., A. Stohl, A. J. Prata, A. Richter, S. Eckhardt, P. Seibert, A. Hoffmann, C. Ritter, L. Bitar, T. J. Duck, and K. Stebel (2010), Remote sensing and inverse transport modeling of the Kasatochi eruption sulfur dioxide cloud, *J. Geophys. Res.*, *115*(D2); doi:10.1029/2009JD013286.

Kristiansen, N. I., A. Stohl, A. J. Prata, N. Bukowiecki, H. Dacre, S. Eckhardt, S. Henne, M. C. Hort, B. T. Johnson, F. Marenco, B. Neininger, O. Reitebuch, P. Seibert, D. J. Thomson, H. N. Webster, and B. Weinzierl (2012), Performance assessment of a volcanic ash transport model mini-ensemble used for inverse modeling of the 2010 Eyjafjallajökull eruption, *J. Geophys. Res.*, *117*(D20); doi:10.1029/2011JD016844.

Krotkov, N., S. Carn, A. Kyueger, P. Bhartia, and K. Yang (2006), Band residual difference algorithm for retrieval of SO_2 from the Aura Ozone Monitoring Instrument (OMI), *IEEE Trans. Geosci. Remote Sens.*, *44*; doi:10.1109/TGRS.2005.861932, 5.

Le Dimet, F.-X., and O. Talagrand (1986), Variational algorithms for analysis and assimilation of meteorological observations: theoretical aspects, *Tellus A*, *38A*(2), 97–110; doi:10.1111/j.1600-0870.1986.tb00459.x.

Lorenc, A. C. (1986), Analysis methods for numerical weather prediction, *Quart. J. Roy. Meteor. Soc.*, *112*(474), 1177–1194; doi:10.1002/qj.49711247414.

Mastin, L., M. Guffanti, R. Servranckx, P. Webley, S. Barsotti, K. Dean, A. Durant, J. Ewert, A. Neri, W. Rose, D. Schneider, L. Siebert, B. Stunder, G. Swanson, A. Tupper, A. Volentik, and C. Waythomas (2009), A multidisciplinary effort to assign realistic source parameters to models of volcanic ash-cloud transport and dispersion during eruptions, *J. Volcanol. Geotherm. Res.*, *186*, 10–20; doi:10.1016/j.jvolgeores.2009.01.008.

Miyazaki, K., H. J. Eskes, K. Sudo, M. Takigawa, M. van Weele, and K. F. Boersma (2012), Simultaneous assimilation of satellite no2, o3, co, and hno3 data for the analysis of tropospheric chemical composition and emissions, *Atmos. Chem. Phys.*, *12*, 9545–9579; doi:10.5194/acpd-12-16131-2012.

Molteni, F., R. Buizza, T. N. Palmer, and T. Petroliagis (1996), The ECMWF ensemble prediction system: Methodology and validation, *Quart. J. Roy. Meteor. Soc.*, *122*(529), 73–119.

Moxnes, E., N. I. Kristiansen, A. Stohl, L. Clarisse, A. Durant, K. Weber, and A. Vogel (2014), Separation of ash and sulfur dioxide during the 2011 grmsvtn eruption, *J. Geophys. Res. Atmos.*, *119*(12), 7477–7501; doi:10.1002/2013JD021129.

Palmer, T., G. Shutts, R. Hagedorn, F. Doblas-Reyes, T. Jung, and M. Leutbecher (2005), Representing model uncertainty in weather and climate prediction, *Ann. Rev. Earth Planet. Sci.*, *33*(1), 163–193; doi:10.1146/annurev.earth.33.092203.122552.

Patra, A., M. Bursik, J. Dehn, M. Jones, R. Madankan, D. Morton, M. Pavolonis, E. Pitman, S. Pouget, T. Singh, P. Singla, E. Stefanescu, and P. Webley (2013), Challenges in developing {DDDAS} based methodology for volcanic ash hazard analysis effect of numerical weather prediction variability and parameter estimation, *Procedia Computer Sci.*, *18*(0), 1871–1880; doi:10.1016/j.procs.2013.05.356, 2013 International Conference on Computational Science.

Pelley, R., M. Cooke, A. Manning, D. Thomson, C. Witham, and M. Hort (2014), Inversion technique for estimating emissions of volcanic ash from satellite imagery, *Geophys. Res. Abstr.*, *16*, EGU2014-12950, eGU General Assembly 2014.

Prata, A. J. (1989), Infrared radiative transfer calculations for volcanic ash clouds, *Geophys. Res. Lett.*, *16*(11), 1293–1296; doi:10.1029/GL016i011p01293.

Prata, A. J. (2013), Detecting and retrieving volcanic ash from Seviri measurements, Algorithm Theoretical Basis Document Version 1.0, 22.05.2013, *Tech. Rep.*, Climate and Atmosphere Department, Norwegian Institute for Air Research, Kjeller, Norway.

Prata, A. J., and A. T. Prata (2012), Eyjafjallajökull volcanic ash concentrations determined using Spin Enhanced Visible and Infrared Imager measurements, *J. Geophys. Res.*, *117*(D20); doi:10.1029/2011JD016800.

Prata, A. J., and A. Tupper (2009), Aviation hazards from volcanoes: The state of the science, *Nat. Hazards*, *51*(2), 239–244; doi:10.1007/s11069-009-9415-y.

Prata, A. J., and C. Bernardo (2007), Retrieval of volcanic SO_2 column abundance from Atmospheric Infrared Sounder data, *J. Geophys. Res. Atmos.*, *112*(D20); doi:10.1029/2006JD007955.

Prata, A. J., and J. Kerkmann (2007), Simultaneous retrieval of volcanic ash and SO2 using MSG-SEVIRI measurements, *Geophys. Res. Lett.*, *34*(5), n/a–n/a; doi:10.1029/2006GL028691.

Prata, A. J., and P. J. Turner (1997), Cloud-top height determination using ATSR data, *Remote Sens. Environ.*, *59*(1), 1–13; doi:10.1016/S0034-4257(96)00071-5.

Prata, A. J., C. Zehner, and K. Stebel (2014), Earth Observations and Volcanic Ash: A Report of the ESA/Eumetsat Workshop, 4–7 March, 2013, *Tech. Rep.*, ESA; doi:10.5270/atmva-14-04.

Rix, M., P. Valks, N. Hao, D. Loyola, H. Schlager, H. Huntrieser, J. Flemming, U. Koehler, U. Schumann, and A. Inness (2012), Volcanic SO2, bro and plume height estimations using GOME-2 satellite measurements during the eruption of Eyjafjallajökull in May 2010, *J. Geophys. Res. Atmos.*, *117*(D20); doi:10.1029/2011JD016718.

Rix, M., P. Valks, N. Hao, J. van Geffen, C. Clerbaux, L. Clarisse, P.-F. Coheur, R. Loyola, T. Erbertseder, W. Zimmer, and S. E. (2009), Satellite monitoring of volcanic sulfur dioxide emissions for early warning of volcanic hazards, *IEEE J-STARS*, *2*(3), 196–206.

Robock, A. (2000), Volcanic eruptions and climate, in *Volcanism and the Earth's Atmosphere*, edited by A. Robock and C. Oppenheimer, in Geophysical Monograph 139, 191–219, American Geophysical Union, Washington, DC.

Ryall, D., R. Derwent, A. Manning, A. Redington, J. Corden, W. Millington, P. Simmonds, S. O'Doherty, N. Carslaw, and G. Fuller (2002), The origin of high particulate concentrations over the United Kingdom, March 2000, *Atmos. Environ.*, *36*(8), 1363–1378; doi:10.1016/S1352-2310(01)00522-2.

Skamarock, W., J. B. Klemp, J. Dudhia, D. O. Gill, D. Barker, M. G. Duda, X.-Y. Huang, and W. Wang(2008), A Description of the Advanced Research WRF Version 3., *NCAR Technical Note NCAR/TN-475+STR*; doi:10.5065/D68S4MVH.

Sofiev, M., M. Galperin, and E. Genikhovich (2008), A construction and evaluation of Eulerian dynamic core for the air quality and emergency modelling system silam, in *Air Pollution Modeling and Its Application XIX*, edited by C. Borrego and A. Miranda, NATO Science for Peace and Security Series Series C: Environmental Security, 699–701, Springer Netherlands; doi:10.1007/978-1-4020-8453-9_94.

Solazzo, E., and S. Galmarini (2015), A science-based use of ensembles of opportunities for assessment and scenario studies, *Atmos. Chem. Phys.*, *15*(5), 2535–2544; doi:10.5194/acp-15-2535-2015.

Sparks, R. S. J., M. I. Bursik, S. N. Carey, J. S. Gilbert, L. S. Glaze, H. Sigurdsson, and A. W. Woods (1997), *Volcanic Plumes*, John Wiley, Chichester, UK.

Stefanescu, E., A. Patra, M. Bursik, M. Jones, R. Madankan, E. Pitman, S. Pouget, T. Singh, P. Singla, P. Webley, and D. Morton (2014), Fast construction of surrogates for {UQ} central to dddas application to volcanic ash transport, *Procedia Computer Sci.*, *29*(0), 1227–1235; doi: 10.1016/j.procs.2014.05.110, 2014 International Conference on Computational Science.

Stohl, A., A. J. Prata, S. Eckhardt, L. Clarisse, A. Durant, S. Henne, N. I. Kristiansen, A. Minikin, U. Schumann, P. Seibert, K. Stebel, H. E. Thomas, T. Thorsteinsson, K. Tørseth, and B. Weinzierl (2011), Determination of time- and height-resolved volcanic ash emissions and their use for quantitative ash dispersion modeling: The 2010 Eyjafjallajökull eruption, *Atmos. Chem. Phys.*, *11*(9), 4333–4351; doi:10.5194/acp-11-4333-2011.

Stohl, A., M. Hittenberger, and G. Wotawa (1998), Validation of the lagrangian particle dispersion model flexpart against large scale tracer experiments, *Atmos. Environ.*, *32*, 4245–4264.

Stuefer, M., S. R. Freitas, G. Grell, P. Webley, S. Peckham, S. A. McKeen, and S. D. Egan (2013), Inclusion of ash and SO_2 emissions from volcanic eruptions in WRF-Chem: Development and some applications, *Geosci. Mod. Dev.*, *6*(2), 457–468; doi:10.5194/gmd-6-457-2013.

Theys, N., R. Campion, L. Clarisse, H. Brenot, J. van Gent, B. Dils, S. Corradini, L. Merucci, P. F. Coheur, M. Van

Roozendael, D. Hurtmans, C. Clerbaux, S. Tait, and F. Ferrucci (2013), Volcanic SO_2 fluxes derived from satellite data: a survey using OMI, GOME-2, IASI and MODIS, *Atm. Chem. Phys.*, *13*(12), 5945–5968; doi:10.5194/acp-13-5945-2013.

Thomas, H., and I. Watson (2010), Observations of volcanic emissions from space: Current and future perspectives, *Nat. Hazards*, *54*(2), 323–354; doi: 10.1007/s11069-009-9471-3.

Vernier, J. P., T. D. Fairlie, J. Murray, A. Tupper, C. Trepte, D. Winker, J. Pelon, A. Garnier, J. Jumelet, M. Pavolonis, A. H. Omar, and K. A. Powell (2013), An advanced system to monitor the 3d structure of diffuse volcanic ash clouds, *J. App. Meteor. Clim.*, *52*(9), 2125–2138; doi:10.1175/JAMC-D-12-0279.1.

Vira, J., and M. Sofiev (2012), On variational data assimilation for estimating the model initial conditions and emission fluxes for short-term forecasting of SOx concentrations, *Atmos. Environ.*, *46*, 318–328; doi:10.1016/j.atmosenv.2011.09.066.

Vogel, H., J. Förstner, B. Vogel, T. Hanisch, B. Mühr, U. Schättler, and T. Schad (2014), Time-lagged ensemble simulations of the dispersion of the Eyjafjallajökull plume over Europe with cosmo-art, *Atmos. Chem. Phys.*, *14*(15), 7837–7845; doi:10.5194/acp-14-7837-2014.

Walker, J. C., E. Carboni, A. Dudhia, and R. G. Grainger (2012), Improved detection of sulphur dioxide in volcanic plumes using satellite-based hyperspectral infrared measurements: Application to the Eyjafjallajökull 2010 eruption, *J. Geophys. Res.*, *117*(D20); doi:10.1029/2011JD016810.

Wang, J., S. Park, J. Zeng, C. Ge, K. Yang, S. Carn, N. Krotkov, and A. H. Omar (2013), Modeling of 2008 Kasatochi volcanic sulfate direct radiative forcing: Assimilation of omi SO_2 plume height data and comparison with MODIS and CALIOP observations, *Atmos. Chem. Phys.*, *13*(4), 1895–1912; doi:10.5194/acp-13-1895-2013.

Webley, P. W., and T. Steensen (2013), *Operational Volcanic Ash Cloud Modeling: Discussion on Model Inputs, Products, and the Application of Real-Time Probabilistic Forecasting*, 271–298, American Geophysical Union; doi:10.1029/2012GM001271.

Wen, S., and W. I. Rose (1994), Retrieval of sizes and total masses of particles in volcanic clouds using AVHRR bands 4 and 5, *J. Geophys. Res.*, *99*(D3), 5421–5431; doi:10.1029/93JD03340.

Wilkins, K., M. Watson, H. Webster, D. Thomson, H. Dacre, S. Mackie, and N. Harvey (2014), Volcanic ash cloud forecasting: combining satellite observations and dispersion modelling, *Geophys. Res. Abstr.*, *16*, EGU2014-1615, EGU General Assembly 2014.

Yang, K., N. A. Krotkov, A. J. Krueger, S. A. Carn, P. K. Bhartia, and P. F. Levelt (2007), Retrieval of large volcanic SO2 columns from the Aura Ozone Monitoring Instrument: Comparison and limitations, *J. Geophys. Res.*, *112*(D24), D24S43; doi:10.1029/2007JD008825.

Yang, K., X. Liu, N. A. Krotkov, A. J. Krueger, and S. A. Carn (2009), Estimating the altitude of volcanic sulfur dioxide plumes from space borne hyper-spectral uv measurements, *Geophys. Res. Lett.*, *36*(10); doi:10.1029/2009GL038025.

9

Uncertainty Assessment of Pyroclastic Density Currents at Mount Vesuvius (Italy) Simulated Through the Energy Cone Model

Pablo Tierz,[1] Laura Sandri,[1] Antonio Costa,[1] Roberto Sulpizio,[2] Lucia Zaccarelli,[1] Mauro Antonio Di Vito,[3] and Warner Marzocchi[4]

ABSTRACT

Pyroclastic density currents (PDCs) are extremely dangerous phenomena so their modeling is essential for hazard and risk purposes. However, PDCs are governed by very complex processes, making their deterministic prediction impossible. Probabilistic approaches are in a pioneering phase and feature large (and still unknown) uncertainties, from the natural variability of PDCs (aleatory uncertainty) to the main sources of epistemic uncertainty (input, parametric, theoretical and structural). In this chapter, we quantify these uncertainties by using the Energy Cone Model (ECM) in a Monte Carlo scheme applied to Mount Vesuvius. According to our results, theoretical uncertainty has the largest impact, 5 to 100 times bigger than input uncertainty, which seems to play a minor role. We find that conditional probabilities of PDC arrival (given an eruption of a specific size) show spatial distributions related to the surrounding topography. In particular, for medium and large eruptions, the conditional probability of PDCs traveling beyond Mount Somma is 1%–15% and 50%–60%, while they reach the Napoli airport in about 0%–1% and 0%–15% of the simulations, respectively. Small-eruption PDCs remain restricted to the south flank and summit area. These results may guide future research devoted to reduce epistemic uncertainties and improve volcanic hazard analyses associated with PDCs.

9.1. INTRODUCTION

Phenomena related to explosive volcanic systems threaten life and property of many millions of people around the world. In particular, Pyroclastic density currents (PDCs) are one the most destructive physical phenomena, both in terms of structural damage [*Valentine*, 1998] and threat to life [*Baxter et al.*, 1998]. PDCs are composed of a hot gravity-driven mixture of gas and solid particles, which travels at high speed along the area surrounding the volcanic vent [*Cas and Wright*, 1987; *Druitt*, 1998; *Branney and Kokelaar*, 2002; *Sulpizio et al.*, 2014]. Their high potential damage is due to several causes, such as lateral impact, elevated temperature, fine- and coarse-particle concentration, toxic gases, and so on [*Baxter et al.*, 1998].

During the last 400 yr, PDCs have been responsible for the largest number of fatalities related to volcanic eruptions (≈100,000 lives, 33% of all fatalities; *Auker et al.* [2013]). Among the most devastating events, we recall Mount Pelée, 1902, Martinique [*Fisher et al.*, 1980]; Mount Lamington, 1951, Papua, New Guinea [*Taylor*, 1958]; and El Chichón, 1982, Mexico [*Sigurdsson et al.*, 1984].

[1]*Istituto Nazionale di Geofisica e Vulcanologia, INGV, Sezione di Bologna, Bologna, Italy*

[2]*Dipartimento di Scienze della Terra e Geoambientali, Università di Bari, Bari, Italy and Istituto per la Dinamica dei Processi Ambientali, Consiglio Nazionale delle Ricerche, IDPA-CNR, Milano, Italy*

[3]*Istituto Nazionale di Geofisica e Vulcanologia, INGV, Osservatorio Vesuviano, Napoli, Italy*

[4]*Istituto Nazionale di Geofisica e Vulcanologia, INGV, Sezione di Roma1, Roma, Italy*

Natural Hazard Uncertainty Assessment: Modeling and Decision Support, Geophysical Monograph 223, First Edition.
Edited by Karin Riley, Peter Webley, and Matthew Thompson.

The extreme complexity of PDC generation, transport, and deposition processes makes PDC modeling extremely challenging. For this reason, the first attempts to evaluate the hazard posed by PDCs were mostly based on maps describing PDC deposits from past eruptions [e.g., *Hall et al.*, 1999; *Orsi et al.*, 2004; *Gurioli et al.*, 2010]. On the other hand, Probabilistic Volcanic Hazard Assessment (PVHA) [*Newhall and Hoblitt*, 2002; *Aspinall et al.*, 2003; *Marzocchi et al.*, 2004] requires proper quantitative PDC modeling strategies that are able to describe the inherent complexity of the process (the aleatory uncertainty), and the incomplete knowledge of the system (the epistemic uncertainty). Although the distinction between aleatory and epistemic uncertainties has been often considered inherently ambiguous, we adopt the taxonomy of uncertainties proposed by *Marzocchi and Jordan* [2014] in which aleatory variability and epistemic uncertainty can be unambiguously distinguished. The inclusion of aleatory and epistemic uncertainties is essential for any reliable PVHA and it allows scientists to provide quantitative assessments that can be used in rational decision making [e.g., *Marzocchi et al.*, 2012].

Presently, PVHA has been carried out already for tephra fallout [e.g., *Bonadonna et al.*, 2005], lava flows [e.g., *Del Negro et al.*, 2013], or lahars [e.g., *Sandri et al.*, 2014]. Conversely, only a few studies have looked into a systematic characterization of the uncertainties associated with numerical modeling of PDCs [e.g., *Wadge et al.*, 1998; *Calder et al.*, 2009; *Stefanescu et al.*, 2012a, b] or have produced PVHA of this phenomenon [e.g., *Dalbey et al.*, 2008; *Bayarri et al.*, 2009; *Sheridan et al.*, 2010; *Neri et al.*, 2015]. This is mainly owing to the difficulty to simulate PDCs, in terms of numerical algorithms and computational resources. PDC simulators that aim to reproduce the most detailed physics of the process are computationally expensive [e.g., *Esposti Ongaro et al.*, 2008] and, therefore, not suitable for the exploration of the large uncertainties involved in the hazard assessment.

In this chapter, we opt for a widely used PDC model, the Energy Cone Model (ECM), applied to Mount Vesuvius, in Italy. Explosive eruptions at the volcano have generated dense and dilute PDCs as a result of transient eruption-column collapses, continuous collapses (pyroclastic fountaining), or very energetic phreato-magmatic explosions [e.g., *Cioni et al.*, 2008]. By means of extensive Monte Carlo sampling of the model parameters, we investigate the role of different types of uncertainty in describing the PDC invasion around the volcano. The aleatory uncertainty is addressed by building and sampling probability density functions (PDFs) of the model eruptive parameters (collapse height, H_0, and PDC mobility, ϕ). Epistemic uncertainty is examined by the specific contribution of four distinct sources [after *Rougier and Beven*, 2013]: input, parametric, theoretical, and structural uncertainties.

Input uncertainty refers to the lack of knowledge about boundary conditions, for instance, the real terrain over which PDCs propagate; here we quantify input uncertainty by running equivalent sets of simulations over Digital Elevation Models (DEMs) with different horizontal spatial resolutions. Parametric uncertainty stems from the fact that we do not know exactly the PDFs for sampling the eruptive parameters describing the aleatory uncertainty; here we describe the effects on model output when we use different types of PDFs. Theoretical uncertainty is linked to the assumptions adopted in the simulation strategy, for example, whether considering the model parameters as independent or not; this source of epistemic uncertainty is addressed by testing several possible relationships between the ECM parameters. Finally, structural uncertainty derives from all the simplifications of the model itself; in other words, it is the uncertainty that remains after having run the model using perfect-known boundary conditions and the 'best' parameter values [*Rougier and Beven*, 2013]. We evaluate structural uncertainty by using computed values of misfit between the best set of ECM simulations and past PDC deposits at Mount Vesuvius.

Exploring all these sources of uncertainty allows us to quantify their specific contribution as recorded in the model outputs. We express such a quantification through: (1) empirical cumulative distribution functions (ECDFs) of two important variables in terms of volcanic hazard posed by PDCs: area of invasion and maximum runout; and (2) conditional-probability maps of PDC invasion in the area around Mount Vesuvius (given the occurrence of an eruption of a specific size).

The implications of this study are twofold: on the one hand, it allows for ranking the different types of uncertainty and checking, quantitatively, their effect on the model outputs. On the other hand, it provides a detailed and structured quantification of epistemic uncertainty associated with the simulation of PDCs through the ECM. This can be applied to assess epistemic uncertainty within PVHA tools and, in the end, may help to improve quantitative volcanic risk assessments.

9.2. METHODS

9.2.1. Energy Cone Model

The Energy Cone Model, ECM [*Malin and Sheridan*, 1982], is an analytical model based upon a simplified formulation of the energy conservation equation. The model assumes that PDCs have an initial potential energy (dependent on the initial vertical position of the mass that will form the PDCs), which is transformed into kinetic energy and dissipated by the effective friction with the ground. The pyroclastic current behaves as a gravity-driven, cohesionless suspension of particles and gas with mass remaining constant [*Sheridan*, 1979]. Under these assumptions, the energy loss is simplified into a linear decay, that is, the energy line [*Heim,* 1932]:

$$\left[(H_v + H_0) - h(x)\right] \cdot g = 1/2 \cdot v^2(x) + gx \cdot \tan\phi \quad (9.1)$$

where H_v is the height of the volcanic vent (a.s.l.), H_0 is the height of column collapse (above H_v), $h(x)$ is the topographic height at distance x from the vent, g is the acceleration of gravity, $v(x)$ is the velocity of the PDC at distance x, and $\tan\phi$ denotes the equivalent coefficient of friction, in other words, the PDC mobility (the smaller the ϕ, the more mobile the PDC).

Along a given direction, PDC terminates when all energy has been dissipated, that is, when the energy line cuts the topographic surface (Fig. 9.1). At the point where the flow stops (x_{stop}), located at a horizontal distance from the vent equal to L and at a height, $h(x_{stop}) = H_{stop}$, we have:

$$v^2(x_{stop}) = 0 \Rightarrow (H_v + H_0) - H_{stop} = L \cdot \tan\phi \quad (9.2)$$

Note that L is the PDC runout in a given direction, and so the maximum runout (MR) is equal to max(L) among all the simulated azimuths. In our case, ϕ and H_0 represent the ECM parameters that, besides the starting position for the PDCs and the DEM, are used to perform the simulations. In this study, every simulation starts from the center of the Mount Vesuvius crater since the vent-opening probability over the crater area is considerably much higher than outside it [*Sandri et al.*, 2009]. Finally, our ECM is built through the combination of 360 energy lines, one every 1° of azimuth, covering the whole area around the volcano.

9.2.2. Case Study: Mount Vesuvius Volcano

Mount Vesuvius is a medium-sized stratovolcano located about 15 km away from the city of Napoli in southern Italy (Fig. 9.2). Its present-day edifice displays a complex shape built during the last 40 ka and is highlighted by the presence of Mount Somma (Fig. 9.2), a topographic remnant of several volcanic-edifice collapses that occurred during the last 20 ka [*Cioni et al.*, 1999]. For hazard purposes, we follow the assumption by *Cioni et al.* [2008] and consider the last 20 ka as representative of

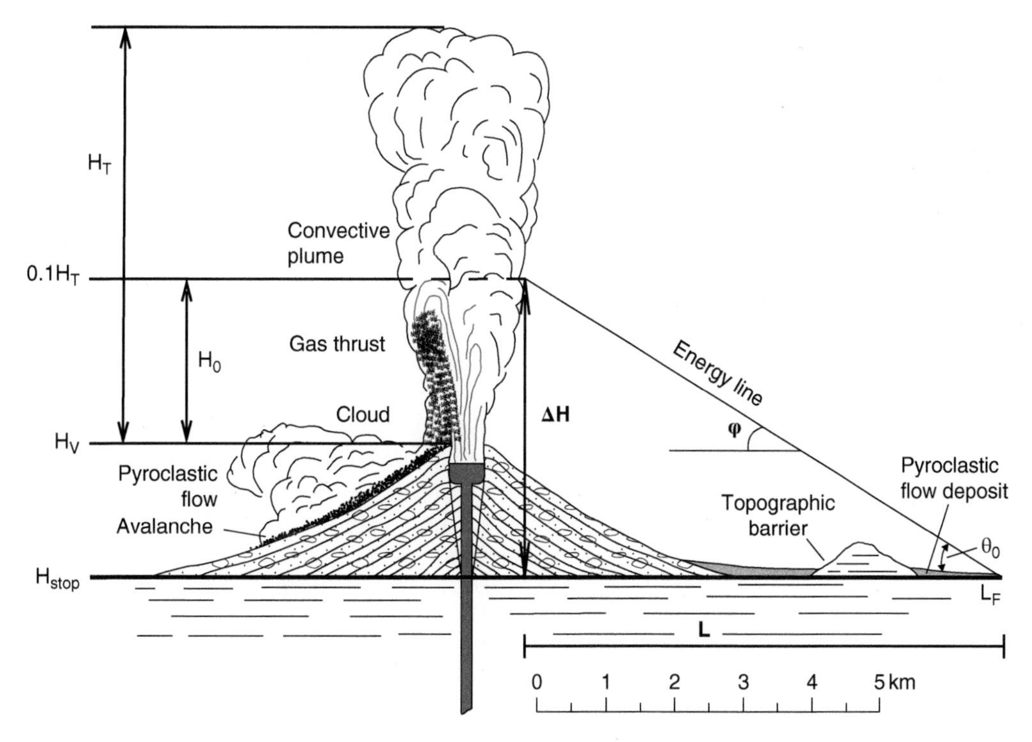

Figure 9.1 The energy line concept. PDC generation and propagation are conditioned by the initial height of collapse (H_0) and the angle between the energy line and the horizontal (ϕ), the latter representing a proxy for PDC mobility. Note that, in the diagram, H_0 coincides with the top of the gas-thrust region, $H_0 \approx 0.1 H_T$ [*Wilson et al.*, 1978]; not to scale [modified from *Wohletz and Heiken*,1992].

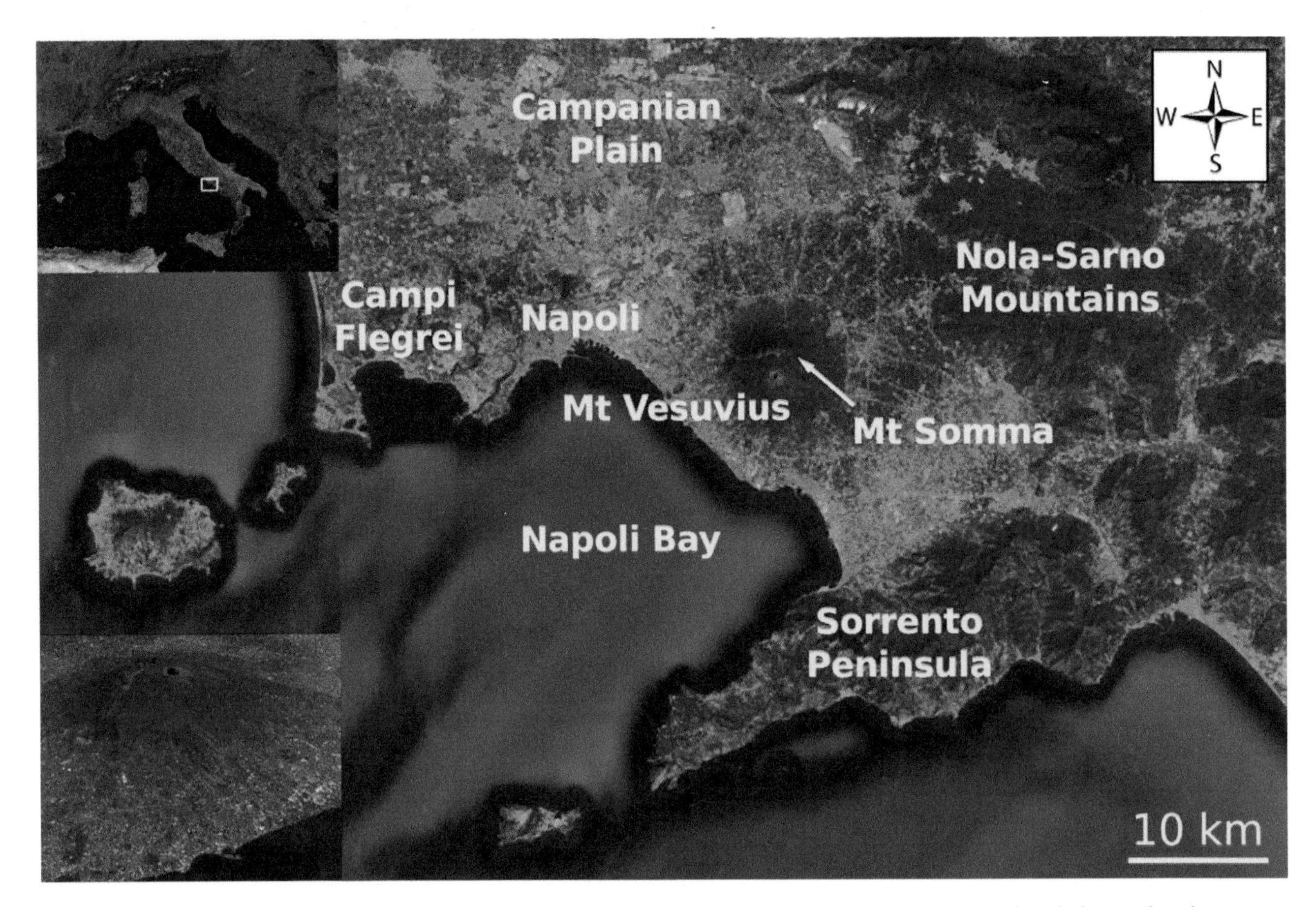

Figure 9.2 Geographical setting of the study area. (Top left) Map of south-central Europe with Italy located in the middle. The yellow square denotes boundaries of the main image where different key locations are identified. The city of Napoli stands on or is surrounded by two principal volcanic systems: Campi Flegrei to the west, and Mount Vesuvius, our target volcano, to the east. The Campanian plain is the tectonic basin where the two volcanic systems originated. Some geomorphological highs, which limit the Campanian plain, are the Nola-Sarno mountains (top right of map) and the Sorrento Peninsula (bottom right of map). Northward from Vesuvius, a remnant from previous edifice collapses is present: Mount Somma. A 3D view from the southwest (Torre del Greco is in the bottom-center of the caption) of Mount Vesuvius cone and Mount Somma rim is given on the bottom left of the figure [Source: Google, DigitalGlobe, 11 December 2014].

what to expect in future eruptions. *Cioni et al.* [2008] defined four eruptive scenarios: Plinian, sub-Plinian I, sub-Plinian II, violent Strombolian eruptions and ash emission events. The first two classes would represent eruptions with Volcanic Explosivity Index, VEI ≥ 5 and VEI4, respectively; while the last three classes could be related to VEI3 eruptions [*Cioni et al.*, 2008]. These VEI classes were proposed by *Marzocchi et al.* [2004] as an exhaustive set of the possible sizes in case of renewal of explosive activity at Mount Vesuvius, given the minimum energy required to reopen the conduit after the current repose time of about 70 yr.

The expected PDCs to occur during Plinian and sub-Plinian eruptions consist of more or less radial PDCs produced by partial and total column collapses, pyroclastic fountaning, or phreatomagmatic explosions. In the case of sub-Plinian II and violent Strombolian eruptions (approximately equivalent to VEI3 and VEI2-3 eruptions, respectively [*Cioni et al.*, 2008]), PDC events can vary from small-volume PDCs formed by column collapse to low-mobility hot avalanches. This variability influences our parametrization of the ECM (see Section 9.2.5).

We hold the same assumption of *Marzocchi et al.* [2004] and consider their three classes in our study (note, we do not consider ash emission events). However, we express eruption size in terms of magnitude (total erupted mass) instead of VEI, since the latter classification assumes there is a correlation between magnitude and intensity (e.g., mass eruption rate [MER], eruption column height), which is not generally valid [*Orsi et al.*, 2009].

9.2.3. Quantification of Aleatory Uncertainty: Intrinsic Randomness

Given a specific eruption size, the randomness in PDC generation (see PDC phenomenology above) can be simulated through a set of possible values for the ECM parameters, ϕ and H_0 (Fig. 9.3). Again, we do not take

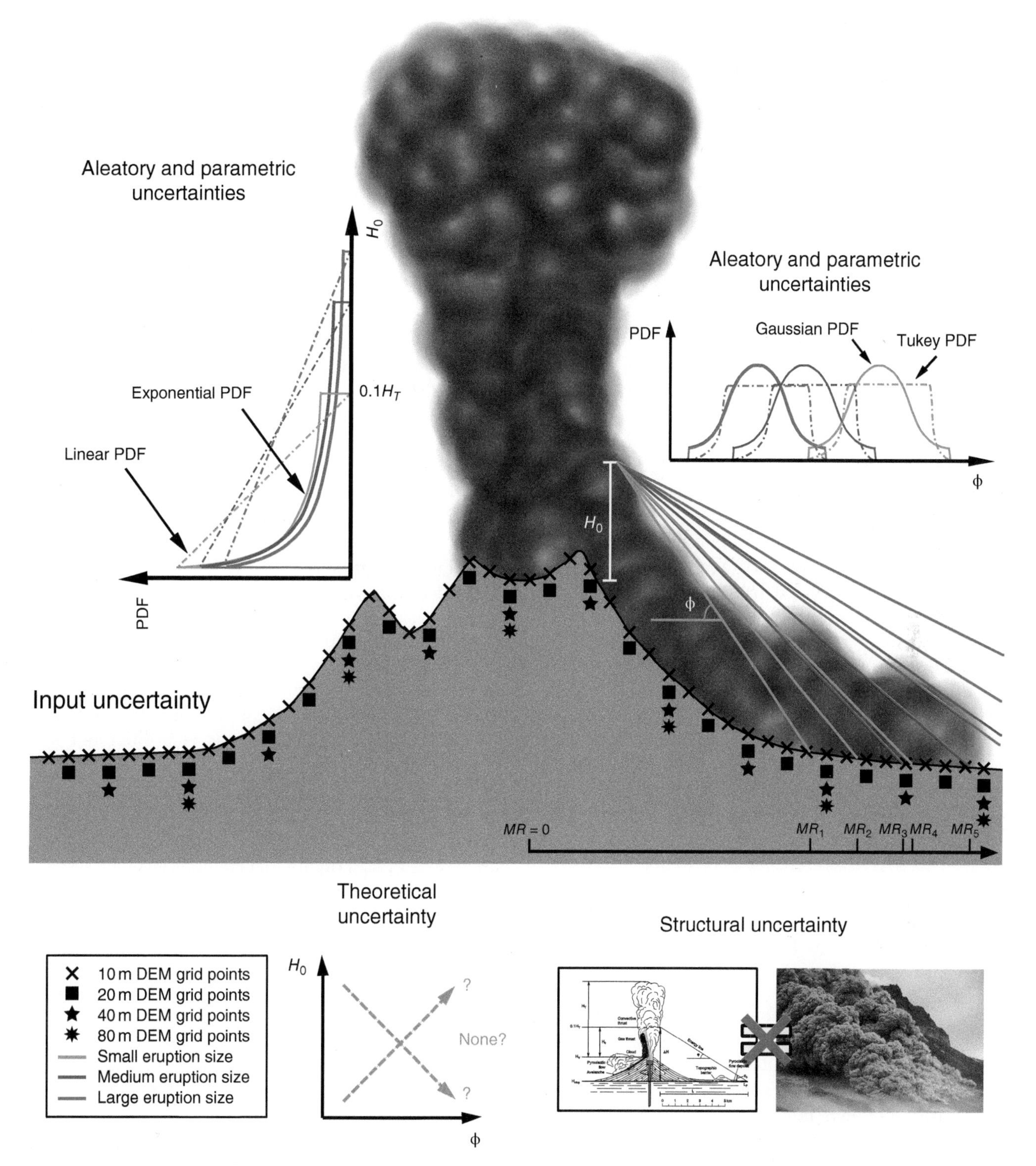

Figure 9.3 Schematic representation of the origin and way of addressing the different types of uncertainty (aleatory and epistemic) quantified in this chapter. Aleatory uncertainty (solid lines) is described through probability density functions (PDFs) of the Energy Cone Model (ECM) parameters: collapse height, H_0, and PDC mobility, ϕ (note that H_0 values are not to scale and ϕ values might seem greater than the actual values used in the chapter). Every simulation provides a value of area of PDC invasion (not shown here) and maximum runout (MR; bottom right of the cartoon). Input uncertainty is explored by running the ECM over Digital Elevation Models (DEMs) with diverse spatial resolutions (also notice that only the horizontal position of the 20, 40, and 80 m DEM grid points has to be considered). Parametric uncertainty is characterized by means of alternative choices for the PDFs (e.g., Tukey or linear PDFs; dotted-dashed lines). Theoretical uncertainty arises from the fact that possible relationships between H_0 and ϕ are not known. Finally, structural uncertainty derives from all the simplifications adopted by simulating the real phenomenon, PDCs, via the ECM (see text for more details).

Table 9.1 Shape Parameters of the Selected Probability Density Functions, PDFs

ECM par →	ϕ (degrees)								H_0 (meters)				
PDF type →	Truncated Gaussian				Asymmetric Tukey Window				Truncated Exponential			Linear Decay	
Eruption size	μ	σ	min	max	a	b	min	max	λ^{-1}	min	max	min	max
Small	23	6	15	36	18	28	15	36	173	20	1000	20	1000
Medium	17	8	8	28	10	25	8	28	441	20	2000	20	2000
Large	12	4	2	20	10	15	2	20	807	20	3500	20	3500

Note: λ^{-1} are mean values for each eruption size. NB. All values are ceil-rounded.
ECM par: ECM parameter; *a*: lower-end of the Tukey plateau; *b*: upper-end of the Tukey plateau.

into account vent locations outside the current crater owing to the very high vent-opening probability over the crater. Nonetheless, a complete PVHA procedure should include the possibility of vents opening on the flanks of the volcano.

In our approach, we account for aleatory uncertainty by describing the ECM parameters through bounded PDFs (Table 9.1; negative or infinity values of the parameters are not physically possible) for each eruption size: Truncated-Gaussian (ϕ) and Truncated-Exponential (H_0) PDFs (see the motivation for these choices in Section 9.2.5).

We propagate aleatory uncertainty by sampling from such PDFs, through a Monte Carlo inversion method [*Tarantola*, 2005], 10,000 pairs of values for ϕ and H_0 per each eruption size. These 30,000 pairs are finally used to run the ECM over a 40 m resolution DEM. This choice is motivated by a potentially wider applicability of the obtained results (high-resolution DEMs are not always available). Nonetheless, as we shall see in Section 9.3 and 9.4, our results indicate that input uncertainty is, by far, the smallest source of epistemic uncertainty.

9.2.4. Quantification of Input Uncertainty: Incomplete Knowledge on Boundary Conditions

Input uncertainty is here explored by running the same set of ECM simulations (as described above in Section 9.2.3) over DEMs with varied horizontal resolutions, namely: 10, 20, 40, and 80 m (Fig. 9.3). This totals 120,000 ECM simulations. The mean vertical errors of the DEMs, although not treated in the analysis, are, respectively, 3.5 [*Tarquini et al.*, 2007], 10, 10, and 10 m. Thus, we assess input uncertainty linked only to sparse topographic data. A more complete quantification of input uncertainty may include the DEM vertical errors, spatial correlation between errors, and could even model the entire DEM as an uncertain variable [e.g., *Stefanescu et al.*, 2012b].

The 10 m DEM is constructed with data downloaded from OpenMap (http://openmap.rm.ingv.it/openmap/) and it is based on data derived from the Italian regional topographic maps, GPS points, ground-based and radar altimetry data [*Tarquini et al.*, 2007]. The 20 m DEM is derived from interpolation of contour lines and spot heights present in the 1:25,000 Italian topographic maps. Finally, the 40 m and 80 m DEMs are derived from resampling of the 20 m DEM. The 10 m DEM is taken as the reference model (i.e., the closest representation of the real terrain) and its associated input uncertainty is assumed to be negligible [*Stefanescu et al.*, 2012a].

9.2.5. Quantification of Parametric Uncertainty: Incomplete Knowledge on Model Parameterization

Parametric uncertainty is examined by imposing different PDFs for ϕ and H_0. In principle, and given the scarce amount of real data available for both parameters, several PDFs could be used. Here, we quantitatively assess how much the ECM outputs change when different PDFs are explored.

9.2.5.1. PDFs for ϕ

Concerning the ϕ parameter, we rely on a worldwide database of PDC mobility that contains volcanic systems morphologically similar to Mount Vesuvius [*Ogburn*, 2013]. We initially explore three different PDFs (in brackets we list their parameters):

1. Uniform (min and max limits). This PDF represents the "maximum ignorance" as no value, within the selected domain, is preferred.

2. Truncated Gaussian (mean, μ; standard deviation, σ; min and max limits). This PDF is chosen as previous works have indicated that Gaussian PDF may be consistent with ϕ data from Volcán de Colima, Mexico [*Sheridan and Macías*, 1995].

3. Asymmetric-Tukey window (high-probability plateau, *a*, *b*; min and max limits; Fig. 9.3 and Table 9.1). This PDF is chosen following the idea that ϕ could follow a distribution characterized by a higher likelihood in its central values.

Then, we perform one-sample Kolmogorov-Smirnov tests [*Massey*, 1951], using ϕ data in *Ogburn* [2013], to test the null hypothesis of such data having been sampled from each PDF. We find that the null hypothesis can only be rejected, at the 1% level of significance, for the Uniform PDF. Therefore, we discard this PDF and use Truncated-Gaussian and Asymmetric-Tukey window PDFs to assess how much they influence the ECM outputs. Their parameters (Table 9.1) are derived from *Ogburn* [2013] after applying some constraints. First, ϕ values are partitioned into VEI ≤ 3, VEI4, VEI ≥ 5 eruptions. As we mentioned above, expected PDCs during VEI ≈ 3 at Mount Vesuvius can show large variability. For this reason, we decide not to exclude the ϕ values related to VEI ≤ 3 eruptions since we consider they are compatible with the PDC phenomenology (e.g., low-mobility hot avalanches) attainable during small eruptions.

Second, data from block and ash flows (BAFs) and ignimbrites are discarded since the occurrence of such types of PDC is unlikely (i.e., their deposits have not been observed in the last 20 ka of stratigraphic record at Mount Vesuvius; *Cioni et al.*[2008]; *Gurioli et al.* [2010]). Thus, we use data only from pumice flows (formed by column collapse) in *Ogburn* [2013].

For each eruption size, Truncated-Gaussian μ and σ are constrained using data from *Ogburn* [2013] and, similarly, Tukey high-probability plateaus are placed, subjectively, between the 20th and 80th percentiles of these samples. The latter implies that 60% of the ϕ values are placed inside the plateau (approximately the same density of probability [$\approx 68\%$] is located within $\mu \pm \sigma$ in a Gaussian PDF). Minimum and maximum ϕ values, for Gaussian and Tukey PDFs, are those found in *Ogburn* [2013].

9.2.5.2. PDFs for H_0

In the absence of real data for H_0, our choice for the PDFs is based on the assumption that high column collapses are less likely than low ones and that column collapse will occur within the gas-thrust region whose top is roughly estimated as 10% of the total height of the eruption column, H_T [*Wilson et al.*, 1978].

The PDFs used are the following (in brackets we list their parameters):

1. Truncated Exponential (mean, λ^{-1}; min and max limits)
2. Linear-decaying (min and max limits)

For each eruption size, we set the minimum and maximum limits of the Linear PDF at 20 m and 0.1 H_T, respectively (Table 9.1). Nonetheless, alternative choices of the minimum limit, at $H_0 = 10$ m and $H_0 = 50$ m, are tested for consistency and no significant differences can be recognized in the ECM outputs. The distribution for H_T is derived from eruption column simulations at Mount Vesuvius [*Selva et al.*, 2014]. In that work, H_T values were calculated from MER [*Mastin et al.*, 2009], after having sampled, from proper PDFs, values of total erupted mass and eruption duration.

The parameter λ defining the Truncated Exponential PDF is inferred by assuming that the value 0.1 H_T (i.e., the top of the gas-thrust region) marks the 95th percentile[1] of the corresponding non-Truncated Exponential PDF (i.e., $f(x) = \lambda e^{-\lambda x}$). The obtained Exponential PDFs are then truncated and renormalized between $H_0 = 20$ m and $H_0 = 0.1 H_T$.

9.2.5.3. Combinations of ϕ and H_0 Sampled Pairs

For each eruption size, we test three different combinations of PDFs. The combination Gaussian-Exponential is taken as reference since: (1) Gaussian PDFs have been previously proposed as the PDF for ϕ [*Sheridan and Macías*, 1995]; and (2) the consideration that many natural phenomena follow exponential-like distributions.

In order to highlight the deviations from this reference combination, we test only its alternatives: (1) Tukey-Exponential and (2) Gaussian-Linear combinations. Again, 10,000 pairs of ϕ-H_0 (per eruption size) are sampled from each combination via the Monte Carlo inversion method. In total, 90,000 simulations of the ECM are run.

9.2.6. Quantification of Theoretical Uncertainty: Incomplete Knowledge on Theoretical Assumptions

PDC mobility stands as an important variable to assess the potential extent of PDCs at a given volcanic system. However, there is not a complete agreement upon either, which is the principal factor controlling ϕ or what might be the relationship (if any) with other variables: for example, PDC volume [*Siebert*, 1984; *Calder et al.*, 1999; *Toyos et al.*, 2007], ground surface characteristics [*Sheridan et al.*, 2005], ground slope [*Charbonnier and Gertisser*, 2012], or column collapse height [*Esposti Ongaro et al.*, 2008; *Doyle et al.*, 2010].

Here, we estimate theoretical uncertainty by exploring three different hypotheses on the relationship between ϕ and H_0 within each eruption size[2]:

1. Direct pattern: column collapse height and PDC mobility are directly related, that is, the higher the collapse, the smaller the value of ϕ. For example, the dilute part of PDCs in the model developed by *Doyle et al.* [2010] would follow this direct pattern.
2. Inverse pattern: column collapse height and PDC mobility are inversely related, that is, the higher the collapse, the larger the value of ϕ. *Esposti Ongaro et al.* [2008] simulations for a VEI4 scenario at Mount Vesuvius would agree with this inverse pattern.
3. Independent pattern: no relationship exists between collapse height and PDC mobility (at least within a given eruption size). Data in *Ogburn* [2013] as well as in *Siebert* [1984] would support the independent pattern (i.e., the ECM parameters are not related).

Our description of aleatory uncertainty comes from the independent pattern. Direct- and inverse-pattern are obtained as subsets of such description. In order to obtain reasonable subsets sizes, the parameter space (ϕ-H_0) is divided into the nine sectors shown in Figure 9.4. Boundaries are, subjectively, placed at the 20th and 80th percentiles of the ϕ-H_0 sample distributions in order to ensure that the parameter space is properly separated into regions of low ϕ-high H_0, high ϕ-high H_0, and so forth.

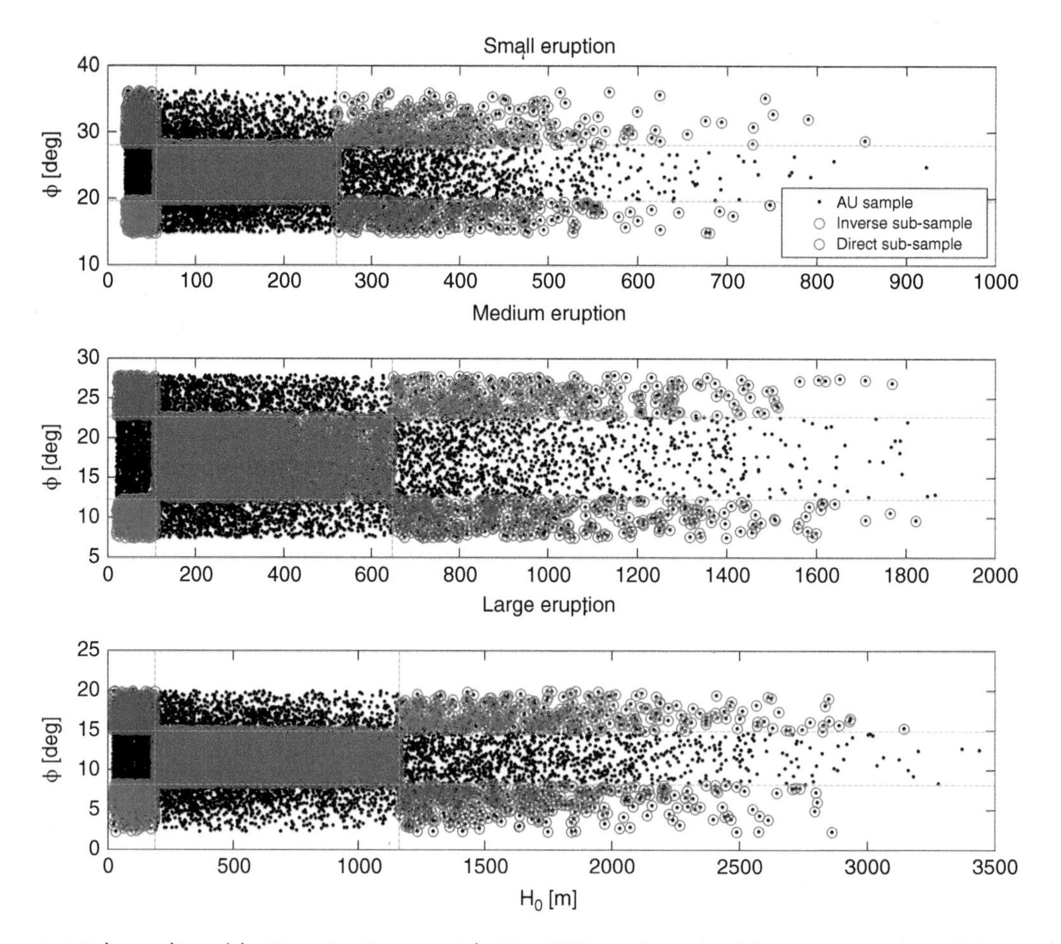

Figure 9.4 Subsampling of the Gaussian-Exponential, 40 m DEM configuration (aleatory uncertainty, AU) to explore theoretical uncertainty. Each graph delineates the ECM parameter space according to three eruption sizes: small (top), medium (middle), and large (bottom). Black dots indicate pairs of ϕ and H_0 as sampled from the aleatory-uncertainty configuration. Red open circles denote the inverse-pattern subsample and purple open circles denote the direct-pattern subsample. Magenta circles represent points shared by both inverse and direct subsamples.

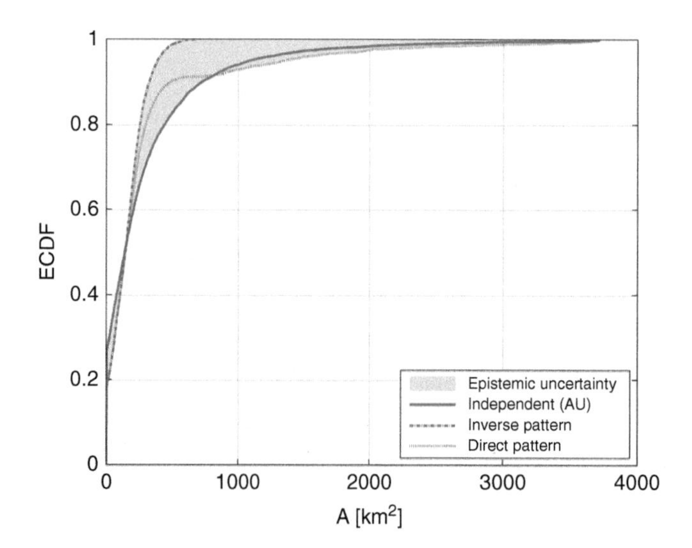

Figure 9.5 Epistemic uncertainty description taking as example the case of theoretical uncertainty and area of PDC invasion. Aleatory uncertainty (AU) corresponds to the solid red line (independent pattern). Epistemic uncertainty is defined as the pale red area between the three empirical cumulative distribution functions (ECDFs). Horizontal distances between the curves demark possible ranges of the output variable considering both aleatory and epistemic uncertainties (see text for more details).

However, both the direct- and inverse-pattern ϕ-H_0 samples are continuous throughout the parameter space since the magenta areas in Figure 9.4 indicate ϕ-H_0 values that belong to both patterns. Finally, theoretical uncertainty is quantified using the ECM outputs from each pattern and building output ECDFs accordingly (Fig. 9.5).

9.2.7. Quantification of Structural Uncertainty: Incomplete Knowledge Reflected in the Model

Structural uncertainty is quantified by comparing the best ECM outputs against real data that, in our case, are the PDC deposits preserved from past eruptions. Here, six eruptions of Mount Vesuvius are used: the 1944 AD eruption [*Cole and Scarpati*, 2010] for the small eruption size; Pollena eruption, 472 AD [*Sulpizio et al.*, 2007], and 1631 AD eruption [*Rolandi et al.*, 1993] for the medium eruption size; and Mercato eruption, 8540 ± 50 cal. yr BP [*Mele et al.*, 2011], Avellino eruption, 3945 ± 10 BP [*Sulpizio et al.*, 2010a; *Sevink et al.*, 2011], and Pompeii eruption, 79 AD [*Carey and Sigurdsson*, 1987], as representative of large eruptions.

In a similar way to the validation carried out by *Tierz et al.* [2016], we first compute a measure of misfit on area of PDC invasion, $\mathcal{M}_{A_{ij}}$, associated with the best 250 simulations (5% of the total[3]), that is, those with highest areal fit, $AF = A_{inter}/A_{tot}$ (where A_{inter} is the intersection area between the simulated area, A_{sim}, and the area of the PDC deposit, A_{obs}; and A_{tot} is the union area between A_{sim} and A_{obs}), per each selected eruption[3]:

$$\mathcal{M}_{A_{ij}} = A_{sim_{ij}} - A_{obs_j} \quad i = 1,\ldots,250 \quad j = 1,\ldots,N_{erup} \qquad (9.3)$$

where $A_{sim_{ij}}$ is the area recorded in the i th best simulation of a specific eruption size and eruption j, A_{obs_j} is the area of the PDC deposits preserved from the j th eruption of this size and N_{erup} is the number of eruptions for this size.

The simulations are extracted from the configuration: Gaussian-Exponential PDFs run over 10 m DEM (i.e., the other sources of epistemic uncertainty are minimized, *Rougier and Beven* [2013]). We repeat the same process for the misfit on maximum runout ($\mathcal{M}_{MR_{ij}}$), selecting the 250 simulations, per eruption, with the closest maximum runout to the real observations.

We then build ECDFs based upon the obtained misfits of area and maximum runout according to the number of eruptions in each eruption size (e.g., the misfit distribution for the large eruption size contains 750 values of misfit). We add the misfit values (which can be negative[4]) along the whole aleatory uncertainty output ECDF and compute the final quantification of structural uncertainty from the minimum and maximum alternate ECDFs (Fig. 9.6).

This way of quantifying structural uncertainty is neither unique nor exhaustive. We note that the use of alternative PDC simulators for characterization of processes not accounted for in the ECM would improve the structural uncertainty quantification. In addition, our misfit distributions are based upon a reduced number of realizations of the "true model" (the PDC deposits) and therefore we are adding a source of aleatory uncertainty within our quantification. Nevertheless, we consider that even a preliminary quantification of structural uncertainty is still better than assuming that the model is perfect [*Rougier and Beven*, 2013].

9.3. RESULTS

Our results are visualized in two different ways: (1) plots of ECDFs for invaded area and maximum runout; and (2) conditional-probability maps of PDC arrival (given an eruption of a specific size) over the surroundings of Mount Vesuvius. As introduced before, aleatory uncertainty is described by the range in model outputs, in terms of invaded area and maximum runout, derived from the configuration: "Gaussian-Exponential, 40 m DEM, independent pattern" (Fig. 9.7). Epistemic uncertainty (Fig. 9.8) is quantified by the range in model outputs obtained by: (1) running the ECM with "Gaussian-Exponential, independent" samples over different DEMs (input uncertainty); (2) sampling the ECM parameters from alternate configurations (parametric uncertainty); and (3) analyzing specific subsets of the aleatory uncertainty simulations (theoretical and structural uncertainties).

Figure 9.5 exemplifies our uncertainty quantification in terms of ECDFs. Aleatory uncertainty (due to randomness in the process) is given as the probability: Prob($X \leq x$); where X represents the ECM output variable (e.g., invaded area). Hence: Prob ($X_{large} \leq 450 \text{ km}^2$) = 0.8 is a measure of aleatory uncertainty. In other words, the quantification of aleatory uncertainty is provided by different percentiles in the model output variable (in the above example, 450 km^2 is the 80th percentile). Epistemic uncertainty (due to incomplete knowledge) is measured by the range in model outputs for any given percentile. In the example above, the 80th percentile is in the range [250,450] km^2, when accounting for epistemic uncertainty.

Figure 9.9 shows the spatial distribution of aleatory and epistemic uncertainties over the surroundings of Mount Vesuvius. In the leftmost column, we plot the conditional probability of PDC arrival, at each grid point, given the occurrence of an eruption of a specific size ($CP = Prob\ [PDCs(x,y) \mid Eruption,\ Size]$); as computed from the aleatory uncertainty simulations. Examples of epistemic uncertainty (Fig. 9.9; right-hand side columns) are given as differences in conditional probability, ΔCP, among aleatory and epistemic uncertainty simulations. For the sake of brevity, we focus on the configurations that lead to the highest absolute ΔCP values (Fig. 9.9).

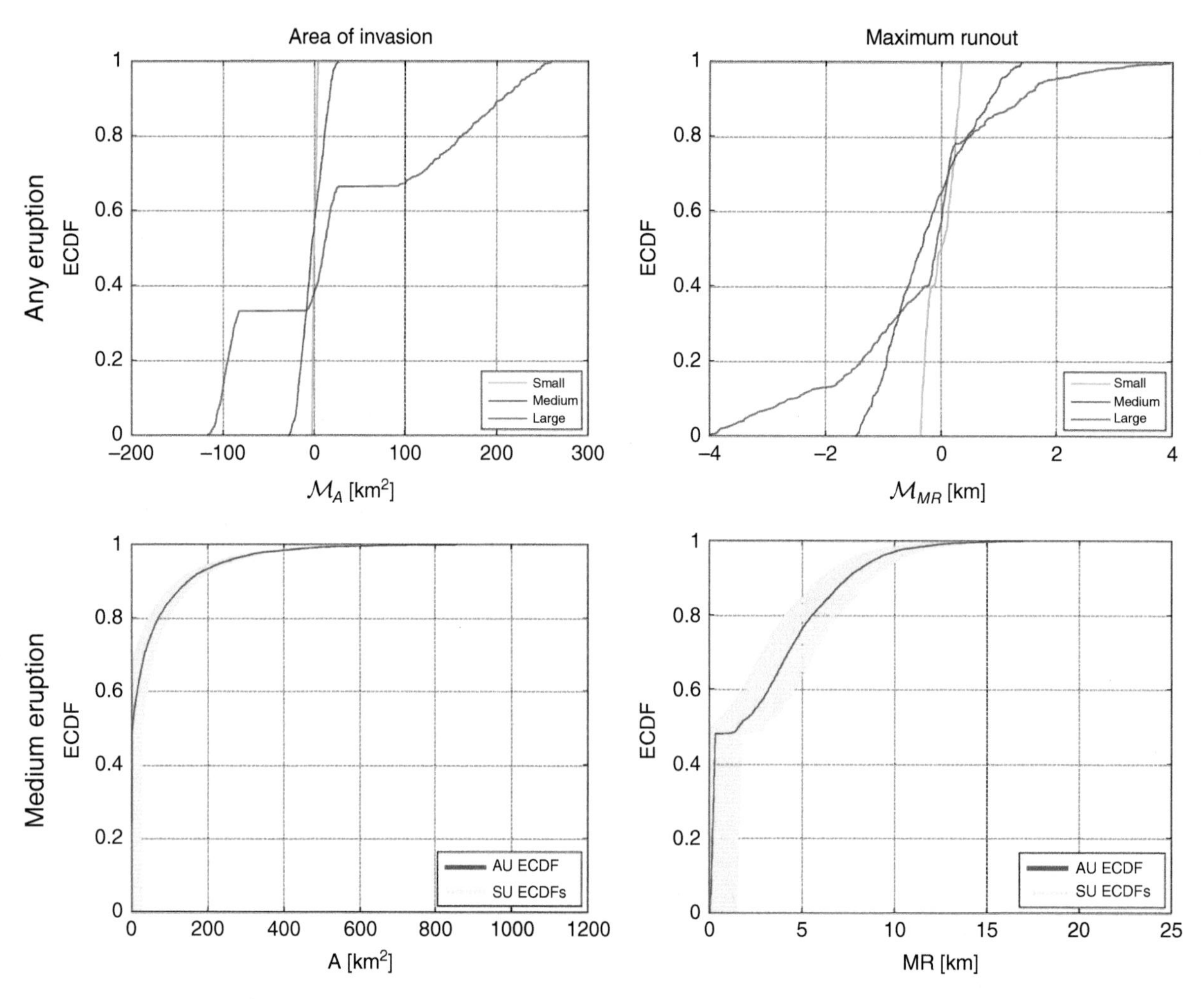

Figure 9.6 Structural uncertainty (SU) description. (Top) Misfit distributions of area of invasion (top left) and maximum runout (top right) according to three different eruption sizes at Mount Vesuvius (Italy): small (green), medium (purple), and large (red). (Bottom) Medium-sized structural-uncertainty representation in the form of SU ECDFs (dashed cyan lines) computed from aleatory uncertainty (AU ECDF, solid purple line) by adding distinct values of areal misfit, M_A (bottom left), and maximum-runout misfit, M_{MR} (bottom right), picked up from the graphs on top (see text for details). Note that the horizontal distance between the minimum-maximum SU ECDFs in the bottom graphs corresponds to the domain of M_A and M_{MR} (top graphs) (i.e., around 50 km^2 and 3 km, respectively, in the medium eruption size).

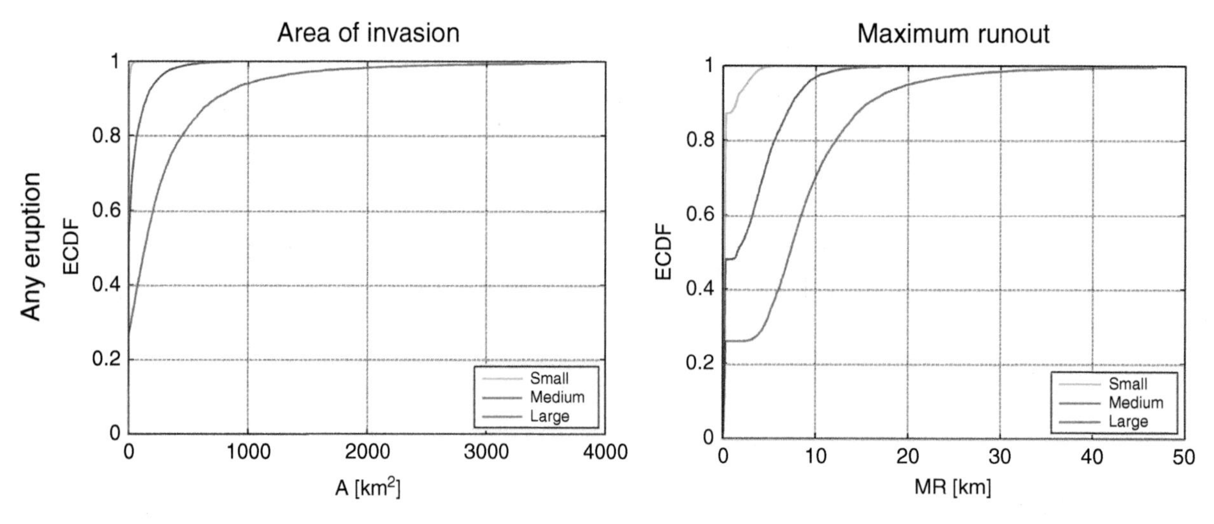

Figure 9.7 Aleatory uncertainty description, in terms of output empirical cumulative distribution functions (ECDF) for area of invasion and maximum runout (*A* and *MR*) of pyroclastic density currents simulated with ECM at Mount Vesuvius (Italy), according to three different eruption sizes: small (green), medium (purple), and large (red).

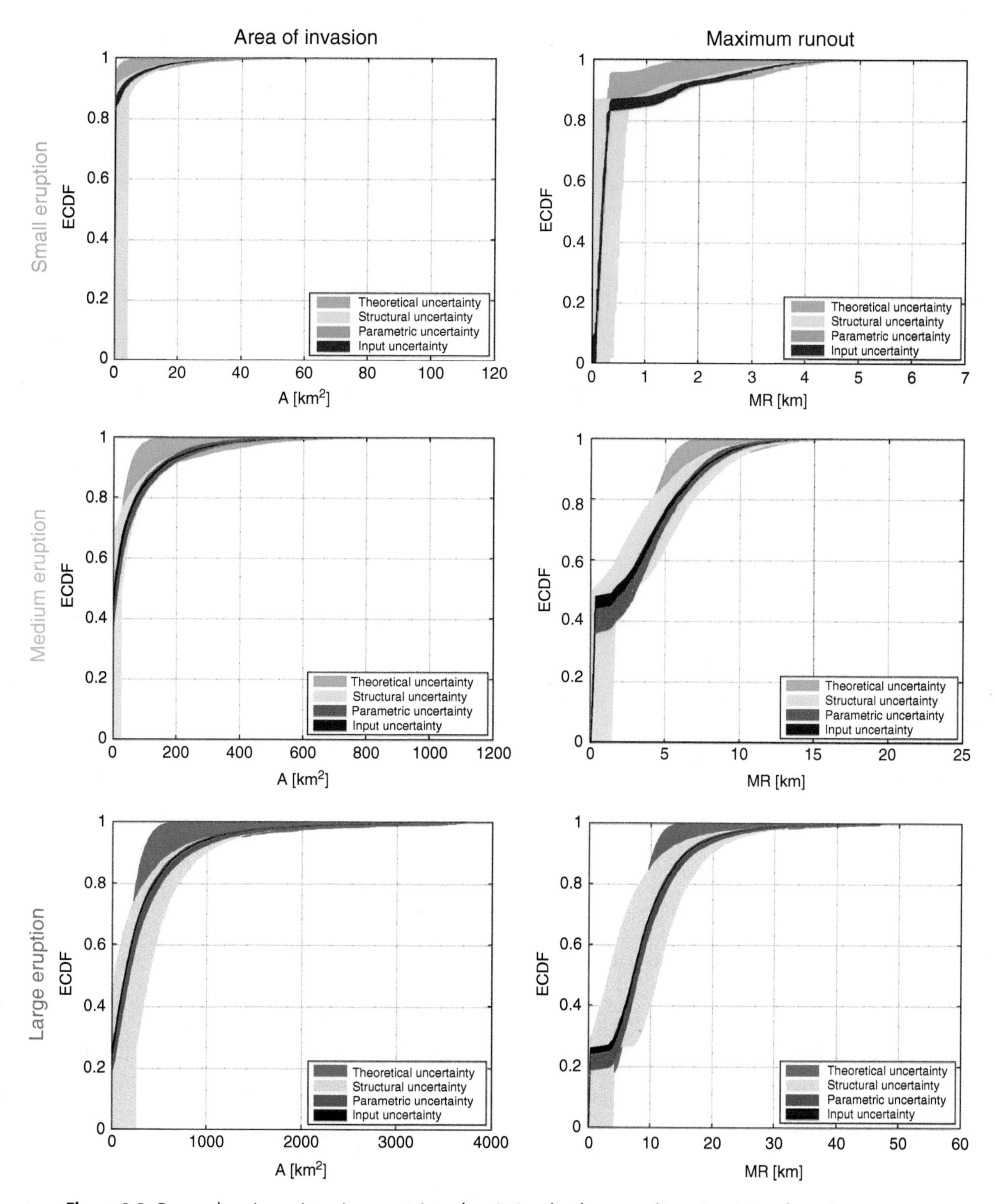

Figure 9.8 Comprehensive epistemic uncertainty description for the area of invasion (A) and maximum runout (MR) of pyroclastic density currents at Mount Vesuvius (Italy) according to three different eruption sizes: small (green), medium (blue), and large (red). Aleatory uncertainty (Fig. 9.7) lies inside the band of input uncertainty. Note how the contribution of each type of epistemic uncertainty to the total uncertainty changes along the graphs.

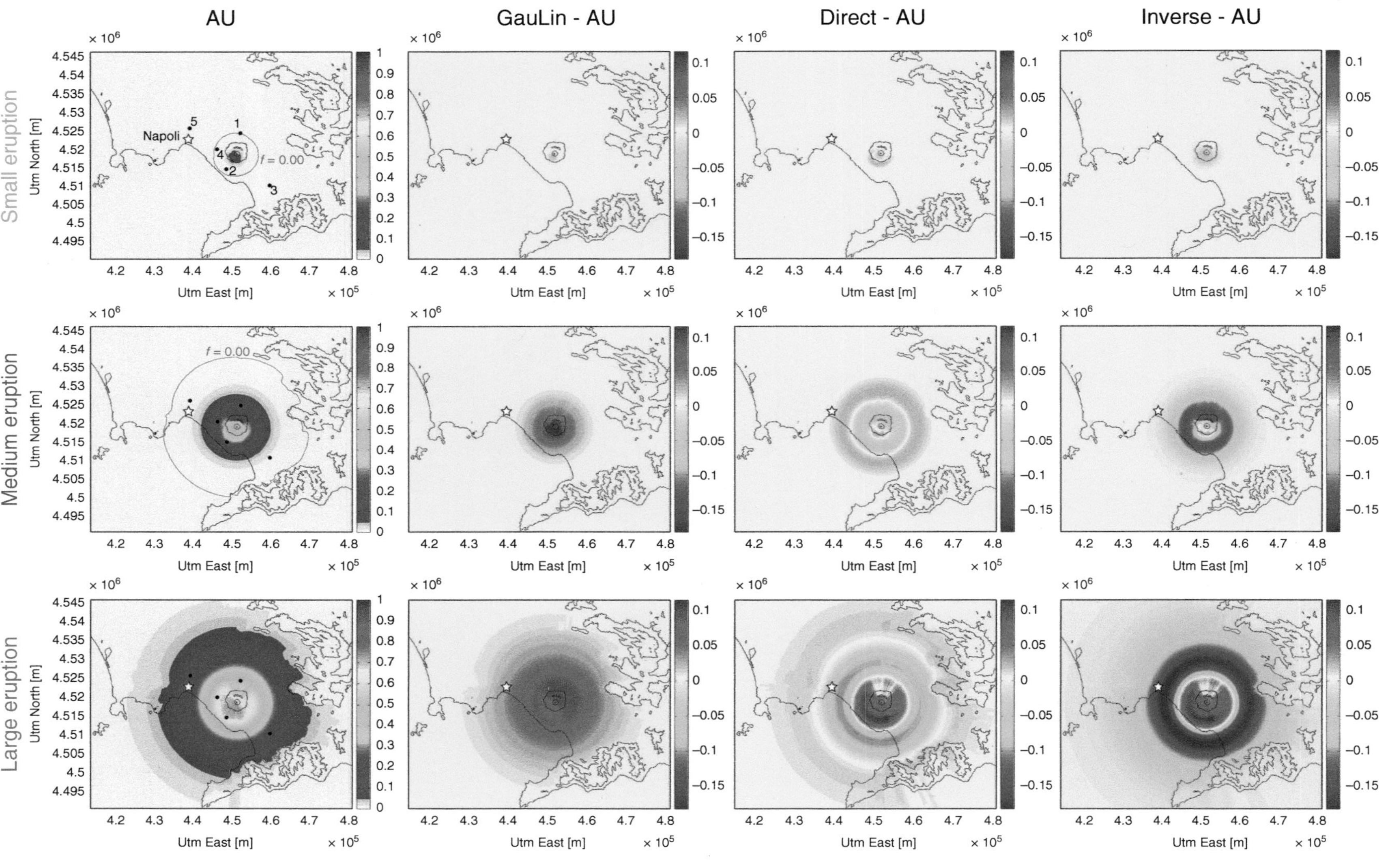

Figure 9.9 Conditional-probability (*CP*) maps of PDC arrival (given the occurrence of an eruption of a specific size) computed by running the Energy Cone Model at Mount Vesuvius, Italy, according to three different eruption sizes: small (green text), medium (blue text), and large (red text). Leftmost-side maps are actual conditional-probabilities of PDC arrival (expressed between 0 and 1) while the other three maps in each eruption size show the differences in conditional probability, at each grid point, between alternate configurations (Gaussian-Linear, 40 m DEM, independent pattern -second column-; Gaussian-Exponential, 40 m, direct pattern -3rd column-; and Gaussian-Exponential, 40 m, inverse pattern -4th column-) and the aleatory-uncertainty (AU) configuration (see text for more details). Colored zones in aleatory-uncertainty maps indicate conditional-probabilities greater than 0.05 and the solid red line displays the limit of $CP > 0$ (notice there is no red line in the large aleatory-uncertainty map). The white star indicates the location of the city of Napoli (1: Somma Vesuviana; 2: Torre del Greco; 3: Scafati; 4: hospital in Massa di Somma; 5: Napoli-Capodichino airport).

9.3.1. Aleatory Uncertainty

The aleatory uncertainty ECDFs (Fig. 9.7) show that, respectively for small, medium and large eruptions, 85%, 50%, and 25% of the ECM simulations correspond to PDCs with negligible area of invasion and maximum runout (i.e., they remain almost inside the crater). The simulations that escape the crater provide areal values rapidly increasing to reach maxima around 100, 1000, and 3700 km^2, respectively. In the case of maximum runout, the increase is not continuous. A lack of values (nearly horizontal ECDFs) occurs between 300 m and 1, 1.5, and 3 km of maximum runout in small, medium, and large eruption sizes, respectively. This is due to heterogeneities in crater-rim altitude and the fact that, for a given simulated energy cone, a small number of energy lines might escape the crater while others remain inside it. This produces an abrupt increase in the maximum runout (but not in the invaded area) of the simulation. Maximum runout values reach maxima around 6, 19, and 47 km, for small, medium, and large eruptions, respectively.

Extreme values of both output variables are restricted to very high percentiles. For instance, considering large eruptions, $MR \approx 30$ km occurs around the 99th percentile while $A \approx 1000$ km^2 occurs around the 95th percentile. In other words, 1% of the simulations record $MR \geq 30$ km while 5% of the simulations cover $A \geq 1000$ km^2.

Regarding the conditional probability, the contrast among eruption sizes is obvious (Fig. 9.9). Small-size simulations are restricted to the southern flank of the volcano, cover small areas, and have modest runouts. Medium-size simulations cover larger areas with moderate conditional probabilities ($CP > 0.4$) remaining over the south, east, and west flanks of Mount Vesuvius. The area with $CP \geq 0.05$ (colored area) has a radius of about 9 km around the volcano. An area similar to this is covered by $CP \geq 0.4$ in large-size simulations where, in turn, the $CP \geq 0.05$ area shows a radius of ≈ 22 km. Nonetheless, distal topographic highs such as the Sorrento Peninsula and the Nola-Sarno Mountains influence the runout blocking the simulated PDCs. Almost the entire city of Napoli lies inside the $CP \geq 0.05$ area. The topographic effect of the natural barrier of Mount Somma is recognizable for all eruption sizes (Fig. 9.9).

9.3.2. Input Uncertainty

Input uncertainty, investigated using differing DEM resolutions, is the source of epistemic uncertainty with the smallest contribution, taking into account all sizes and both output variables. Quantitatively, its maximum absolute values are, respectively, $\approx$ 3, 7, 25 km^2 for area of invasion and $\approx$ 1, 1.5, 3.5 km for maximum runout in small, medium, and large eruptions. These values are hardly distinguishable, particularly in the medium- and large-size ECDFs of area of invasion. In terms of maximum runout, input uncertainty has a larger effect on total epistemic uncertainty. Its maximum extent coincides with the first simulations exiting the crater and this is recognized in all eruption sizes (Fig. 9.8).

Due to its very small contribution, we do not show any conditional-probability map for input uncertainty, since these maps are very similar to those calculated from aleatory uncertainty (i.e., $\Delta CP \sim 0$).

9.3.3. Parametric Uncertainty

Parametric uncertainty, investigated using different PDFs for the model parameters, has a larger impact than input uncertainty on both output variables and, mostly, in medium and large eruptions. In small eruptions, parametric and input uncertainties are superposed (Fig. 9.8). Maximum parametric-uncertainty absolute values are about 20, 300, 700 km^2 for area of invasion and 1, 4, 8 km, for maximum runout in small, medium, and large sizes, respectively.

Regarding the conditional-probability maps, we display only the Gaussian-Linear (40 m, independent pattern) simulations, which show clear positive ΔCP values over the whole map (Fig. 9.9). These ΔCP values are bigger in the vicinity of the volcanic vent and get smaller at increasing distances. For large eruption simulations, ΔCP is approximately +0.1 over the whole north sector (Fig. 9.9), which implies that most of the Gaussian-Linear large-size simulations produce PDCs overcoming Mount Somma (i.e., the conditional probability over the north flank is similar to that on the other flanks of the volcano).

9.3.4. Theoretical Uncertainty

Theoretical uncertainty, investigated analyzing different relationships between the model parameters, gives the largest contribution to total epistemic uncertainty. Maximum theoretical uncertainty, in absolute terms, reaches around 90, 850, 3000 km^2 for area of invasion and 3, 11, 33 km for maximum runout in small, medium, and large eruptions, respectively. This is due to two reasons: (1) the inverse-pattern ECDF has a much narrower range of possible (smaller) values of area of invasion and maximum runout than the aleatory-uncertainty output ECDF; and (2) the direct-pattern ECDF shows a significant break in slope (Fig. 9.5) and its tail is slightly heavier than that seen in the aleatory-uncertainty output ECDF.

Considering the conditional-probability maps, the direct and inverse patterns display remarkably different features as well. Direct-pattern difference maps are characterized by a common spatial distribution of ΔCP, independent of eruption size (Fig.9.9): proximal areas display positive ΔCP values; in medial areas (e.g., about 4 km and 7 km from the crater in medium and large sizes, respectively), we observe negative ΔCP values; and, then, at more distal locations (e.g., beyond the $CP \geq 0.05$ limit in aleatory uncertainty), ΔCP changes sign again, although the positive differences here are smaller than in proximal areas. By and large, Mount Somma barrier seems to prevent PDC propagation northward. The inverse-pattern maps also display positive ΔCP values over proximal sectors. However, beyond the positive-negative ΔCP boundary (Fig. 9.9), ΔCP gets strongly negative, decaying to smaller negative ΔCP values only at distal locations. This reflects the fact that no inverse-pattern simulation reaches medial and distal areas.

9.3.5. Structural Uncertainty

In regard to structural uncertainty, we present only ECDFs since this type of epistemic uncertainty cannot be directly mapped in this study. Here, we use the "best" ECM simulations to characterize the misfit distributions and we apply the latter to build the final quantification of structural uncertainty (see Section 9.2.7).

However, estimating the effect of structural uncertainty on conditional-probability maps would require a procedure able to incorporate this source of epistemic uncertainty in the model itself, maybe as some sort of asymmetric buffer depending on the surrounding topography [e.g., *Widiwijayanti et al.*, 2009]. This propagation of structural uncertainty to the conditional-probability maps should rely upon misfit distributions, as previously estimated.

In terms of ECDFs, structural uncertainty is the most uniform source of epistemic uncertainty (Fig. 9.8). The maximum absolute values can get as high as ≈ 6, 50, 380 km^2 for area of invasion and ≈ 0.7, 3, 8 km for maximum runout in small, medium, and large eruptions, respectively.

9.4. DISCUSSION

Computing PVHA is time- and resources-consuming and not all numerical simulators or uncertainty quantification techniques are suitable for this purpose. The combined use of the ECM and Monte Carlo sampling supplies a comprehensive description of aleatory and epistemic uncertainties for PDCs at Mount Vesuvius, Italy (Figs. 9.8, 9.9 and 9.10), that may be incorporated into PVHA.

9.4.1. A Comprehensive Uncertainty Description

In order to compare the different sources of epistemic uncertainty among them and across eruption sizes, we calculate the relative maximum expected deviation as: $\delta_{ij} = \Delta_{ij} / x_{50_i}$, where, for a given size i, Δ_{ij} is the maximum horizontal distance between the aleatory-uncertainty ECDF and the alternate ECDFs of the j th source of epistemic uncertainty; and x_{50_i} is a common reference value: the 50th percentile (or median) of the aleatory-uncertainty ECDF, again for size i. Note that $\delta_{ij} > 1$ implies that maximum expected deviation is greater than the median, x_{50_i}.

Concerning area of invasion, $\delta_{A_{ij}}$ spans from 10^{-1} to 10^3 considering all sources. The largest deviations occur on parametric and theoretical uncertainties in all eruption sizes but especially in the small and medium sizes where $\delta_{A_{ij}}$ reaches values on the order of 10^2–10^3 (Table 9.2). Among all types of epistemic uncertainty, input uncertainty has the smallest effect on the areal outputs.

In the case of maximum runout, the obtained values are much more homogeneous, spanning only two orders of magnitude ($\delta_{MRij} \sim 10^{-1}$ to 10^1), considering all sources. Again, small and medium sizes show greater deviations in the outputs with respect to the large size. Theoretical uncertainty is confirmed to give the largest contribution. Input and structural uncertainties exhibit the smallest deviations even though the former plays a role in the small-size epistemic uncertainty, probably due to a stronger interaction of its simulated PDCs with the proximal topography (e.g., the crater and Mount Somma caldera rim). Figure 9.10 shows a south-north profile

Table 9.2 Relative Maximum Expected Deviations From Aleatory Uncertainty Considering Area of PDC Invasion and Maximum Runout (δ_A, δ_{MR}) and Four Different Sources of Epistemic Uncertainty

Variable →	δ_A				δ_{MR}			
Eruption size	IU	PU	TU	SU	IU	PU	TU	SU
Small	32.34	432.6	1168	53.81	5.595	5.404	18.37	1.921
Medium	3.569	130.7	402.2	12.87	0.927	2.073	7.312	0.926
Large	0.201	4.312	22.48	1.889	0.491	0.993	4.529	0.550

Note: IU = input, PU = parametric, TU = theoretical, SU = structural.

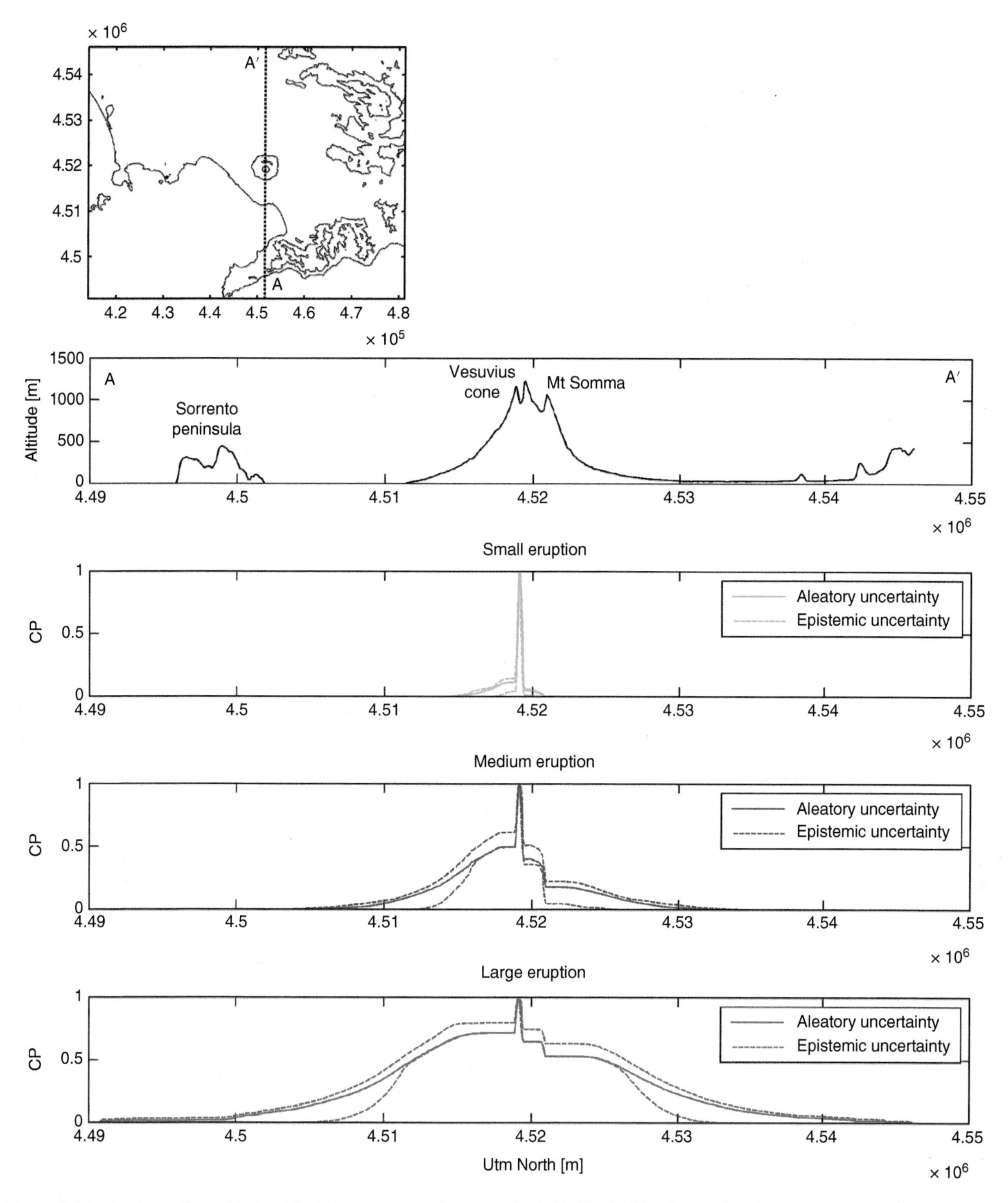

Figure 9.10 South-north profile (A-A′) of the study area (map on the left). (Top) Altitude profile where some regional topographic highs are identified. The three other profiles, from top to bottom, show the conditional probabilities of PDC invasion computed from ECM simulations at Mount Vesuvius and taking into consideration three eruption sizes: small (green), medium (purple), and large (red). Solid lines indicate the value of conditional probability as calculated from aleatory-uncertainty simulations. Dashed lines denote the minimum and maximum values of conditional probability taking into account all sources of epistemic uncertainty but the structural uncertainty (see text for more details). The link between topography and PDC invasion can be visualized in all three eruption sizes.

Table 9.3 Two-Sample Kolmogorov-Smirnov Tests Performed by Comparing the Aleatory-Uncertainty Output Empirical Cumulative Distribution Function (ECDF) Against Diverse Alternate Output ECDFs (Epistemic-Uncertainty Configurations) for Two Different Output Variables (Area of Invasion and Maximum Runout) and Three Eruption Sizes at Mount Vesuvius (Italy)

ECM output →	Area of PDC invasion						Maximum runout					
Eruption →	Small		Medium		Large		Small		Medium		Large	
ECDF	Null	*p*	Null	*p*	Null	*p*	Null	*p*	Null	*p*	Null	*p*
gauexp10ind	1	<1e-5	1	<1e-5	1	0.026	1	<1e-5	1	<1e-5	1	<1e-5
gauexp20ind	1	<1e-5	1	0.003	0	0.144	1	<1e-5	1	<1e-5	1	<1e-5
gauexp80ind	1	<1e-5	1	<1e-5	1	<1e-5	1	<1e-5	1	<1e-5	1	<1e-5
tukexp40ind	0	0.927	0	0.287	0	0.374	0	0.871	0	0.058	0	0.374
gaulin40ind	1	<1e-5	1	<1e-5	1	<1e-5	1	<1e-5	1	<1e-5	1	<1e-5
gauexp40inv	1	<1e-5	1	<1e-5	1	<1e-5	1	<1e-5	1	<1e-5	1	<1e-5
gauexp40dir	1	<1e-5	1	<1e-5	1	<1e-5	1	<1e-5	1	<1e-5	1	<1e-5
SUmin	1	0.000	1	0.000	1	0.000	1	0.000	1	0.000	1	0.000
SUmax	1	0.000	1	0.000	1	0.000	1	0.000	1	0.000	1	0.000

Note: Null: null hypothesis decision (0: not to reject; 1: reject); *p*: p-value; gauexp: Gaussian-Exponential; tukexp: Tukey-Exponential; gaulin: Gaussian-Linear; 10, 20, 40, 80: DEM horizontal resolution (in meters); ind: independent pattern; inv: inverse pattern; SUmin: minimum structural-uncertainty; SUmax: maximum structural-uncertainty.

covering our study area, where topographic altitude is plotted against conditional probabilities of PDC arrival taking into account both aleatory and epistemic uncertainties. The smaller the eruption size, the greater the influence of Mount Somma. The different shape of the conditional-probability profiles over the north flank (Mount Somma) compared to the south flank of the volcano is recognizable for all sizes. Nonetheless, the profile is more homogeneous in the case of a large eruption. This is in agreement with the general idea that large PDCs are less controlled by topography than smaller ones [e.g., *Druitt*, 1998].

The very small contribution of input uncertainty to simulation results of medium and large sizes could be related to two main factors. First, our case study is based on a topographic setting controlled by a high-standing stratocone (Mount Vesuvius) surrounded by a quite flat area (Campanian plain, Fig. 9.2). Assessing the importance of input uncertainty linked to ECM simulations on other volcanic systems featured by more complex topographies (e.g., Campi Flegrei, Italy) could be of particular interest for PVHA. The example of Campi Flegrei is relevant here as PDC propagation could be controlled by (1) the complex morphology of the volcanic system (where caldera- and edifice-collapse structures are spatially combined with preserved eruptive cones [e.g., *Di Vito et al.*, 1999; *Orsi et al.*, 2004]); and (2) the vent-opening spatial variability that can be expected from future eruptions [e.g., *Selva et al.*, 2012].

Second, it should be stressed that the topographic control on PDC propagation that can be captured using the ECM is limited. Some effects such as channelization of PDCs [*Esposti Ongaro et al.*, 2008; *Komorowski et al.*, 2013] or dense-dilute PDC decoupling [*Fisher*, 1995] are not captured by such a simple model. Still, these effects can be crucial for PVHA and adequate scientific support for risk management. On the one hand, single-simulation or single-scenario (composed of few ECM simulations) approaches will likely fail in describing the possible extent of surge-decoupling processes as one might expect to occur, for instance, when PDCs encounter Mount Somma (there are other situations in which surge decoupling is favored, such as sharp variations in channel capacity and/or sinuosity [e.g., *Ogburn et al.*, 2014]). On the other hand, structured analyses quantifying both aleatory and epistemic uncertainties (where many thousands of simulations, covering coherent ranges of ϕ and H_0, are run) may be able to somehow assess this extent, since they generate a statistical sample of ECM outputs, which in part end before/at Mount Somma but in part go beyond it (Figs. 9.9, 9.10). As refers to PDC channelization, little can be done with the ECM and more sophisticated simulators are needed to evaluate the importance of such a phenomenon in the final PDC extent (see, for instance, Fig. 7 in *Esposti Ongaro et al.* [2008]).

In order to statistically test which alternate configurations significantly change the output ECDFs of area of invasion and maximum runout from the aleatory-uncertainty output ECDF, we perform two-sample Kolmogorov-Smirnov tests, one per each epistemic-uncertainty ECDF tested against the aleatory-uncertainty ECDF (Table 9.3). This test evaluates the null hypothesis that both ECDFs (each epistemic one and the common aleatory) have been sampled from the same underlying distribution, that is, a theoretical CDF [e.g., *Davis*, 2002]. By means of the K-S tests, we want to find

Table 9.4 Conditional Probabilities of PDC Invasion (*CP*, in Percentage) on Five Selected Points Over the Surroundings of Mount Vesuvius (Italy)

Eruption →		Small			Medium			Large		
Location	*d*[km]	minEU	AU	maxEU	minEU	AU	maxEU	minEU	AU	maxEU
SV[a]	5.8	0	0	0	1.4	13	15	51	51	60
TdG	4.8	0	0.1	0.3	12	25	31	66	67	77
Scf	12	0	0	0	0.0	1.2	2.8	2.9	20	25
MdSH	5.2	0	0	0	4.1	19	23	60	61	70
NCA	14	0	0	0	0	0.3	0.7	0	12	15

Note: Calculated from ECM simulations and taking into account both aleatory and epistemic uncertainties for three different eruption sizes: small, medium and large. For the sake of consistency, structural uncertainty is not included in the calculations, although this type of uncertainty is one of the less efficient in modifying the aleatory-uncertainty values (see text for more details); *d*: distance from the vent; minEU, maxEU: minimum and maximum *CP* values taking into account epistemic uncertainty; AU: *CP* value considering only aleatory uncertainty; SV: Somma Vesuviana; TdG: Torre del Greco; Scf: Scafati; MdSH: Massa di Somma Hospital; NCA: Napoli-Capodichino Airport.
[a] Minimum critical values of ϕ-H_0, which allow the ECM simulations to surpass Mount Somma (arriving to Somma Vesuviana) are $\phi \leq 13^\circ$ for $H_0 \approx$ 300–500 m up to $\phi \approx 23^\circ$ for $H_0 \approx$ 1700 m.

out whether the epistemic ECDFs are significantly different from the aleatory ECDF and, hence, quantifying epistemic uncertainty provides useful information about the ECM outputs. The only alternate configuration for which we cannot reject the null hypothesis, at the 5% significance level, is the Tukey-Exponential combination (parametric uncertainty). This fact reaffirms the choice of not exploring the Tukey-Linear configuration as we may assume their outputs would not be significantly different from Gaussian-Linear ones.

By taking into account all other epistemic uncertainties, instead, we observe a statistically significant variation of the final distribution for both area and maximum-runout values (Table 9.3). This means that, although input uncertainty changes these values by a small δ_{ij} amount, it is sufficient to modify the output ECDFs. More generally, it suggests that all types of epistemic uncertainty analyzed in this study are able to significantly change the resulting output values compared to those obtained when considering only aleatory uncertainty.

Finally, we can provide the probability ranges (considering both aleatory and epistemic uncertainties), conditional to the generation of PDCs and given the occurrence of an eruption of a specific size, for different locations being invaded by PDCs in case of small, medium, and large eruptions at Mount Vesuvius. For illustrating this, we report some examples of conditional probabilities (Fig. 9.9) for PDCs arriving (1) beyond Mount Somma (e.g., Somma Vesuviana), (2) at Torre del Greco (Napoli-Reggio Calabria highway and train station), (3) at Scafati (E-SE from Mount Vesuvius), (4) at the hospital in Massa di Somma, and (5) at the Napoli-Capodichino airport (Table 9.4).

9.4.2. Applicability to Other Volcanoes

As indicated before, quantifying uncertainties is often a computationally expensive procedure. In spite of this, uncertainty is unavoidably linked to volcanic hazards [e.g., *Hincks et al.*, 2014]. A better understanding of its sources, and how it can be reduced or, when not possible, quantified properly, will be beneficial for decision making and risk mitigation [*Marzocchi et al.*, 2012]. Moreover, exploring the best ways of visualizing and communicating uncertainty requires close collaboration between scientists and decision makers [e.g., *Doyle et al.*, 2014].

Accordingly, uncertainty characterization should be gradually included in volcanic hazard products as it has become a routine procedure in other natural hazards, such as earthquakes [e.g., *Abrahamson and Bommer*, 2005; *Stucchi et al.*, 2011].

The ECM has been one of the most-used models in volcanic hazard mapping studies [*Malin and Sheridan*, 1982; *Wadge and Isaacs*, 1988; *Barberi et al.*, 1992; *Sheridan and Macías*, 1995; *Alberico et al.*, 2002; *Sulpizio et al.*, 2010b] because of its capability to capture some first-order characteristics of PDCs and its computational efficiency. The results presented in this chapter may be used to include preliminary uncertainty quantifications into single-scenario or even single-simulation hazard studies. Despite being an order-of-magnitude estimation of aleatory and/or epistemic uncertainty, this can enrich the information provided in the hazard analysis. Besides, getting an understanding of where or why different sources of uncertainty may arise, can guide future research efforts to better describe and/or reduce them.

For these reasons, the application of the presented approach to other volcanic systems (as well as to other

PDC simulators) might serve to bridge the gap between some current practices, which lack of any uncertainty description, and an eventual common framework for uncertainty quantification in PVHA of PDCs.

9.4.3. Applicability to PVHA

Epistemic uncertainty represents an essential part of uncertainty assessment procedures. Quantifying the extent to which our knowledge is limited requires, first of all, to have an idea of what the missing processes are, the ill-constrained data, and so on. In other words, we need to be aware of what we *don't know* [*Spiegelhalter and Riesch*, 2011].

This implies that epistemic uncertainty may increase or decrease as knowledge about volcanic systems changes over time. For instance, before the tragic eruption of Mount Saint Helens (USA) in 1980, little was known about the possibility of a gravitational failure of part of the volcanic edifice, which triggered the devastating lateral blast [e.g., *Crandell and Mullineaux*, 1978]. Consequently, at that time, the epistemic uncertainty associated with a nearly unknown outcome would have been very low. After the eruption, we discover an ontological error [*Marzocchi and Jordan*, 2014]. Many horseshoe-shaped morphologies on different volcanoes around the world were recognized afterward and volcanic hazard scientists would start including the possibility of flank failure in their assessments [e.g., *Martí et al.*, 2008]. However, the causal links between magma ascent, emplacement or eruption, and the triggering of a flank collapse are not perfectly known. This means that epistemic uncertainty about the probability of flank failure has increased over time even though the knowledge about volcanic systems has clearly increased during the last 35 yr.

This changing nature of epistemic uncertainty copes well with Bayesian methodologies, where new information can be incorporated to previous states of information in a structured manner [e.g., *Tarantola*, 2005], and with the taxonomy of uncertainties proposed by *Marzocchi and Jordan* [2014].

PVHA has been connected to Bayesian theory since its beginnings [*Newhall and Hoblitt*, 2002; *Aspinall et al.*, 2003; *Marzocchi et al.*, 2004]. Some development in formal frameworks to evaluate epistemic uncertainty was published afterward [e.g., *Marzocchi et al.*, 2008]. In these works, a necessarily subjective (although expert) choice must be made in order to assign a degree of confidence on the proposed model with respect to other possible models and observed data. This is ultimately linked to the selected modeling procedure, including the simulator itself, the underlying assumptions, the DEM, the number of simulations, and so on [e.g., *Selva et al.*, 2010]. In this respect, we think that our study (and similar studies) can provide useful insights for assigning the subjective degree of confidence based upon a more robust, transparent, and justifiable procedure, taking into account all the information about the morphology and magnitude of epistemic uncertainty gathered from the presented results.

ACKNOWLEDGMENTS

The research leading to these results has received funding from the EU FP7 projects NEMOH (Numerical, Experimental, and stochastic Modeling of vOlcanic processes and Hazard, grant agreement no. 289976) and MED-SUV (MEDiterranean SUpersite Volcanoes, grant agreement no. 308665), from the Futuro in Ricerca 2008 FIRB Project ByMur (RBFR0880SR) financed by MIUR, the Italian Ministry for Research and Education, and from the Italian project DPC-INGV "V1: Probabilistic Volcanic Hazard Analysis," funded by Dipartimento della Protezione Civile. Some results shown here were obtained through computational resources provided by the Center for Computational Research (University at Buffalo, New York, USA). The 20 m and 40 m DEMs are courtesy of the Laboratory of Geomatics (INGV, Osservatorio Vesuviano). We warmly thank Jacopo Selva for valuable discussions about some topics treated in the chapter, Giuseppe Vilardo for providing some DEM products used in this chapter, Fawzi Doumaz for gathering and making available the OpenMap cartographic data, and Andrea Di Vico for GIS assistance. We are very thankful to Sylvain Charbonnier and two anonymous reviewers for their thoughtful reviews, from which the quality of our chapter has greatly benefited. Likewise, we are very grateful to the editors for all their help and their excellent editorial handling of this publication.

NOTES

1. The selected percentile is a subjective choice guided by a precautionary principle, since assigning a higher percentile (e.g., 99th) would produce PDFs with higher density at the smaller H_0 values and lower density at larger values.
2. The eruption column heights [*Selva et al.*, 2014] and ϕ values [*Ogburn*, 2013] determine a direct pattern among eruption sizes.
3. The total number of simulations utilized to quantify structural uncertainty (whether for area of invasion or maximum runout) is 250 sims × 6 eruptions = 1500 simulations. This corresponds to the 5% of the total number of simulations: 10,000 sims × 3 = 30,000 simulations. Therefore, our quantification of structural uncertainty gives equal weight to the misfit distributions estimated from each eruption, as far as there is not evidence to proceed differently.
4. N.B.: Nonetheless, negative values of area and maximum runout are not allowed.

REFERENCES

Abrahamson, N. A., and J. J. Bommer (2005), Probability and uncertainty in seismic hazard analysis, *Earthquake Spectra*, *21*(2), 603–607.

Alberico, I., L. Lirer, P. Petrosino, and R. Scandone (2002), A methodology for the evaluation of long-term volcanic risk from pyroclastic flows in Campi Flegrei (Italy), *J. Volcanol. Geotherm. Res.*, *116*(1), 63–78.

Aspinall, W., G. Woo, B. Voight, and P. Baxter (2003), Evidence-based volcanology: Application to eruption crises, *J. Volcanol. Geotherm. Res.*, *128*(1), 273–285.

Auker, M., R. Sparks, L. Siebert, H. Crosweller, and J. Ewert (2013), A statistical analysis of the global historical volcanic fatalities record, *J. Appl. Volcanol.*, *2*(1), 2; doi:10.1186/2191-5040-2-2.

Barberi, F., M. Ghigliotti, G. Macedonio, H. Orellana, M. Pareschi, and M. Rosi (1992), Volcanic hazard assessment of Guagua Pichincha (Ecuador) based on past behaviour and numerical models, *J. Volcanol. Geotherm. Res.*, *49*(1), 53–68.

Baxter, P., A. Neri, and M. Todesco (1998), Physical modelling and human survival in pyroclastic flows, *Nat. Hazards*, *17*(2), 163–176; doi:10.1023/A:1008031004183.

Bayarri, M., J. O. Berger, E. S. Calder, K. Dalbey, S. Lunagomez, A. K. Patra, E. B. Pitman, E. T. Spiller, and R. L. Wolpert (2009), Using statistical and computer models to quantify volcanic hazards, *Technometrics*, *51*(4).

Bonadonna, C., C. B. Connor, B. Houghton, L. Connor, M. Byrne, A. Laing, and T. Hincks (2005), Probabilistic modeling of tephra dispersal: Hazard assessment of a multiphase rhyolitic eruption at Tarawera, New Zealand, *J. Geophys. Res. Sol. Earth (1978–2012)*, *110*(B3).

Branney, M. J., and B. P. Kokelaar (2002), Pyroclastic density currents and the sedimentation of ignimbrites, Geological Society of London.

Calder, E., B. Pitman, R. Wolpert, S. Bayarri, E. Spiller, and J. Berger (2009), Towards a new approach for generating probabilistic hazard maps for pyroclastic flows during lava dome eruptions, *AGU Spring Meeting Abstracts*, *1*, 7.

Calder, E., P. Cole, W. Dade, T. Druitt, R. Hoblitt, H. Huppert, L. Ritchie, R. Sparks, and S. Young (1999), Mobility of pyroclastic flows and surges at the Soufriere Hills volcano, Montserrat, *Geophys. Res. Lett.*, *26*(5), 537–540.

Carey, S., and H. Sigurdsson (1987), Temporal variations in column height and magma discharge rate during the 79 AD eruption of Vesuvius, *Geological Society of America Bulletin*, *99*(2), 303–314.

Cas, R. A., and J. V. Wright (1987), *Volcanic Successions, Modern and Ancient: A Geological Approach to Processes, Products, and Successions*, Allen & Unwin.

Charbonnier, S. J., and R. Gertisser (2012), Evaluation of geophysical mass flow models using the 2006 block-and-ash flows of Merapi Volcano, Java, Indonesia: Towards a short-term hazard assessment tool, *J. Volcanol. Geotherm. Res.*, *231*, 87–108.

Cioni, R., A. Bertagnini, R. Santacroce, and D. Andronico (2008), Explosive activity and eruption scenarios at Somma-Vesuvius (Italy): Towards a new classification scheme, *J. Volcanol. Geotherm. Res.*, *178*(3), 331–346.

Cioni, R., R. Santacroce, and A. Sbrana (1999), Pyroclastic deposits as a guide for reconstructing the multi-stage evolution of the Somma-Vesuvius Caldera, *Bull. Volcanol.*, *61*(4), 207–222.

Cole, P. D., and C. Scarpati (2010), The 1944 eruption of Vesuvius, Italy: combining contemporary accounts and field studies for a new volcanological reconstruction, *Geolog. Mag.*, *147*(03), 391–415.

Crandell, D. R., and D. R. Mullineaux (1978), *Potential hazards from future eruptions of Mount St. Helens volcano, Washington*, 1383, Dept. of the Interior, Geological survey, US Govt. Print. Off., Washington, D.C.

Dalbey, K., A. Patra, E. Pitman, M. Bursik, and M. Sheridan (2008), Input uncertainty propagation methods and hazard mapping of geophysical mass flows, *J. Geophys. Res. Sol. Earth (1978–2012)*, *113*(B5).

Davis, J. C. (2002), *Statistics and Data Analysis in Geology*, 3rd ed., John Wiley & Sons, New York.

Del Negro, C., A. Cappello, M. Neri, G. Bilotta, A. Hérault, and G. Ganci (2013), Lava flow hazards at Mount Etna: Constraints imposed by eruptive history and numerical simulations, *Sci. Reports*, *3*.

Di Vito, M., R. Isaia, G. Orsi, J. Southon, S. De Vita, M. d'Antonio, L. Pappalardo, and M. Piochi (1999), Volcanism and deformation since 12,000 years at the Campi Flegrei caldera (Italy), *J. Volcanol. Geotherm. Res.*, *91*(2), 221–246.

Doyle, E., A. Hogg, H. Mader, and R. Sparks (2010), A two-layer model for the evolution and propagation of dense and dilute regions of pyroclastic currents, *J. Volcanol. Geotherm. Res.*, *190*(3), 365–378.

Doyle, E. E., J. McClure, D. Paton, and D. M. Johnston (2014), Uncertainty and decision making: Volcanic crisis scenarios, *Int. J. Dis. Risk Reduct.*, *10*, 75–101.

Druitt, T. (1998), Pyroclastic density currents, *Geological Society, London, Special Publications*, *145*(1), 145–182.

Esposti Ongaro, T., A. Neri, G. Menconi, M. De'Michieli Vitturi, P. Marianelli, C. Cavazzoni, G. Erbacci, and P. Baxter (2008), Transient 3D numerical simulations of column collapse and pyroclastic density current scenarios at Vesuvius, *J. Volcanol. Geotherm. Res.*, *178*(3), 378–396.

Fisher, R. V. (1995), Decoupling of pyroclastic currents: Hazards assessments, *J. Volcanol. Geotherm. Res.*, *66*(1), 257–263.

Fisher, R. V., A. L. Smith, and M. J. Roobol (1980), Destruction of St. Pierre, Martinique, by ash-cloud surges, May 8 and 20, 1902, *Geology*, *8*(10), 472–476.

Gurioli, L., R. Sulpizio, R. Cioni, A. Sbrana, R. Santacroce, W. Luperini, and D. Andronico (2010), Pyroclastic flow hazard assessment at Somma-Vesuvius based on the geological record, *Bull. Volcanol.*, *72*(9), 1021–1038.

Hall, M. L., C. Robin, B. Beate, P. Mothes, and M. Monzier (1999), Tungurahua Volcano, Ecuador: structure, eruptive history and hazards, *J. Volcanol. Geotherm. Res.*, *91*(1), 1–21.

Heim, A. (1932), *Bergsturz und menschenleben*, Vol. *30*, Fretz & Wasmuth.

Hincks, T. K., J.-C. Komorowski, S. R. Sparks, and W. P. Aspinall (2014), Retrospective analysis of uncertain eruption precursors at La Soufrière volcano, Guadeloupe, 1975–77: Volcanic hazard assessment using a Bayesian Belief Network approach, *J. Appl. Volcanol.y*, *3*(1), 1–26.

Komorowski, J.-C., S. Jenkins, P. J. Baxter, A. Picquout, F. Lavigne, S. Charbonnier, R. Gertisser, K. Preece, N. Cholik, A. Budi-Santoso, and Surono (2013), Paroxysmal dome explosion during the Merapi 2010 eruption: Processes and facies relationships of associated high-energy pyroclastic density currents, *J. Volcanol. Geotherm. Res.*, *261*, 260–294.

Malin, M. C., and M. F. Sheridan (1982), Computer-assisted mapping of pyroclastic surges, *Science*, *217*(4560), 637–640.

Martí, J., W. Aspinall, R. Sobradelo, A. Felpeto, A. Geyer, R. Ortiz, P. Baxter, P. Cole, J. Pacheco, M. Blanco, et al. (2008), A long-term volcanic hazard event tree for Teide-Pico Viejo stratovolcanoes (Tenerife, Canary Islands), *J. Volcanol. Geotherm. Res.*, *178*(3), 543–552.

Marzocchi, W., and T. H. Jordan (2014), Testing for ontological errors in probabilistic forecasting models of natural systems, *Proc. Nat. Acad.Sci.*, *111*(33), 11,973–11,978.

Marzocchi, W., C. Newhall, and G. Woo (2012), The scientific management of volcanic crises, *J. Volcanol. Geotherm. Res.*, *247*, 181–189.

Marzocchi, W., L. Sandri, and J. Selva (2008), BET_EF: a probabilistic tool for long-and short-term eruption forecasting, *Bull. Volcanol.*, *70*(5), 623–632.

Marzocchi, W., L. Sandri, P. Gasparini, C. Newhall, and E. Boschi (2004), Quantifying probabilities of volcanic events: the example of volcanic hazard at Mount Vesuvius, *J. Geophys. Res. Sol. Earth (1978–2012)*, *109*(B11).

Massey, F. J. (1951), The Kolmogorov-Smirnov test for goodness of fit, *J. Amer. Stat. Assoc.*, *46*(253), 68–78.

Mastin, L., M. Guffanti, R. Servranckx, P. Webley, S. Barsotti, K. Dean, A. Durant, J. Ewert, A. Neri, W. Rose, et al. (2009), A multidisciplinary effort to assign realistic source parameters to models of volcanic ash-cloud transport and dispersion during eruptions, *J. Volcanol. Geotherm. Res.*, *186*(1), 10–21.

Mele, D., R. Sulpizio, P. Dellino, and L. La Volpe (2011), Stratigraphy and eruptive dynamics of a pulsating Plinian eruption of Somma-Vesuvius: the Pomici di Mercato (8900 years BP), *Bull. Volcanol.*, *73*(3), 257–278.

Neri, A., A. Bevilacqua, T. Esposti Ongaro, R. Isaia, W. P. Aspinall, M. Bisson, F. Flandoli, P. J. Baxter, A. Bertagnini, E. Iannuzzi, et al. (2015), Quantifying volcanic hazard at Campi Flegrei caldera (Italy) with uncertainty assessment: 2. pyroclastic density current invasion maps, *J. Geophys. Res. Sol. Earth*, *120*(4), 2330–2349.

Newhall, C., and R. Hoblitt (2002), Constructing event trees for volcanic crises, *Bull. Volcanol.*, *64*(1), 3–20.

Ogburn, S. E. (2013), FlowDat: VHub Mass Flow Database.

Ogburn, S. E., E. S. Calder, P. D. Cole, and A. J. Stinton (2014), The effect of topography on ash-cloud surge generation and propagation, *Geological Society, London, Memoirs*, *39*(1), 179–194.

Orsi, G., M. A. Di Vito, and R. Isaia (2004), Volcanic hazard assessment at the restless Campi Flegrei caldera, *Bull. Volcanol.*, *66*(6), 514–530.

Orsi, G., M. A. Di Vito, J. Selva, and W. Marzocchi (2009), Long-term forecast of eruption style and size at Campi Flegrei caldera (Italy), *Earth Planet. Sci.Lett.*, *287*(1), 265–276.

Rolandi, G., A. Barrella, and A. Borrelli (1993), The 1631 eruption of Vesuvius, *J. Volcanol. Geotherm. Res.*, *58*(1), 183–201.

Rougier, J., and K. Beven (2013), Model limitations: the sources and implications of epistemic uncertainty, *Risk and Uncertainty Assessment for Natural Hazards*, Cambridge University Press, Cambridge, 40–63.

Sandri, L., E. Guidoboni, W. Marzocchi, and J. Selva (2009), Bayesian event tree for eruption forecasting (BET_EF) at Vesuvius, Italy: a retrospective forward application to the 1631 eruption, *Bull. Volcanol.*, *71*(7), 729–745.

Sandri, L., J.-C. Thouret, R. Constantinescu, S. Biass, and R. Tonini (2014), Long-term multi-hazard assessment for El Misti volcano (Peru), *Bull. Volcanol.*, *76*(2), 1–26.

Selva, J., A. Costa, W. Marzocchi, and L. Sandri (2010), BET_VH: exploring the influence of natural uncertainties on long-term hazard from tephra fallout at Campi Flegrei (Italy), *Bull. Volcanol.*, *72*(6), 717–733.

Selva, J., G. Orsi, M. A. Di Vito, W. Marzocchi, and L. Sandri (2012), Probability hazard map for future vent opening at the Campi Flegrei caldera, Italy, *Bull. Volcanol.*, *74*(2), 497–510.

Selva, J., L. Sandri, A. Costa, R. Tonini, A. Folch, and G. Macedonio (2014), Exploring the full natural variability of eruption sizes within probabilistic hazard assessment of tephra dispersal, in *EGU General Assembly Conference Abstracts*, *16*, 7612.

Sevink, J., M. J. van Bergen, J. van der Plicht, H. Feiken, C. Anastasia, and A. Huizinga (2011), Robust date for the Bronze Age Avellino eruption (Somma-Vesuvius): 3945±10 calBP (1995±10 calBC), *Quaternary Sci. Rev.*, *30*(9), 1035–1046.

Sheridan, M., A. Stinton, A. Patra, E. Pitman, A. Bauer, and C. Nichita (2005), Evaluating Titan2D mass-flow model using the 1963 Little Tahoma Peak avalanches, Mount Rainier, Washington, *J. Volcanol. Geotherm. Res.*, *139*(1), 89–102.

Sheridan, M. F. (1979), Emplacement of pyroclastic flows: a review, *Geological Society of America Special Papers*, *180*, 125–136.

Sheridan, M. F., and J. Macías (1995), Estimation of risk probability for gravity-driven pyroclastic flows at Volcan Colima, Mexico, *J. Volcanol. Geotherm. Res.*, *66*(1), 251–256.

Sheridan, M. F., A. K. Patra, K. Dalbey, and B. Hubbard (2010), Probabilistic digital hazard maps for avalanches and massive pyroclastic flows using TITAN2D, *Geological Society of America Special Papers*, *464*, 281–291.

Siebert, L. (1984), Large volcanic debris avalanches: Characteristics of source areas, deposits, and associated eruptions, *J. Volcanol. Geotherm. Res.*, *22*(3), 163–197.

Sigurdsson, H., S. Carey, and J. Espindola (1984), The 1982 eruptions of El Chichón volcano, Mexico: stratigraphy of pyroclastic deposits, *J. Volcanol. Geotherm. Res.*, *23*(1), 11–37.

Spiegelhalter, D. J., and H. Riesch (2011), Don't know, can't know: embracing deeper uncertainties when analysing risks, *Philosophical Transactions of the Royal Society A: Mathematical, Physical and Engineering Sciences*, *369*(1956), 4730–4750.

Stefanescu, E., M. Bursik, and A. Patra (2012a), Effect of digital elevation model on Mohr-Coulomb geophysical flow model output, *Nat. Hazards*, *62*(2), 635–656.

Stefanescu, E., M. Bursik, G. Cordoba, K. Dalbey, M. Jones, A. Patra, D. Pieri, E. Pitman, and M. Sheridan (2012b), Digital elevation model uncertainty and hazard analysis using a geophysical flow model, *Proceedings of the Royal Society A: Mathematical, Physical and Engineering Science*, *468*(2142), 1543–1563.

Stucchi, M., C. Meletti, V. Montaldo, H. Crowley, G. M. Calvi, and E. Boschi (2011), Seismic Hazard Assessment (2003–2009) for the Italian Building Code, *Bull. Seism. Soc. Amer.*, *101*(4), 1885–1911.

Sulpizio, R., D. Mele, P. Dellino, and L. La Volpe (2007), Deposits and physical properties of pyroclastic density currents during complex Subplinian eruptions: the AD 472 (Pollena) eruption of Somma-Vesuvius, Italy, *Sedimentology*, *54*(3), 607.

Sulpizio, R., L. Capra, D. Sarocchi, R. Saucedo, J. Gavilanes-Ruiz, and N. Varley (2010b), Predicting the block-and-ash flow inundation areas at Volcán de Colima (Colima, Mexico) based on the present day (February 2010) status, *J. Volcanol. Geotherm. Res.*, *193*(1), 49–66.

Sulpizio, R., P. Dellino, D. M. Doronzo, and D. Sarocchi (2014), Pyroclastic density currents: state of the art and perspectives, *J. Volcanol. Geotherm. Res.*, *283*(0), 36–65, doi:http://dx.doi.org/10.1016/j.jvolgeores.2014.06.014.

Sulpizio, R., R. Bonasia, P. Dellino, D. Mele, M. Di Vito, and L. La Volpe (2010a), The Pomici di Avellino eruption of Somma-Vesuvius (3.9 ka BP), Part II: Sedimentology and physical volcanology of pyroclastic density current deposits, *Bull. Volcanol.*, *72*(5), 559–577.

Tarantola, A. (2005), *Inverse Problem Theory and Methods for Model Parameter Estimation*, siam, Philadelphia, PA.

Tarquini, S., I. Isola, M. Favalli, F. Mazzarini, M. Bisson, M. T. Pareschi, and E. Boschi (2007), Tinitaly/01: A new triangular irregular network of Italy, *Ann. Geophys.*, *50*(3).

Taylor, G. A. (1958), The 1951 eruption of Mount Lamington, Papua, *Australian Bureau of Mineral Resources, Geology and Geophysics Bulletin*, *38*, 1–117.

Tierz, P., Sandri, L., Costa, A., Zaccarelli, L., Di Vito, M. A., Sulpizio, R., Marzocchi, W. (2016), Suitability of energy cone for probabilistic volcanic hazard assessment: validation tests at Somma-Vesuvius and Campi Flegrei (Italy), *Bull. Volcanol.*, *78*(79), doi:10.1007/s00445-016-1073-9.

Toyos, G., P. Cole, A. Felpeto, and J. Marti (2007), A GIS-based methodology for hazard mapping of small volume pyroclastic density currents, *Nat. Hazards*, *41*(1), 99–112.

Valentine, G. A. (1998), Damage to structures by pyroclastic flows and surges, inferred from nuclear weapons effects, *J. Volcanol. Geotherm. Res.*, *87*(1), 117–140.

Wadge, G., and M. Isaacs (1988), Mapping the volcanic hazards from Soufriere Hills Volcano, Montserrat, West Indies using an image processor, *J. Geolog. Soc.*, *145*(4), 541–551.

Wadge, G., P. Jackson, S. Bower, A. Woods, and E. Calder (1998), Computer simulations of pyroclastic flows from dome collapse, *Geophys. Res. Lett.*, *25*(19), 3677–3680.

Widiwijayanti, C., B. Voight, D. Hidayat, and S. Schilling (2009), Objective rapid delineation of areas at risk from block-and-ash pyroclastic flows and surges, *Bull. Volcanol.*, *71*(6), 687–703.

Wilson, L., R. Sparks, T. Huang, and N. Watkins (1978), The control of volcanic column heights by eruption energetics and dynamics, *J. Geophys. Res. Sol. Earth (1978–2012)*, *83*(B4), 1829–1836.

Wohletz, K., and G. Heiken (1992), Volcanology and Geothermal Energy, 432 pp., Univ. of Calif. Press, Berkeley.

10

Earthquake Loss Estimation in the Gyeongju Area, Southeastern Korea, Using a Site Classification Map

Su Young Kang and Kwang-Hee Kim

ABSTRACT

Understanding the potential threats due to a disaster is the first step toward the prevention of future damage and losses from such events. HAZUS is a damage- and loss-estimation software package used in the management of natural disasters. The software enables federal and local authorities to provide rapid and effective recovery measures under actual earthquake situations. It also provides the information required to establish a realistic and comprehensive earthquake risk mitigation plan for future potentially damaging earthquakes. This study examined the HAZUS input parameters for earthquake loss estimations in Korea. Historic records for 1900 years, as well as a recent instrument-based earthquake catalog, were examined in detail to identify areas where a high probability of large earthquakes was indicated. We particularly considered southeastern Korea, where the largest historic earthquake was reported, and small- to moderate-sized earthquakes have frequently occurred during the modern period of instrument observations. The area is also located at the center of a potentially active fault identified in previous studies. The Gyeongju area was chosen after a careful review. Because the extent of ground shaking is strongly influenced by local site conditions, a site classification map of the study area was prepared to provide a realistic simulation of estimated earthquake losses. We then estimated losses due to a magnitude 6.7 earthquake scenario in the Gyeongju area by applying the site classification map. The results indicated large spatial variations in damage estimates in the study area. As the epicenter of the scenario earthquake, the Gyeongju area was predicted to experience significant damage. Despite the large distance from the epicenter, the results indicated that Pohang Nam-gu was expected to experience levels of damage comparable to those in the Gyeongju area. Comparisons of results with and without the use of the site classification map indicated that the large estimated losses were caused mainly by the soft soil conditions that occur widely in the area.

10.1. INTRODUCTION

The Korean Peninsula is considered a largely stable continental region because it is located far from major tectonic boundaries and exhibits relatively low seismicity [*Koh et al.*, 2005; *Kyung*, 2012; *Kim and Kim*, 2014]. However, historic records for Korea reveal that several large earthquakes have occurred during its history, including a period of earthquake activity between the fifteenth and eighteenth centuries, during which exceptional high seismicity was reported [*Lee and Lee*, 1997; *Koh et al.*, 2005; *Kyung*, 2012]. Modern seismic observations in the peninsula also indicate that the peninsula has experienced several moderate earthquakes. Therefore, earthquake hazard and the subsequent risks due to large earthquakes with a long recurrence interval should not be ignored or underestimated.

Because short-term earthquake prediction is not viable in the foreseeable future and current earthquake mitigation strategies focus on long-term ground-shaking forecasts that can be used in building design, the best practice

Department of Geological Sciences, Pusan National University, Busan, Korea

Natural Hazard Uncertainty Assessment: Modeling and Decision Support, Geophysical Monograph 223, First Edition.
Edited by Karin Riley, Peter Webley, and Matthew Thompson.

to minimize losses due to earthquake hazard is to make ourselves resilient from vulnerability by being well prepared for and responding effectively to disastrous events. This can be achieved by scientific risk assessments and informed policy making and decisions.

Simulations to predict damage areas and losses due to natural disasters and visualizations of the results can provide invaluable information for decision makers and emergency crews [*Kim*, 2010; *Kang et al.*, 2008]. HAZUS, a geographic information system (GIS)-based natural hazard loss estimation software package developed by the US Federal Emergency Management Agency (FEMA), is a well-known tool for that purpose. HAZUS not only calculates and tabulates the potential losses in terms of social, physical, and economic losses but also visualizes them on maps [*FEMA*, 2009]. The results of the loss estimation reveal the vulnerability of the target area to the specific disaster. Initially, the software estimates the exposure of the study area (e.g., population, buildings and their vulnerability, infrastructures, various assets, etc.). Then, it characterizes the level of hazard and calculates the potential losses by comprehensively considering the exposure and level of hazard. HAZUS can model multihazard scenarios including flooding, hurricanes, tsunami, and earthquakes. Although it was developed primarily for the US continent, the HAZUS toolset can be adopted in other countries including Korea [*Kang et al.*, 2007a; *Kang et al.*, 2008].

Ground shaking due to an earthquake is frequently amplified by local site conditions. It is also frequently observed that large structural damage is accompanied by land subsidence, liquefaction, and landslides. For similar-magnitude earthquakes, less damage is observed for those occurring in areas of firm ground [*Sun and Chung*, 2008]. Realistic loss estimation is more likely when the topographical characteristics and soil conditions are included in the input parameters. Characterizing the site amplification for earthquake ground motions requires geological and geotechnical information [*Sun and Chung*, 2008]. In this study, a site condition map of the study area was prepared by combining information from geological and topographical maps.

We also carefully reviewed both the historic earthquake catalog and instrument earthquake records in the Korean peninsula to identify the earthquake-prone area. Then, we estimated potential losses in terms of physical damage and social losses in different areas of the study region using HAZUS.

10.2. METHODS

Earthquakes tend to occur along active faults. However, in the stable continental region, including the Korean Peninsula, the situation is not as clear as that along the plate boundaries, primarily due to the low seismicity and long recurrence interval of major earthquakes. We examined both historic earthquake records (between AD 2 and 1904) compiled by the Korea Meteorological Administration [*KMA*, 2010] and the modern instrument-based earthquake catalog (from 1978 to 2013) held by the Korea Meteorological Administration [*KMA*, 2013].

Local site conditions may amplify ground motion. Information such as site characteristics therefore plays an important role in the risk assessment of ground shaking, liquefaction, landslides, and other geological hazards. Sites can be classified by a direct or indirect approach. The direct approach measures the physical properties of the medium in a borehole, which is expensive and time consuming. The indirect approach uses seismological, geological, and geomorphological information for site classification. Information about surface geology is readily available from the digital geological map of Korea and can be correlated with discrete velocity measurements. These correlations have been extrapolated to a larger area [*Park and Elrick*, 1998]. Information available from the geological map includes geological ages, strata, related field information, and so on. To overcome the limitations posed by the biased spatial distribution of available direct measurements from boreholes, we used supplementary information from geological, topographical, and seismic data to construct a site classification map of the study area. Geological age and strata from the geological map along with the elevation and slope from the topographic map were incorporated using GIS tools [*Kang and Kim*, 2012].

Losses due to an earthquake in the Gyeongju area were estimated using the deterministic method in HAZUS. The first step involved preparing inventory data for the loss estimation system. This included determining the population, classifying the occupancy of buildings and facilities, classifying building structure and occupancy, and collecting information regarding essential facilities. A site classification map of the study area was also prepared and used as input data. For earthquake information, the location, magnitude, and ground motion attenuation in relation to distance were selected and provided as input parameters. Further details are given in the following sections. They are also available in our previous studies [*Kang and Kim*, 2009; 2011; 2012; *Kang et al.*, 2007a; *Kang et al.*, 2007b; *Kang et al.*, 2008].

The primary hazardous effect of an earthquake is what happens during and just after the earthquake, which may include the collapse or toppling of buildings, people being buried in the rubble, and disorder due to the malfunction of critical facilities in the affected area. Secondary hazardous effects are the consequences of the primary effects or what happens after the earthquake;

these may include fire, presence and disposal of hazardous materials and the subsequent risks of disease spread, landslides, liquefaction, and so on. Emergency shelters and hospitals for affected residents are required. In the long term, some residents may need to be relocated. Assuming an earthquake hazard and the fragility of structures in the region studied, HAZUS estimates the potential loss in terms of the direct physical damage, secondary physical damage, and social/economic losses. Direct damage may include the physical damage to residential and commercial buildings, schools, critical facilities, and infrastructure. Secondary physical damage may include losses due to fire, the disposal of hazardous materials and the subsequent spread of disease, and debris from collapsed buildings. Social/economic losses include lost jobs, shelter requirements, displaced households, business interruptions, repairs, and reconstruction costs.

10.3. SEISMICITY IN THE KOREAN PENINSULA

Generally, the spatial and temporal distributions of large earthquakes can be determined by careful inspection of earthquake records based on instrument measurements. Unfortunately, this is not true in Korea because instrument earthquake records in Korea are relatively short, and the recurrence interval of large earthquakes in Korea is very long. Thus, in addition to reviewing instrument-measured seismicity since 1978, we also reviewed a historic earthquake catalog. The most recently compiled historic records by the *KMA* [2010] were reviewed in this study. There were 2139 and 1109 earthquakes in the historic records and in the instrument-based earthquake catalog, respectively. Their magnitudes ranged from 3.5 to 6.7 for the historic earthquakes and from 2.0 to 5.3 for the instrumental earthquakes. Given the lack of seismicity information in the modern seismic observation period (between 1978 and 2013), seismicity and the likelihood of high-magnitude earthquakes in the Korean Peninsula are better estimated using both sets of records.

The GIS spatial analysis function is frequently used to visualize geographical phenomena [*Clarke*, 2010]. The point density tool in the spatial analysis function of a GIS counts and visualizes the number of earthquakes in a unit area. The tool was applied to historic and instrumental earthquake data to determine spatial variation in seismicity. Any area with a high point density is regarded as an area of high seismicity, and it therefore has a relatively high earthquake hazard.

The historic and instrumental data used in the study also included information regarding the magnitude of earthquakes. In Korea, it is generally acknowledged that earthquakes with a magnitude greater than 5 can result in serious damage to the source area [*Kyung*, 2012]. It is also desirable to reflect the hazard generated by large-magnitude earthquakes in the results of the spatial analyses. Weightings according to the corresponding earthquake magnitude were given in the point density analysis to reflect the significance of earthquake size. Thus, the results of spatial analyses can identify areas of high seismic energy release. Note that the results still do not represent a seismic risk map because the seismic risk can vary due to many factors that were not addressed in the study, including the location, type and length of faults, and local site conditions.

Results from the point density analysis with the magnitude weighting, using the GIS spatial function are presented in Figure 10.1. The Pyungyang area and its vicinity were distinguished as an area with high seismic hazard based on analysis of the historic records. Seoul and Gaeseong were also identified as areas with a high seismic hazard (Fig. 10.1a). Results using records of historic earthquakes with a magnitude >5.0, which are likely to cause earthquake damage, produced slightly different results, with the Gyeongju and Seoul areas and their vicinities experiencing higher seismic hazards than any other area in the Korean Peninsula (Fig. 10.1b). Results obtained using another dataset, the instrumental earthquake data for the Korean Peninsula and its surrounding regions between 1978 and 2013 (Fig. 10.2), identified Pyungyang, Chungchung, and Gyeongsang provinces as high seismic areas. The Yellow Sea, the eastern offshore, and the Jeju offshore areas were also affected by high seismicity. Both historic and recent earthquake records indicated that Pyungyang and Gyeongju have experienced relatively high seismicity. Considering the accessibility to the study area, the Gyeongju and its vicinity were selected for this study. They are shown as a thick line on Figure 10.2.

10.4. SITE CLASSIFICATION MAP

The mean shear-wave velocity in the top 30 m of the surface (Vs_{30}) is frequently used to characterize the site response to an earthquake, that is, as a proxy for the amplification of ground motion [*Collins et al.*, 2006; *Castellaro et al.*, 2008; *Allen and Wald*, 2009]. Although it is preferable to use well-logging data with direct measurements of shear-wave velocity to construct a site classification map, these data are expensive and time consuming to collect. As an alternative to the direct measurements, it has been proposed that geological and geomorphological information can be used to construct a regional-scale site classification map [*Lee et al.*, 2001; *Allen and Wald*, 2009]. We used this technique to construct a site classification map of Gyeongju and its vicinity using GIS tools. The detailed procedures and the resultant site classification map are given in *Kang and Kim* [2012]. Only a brief summary is provided here.

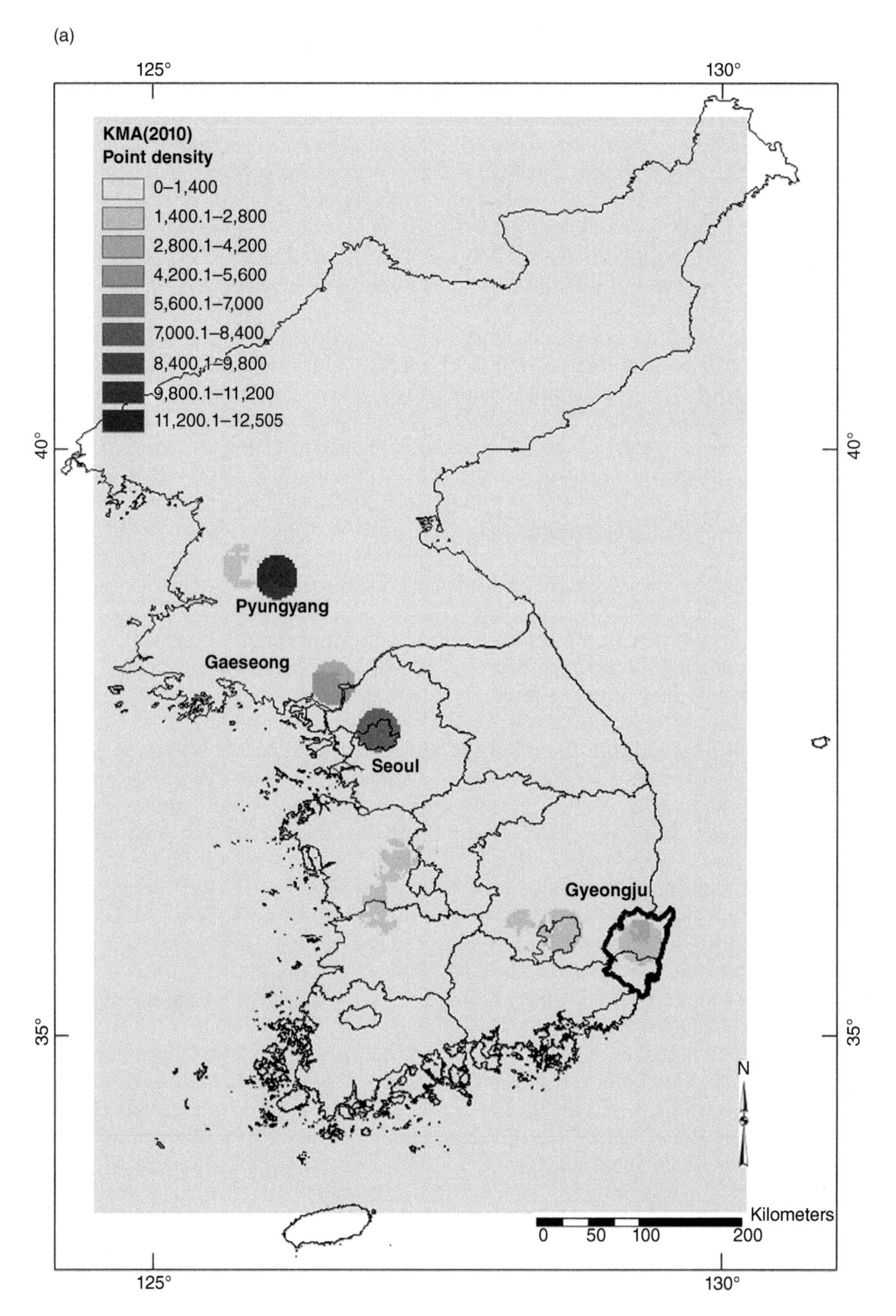

Figure 10.1 Point density analysis of (a) historic earthquakes with a magnitude greater than 3.5 and (b) historic earthquakes with a magnitude greater than 5.0.The Gyeongju area in the southeast of the Korea peninsula is marked by thick solid line.

The primary data used to construct the site classification map of the study area were derived from a geological map of Korea (1:250,000 scale), which is publicly available from the Korea Institute of Geoscience and Mineral Resources [*KIGAM*, 2010], and a digital elevation model (DEM at 1- and 3-arcsecond resolution) [*USGS*, 2008]. Using the information provided by the geological map of Korea and the DEM, the study area was classified based

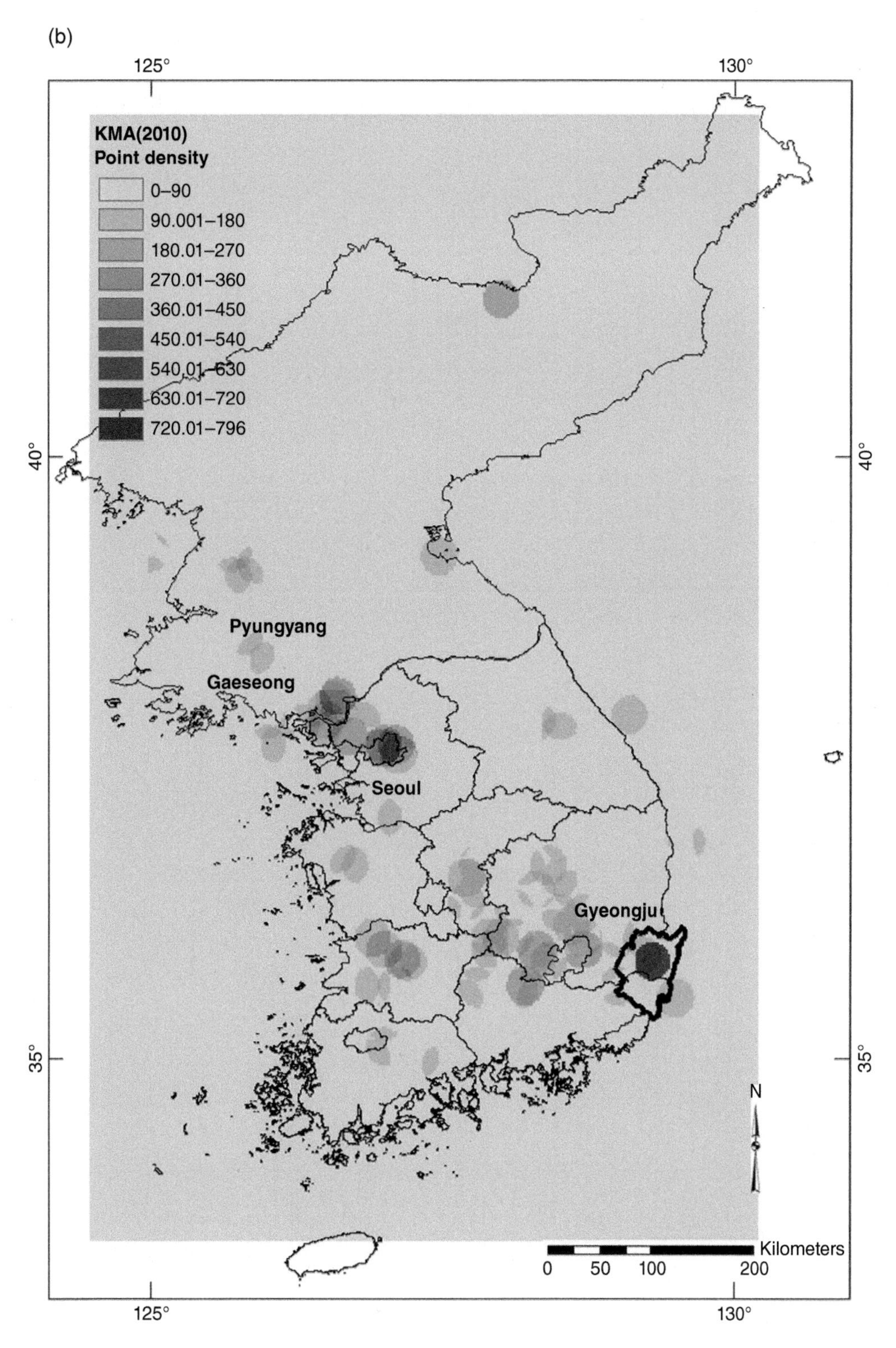

Figure 10.1 (Continued)

upon rock types, ages, and topographic slope following the criteria proposed by *Lee et al.* [2001] and *Wald and Allen* [2007]. Then, results were associated with shear wave velocity values bounded by the National Earthquake Hazard Reduction Program (NEHRP). The site classification scheme based on the range of slopes, rock types, and shear wave velocities is summarized in Table 10.1 and is discussed in detail by *Kang and Kim* [2012] and references therein.

The Gyeongju area and its vicinity are characterized by site classes between B (rock) and D (stiff soil). Site class B is dominant in the study area, occupying 62% of the

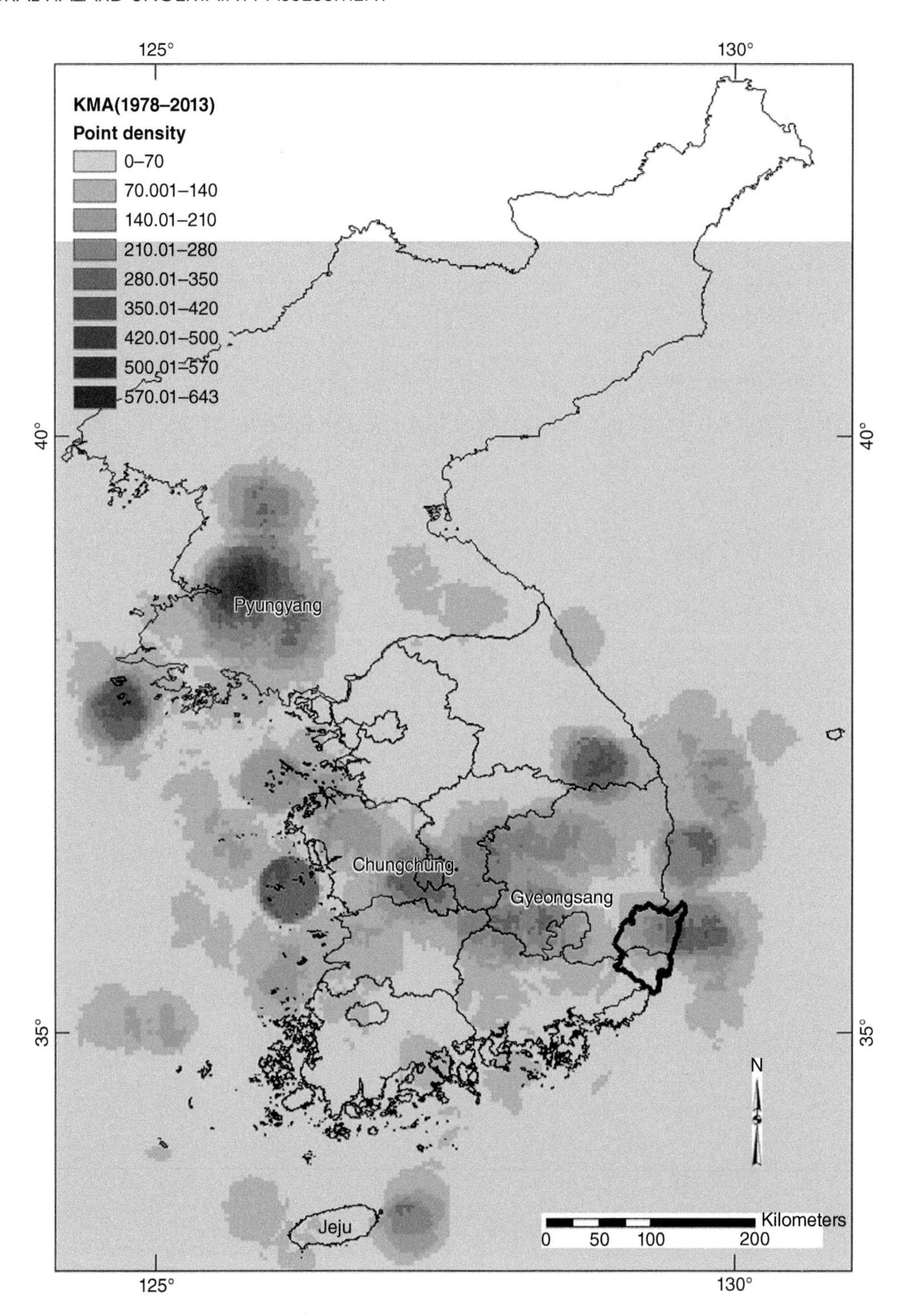

Figure 10.2 Point density analysis of instrument-detected earthquakes between 1978 and 2013 with a magnitude greater than 2.0.

total area. In the eastern part of the study area, site class C (very dense soil/soft rock) is frequently observed and covers 17% of the total area. Site class D occupies 22% of the total area and extends in an elongated north-south pattern across the study area (Fig. 10.3).

In Gyeongju, which is located in the center of the study area, site class B is dominant, although site classes D and C are observed in a NNW-SSE trend and in the eastern part of the city, respectively. Approximately half of Pohang Nam-gu, which is located in the northern

Table 10.1 Summary of Correlations Among the Earthquake Hazard Reduction Program (NEHRP) Site Classifications, Geological and Topographical Information, and Shear-Wave Velocity

Site class	Description	Range of Vs30 (m/sec)	Slope (m/m)
A	Hard rock	>1500	
B	Rock	760–1500	>0.025
C	Very dense soil and soft rock	360–760	7.2E-3~0.025
D	Stiff soils	180–360	2.0E-5~7.2E-3
E	Soft soils	~180	<2.0E-5
F	Requires site-specific evaluation		

Source: Wald and Allen [2007].

part of the study area, is classified as site class C. Large areas of Pohang Nam-gu are also classified as site class D. Located in the southern part of the study area, most of Ulsan is classified as site class B, except Ulsan Nam-gu, where site class D is observed. In the study area, soft soils are frequently found in Pohang Nam-gu and Ulsan Nam-gu.

10.5. SCENARIO EARTHQUAKE AND INVENTORY DATA FOR THE EARTHQUAKE LOSS ESTIMATION

A scenario earthquake of magnitude 6.7 located at Gyeongju was specified for the calculation of ground shaking and damage in the study area. The location and magnitude of the scenario earthquake were

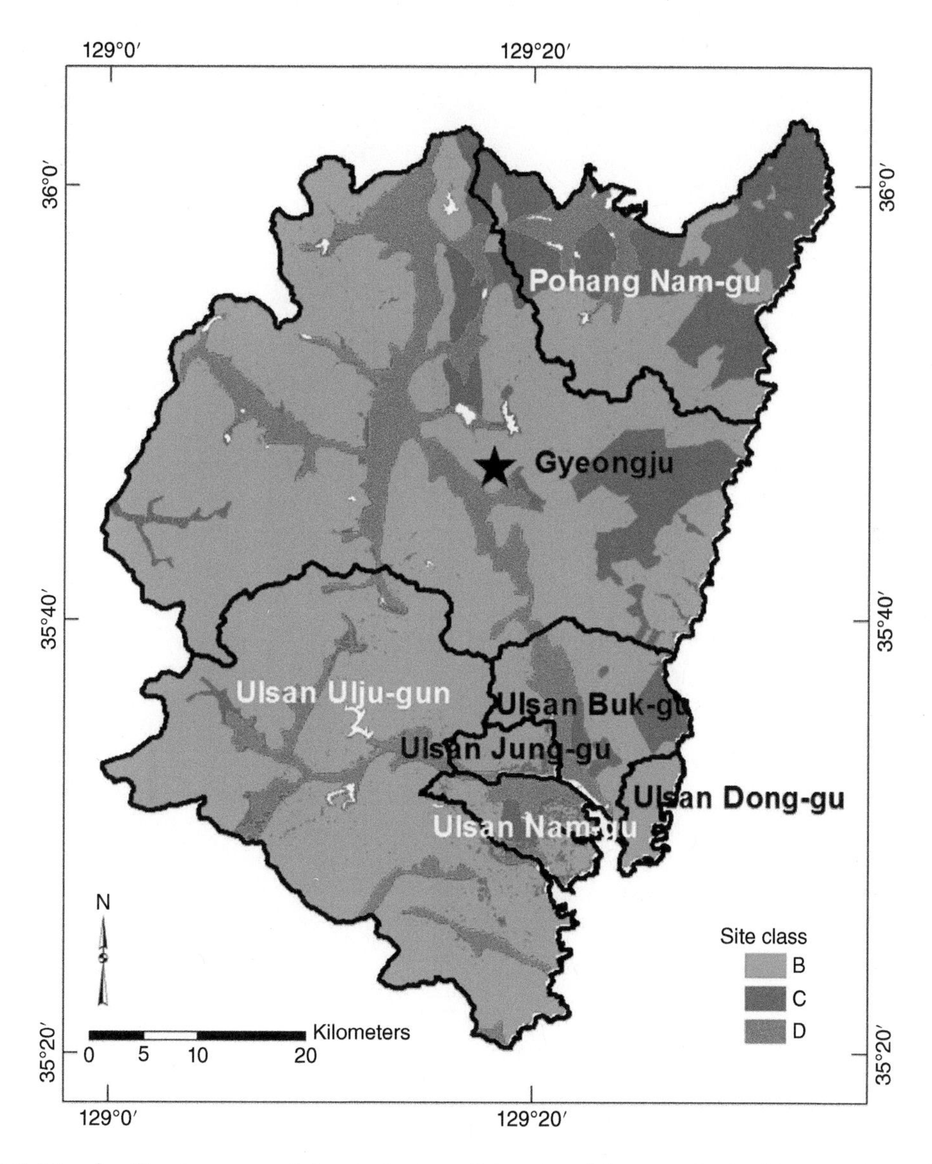

Figure 10.3 Site classification map of the study area. Epicenter of the magnitude 6.7 scenario earthquake is shown by a black star.

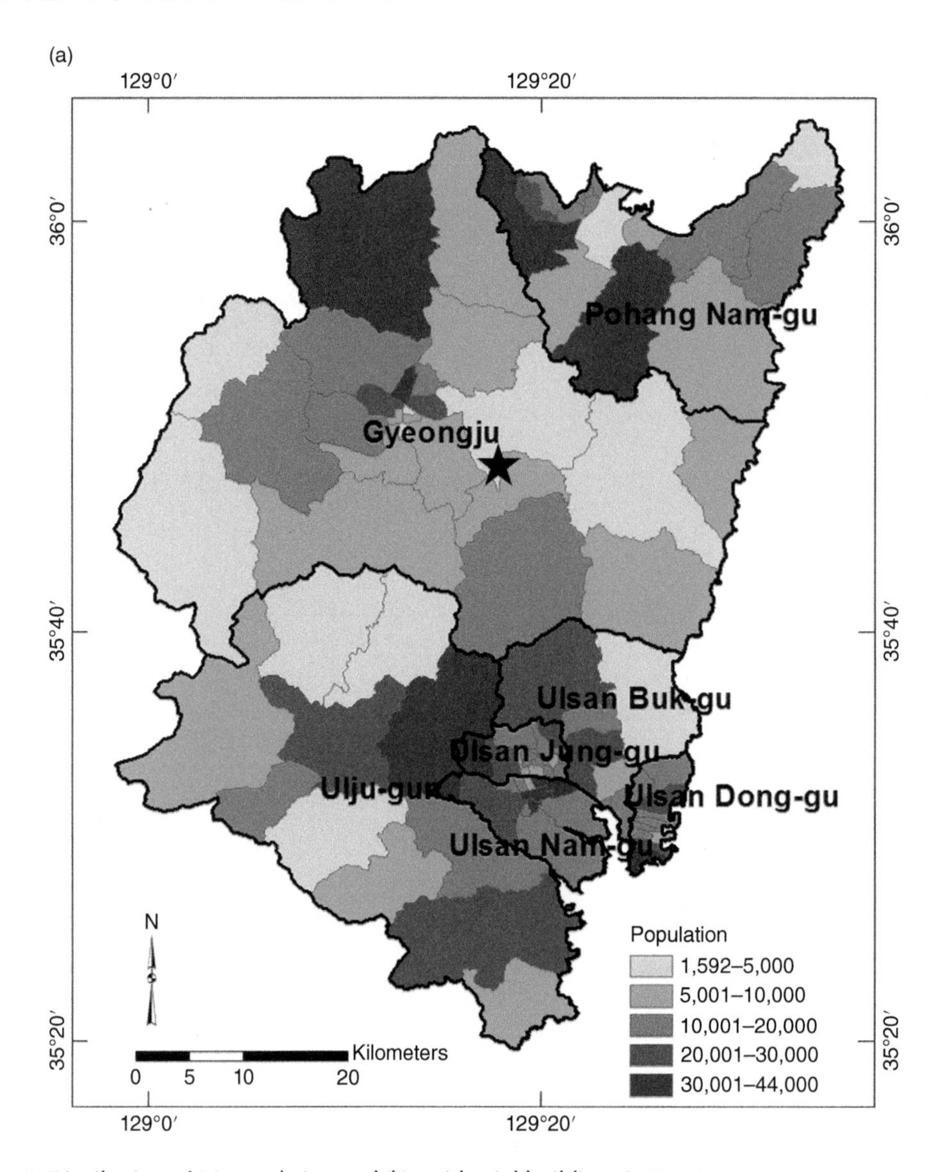

Figure 10.4 Distribution of (a) population and (b) residential buildings in 2005.

selected from the historic earthquake database recently compiled by *KMA* [2010]. The earthquake with a magnitude of 6.7, one of the largest earthquakes recorded in Korea, occurred at Gyeongju in 779. It was assumed that the scenario earthquake was nucleated at a depth of 10 km and that it ruptured a strike-slip fault. We limited the loss estimation to the Gyeongju and its vicinity, where most damage due to the scenario earthquake was assumed to occur. We also selected the attenuation relation proposed by *Sadigh et al.* [1997], which was based on a previous study that compared observed ground motion from a moderate earthquake in Korea with the attenuation relations available for HAZUS [*Kang et al.*, 2007b].

Inventory data required for loss estimations in the study area were collected primarily from the census report of 2005, which is publicly available from Korean Statistical Information Service [*KOSIS*, 2007]. The study area had a population of 1,564,765 in 2005 (Fig. 10.4). As of 2005, it also contained 436,282 residential buildings, which are typically houses, apartment blocks, and multipurpose buildings (Fig. 10.4b). The building materials were inferred from a report published in 1990 [*KOSIS*, 2007]. The majority of the buildings were constructed from reinforced concrete (78% of the whole building stock). Approximately 10% and 12% of the rest were made from bricks and wood, respectively [*KOSIS*, 2007]. Essential facilities in the area include 396 medical facilities (which include hospitals, clinics, and public health centers), 76 police stations, 32 fire stations, and 352 schools. We paid extra attention to these essential facilities because they provide services to the community and their functionality is critical in the aftermath of an earthquake.

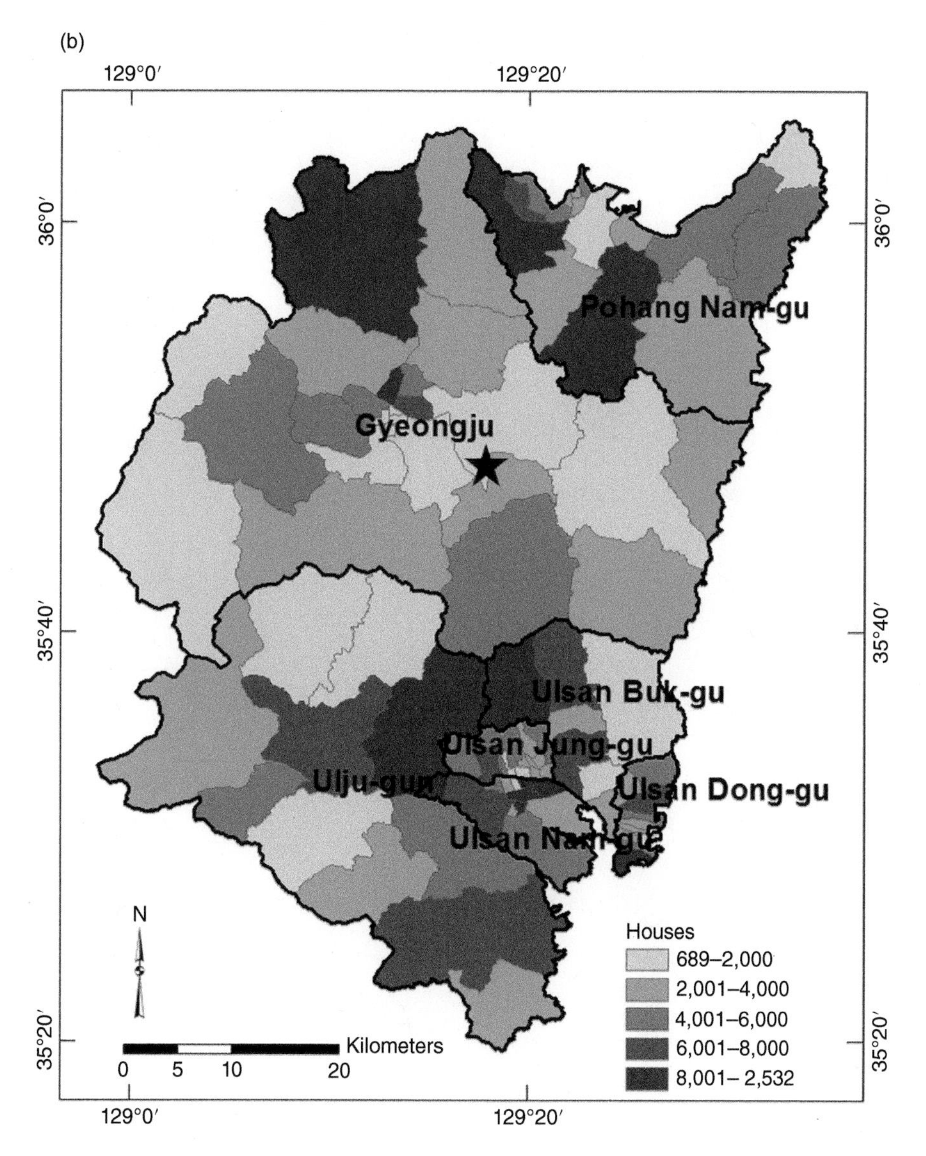

Figure 10.4 (Continued)

10.6. LOSS ESTIMATION RESULTS

Our loss estimation results were derived using the deterministic method in HAZUS, assuming a scenario earthquake of magnitude 6.7 in Gyeongju and the attenuation relation proposed by *Sadigh et al.* [1997]. The study area includes Gyeongju, Pohang Nam-gu, and Ulsan Metropolitan Cities. We estimated the losses with and without use of the site classification map and compared the results. Regional variations in the loss estimates without the site classification map were mostly affected by the distance from the epicenter. The variations in the loss estimation using the site classification map were heavily influenced by both the distance from the epicenter and local site conditions.

10.6.1. Damage to Residential Buildings

As of 2005, there were 436,282 residential buildings in the study area. Structural damage was dependent on both the distance from the epicenter and site conditions. For example, HAZUS estimated that about 98,163 residential buildings were at least moderately damaged in the scenario earthquake (Fig. 10.5a). This represents approximately 23% of the residential building stock in the study area. Moderate damage means that cracks would be observed in building walls. The estimates also include 6171 buildings that collapsed (Fig. 10.5b).

We compared the damage to residential buildings by jurisdiction. The inventory data indicated that there were 86,479 residential buildings in Gyeongju, 73,225 in Pohang Nam-gu, 39,821 in Ulsan Buk-gu, 52,482 in

(a)

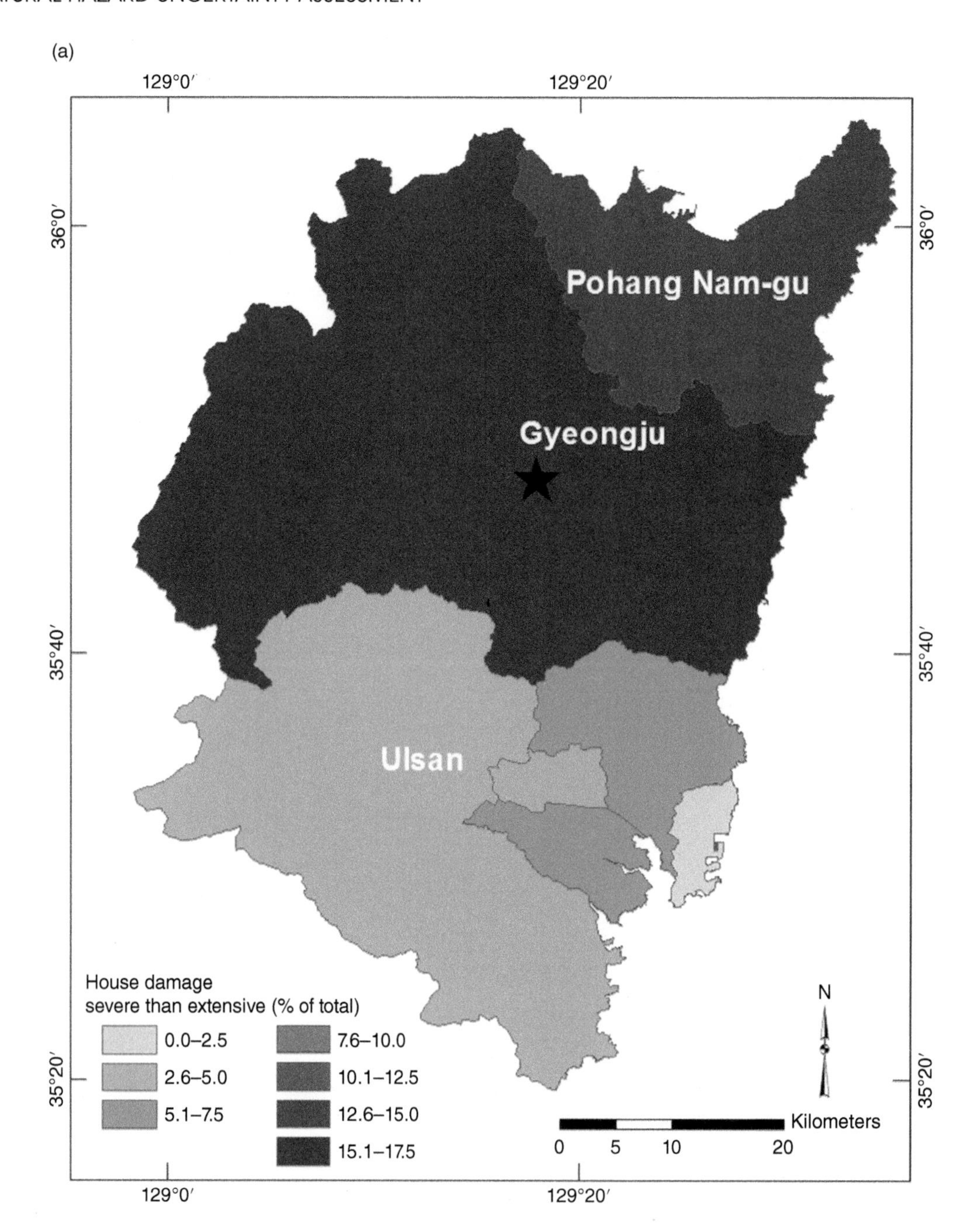

Figure 10.5 Estimates of damage to residential buildings: (a) Distribution of residential buildings with moderate or severe damage and (b) damage to residential buildings by jurisdictions.

Ulsan Jung-gu, 52,865 in Ulsan Ulju-gun, 86,397 in Ulsan Nam-gu, and 45,013 in Ulsan Dong-gu. The number of damaged residential buildings was largest in Gyeongju. HAZUS estimated that 35% (30,186 buildings) of the residential building stock in Gyeongju would experience at least moderate damage. In comparison, 34% (24,749 buildings) of the residential building stock in Pohang Nam-gu was estimated to experience at least moderate damage. In Ulsan Nam-gu, there were 15,994 moderately or severely damaged buildings, corresponding to 19% of the residential building stock. This indicates that within Ulsan Metropolitan City, the greatest damage was expected in Ulsan Nam-gu. The loss estimation results also predicted the complete destruction of 3014 buildings in Gyeongju and 1733 buildings in Pohang Nam-gu. Although the overall damage in Ulsan Nam-gu was not as severe as that in Gyeongju and Pohang Nam-gu, 625 buildings were also expected to be completely

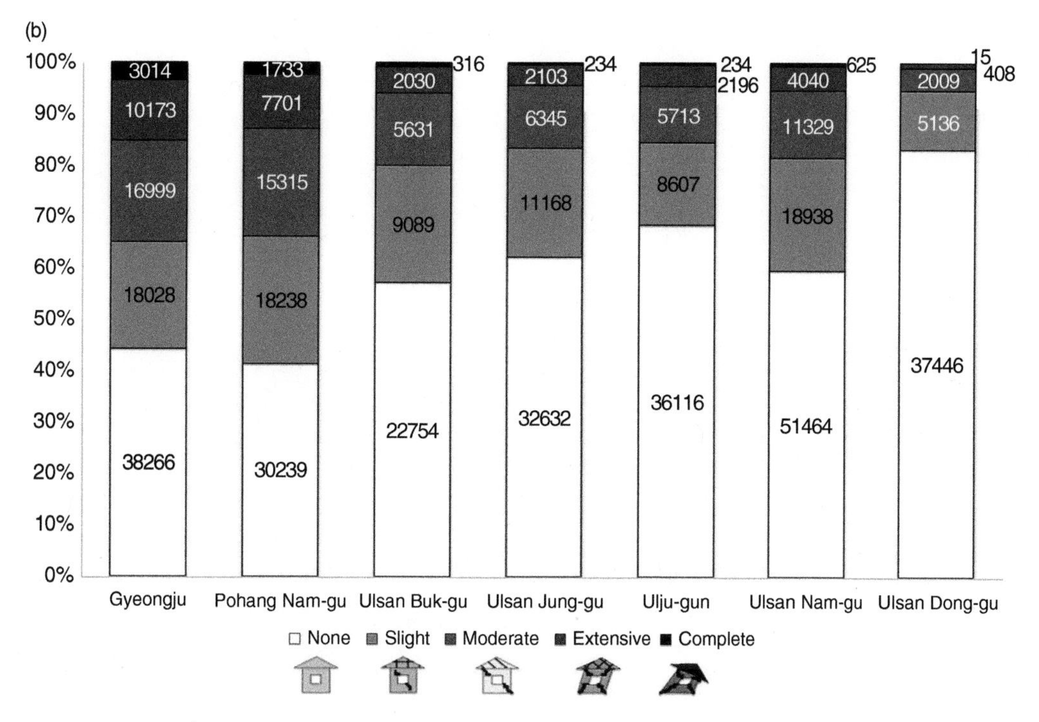

Figure 10.5 (Continued)

destroyed. The least damage to residential buildings was expected in Ulsan Dong-gu, where only 15 buildings were expected to collapse.

10.6.2. Damage to Essential Facilities

Essential facilities include hospitals, schools, police stations, and fire stations. Emergency countermeasures and their operational readiness are dependent upon their functionality following an earthquake. As reported in the previous section, damage to residential buildings and other losses are expected to increase if essential facilities are located near the epicenter and on soft soil. Generally, if any facility experiences a functionality loss of about 50%, it cannot perform well. In an emergency situation, field crews should be advised to consider any alternatives. Mapping and displaying the functionality of the essential facilities on the day of earthquake and over time under scenario situations are therefore critical for decision makers to develop emergency action plans. Figures 10.6 and 10.7 display the functionality of essential facilities in the study area after the scenario earthquake. The details of the results are as follows.

10.6.2.1. Hospitals

Hospitals play a critical role during the aftermath of an earthquake. For the purposes of this study, hospitals in the study area included any facility that could offer emergency medical care (i.e., general hospitals, clinics, and community health centers). There are 396 hospitals in the area, of which 211 were expected to be 50% functional on the day of the earthquake, corresponding to 53% of the hospital stock. Although each hospital's functionality and recovery would vary due to the extent of ground shaking and subsequent impacts, within 7 days of the earthquake, an average 30% loss of hospital functionality was expected. By 30 days after the earthquake, this figure was predicted to have decreased to 7%.

There are 76 hospitals in Gyeongju, 64 in Pohang Nam-gu, 14 in Ulsan Buk-gu, 88 in Ulsan Dong-gu, 44 in Ulsan Nam-gu, 7 in Ulsan Jung-gu, and 103 in Ulsan Ulju-gun (256 hospitals in total in Ulsan Metropolitan City). Many hospitals near the epicenter were expected to experience significant damage and functionality loss. The number of hospitals affected decreased with distance from the epicenter. Affected hospitals would gradually recover their functionality over time. Figure 10.6 shows the large degree of spatial variability in the predicted damage to hospitals in the study area. Hospitals in Pohang Nam-gu and Gyeongju were expected to experience the most serious damage. On the day of the earthquake, 47 of the 64 hospitals in Pohang Nam-gu were likely to be operating at less than 50% of their functionality. Although they gradually recovered over time, 16 did not recover to the 50% functionality level until 30 days after the earthquake. Fifty-one hospitals in Gyeongju

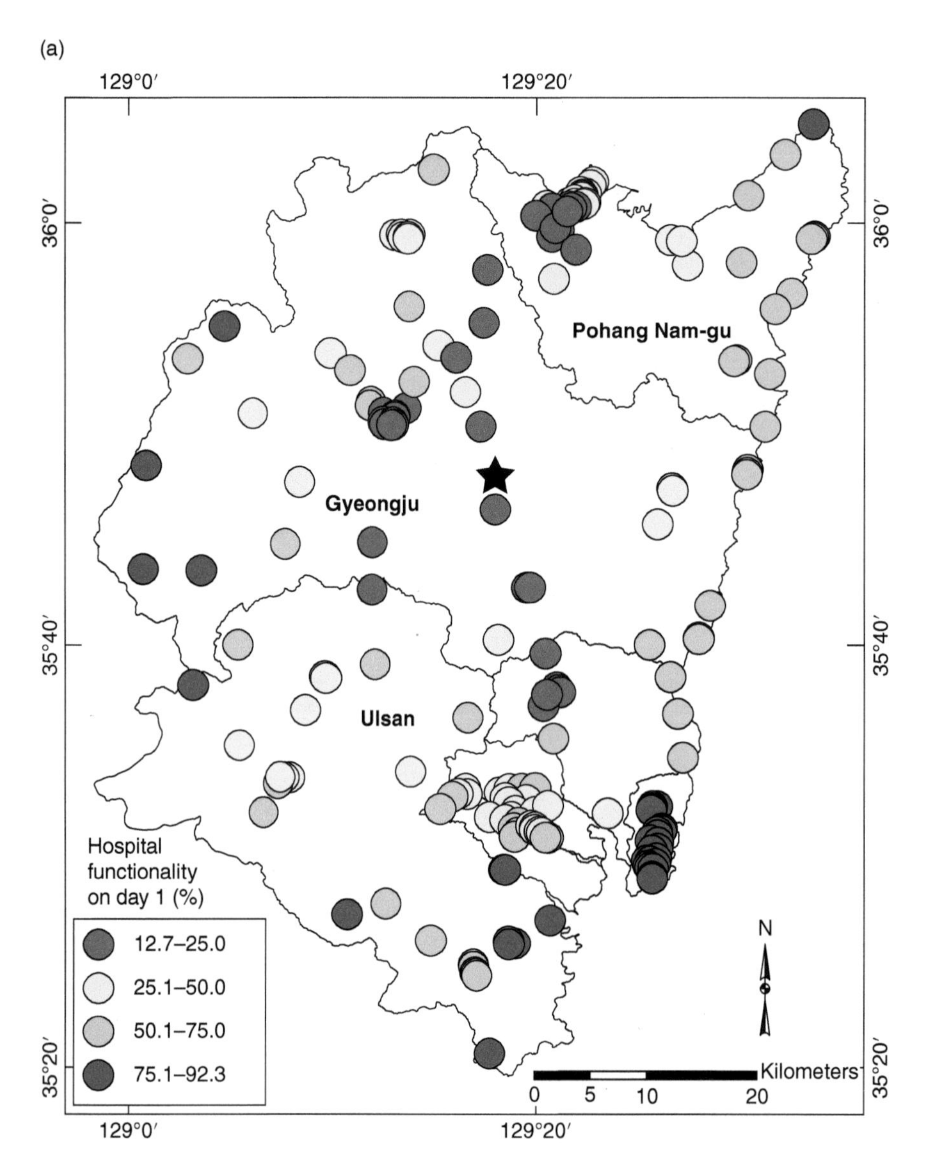

Figure 10.6 Hospital functionality at (a) day 1 and (b) day 7 after the earthquake.

were expected to lose more than 50% of their functionality on the day of the earthquake. The number of hospitals with less than 50% functionality in Ulsan Nam-gu was 34, although all were expected to regain at least 50% of their functionality during the first week after the earthquake. In Ulsan Dong-gu, all of the hospitals were expected to retain more than 50% of their functionality on the day of the earthquake.

10.6.2.2. Schools, Police Stations, and Fire Stations

Schools are generally located near residential areas and are considered to be useful as emergency shelters for affected residents. HAZUS was used to estimate the damage to school buildings and the degree of their subsequent loss of functionality as emergency shelters following the scenario earthquake. The results indicated large spatial variations over the study area. Fifty of the 352 schools in the study area were expected to lose more than 50% of their functionality. Only 302 schools would be available as emergency shelters in the proposed scenario.

There are 116 schools in Gyeongju, 62 in Pohang Nam-gu, 21 in Ulsan Buk-gu, 28 in Ulsan Dong-gu, 44 in Ulsan Nam-gu, 29 in Ulsan Jung-gu, and 52 in Ulsan Ulju-gun (174 schools in total in Ulsan Metropolitan City). As with other essential facilities, schools in Gyeongju had the highest probability of damage. In the analysis, it was assumed that school buildings with less than 50% functionality were not able to perform well as emergency shelters. Forty of 116 schools were expected to be operating at less than 50% functionality on the day of the earthquake. This number was expected to decrease to 35 by 3 days after the earthquake. All schools were expected to recover at least 50% of their functionality

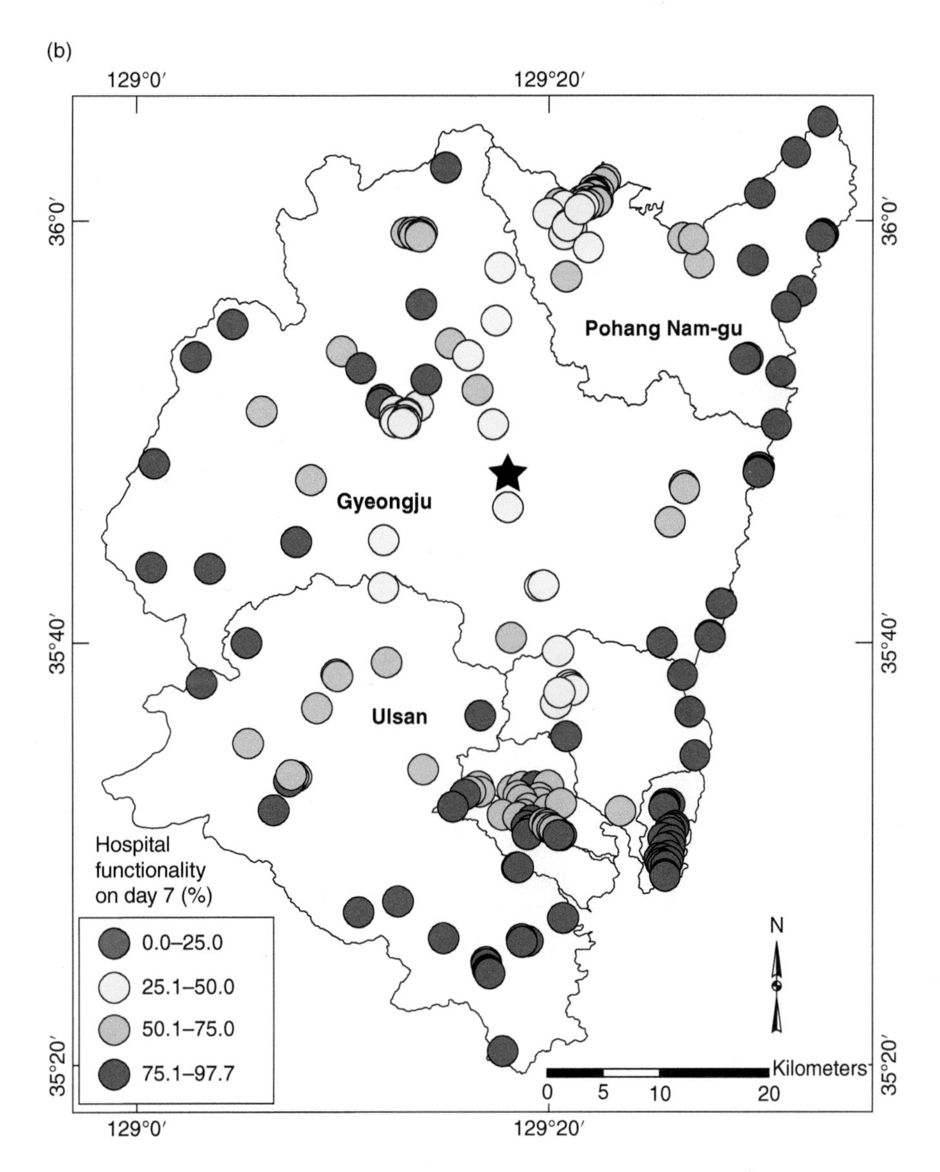

Figure 10.6 (Continued)

within 7 days. For Pohang Nam-gu, 6 of the 62 schools were still operating below 50% functionality 3 days after the earthquake. All schools in Pohang Nam-gu were expected to recover at least 50% functionality within 7 days. In Ulsan Metropolitan City, 3 schools in Ulsan Buk-gu and 1 in Ulsan Ulju-gun were expected to be operating at less than 50% functionality 3 days after the earthquake (Fig. 10.7a).

Police stations and fire stations are the control centers where decision makers can prioritize and allocate emergency resources under panic conditions. There are 76 police stations in the study area, with 17 in Gyeongju, 18 in Pohang Nam-gu, 6 in Ulsan Buk-gu, 6 in Ulsan Dong-gu, 13 in Ulsan Nam-gu, 5 in Ulsan Jung-gu, and 11 in Ulsan Ulju-gun. Sixty-three of them remained operational at above 50% functionality. Eight police stations in Gyeongju and 5 in Pohang Nam-gu, were expected to be operating at less than 50% functionality. HAZUS estimated that all of the police stations in Ulsan Metropolitan City maintained at least 50% functionality on the day of the earthquake (Fig. 10.7b).

There are 32 fire stations in the study area, with 12 in Gyeongju, 5 in Pohang Nam-gu, 2 in Ulsan Buk-gu, 4 in Ulsan Nam-gu, 1 in Ulsan Jung-gu, and 8 in Ulsan Ulju-gun. Twenty-three of them were expected to retain a functionality of above 50% on the day of the earthquake. All of the fire stations were predicted to recover 50% functionality within 7 days of the earthquake. Estimates of fire station functionality indicated that 7 fire stations in Gyeongju and 2 in Pohang Nam-gu, would be less than 50% functional on the day of the earthquake. All of the fire stations in Ulsan Metropolitan City were expected to have more than 50% functionality on the day of the earthquake (Fig. 10.7c).

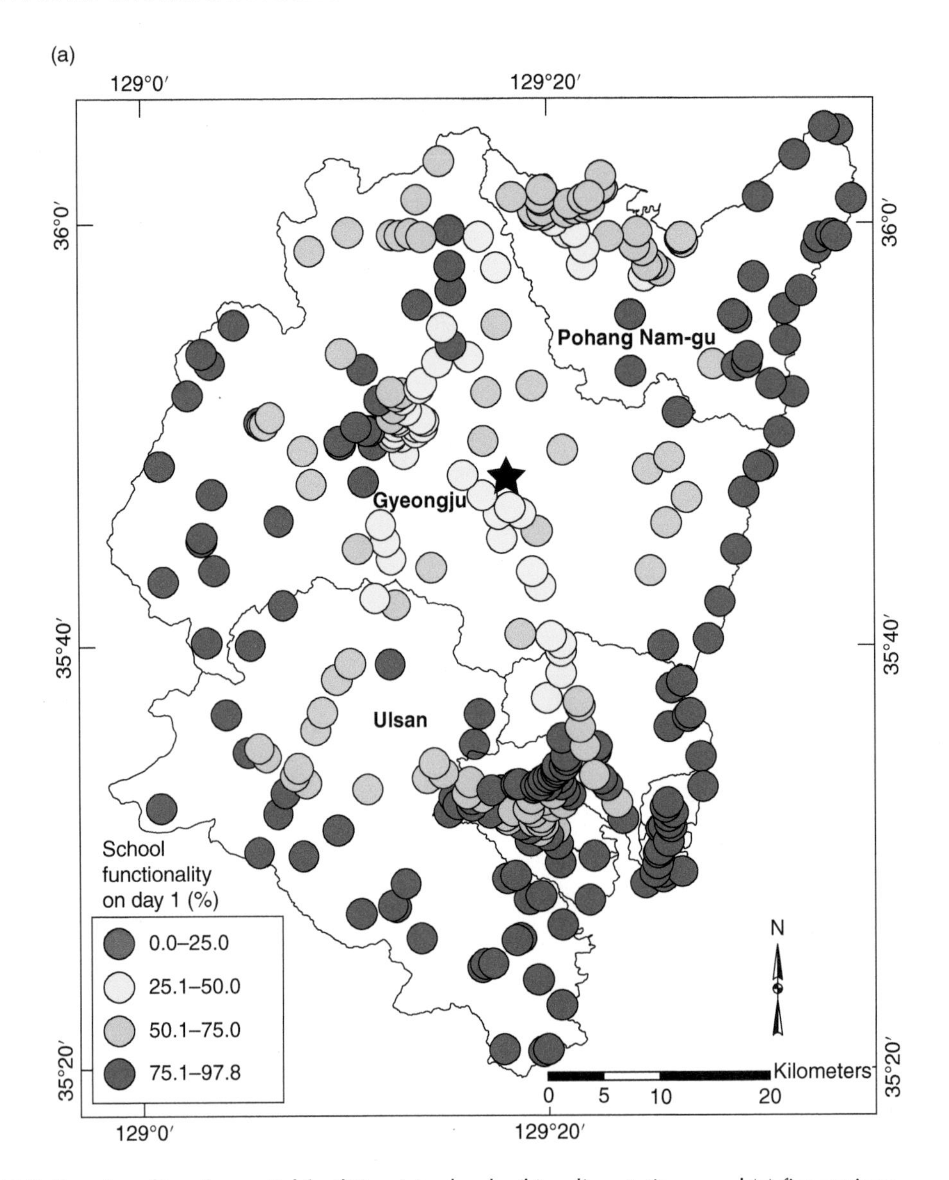

Figure 10.7 Functionality of essential facilities: (a) schools, (b) police stations, and (c) fire stations.

10.6.2.3. Fires, Casualties, and People Seeking Temporary Shelter

Fires may cause significant damage following an earthquake. The scenario results estimated that approximately 1000 fires would break out in the study region, with 79,132 people (5.1% of the population of the study area) exposed to fire. A high risk of 470 and 140 fire breakouts was estimated in Gyeongju and Pohang Nam-gu, respectively, and Ulsan Jung-gu was estimated to have more people exposed to fires. The number of people exposed to the fires were 19,197, 18,492, and 14,203 in Ulsan Nam-gu, Ulsan Jung-gu, and Gyeongju, respectively. Note that Ulsan Nam-gu was estimated to have the largest number of people exposed to fires.

The scenario results estimated that 19,765 people (1.3% of the population) would seek temporary shelter in the study region (Fig. 10.8), with smaller figures of 6196 in Pohang Nam-gu, 5322 in Gyeongju, and 4024 in Ulsan Nam-gu.

The casualty estimates varied according to the time of day. HAZUS estimates the casualty numbers for 2:00 a.m., 2:00 p.m., and 5:00 p.m. These times represent the periods of the day when different sectors of the community are at their peak occupancy levels. The casualties in HAZUS are broken down into four severity levels that describe the extent of injuries [*FEMA*, 2009]: injuries that require medical attention but not hospitalization (Severity Level 1), injuries that require hospitalization but that are not considered life threatening (Severity Level 2), injuries that require hospitalization and can become life threatening if not treated promptly (Severity Level 3), and victims who are killed by the earthquake (Severity Level 4).

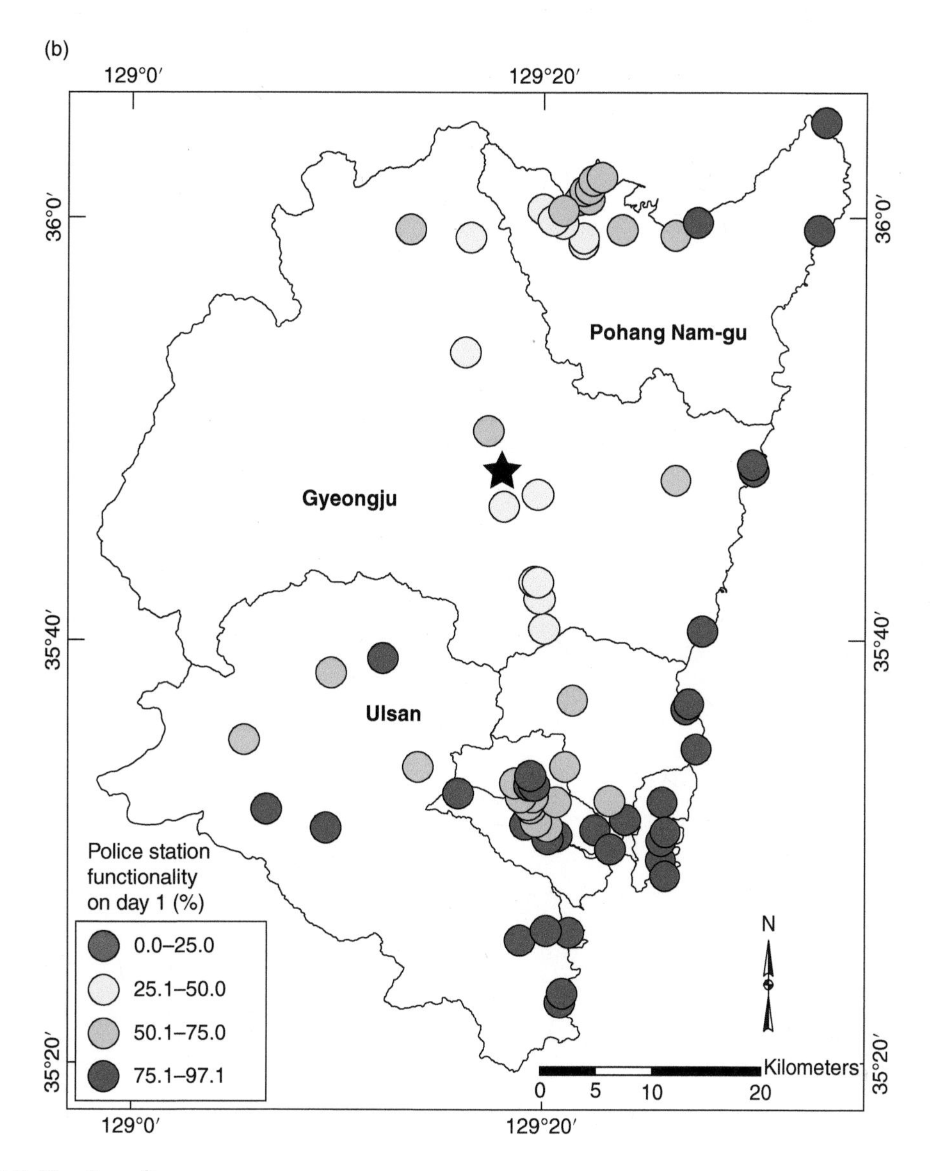

Figure 10.7 (Continued)

Casualties were predicted to be highest when the earthquake occurred in the early morning (2:00 a.m.) because residential occupancy would be at its maximum. At this time, the estimated numbers of casualties were 3566, 894, 145, and 288 for Severity Levels 1, 2, 3, and 4, respectively (Fig. 10.9). The 288 fatalities included 128 in Gyeongju, 85 in Pohang Nam-gu, and 35 in Ulsan Nam-gu. The 2:00 p.m. estimate considers that the occupancy level of educational, commercial, and industrial buildings would be at a maximum. The numbers of casualties were predicted to be 684, 169, 26, and 53 for Severity Levels 1, 2, 3, and 4, respectively. There were 23 and 16 fatalities predicted in Gyeongju and Pohang Nam-gu, respectively. More casualties were predicted for the 5:00 p.m. time, which represents the peak commuting time, than for 2:00 p.m., but fewer than for 2:00 a.m. The numbers of casualties at 5:00 p.m. were predicted to be 1366, 337, 55, and 106 for Severity Levels 1, 2, 3, and 4, respectively. There were 47 and 31 fatalities predicted in Gyeongju and Pohang Nam-gu, respectively, and only 13 in Ulsan Nam-gu.

10.6.3. Comparisons of Loss Estimates for Residential Buildings with and without the Use of the Site Classification Map

Damage to residential buildings as revealed using the site classification map was considered in a previous section. The results indicated that Pohang Nam-gu would experience a large amount of damage that was comparable to the damage predicted in Gyeongju despite its relatively large distance from the epicenter

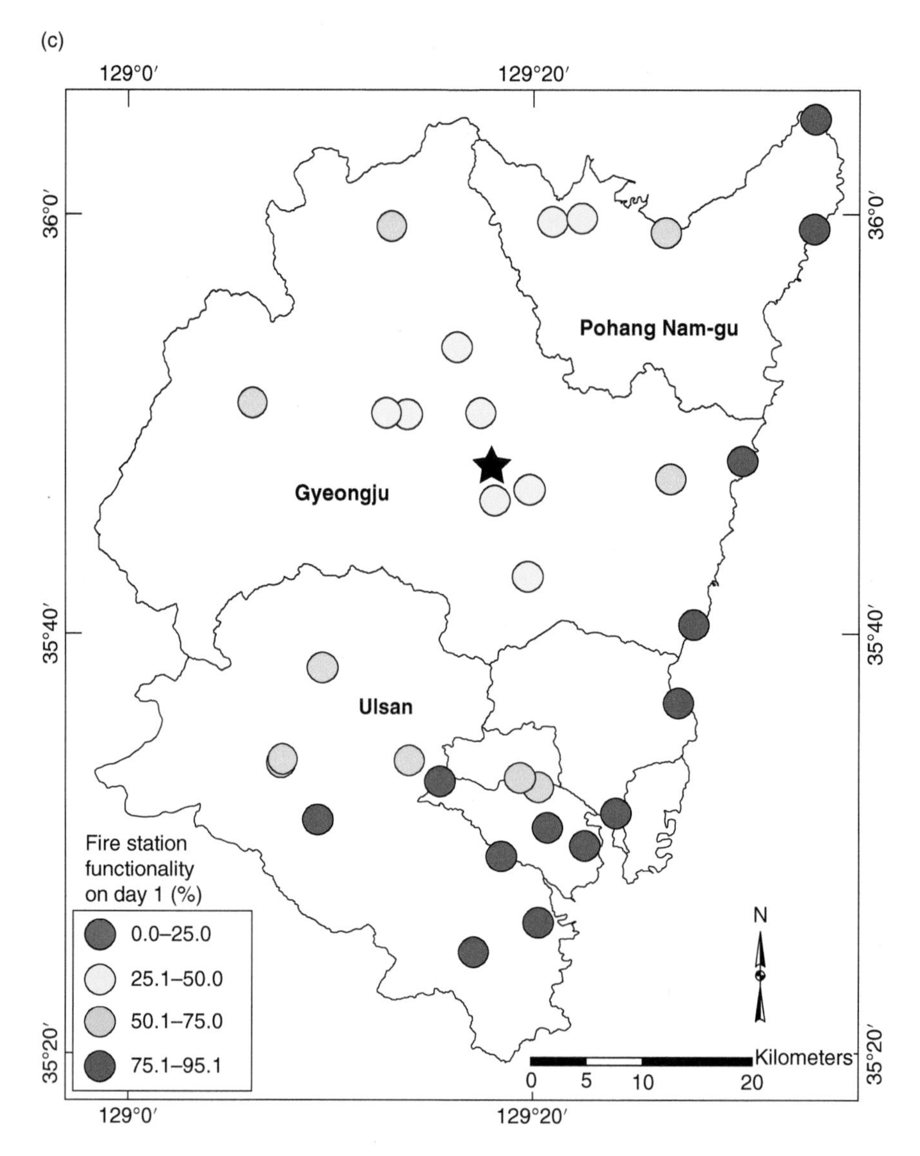

Figure 10.7 (Continued)

of the scenario earthquake. The large amount of damage partially represents the effect of the soft soil conditions in the shallow subsurface of Pohang Nam-gu. Without the use of site information, the estimate of damage to the residential buildings in Gyeongju was much greater than that for the residential buildings in Pohang Nam-gu. Similar predictions were made for Ulsan Nam-gu and Ulsan Ulju-gun. Damage to residential buildings in Ulsan Nam-gu was estimated to be greater than that in Ulsan Ulju-gun. The soft soils that are prevalent in Pohang Nam-gu would amplify the ground motion, resulting in greater damage. However, estimates made without the use of site conditions, indicated greater damage in Ulsan Ulju-gun than in Ulsan Nam-gu, mainly due to Ulsan Ulju-gun's proximity to the epicenter of the scenario earthquake. In general, damage estimates using the site classification were intermediate between the rough estimates of site classes B and C (Fig. 10.10).

10.7. CONCLUSIONS

The prediction of potential losses from natural disasters helps authorities to make informed decisions. Knowledge and understanding of potential losses and their variations across a region are always important for decision makers who have to develop emergency plans. They are also critical in deciding priorities and allocating resources under crisis conditions. Emergency crews and the general public are more able to mitigate the damage caused by earthquakes through carefully orchestrated operations based on informed decisions.

The extent of ground shaking and the subsequent losses can vary dramatically across a region due to

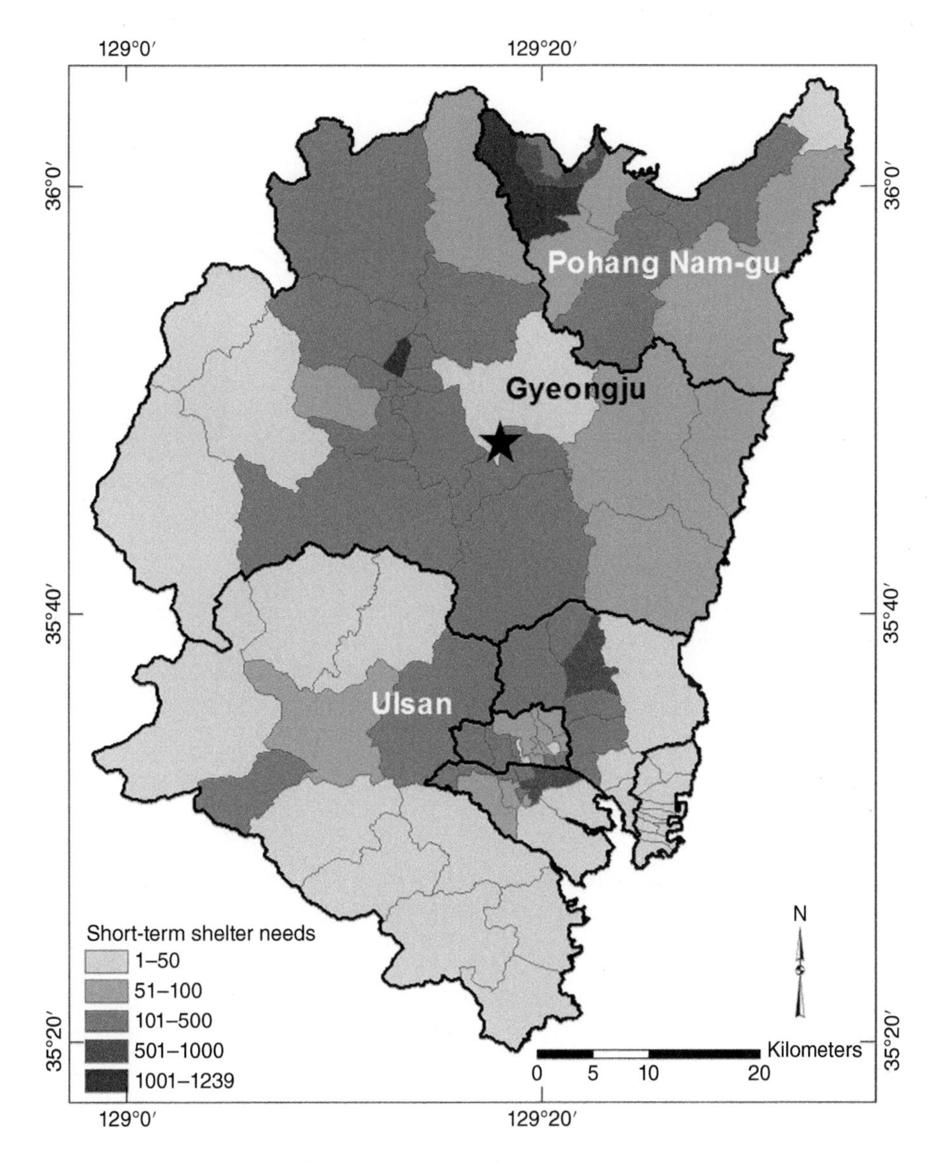

Figure 10.8 Distribution of people seeking temporary shelter.

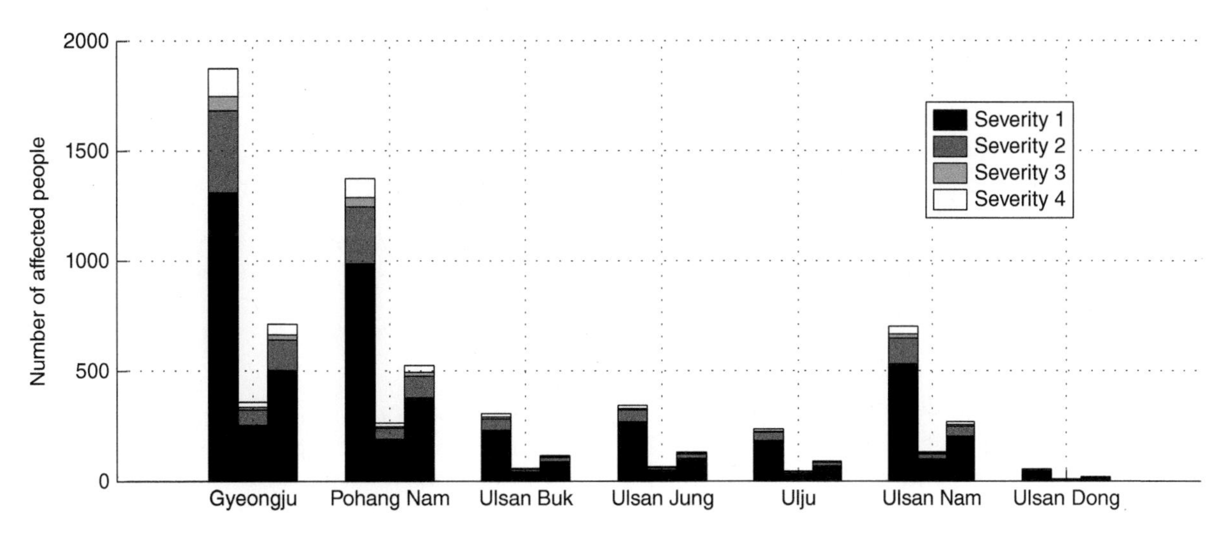

Figure 10.9 Estimated casualties and fatalities for an earthquake occurring at 2 a.m., 2 p.m., and 5 p.m. from left to right. Estimates are also grouped by jurisdiction.

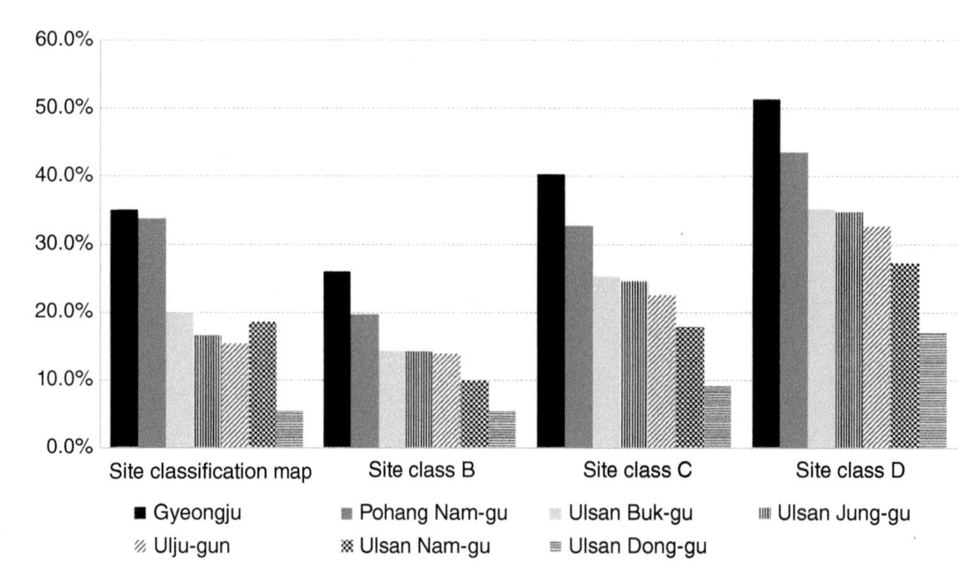

Figure 10.10 Comparison of moderate or severe damage to residential buildings with and without the use of a site classification map by jurisdiction.

variations in site conditions and distance from the source. The use of detailed local data is therefore essential for obtaining accurate loss estimations. Loss estimation associated with a scenario earthquake using HAZUS and a site classification map were used to provide a picture of potential earthquake damage. The losses due to the scenario earthquake were estimated and compared with and without the use of a site classification map of the study region. Various types of damage were addressed.

HAZUS predicted that the Gyeongju area would experience a large amount of damage to residential buildings, schools, and fire stations following a magnitude 6.7 scenario earthquake located in Gyeongju. Relatively large amounts of damage to police stations and hospitals were estimated in the Pohang Nam-gu area. If the earthquake occurred at 2:00 a.m., 128 fatalities were predicted in Gyeongju and approximately 6200 people were estimated to require temporary shelter. It is noteworthy that Pohang Nam-gu would experience comparable levels of damage to Gyeongju despite its large distance from the epicenter of the scenario earthquake. Although the distance from the epicenter to Ulsan Nam-gu is larger than the distance to Ulsan Ulju-gun, projected damage to the residential buildings in Ulsan Nam-gu was greater than that for Ulsan Ulju-gun because of the effect of the soft soils in the shallow subsurface of the Ulsan Nam-gu area. In Ulsan Dong-gu, located at distance from the epicenter comparable to Pohang Nam-gu, less damage was estimated due to the firm ground conditions.

ACKNOWLEDGMENTS

The authors would like to express their appreciation for valuable comments and suggestions from two anonymous reviewers. Some aspects of details in this paper were reported in companion papers [*Kang et al.*, 2007b; *Kang and Kim*, 2011, 2012]. This work was supported by Korea Meteorological Administration Research and Development under KMIPA2015-3010. This work was also supported by the National Research Foundation of Korea (NRF) and the Center for Women in Science, Engineering and Technology (WISET) Grant funded by the Ministry of Science, ICT & Future Planning of Korea (MSIP) under the Program for Returners into R&D (KW-2015-PPD-0129).

REFERENCES

Allen, T. I., and D. J. Wald (2009), On the use of high-resolution topographic data as a proxy for seismic site conditions (Vs30), *Bull. Seis. Soc. Amer.*, *99*(2A), 935–943.

Castellaro, S., F. Mulargia, and P. L. Rossi (2008), Vs30: Proxy for seismic amplification? *Seis. Res. Lett.*, *79*(4), 540–543.

Clarke, K. C. (2010), *Getting Started with GIS*, 5th ed., Prentice Hall.

Collins, C., R. Kayen, B. Carkin, T. Allen, P. Cummins, and A. McPherson (2006), Shear wave velocity measurement at Australian ground motion seismometer sites by the spectral analysis of surface waves (SASW) method, *Earthquake Engineering in Australia*, Canberra, 24–26 November 2006, 173–178.

FEMA (2009), *Multi Hazard Loss Estimation Methodology Earthquake Model, User Manual*, Federal Emergency Management Agency, Washington, D.C.

Kang, S. Y., and K. H. Kim (2009), A case study of GIS-based site classification in the Gyeongsang Province constrained by geologic and topographic information, *J. Korean Assoc. Geo. Inf. Stud.*, *12*(4), 136–145 (in Korean with English abstract).

Kang, S. Y., and K. H. Kim (2011), A simulation of earthquake loss estimation with application of a site classification map, *J. Korean Soc. Haz. Mit.*, *11*(6), 67–75 (in Korean with English abstract).

Kang, S. Y., and K. H. Kim (2012), Developing a seismic site classification map of Korea using geologic and topographic maps, *Disaster Adv.*, *5*(2), 16–25.

Kang, S. Y., K. H. Kim, D. C. Kim, H. S. Yoo, D. J. Min, and B. C. Suk (2007a), A preliminary study of the global application of HAZUS and ShakeMap for loss estimation from a scenario earthquake in the Korean Peninsula, *J. Korean Assoc. Geo. Inf. Stud.*, *10*(1), 47–59 (in Korean with English abstract).

Kang, S. Y., K. H. Kim, B. C. Suk, and H. S. Yoo (2007b), Attenuation relations in HAZUS for earthquake loss estimations in Korea, *J. Earthquake Eng. Soc. Korea*, *11*(6), 15–21 (in Korean with English abstract).

Kang, S. Y., K. H. Kim, B. C. Suk, and H. S. Yoo (2008), A simulation of earthquake loss estimation for a Gyeongju event, *J. Korean Soc. Haz. Mit.*, *8*(3), 95–103 (in Korean with English abstract).

KIGAM (2010), Geology map of Korea, Daejeon, Korea Institute of Geosciences and Mineral Resources

Kim, K. H. (2010), *Spatial Analysis*, Munundang, Seoul (in Korean).

Kim, W. Y., and K. H. Kim (2014), The 9 February 2010 Siheung, Korea, earthquake sequence: Repeating earthquakes in a stable continental region, *Seism. Soc. Amer.*, *104*, 551–559.

KMA (2010), *Historic Earthquakes of Korea (II)*, Korea National University of Education, Cheongju (in Korean).

KMA (2013), 1978–2013 Earthquake Catalogues, last accessed on 15 October 2013, http://www.kma.go.kr/, Seoul.

Koh, J. H., J. H. Kwon, and Y. S. Choi (2005), Error assessment of attitude determination using wireless internet-based DGPS, *J. Korean Soc. Survey Geodesy Photo. Cartog.*, *23*(3), 239–249 (in Korean with English abstract).

KOSIS (2007), Statistical Database, last visited in April 2007, http://www.kosis.kr/, Korean Statistical Information Service, Daejeon.

Kyung, J. B. (2012), Characteristics of damaging earthquakes occurred in Seoul metropolitan area for the last two thousand years, *J. Korean Earth Sci. Soc.*, *33*(7), 637–644 (in Korean with English abstract).

Lee, C. T., C. T. Cheng, C. W. Liao, and Y. B. Tsai (2001), Site classification of Taiwan free-field strong-motion stations, *Bull. Seis. Soc. Amer.*, *91*(5), 1283–1297.

Lee, K. H., and T. G. Lee (1997), An analysis of seismic risk of Seoul area (I), *J. Earthquake Eng. Soc. Korea*, *1*(4), 29–35 (in Korean with English abstract).

Park, S., and S. Elrick (1998), Predictions of shear-wave velocities in Southern California using surface geology, *Bull. Seis. Soc. Amer.*, *88*, 677–685.

Sadigh, K., C. Y. Chang, J. Egan, F. Makdisi, and R. R. Youngs (1997), Attenuation relationships for shallow crustal earthquakes based on California strong motion data, *Seis. Res. Lett.*, *68*(1), 180–189.

Sun, C. G., and C. K. Chung (2008), Regional estimation of site-specific seismic responses at Gyeongju by building a GIS-based geotechnical information system, *J. Korean Assoc. Geo. Inf. Stud.*, *11*(2), 38–50 (in Korean with English abstract).

USGS (2008), Global 1 and 3 arc-second elevation (Shuttle Radar Topography Mission), http://eros.usgs.gov/elevation-products, last accessed on 10 November 2009.

Wald, D. J., and T. I. Allen (2007), Topographic slope as a proxy for seismic site condition and amplification, *Bull. Seis. Soc. Amer.*, *97*(5), 1379–1395.

11

Implications of Different Digital Elevation Models and Preprocessing Techniques to Delineate Debris Flow Inundation Hazard Zones in El Salvador

Eric Ross Anderson,[1] Robert E. Griffin,[2] and Daniel E. Irwin[3]

ABSTRACT

Landslides caused by heavy rains on steep, volcanic slopes pose a persistent threat in El Salvador. Particularly for volcanic debris flows (lahars), limitations in digital elevation model (DEM) spatial resolution, acquisition date of the data, and hydrological preprocessing techniques can contribute to inaccurate hazard-zone delineations. This study investigates these limitations to determine which combination of methods most closely agrees with observed lahar events on San Vicente volcano. A national DEM offers high spatial resolution (10 m horizontal) but may lack recent natural and anthropogenic terrain modifications. Global DEMs are more recent but are of lower spatial resolution (typically, 30 or 90 m). Moreover, traditional hydrological preprocessing may lead to drastic modifications of original elevation values. Optimized pit-filling techniques are designed to minimize such modifications. Satellite image interpretation and field surveying provide observations to validate model simulations. Results show that the national DEM yielded the highest overall accuracy. The hydrologic preprocessing techniques employed affected the extent of the debris-flow zones only under those modeled using global DEMs. Global DEMs persistently missed a historically affected town. This implies that for areas lacking high resolution DEMs, the wrong combination of input data and preprocessing technique could lead to misinformation.

11.1. INTRODUCTION

El Salvador in Central America has documented over 1000 landslide events in its history, many of them causing tens to hundreds of deaths and thousands of buildings and homes damaged [*MARN*, 2013; *OSSO et al.*, 2013]. Often triggered by intense and prolonged rains on steep slopes, landslides pose a threat to populations and infrastructure every year, whether they are shallow landslides or large, volcanic lahars, defined by fast-moving masses of debris, mud, boulders, and water. In 2009, the convergence of a low-pressure system from the Pacific Ocean with Tropical Storm Ida from the Atlantic triggered several lahars on the San Vicente volcano that destroyed entire blocks of the towns downstream [*Government of El Salvador*, 2009; *CEPAL*, 2010]. Neighboring countries in Central America face similar challenges in managing geological and hydrometeorological hazards.

Although potential debris-flow inundation hazard zones had been delineated using digital elevation models (DEMs) [*Major et al.*, 2004], and generally agreed with actual lahars in 2009, some disparities existed between the simulated zones and actual affected areas. Limitations in DEM spatial resolution, publication date of the data, and hydrological preprocessing techniques can contribute to inaccurate hazard zone delineations. These discrepancies

[1] *Earth System Science Center, University of Alabama in Huntsville, Huntsville, Alabama, USA*

[2] *Atmospheric Science Department, University of Alabama in Huntsville, Huntsville, Alabama, USA*

[3] *NASA Marshall Space Flight Center, Huntsville, Alabama, USA*

Natural Hazard Uncertainty Assessment: Modeling and Decision Support, Geophysical Monograph 223, First Edition.
Edited by Karin Riley, Peter Webley, and Matthew Thompson.

could ultimately hinder proper hazard zonation and evacuation decision making.

Landslide risk management is discussed at great length in the literature [*Guzzetti et al.*, 1999; *Dai et al.*, 2002; *Metternicht et al.*, 2005; *van Westen et al.*, 2006]. DEMs are often used as foundational inputs for geospatial analyses of landslide hazards, but numerous uncertainties are associated with such model inputs. First, in practical terms, the DEMs used for landslide hazard characterization are often selected simply by the availability of data, not according to the scale of the hazard. Implications in the resulting hazard maps and decisions made around those outputs must be considered. Moreover, uncertainty can be introduced in hydrological preprocessing techniques, particularly pit or sink filling. DEMs frequently contain erroneous depressions, pits, or sinks, in the surface, often a result of either interpolating contour lines or signals from radar, lidar, or stereo-pair images [*O'Callaghan and Mark*, 1984; *Jenson and Domingue*, 1988].

It is possible to delineate potential debris-flow inundation hazard zones based on DEMs of varying resolutions. In El Salvador, a national DEM of 10 m horizontal spatial resolution offers the finest level of detail, but it was prepared with survey data from the 1970s and 1980s [*CENTA and FAO*, 1998]. Global products from the Shuttle Radar Topography Mission (SRTM) and the Advanced Spaceborne Thermal Emission and Reflection Radiometer (ASTER) also provide elevation data [*Jarvis et al.*, 2008; *Tachikawa et al.*, 2011]; however, their ability to capture the detail of steep valleys and flow paths can be hindered by low spatial resolution and the particular remote sensing techniques for elevation determination (Interferometric Synthetic Aperture Radar [InSAR] and multilook stereoscopy). The confusion of forest canopy for bare earth also comes into play [*Farr and Kobrick*, 2000; *Hirano et al.*, 2003; *Hubbard et al.*, 2007; *Muñoz-Salinas et al.*, 2009].

Regarding errors introduced during preprocessing, false pits must be filled before hydrological or surface analysis can occur [*O'Callaghan and Mark*, 1984; *Jenson and Domingue*, 1988]. Traditionally, pit removal changes meaningful topography in flat areas and can cause surface runoff to flow in paths that differ from reality [*Soille*, 2004]. The combination of improper spatial resolution and different preprocessing techniques could lead to increased levels of uncertainty in model outputs.

This study investigates the effects of using different elevation models and pit-filling techniques in the final debris hazard zone delineations, in an effort to determine which combination of methods most closely agrees with observed landslide events.

11.2. METHODS

11.2.1. Geographic Area, Spatiotemporal Scope

El Salvador is located in the intertropical convergence zone and has roughly a 7–8 mo rainy season [*MARN*, 2015]. Floods and landslides are common, and the magnitude of their impacts is closely related to the exposure of populations. With over six million people inhabiting roughly 21,000 square kilometers, El Salvador has the highest population density in Central America [*CIA*, 2013]. The capital of San Salvador is home to nearly a third of the population [*OPAMSS*, 2010]. The present area of study includes the northern flanks of the San Vicente volcano within the San Vicente department located in the center of the country. The main geological formations of the study area include andesitic and basaltic effusives and pyroclastic rocks at the upper part of the volcano, roughly a 1–2 km radius [*Weber et al.*, 1978]. Further downslope on the northern flanks is "tierra Blanca," or "white earth," a remnant of the catastrophic Tierra Blanca Joven eruption thought to have caused a population crash in Mesoamerica around AD 536 [*Dull et al.*, 2001]. These pyroclastic and epiclastic rocks are known to be poor materials for buildings and roads and are very susceptible to washout during heavy rains [*Schmidt-Thomé*, 1975; *Weber et al.*, 1978]. San Vicente is one in a string of volcanoes that runs through Central America, forming some of the youngest lithology in the region [*Weber et al.*, 1978].

Agricultural activity is widespread throughout the country, permanent crops covering much of the flanks and climbing the steep slopes of the San Salvador volcano [*CATHALAC*, 2011]. The towns of Guadalupe, Verapaz, and Tepetitán, and populations living along Gully los Infiernillos, are all adjacent to rivers and streams originating near the top of the volcano.

The current study focuses on the characterization of spatial uncertainty of debris-flow hazard zones surrounding the November 2009 lahars, for which substantial field and remote sensing observations are available [*Anderson et al.*, 2012]. The scientific community has acknowledged a great challenge in knowing exactly when and where lahars can occur, and this study focuses specifically on characterizing uncertainties in the spatial extent of these events.

11.2.2. Inputs and Tools

The main inputs required to delineate debris-flow inundation hazard zones are a DEM, the location where the lahar initiates, is transported, and subsequently deposited, as well as the overall volume of the lahar. As such, the quality of the DEM and preprocessing techniques are

Table 11.1 DEM Sources and Characteristics

DEM source	Date	Spatial resolution	Strengths	Weaknesses
National	1970s–1980s	10 m	Fine spatial detail; synthesized rivers matched cartographic representation well	Out-dated
ASTER GDEM v2	2003–2011	30 m	Most up to date; greater potential for further updates	Coarse spatial resolution; inconsistent across globe; most error in cloudy regions
SRTM v4.1	2000	90 m	Most consistent across the globe	Coarsest spatial resolution

immediate sources of potential error. Even though the national DEM is of the highest spatial resolution, 10m horizontal grid spacing, it was prepared with survey data from the 1970s and 1980s. This implies that recent surface features and new or modified flow channels are not adequately represented. Two other DEMs provide more recent elevation measurements but lack the spatial detail of the national DEM. The ASTER Global DEM (GDEM) version 2 is composed of individual ASTER scenes from 2003 to 2011, and has a 30m spatial resolution [*Tachikawa et al.*, 2011]. The Shuttle Radar Topography Mission (SRTM) DEM version 4.1 has gone through more rigorous correction processes but has a coarser spatial resolution of 90m [*Jarvis et al.*, 2008]. Others have investigated the effects of different DEM sources and resolutions on lahar delineations and hydrologic and erosion modeling, in other areas. [*Zhang and Montgomery*, 1994; *Stevens et al.*, 2007; *Zhang et al.*, 2008]. A summary of the DEM sources and characteristics we tested for this case in El Salvador is presented in Table 11.1.

The debris-flow simulation tool, LAHARZ, was developed by the US Geological Survey (USGS) [*Schilling*, 1998]. The first model simulations were made with the program as available in ArcInfo Workstation Advanced Macro Language (AML), which is no longer supported by Environmental Systems Research Institute (Esri). More recently, USGS has made this program available in Python scripts for more direct incorporation into an ArcGIS working environment, which we used to rerun simulations and calculate confidence intervals [*Schilling*, 2014]. LAHARZ is an appropriate tool for this study for the primary reason that it is free and openly available and thus has a high level of accessibility as a tool for decision makers across many countries with lahar prone areas. By default, LAHARZ and many other hydrological toolkits rely on the default ArcInfo Spatial Analyst (SA) "Fill" tool to remove false depressions or pits in the DEMs [*O'Callaghan and Mark*, 1984; *Jenson and Domingue*, 1988].

Besides testing different DEMs, the other objective of this study is to test different pit filling techniques. The Optimized Pit Removal (OPR) technique is designed to provide a hydrologically conditioned surface without modifying original elevation values so drastically. Instead of raising any and all false depressions, OPR also cuts and fills surrounding artifacts (i.e., false peaks) that might have inadvertently created a false depression [*Soille*, 2004; *Jackson*, 2013]. We tested a total of three hydrological conditioning techniques, the ArcInfo SA Fill Tool, and two approaches for optimizing pit removal with OPR. A summary of these techniques is shown in Table 11.2.

For this study, we focused on the five major lahars that were triggered by approximately 355mm of rain in 24hr between 7 and 8 November 2009, on the San Vicente volcano [*Government of El Salvador*, 2009]. We obtained the initiation points from landslide catalog entries (Fig. 11.1) [*MARN*, 2013] and verified them with high resolution satellite imagery. These lahars that devastated the towns surrounding the volcano had field-estimated volumes that ranged from 240,000 to 370,000 m^3, mostly containing soil, rocks, and trees [*MARN*, 2013]. A detailed field report on the lahars considered in this study can be found in [*SNET*, 2009]. This experiment setup resulted in nine combinations of input DEM and hydrological conditioning technique, each simulated from five lahars initiation points, totaling 45 lahar simulations.

In order to validate debris-flow simulations generated with LAHARZ, we used observations derived from a Formosat-2 visible and near infrared satellite image, and verified in the field. The 2m horizontal resolution of these images offered some of the finest resolution characterization of the 2009 lahars for runout length and width of debris flow and deposit features at the study sites. We conducted basic image processing in ENVI, version 5.0. Finally, we merged simulations and observations in ArcGIS to assess accuracy.

11.2.3. Debris-Flow Simulation

The LAHARZ model requires that input DEMs be hydrologically corrected. By default the SA Fill Tool is employed, but model customization allows for user-generated inputs. For these we used one SA method and two OPR fill methods. The OPR Minimum Absolute Change (MinAbs) method both cuts and fills cells while minimizing

Table 11.2 Hydrological Conditioning Techniques

Technique	Notes	Sample profile view
ArcInfo Spatial Analyst–Fill Tool (SA)	Only fills (no cuts) by raising all pits to create surface with continuous flow	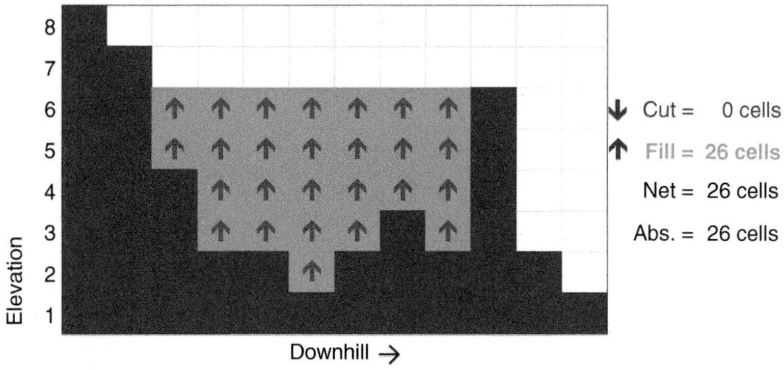 Elevation 1–8; Downhill →; Cut = 0 cells; Fill = 26 cells; Net = 26 cells; Abs. = 26 cells
Optimized Pit Removal Tool–Minimum Absolute Change (OPR MinAbs)	Cuts and fills while minimizing the absolute number of cells changed	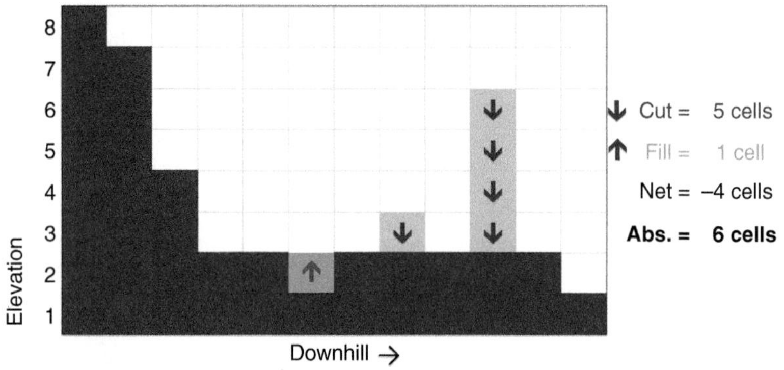 Elevation 1–8; Downhill →; Cut = 5 cells; Fill = 1 cell; Net = –4 cells; **Abs. = 6 cells**
Optimized Pit Removal Tool–Minimum Net Change (OPR MinNet)	Cuts and fills while minimizing the net number of cells changed	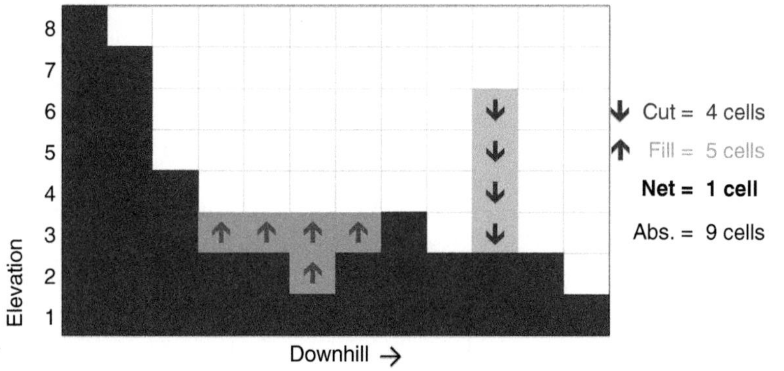 Elevation 1–8; Downhill →; Cut = 4 cells; Fill = 5 cells; **Net = 1 cell**; Abs. = 9 cells

Source: Adapted from *Jackson* [2013].

the absolute number of cells changed. The OPR Minimum Net Change (MinNet) technique also cuts and fills but minimizes the net number of cells changed (see Table 11.2). SA and both OPR techniques were applied to the three test DEMs. All geospatial datasets were projected to the Lambert Conformal Conic system (North American 1927 datum).

Once input DEMs were prepared, the LAHARZ package derived flow direction, flow accumulation, and stream delineation. We assured that flow paths were detected at least as far uphill as the designated lahar initiation points. Using the initiation points shown in Figure 11.1, we ran LAHARZ, simulating debris flows of 370,000 m^3 for each of the two Guadalupe lahars, 240,000 m^3 for the Verapaz lahar, 250,000 m^3 for the Gully los Infiernillos lahar, and 300,000 m^3 for the Tepetitán lahar, referring to the El Salvador Ministry of the Environment and Natural Resources (MARN) General Directorate of the Environmental Observatory (DGOA) landslide inventory for debris-flow volumes [*MARN*, 2013]. As described in [*SNET*, 2009], these estimates were made from field visits and satellite image interpretation. Others have described in greater detail fieldwork methods for lahar modeling in other parts of the world [*Canuti et al.*, 2002; *Johnson*, 1994]. To address uncertainties in field estimations of debris-flow volumes and the difficulty of selecting a single volume in the LAHARZ model, we ran

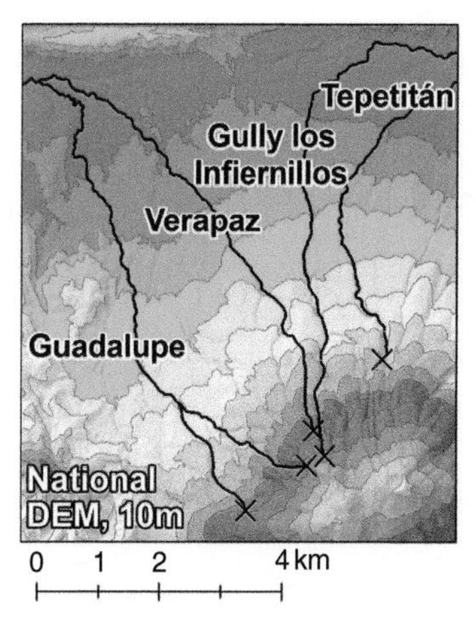

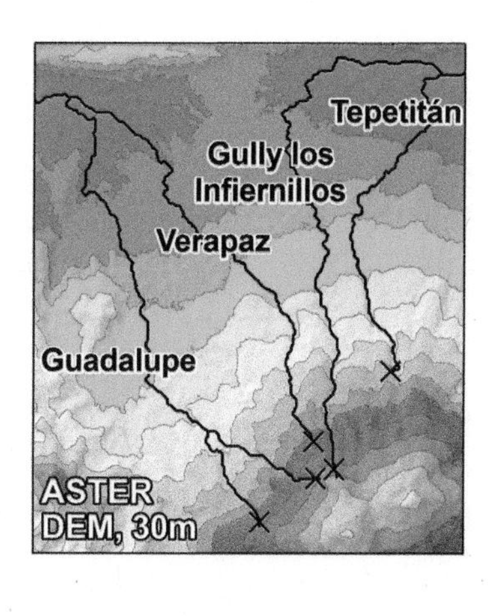

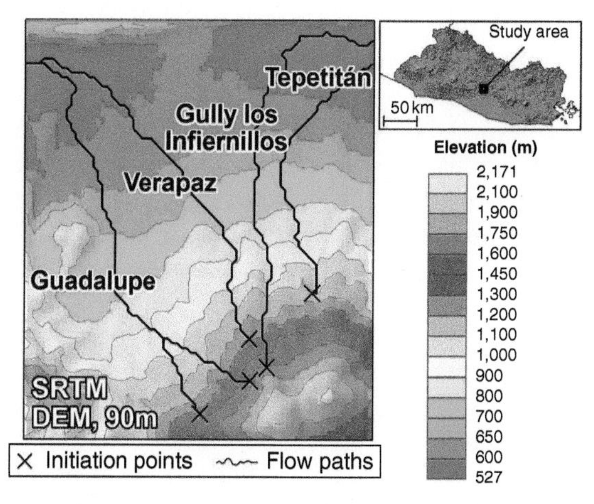

Figure 11.1 The three digital elevation models tested with their respective flow paths for five lahar initiation points on the San Vicente volcano, El Salvador.

additional simulations to calculate ranges of planimetric inundation area with a 95% confidence interval.

We then obtained lahar inundation hazard zones as geospatial outputs of the LAHARZ program, which followed the nearest flow path downhill and distributed the debris volume along and across the valley. An empirically derived relationship between landslide volume and inundated area can be expressed by:

$$A = 0.5 \cdot V^{\frac{2}{3}};\ B = 200 \cdot V^{\frac{2}{3}} \tag{11.1}$$

where A is the maximum cross-sectional area inundated, B is the total planimetric area inundated, for a given volume, V [*Iverson et al.*, 1998; *Griswold and Iverson*, 2008]. The final output of the program was a series of two-dimensional potential inundation hazard zones, defined by the above relationship between cross-sectional and planimetric area and defined volume.

11.2.4. Model Validation

In order to assess the performance of each runout model simulation, we followed the approach demonstrated by *Cepeda et al.* [2010] by calculating "true positive rates" and "false positive rates." The observation dataset principally came from Formosat-2 satellite image interpretation. Due to the expedited delivery format of the image, we applied further georeferencing and orthorectification to align the image correctly with respect to local topography and built features. First, the national DEM served as a guide for topographic referencing along valleys, ridges, and the cone of the volcano. Second, the Esri ArcGIS Imagery basemap provided locations of roads, buildings, and farms on the more gently sloped flanks. Twenty-six control points using the DEM as a reference and 15 control points using the base imagery fed into a spline transformation. This resulted in a more exact representation of the intricate valleys and ridges along the volcano, while also ensuring that built features such as roads and buildings also aligned correctly. Since spline transformations can lead to extreme warping beyond control points, we located control points at a strategic distribution throughout the Formosat-2 image extending to well beyond the reach of lahar inundation zones.

The observation of the extents of the five lahar inundation zones was based on a spectral response of vegetation in the red and near infrared channels. Others have demonstrated a combination of spectral analysis and LAHARZ simulations to delineate lahars from satellite images [*Davila et al.*, 2007]. First, we applied an internal average reflectance (IAR) atmospheric correction to effectively convert radiance data to reflectance [*Exelis*, 2013]. Then we used the infrared band (B_4) and red band (B_3) to calculate the Normalized Vegetation Difference Index (NDVI):

$$NDVI = \frac{B_4 - B_3}{B_4 + B_3} \tag{11.2}$$

where NDVI ranges from −1 to 1 and the higher values indicating greater photosynthetic activity [*Van de Griend and Owe*, 1993]. Lower values can be interpreted as lack of or little vegetation such as bare soil, populated places, water, shadow, or clouds. To minimize the likelihood that lahars might be visually missed, we conducted a first pass threshold and retained cells with NDVI values of 0.3 and lower to begin to separate the lahar signature (and other areas without vegetation) from standing undisturbed vegetation. Postprocessing honed the observation dataset by masking the result to retain only low NDVI pixels in and adjacent to valleys and affected towns. Moreover, parts of the town of Verapaz were retained, based on ground knowledge that debris flowed through many of the streets and beyond the town's inhabited area. The resulting lahar observations reflected runout length, width, and shape, as described in official estimates and mapped in postdisaster damage assessments [*Government of El Salvador*, 2009; *CEPAL*, 2010].

With the simulation and observation datasets complete, we measured to what extent simulations and observations agreed and disagreed. Table 11.3 shows the matrix of possible results. We calculated true positives, *TP*, where model simulations and observations agreed on the presence of a lahar, and false positives, *FP*, where lahars were simulated but not observed. We also calculated false negatives, *FN*, where lahars were observed but not simulated, and true negatives, *TN*, were simulations and observations agreed on the absence of a lahar. The last row of Table 11.3 shows total positives, ΣP, and total negatives, ΣN. We expressed the final metrics of success using true positive rates, *TPR*, and false positive rates, *FPR*.

$$TPR = \frac{TP}{\Sigma P} \tag{11.3}$$

$$FPR = \frac{FP}{\Sigma \mathrm{N}} \tag{11.4}$$

A perfect simulation would yield a *TPR* of 1 and an *FPR* of 0. An overall accuracy assessment was given by:

$$Accuracy = \frac{TP + TN}{TP + TN + FP + FN} \tag{11.5}$$

Following the approach published by *Cepeda et al.* [2010], we limited the measure of negatives to a one pixel buffer around the observed and simulated areas. Given that our analysis used pixels of DEMs with different spatial dimensions, we assessed accuracy with a buffer to 10 m but also discuss how the accuracies would be reported with buffers of 30 and 90 m.

Table 11.3 Description of How Simulated Versus Observed Debris Inundation Zones Are Classified to Determine Overall Accuracy

		Observed (Formosat-2)	
		Positive	Negative
Simulated (LAHARZ)	Yes	True positive, *TP*	False positive, *FP*
	No	False negative, *FN*	True negative, *TN*
Σ		Positives, ΣP	Negatives, ΣN

11.3. RESULTS

We considered three different DEMs and applied three hydrological preprocessing techniques to each, resulting in nine model setups that each considered five lahar initiation points, for a total of 45 lahar simulations. An example of the spatial overlap of simulated and observed debris-flow hazard zones for all three DEMs using the SA fill technique for all five lahars is shown in Figure 11.2. Black areas represent where simulations and observations agreed (*TP*). Green shows where lahars were simulated but not observed (*FP*), and magenta shows where lahars were observed but not simulated (*FN*). We constructed the tables described in the previous section to arrive at true positive rates, false positive rates, and overall accuracy for each simulation.

The true positive rates, *TPR*, false positive rates, *FPR*, and overall accuracy, provided a quantitative assessment of the capability of each combination of DEM and pit-filling method to match the observed lahars. Table 11.4, Table 11.5, and Table 11.6 summarize these results. Overall, simulations with the national DEM consistently yielded the highest *TPR* and best absolute true and false positives, as well as the highest accuracies. The national DEM also offered a lower *FPR* than the other DEMs. The ASTER GDEM had the lowest *TPR* for all hydrological preprocessing techniques. Overall, the different pit-filling techniques did not have a large effect on *TPR*, *FPR*, or accuracy, among simulations, because differences stemming from DEM spatial resolution overpower those based on the several pit-filling techniques examined here.

Considering the performance from a spatial perspective, all simulations using ASTER and SRTM DEMs missed Verapaz. The ASTER GDEM simulations sent the lahar to the west of town (circle B in Fig. 11.2), while the SRTM DEM simulations ended before town. In fact, nearly all simulations using the SRTM DEM resulted in runout lengths more than 1 km too short (circles D in Fig. 11.2). This type of shortcoming of lower resolution DEMs for LAHARZ simulations has been previously identified and discussed in the literature [*Muñoz-Salinas et al.*, 2009]. Note in Figure 11.2 that the ASTER GDEM simulations include a yellow category. This depicts the simulation using the OPR MinAbs pit-filling technique, which resulted in a very early diversion of the Gully los Infiernillos lahar (circle C in Fig. 11.2). This is also reflected by the slightly lower *TPR* and slightly higher *FPR* for this simulation.

The Guadalupe lahar is unique in that it converged roughly 2 km downhill from the initiation point, ran through part of the town, and proceeded to diverge after another 2 km or so and once more just before its northern terminus (circle A in Fig. 11.2). None of the simulations captured this phenomenon. The Verapaz lahar separated

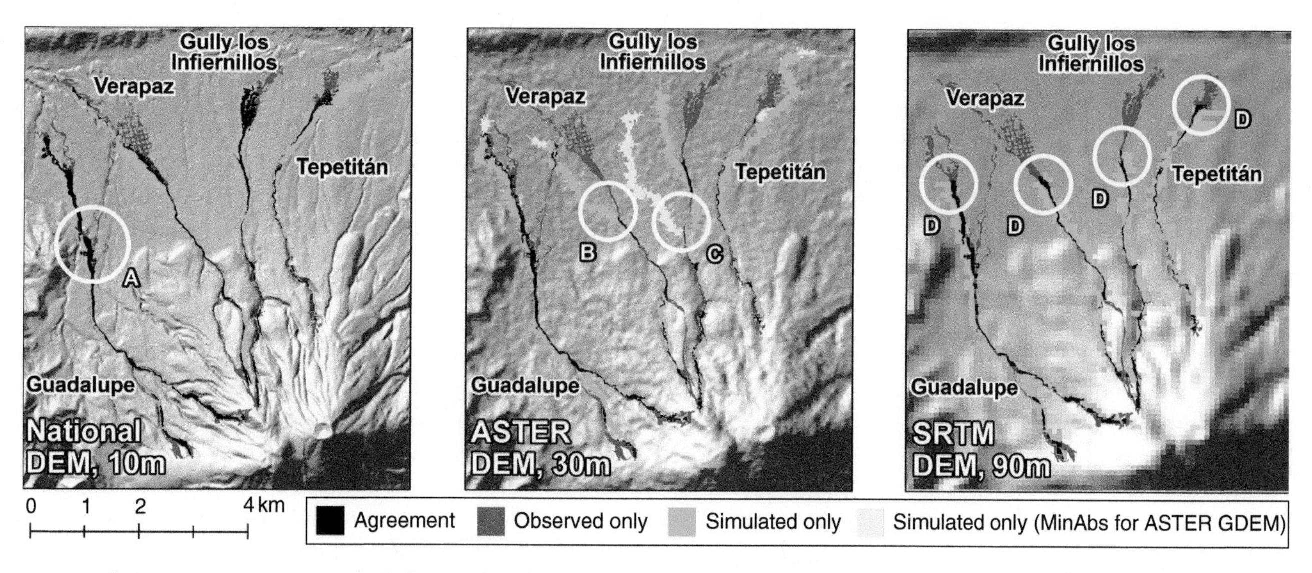

Figure 11.2 Comparisons of observed and simulated debris flows for the three digital elevation models tested (shown with the ArcGIS Spatial Analyst hydrological preprocessing technique, except where noted).

Table 11.4 True Positive Rates, TPR, of Each DEM Versus Fill Method

	DEM		
Fill method	National	ASTER	SRTM
SA	0.54	0.28	0.33
OPR MinAbs	0.54	0.27	0.34
OPR MinNet	0.54	0.28	0.33

Table 11.5 False Positive Rates, FPR, of Each DEM Versus Fill Method

	DEM		
Fill Method	National	ASTER	SRTM
SA	0.30	0.36	0.37
OPR MinAbs	0.30	0.36	0.37
OPR MinNet	0.30	0.36	0.36

Table 11.6 Overall Accuracy of Each DEM Versus Fill Method

	DEM		
Fill Method	National	ASTER	SRTM
SA	0.68	0.57	0.58
OPR MinAbs	0.68	0.57	0.58
OPR MinNet	0.68	0.57	0.58

very close to its source and came back together just over a kilometer later. This braided type of stream network was not resolved in the lahar simulations either. In fact, the national DEM routed debris in one direction, while the other two DEMs preferred the other.

Given that LAHARZ is an empirical model based on documented relationships between lahar volume and cross-sectional and planimetric area inundated [*Iverson et al.*, 1998], we also ran simulations to estimate areas of potential inundation with a 95% confidence interval. The 95% confidence interval test also allowed us to account for uncertainties related to lahar volume and initiation points. We limited these tests to the SA pit-filling technique for all DEMs. Table 11.7 summarizes the ranges of potential inundation areas for these cases, showing the combined surface areas for the five simulated lahars.

Table 11.7 Ranges of Estimated Areas of Potential Inundation Summed for All Debris Flows Tested, at a 95% Confidence Interval

	Range, m^2
National	963,700 to 14,129,300
ASTER	967,613 to 13,482,072
SRTM	1,131,008 to 12,732,676

11.4. DISCUSSION

In very practical terms, the motivation behind lahar simulation is to determine whether or not assets, such as a populated area or road, are in the path of a lahar. In this study, none of the simulations completely captured the inundation of numerous towns. For example, there is a lack of correspondence between simulated and observed lahars within populated areas around the town of Verapaz, shown in Figure 11.2. These results suggest that if much more precise elevation data could resolve built features, more accurate lahar inundation areas within towns may be achievable. Still, regardless of the input data, the LAHARZ model was not designed to simulate bifurcations in debris flows, such as the two flow paths seen after passing through Guadalupe [*Iverson et al.*, 1998].

Focusing on Verapaz, we note that none of the simulations captured the extent to which debris flowed through the town itself. The closest simulation came from the national DEM, but all simulations incorrectly excluded the vast majority of the populated area from the hazard zones. This is most likely due to recent terrain modification that is not reflected in the old DEM or flow path openings a few meters wide that simply cannot be resolved by any DEM used in this study.

The results indicate that for the 45 lahar simulations, the resolution of the input DEM is more important than the hydrologic preprocessing technique, as shown through *FPR* and *TPR* rates in Tables 11.4 and 11.5 and overall accuracy in Table 11.6. Even though SRTM and ASTER datasets are more current (the national DEM being based on elevation data from the 1970s and 1980s), we could not prove that the benefit of timeliness outweighed the cost of spatial resolution.

The method for calculating true negatives influenced the false positive rates and overall accuracy assessments, as discussed in *Cepeda et al.* [2010]. In Tables 11.5 and 11.6, we reported *FPR* and accuracy considering a 10 m buffer around lahar simulations and observations. *TPR* is not influenced by the size of the study area considered in calculating true negatives. Had a buffer of 30 m or 90 m been included, *FPR* would have decreased, and overall accuracy would have increased. For example, we reported an overall accuracy of 67% for simulations with the national DEM when calculating true negatives only within 10 m of the lahar simulations and observations. If a 30 m buffer were considered, *FPR* would have decreased, and the accuracy for the same simulations from the same DEM would have been reported as 74%. With a 90 m buffer, the *FPR* would decrease further, and the accuracy would have been reported as 82%. This trend shows that increasing the study area to calculate true negatives and *FPR* results in a decreased *FPR* and increased overall accuracy calculation. We do not aim to exaggerate the accuracy of these simulations; therefore, we report a more conservative 67% accuracy in the highest performing simulations, those derived from the national DEM.

There was not a direct relationship between DEM resolution and *TPR*. Even though the ASTER GDEM provided medium spatial resolution (relative to the other DEMs tested), it delivered the worst *TPR*. This may be a reflection of the ASTER GDEM being the newest product available, having gone through less rigorous terrain correction than the SRTM DEM or due to the remote sensing technique employed (ASTER GDEM being based on multilook angle stereographic photography and SRTM based on InSAR). Based on stereo images obtained by a passive sensor, the ASTER GDEM is more susceptible to data gaps in cloudy areas than is the SRTM product. From a runout length perspective, it is obvious that the 90 m SRTM DEM consistently falls short of simulating how far lahars of given volumes flow.

We did not calculate the effect of applying different NDVI thresholds on constructing the observation dataset; however, we were able to rely on damage assessment maps from the field to confirm the satellite-based observations. Moreover, the spatial resolution of the observation dataset was 25 times that of the highest resolution DEM tested (4 m^2 pixel versus 100 m^2 pixel). While we acknowledge the inherent uncertainty in measuring the aerial extent and maximum runout from the observation dataset, we do not expect it to have introduced considerable error.

Even though pit-filling techniques did not have a large effect on true or false positive rates, as mentioned in the results it is important to note that individual simulations differed greatly from each other. The clear message is that the wrong combination of DEM and preprocessing technique could lead to drastically wrong debris-flow hazard zone designations, especially as seen in the cases of Gully los Infiernillos and Verapaz (circle C in Fig. 11.2).

In an attempt to compare simulations with a known recent event, we set predefined debris-flow volumes based on field assessments. When potential lahar volumes are unknown (as is generally the case), *Iverson et al.* [1998] suggest that LAHARZ outputs be mapped showing a range of debris-flow volumes. This approach would avoid conveying a black-and-white "hazard," "no hazard" understanding of potential lahars. Our calculations of estimated areas of potential inundation at a 95% confidence interval show wide ranges (Table 11.7). Figure 11.3 shows the inundated areas for the upper and lower limits for the 95% confidence interval tests for each input DEM. Figure 11.3 and Table 11.7 together show how the uncertainties affect the modeling results. The range is narrowest for the SRTM DEM and highest for the national DEM, but the accuracy assessments discussed above support the pursuit of higher resolution DEMs to more successfully simulate lahar runout events on the San Vicente volcano in El Salvador.

San Vicente is not the only volcano in El Salvador with lahar hazards. In 1982, heavy rains triggered a debris flow of 300,000 m^3 that tore through a neighborhood in the capital city San Salvador, killing an estimated 300–500 people [*Moisa and Romano*, 1994; *Finnson et al.*, 1996; *Cepeda et al.*, 2010]. Despite the repeated history of landslides, the same neighborhood that lay in the path of the 1982 lahar has since been repopulated, with the number of residents now greater than in 1982 [*Lopez Paz*, 1985; *SNET*, 2008]. Today, the population in the San Salvador metropolitan area has soared to two million, exposing even more to disasters [*Cepeda et al.*, 2010]. Several years

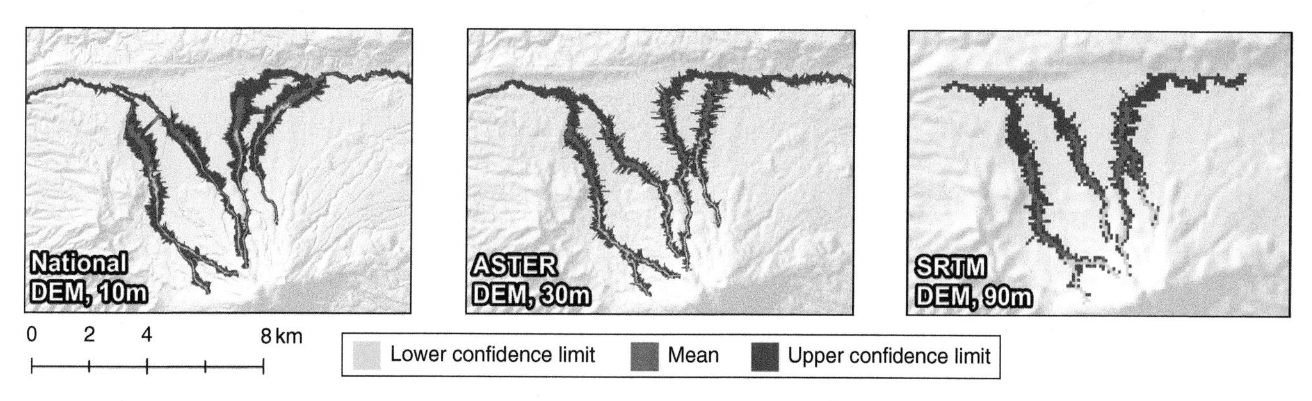

Figure 11.3 Debris flow simulations for the upper and lower limits for a 95% confidence interval for the three digital elevation models tested (shown with the ArcGIS Spatial Analyst hydrological preprocessing technique).

ago, residents and officials were startled when a small landslide grew from the same failure point as the 1982 event. While it has not grown into a full-fledged lahar, today it lurks above the capital city as a reminder of the "Tragedy of Montebello" [*Lopez Paz*, 1985; *SNET*, 2008].

11.5. CONCLUSIONS

Even though global DEMs such as ASTER and SRTM may not be applicable for decision support in El Salvador because of the availability of a 10 m product, this is often not the case for other parts of Central America. Yet, the availability of such a dataset in El Salvador has allowed us to compare model performance with globally available DEMs, giving context to the true positive and false positive rates attained using ASTER and SRTM DEMs. We have shown that decision makers in areas lacking high-resolution DEMs should understand that resolving debris-flow inundation zones relies acutely on the spatial resolution of input data. We have also shown that, due to the distribution of lahar volumes on a cell-by-cell basis, coarser DEMs will likely result in an overestimation of the lateral spreading and an underestimation of runout length, agreeing with similar analyses in Mexico [*Hubbard et al.*, 2007]. Further engagement with decision makers on the maximum allowable uncertainty and required accuracies in lahar modeling would make these analyses more applicable to disaster managers. It would also help to inform applied scientists as to how to invest further research efforts. In this study, we have focused only on the lahar hazard. Combining this information with vulnerability and frequency of flows would allow for further risk analysis [*Scott et al.*, 1995]. Further testing to determine the maximum acceptable spatial resolution to simulate lahars of given volumes is also merited. For more precise lahar mapping, lidar-derived DEMs are often preferred because they can resolve different levels of materials such as vegetation and bare earth and can capture fine-scale anthropogenic and natural topographic features that can push a debris flow one in any given direction [*Griswold*, 2003]. The limited availability of such lidar-based datasets is often a challenge for landslide hazard modeling, particularly in Central America. With the recent release of the SRTM 30 m dataset for other parts of the world, this should open up new opportunities for further assessment [*NASA*, 2015]. Furthermore, data from more recent satellite-based X-band missions, such as Cosmo SkyMed and TanDEM-X, may be able to provide more accurate (less than 10 m horizontal) and timely datasets for inputs into LAHARZ, and thus more appropriate debris-flow modeling.

ACKNOWLEDGMENTS

Thanks go to the Ministry of the Environment and Natural Resources/General Directorate of the Environmental Observatory (MARN/DGOA) in El Salvador, the USGS LAHARZ team, and the US National Aeronautics and Space Administration (NASA) Earth Science Division's Applied Sciences and Capacity Building Programs. This work is an extension of a master's thesis at the University of Alabama in Huntsville's Earth System Science program and was funded by the SERVIR program, a joint initiative led by NASA and the US Agency for International Development (USAID). We also thank the reviewers whose comments helped improve and clarify this manuscript.

REFERENCES

Anderson, E., D. Irwin, F. Delgado, M. Diaz, G. Molina, and E. Cherrington (2012), Evaluation of space-based support through SERVIR during and after the 2009 floods and landslides in El Salvador, in *63rd International Astronautical Congress*, 1–13, International Astronautical Federation, Naples, Italy.

Canuti, P., N. Casagli, F. Catani, F., G. and Falorni (2002), Modeling of the Guagua Pichincha Volcano (Ecuador) lahars, *Physics and Chemistry of the Earth*, *27*, 1587–1599; doi:10.1016/S1474-7065(02)00180-8.

CATHALAC (2011), Final Report: Central American Land Cover and Land Use Map–Land Cover and Land Use Change 1980-1990-2000-2010, Regional Program for the Reduction of Vulnerability and Environmental Degradation/CATHALAC, Panama City, Panama.

CENTA and FAO (1998), Mapa digital: Curvas a nivel cada 100 metros sobre el nivel del mar, San Andrés, El Salvador.

CEPAL (2010), El Salvador: impacto socioeconómico, ambiental y de riesgo por la baja presión asociada a la tormenta tropical Ida en noviembre de 2009, United Nations Economic Commission for Latin America (CEPAL), Mexico City; availablefrom: http://www.cepal.org/es/publicaciones/1382-el-salvador-impacto-socioeconomico-ambiental-y-de-riesgo-por-la-baja-presion (accessed 16 June 2015).

Cepeda, J., J. A. Chávez, and C. Cruz Martínez (2010), Procedure for the selection of runout model parameters from landslide back-analyses: application to the Metropolitan Area of San Salvador, El Salvador, *Landslides*, *7*(2), 105–116; doi:10.1007/s10346-010-0197-9.

CIA (2013), CIA–The World Factbook; available from: https://www.cia.gov/library/publications/the-world-factbook/geos/es.html (Accessed 29 May 2013).

Dai, F., C. Lee, and Y. Ngai (2002), Landslide risk assessment and management: An overview, *Eng. Geol.*, *64*(1), 65–87; doi:10.1016/S0013-7952(01)00093-X.

Davila, N., L. Capra, J. C. Gavilanes-Ruiz, N. Varley, G. Norini, and A. Gomez Vazquez (2007), Recent lahars at Volcan de Colima (Mexico): Drainage variation and spectral classification, *J. Volcanol. Geo. Res.*, *165*(3–4), 127–141; doi:10.1016/j.jvolgeores.2007.05.016.

Dull, R. A., J. R. Southon, and P. Sheets (2001), Volcanism, ecology and culture: A Reassessment of the Volcán Ilopango TBJ Eruption in the Southern Maya Realm, *Lat. Am. Antiq.*, *12*(1), 25–44.

Exelis (2013), Atmospheric Correction (Using ENVI); available from: http://www.exelisvis.com/docs/AtmosphericCorrection.html#IAR (accessed 16 June 2015).

Farr, T. G., and M. Kobrick (2000), Shuttle radar topography mission produces a wealth of data, *Eos, Trans. AGU*, *81*(48), 583–585; doi:10.1029/EO081i048p00583.

Finnson, H., C. Bäcklin, and A. Bodare (1996), Landslide hazard at the San Salvador volcano, El Salvador, in *Landslides, Proceedings of the 7th International Symposium on Landslides*, Trondheim, Norway: Rotterdam, Balkema, vol. 1, 215–220.

Government of El Salvador (2009), El Salvador: Damage, Loss, and Needs Assessment for Disaster Recovery and Reconstruction After the Low Pressure System Associated with Tropical Storm Ida, San Salvador.

Van de Griend, A. A., and M. Owe (1993), On the relationship between thermal emissivity and the normalized difference vegetation index for natural surfaces, *Int. J. Remote Sens.*, *14*(6), 1119–1131.

Griswold, J. (2003), Hazard maps for rapidly moving landslides using GIS and LIDAR, in Geological Society of America Annual Meeting, 2–5 November 2003, Seattle, WA.

Griswold, J. P., and R. M. Iverson (2008), Mobility statistics and automated hazard mapping for debris flows and rock avalanches, *Scientific Investigations Report 2007-5276*, Reston, VA.

Guzzetti, F., A. Carrara, M. Cardinali, and P. Reichenbach (1999), Landslide hazard evaluation: A review of current techniques and their application in a multi-scale study, Central Italy, *Geomorphology*, *31*(1–4), 181–216; doi:10.1016/S0169-555X(99)00078-1.

Hirano, A., R. Welch, and H. Lang (2003), Mapping from ASTER stereo image data: DEM validation and accuracy assessment, *ISPRS J. Photogramm. Remote Sens.*, *57*(5–6), 356–370; doi:10.1016/S0924-2716(02)00164-8.

Hubbard, B. E., M. F. Sheridan, G. Carrasco-Núñez, R. Díaz-Castellón, and S. R. Rodriguez (2007), Comparative lahar hazard mapping at Volcan Citlaltépetl, Mexico, using SRTM, ASTER and DTED-1 digital topographic data, *J. Volcanol. Geotherm. Res.*, *160*(1–2), 99–124; doi:10.1016/j.jvolgeores.2006.09.005.

Iverson, R. M., S. P. Schilling, and J. W. Vallance (1998), Objective delineation of lahar-inundation hazard zones, *Geol. Soc. Am. Bull.*, *110*(8), 972–984; doi:10.1130/0016-7606(1998)110<0972:ODOLIH>2.3.CO;2.

Jackson, S. (2013), Optimized Pit Removal, Cent. Res. Water Resour; available from: http://tools.crwr.utexas.edu/OptimizedPitRemoval/CRWR Tools Optimized Pit Removal.html (accessed 6 January 2014).

Jarvis A., H. I. Reuter, A. Nelson, E. Guevara, (2008), Hole-filled seamless SRTM data V4, International Centre for Tropical Agriculture (CIAT); available from http://srtm.csi.cgiar.org.

Jenson, S. K., and J. O. Domingue (1988), Extracting topographic structure from digital elevation data for geographic information system analysis, *Photogramm. Eng. Remote Sens.*, *54*(11), 1593–1600.

Johnson, A. M. (1994), Debris flow, in *Slope Instability*, ed. D. Brunsden and D. B. Prior, 257–361, Wiley, New York.

Lopez Paz, J. L. (1985), Account of the immediate rescue action developed in the tragedy of Montebello, *Disasters*, *9*(2), 92–95; doi:10.1111/j.1467-7717.1985.tb00918.x.

Major, J. J., S. P. Schilling, C. R. Pullinger, and C. D. Escobar (2004), Debris-flow hazards at San Salvador, San Vicente, and San Miguel volcanoes, El Salvador, in *Natural Hazards in El Salvador*, edited by W. I. Autor Rose, pp. 89–108, Geological Society of America, Boulder, CO.

MARN (2013), Inventario, Geología; available from: http://snet.gob.sv/ver/geologia/inventario/(accessed 20 May 2013).

MARN (2015), Perfiles Climatológicos;. available from http://www.snet.gob.sv/ver/meteorologia/clima/perfiles+climatologicos/ (accessed 11 November 2015).

Metternicht, G., L. Hurni, and R. Gogu (2005), Remote sensing of landslides: An analysis of the potential contribution to geo-spatial systems for hazard assessment in mountainous environments, *Remote Sens. Environ.*, *98*(2–3), 284–303; doi:10.1016/j.rse.2005.08.004.

Moisa, A. M., and L. E. Romano (1994), Caracterización de los desastres en El Salvador: tipología y vulnerabilidad socioeconómica, Centro de Protección para Desastres (CEPRODE), Documento 4711, San Salvador, El Salvador.

Muñoz-Salinas, E., M. Castillo-Rodríguez, V. Manea, M. Manea, and D. Palacios (2009), Lahar flow simulations using LAHARZ program: Application for the Popocatépetl volcano, Mexico, *J. Volcanol. Geotherm. Res.*, *182*(1–2), 13–22; doi:10.1016/j.jvolgeores.2009.01.030.

NASA (2015), NASA Shuttle Radar Topography Mission, Global 1 arc second, NASA EOSDIS Land Processes DAAC, USGS Earth Resources Observation and Science (EROS) Center, Sioux Falls, SD (https://lpdaac.usgs.gov), (accessed 12 September 2015 at http://dx.doi.org/10.5067/MEaSUREs/SRTM/SRTMGL1.003).

O'Callaghan, J. F., and D. M. Mark (1984), The extraction of drainage networks from digital elevation data, *Comput. Vision. Graph. Image Process.*, *28*, 323–344.

OPAMSS (2010), Oficina de Planificación del Area Metropolitana de San Salvador; available from: http://www.opamss.org.sv/ (accessed 5 July 2010).

OSSO, La RED, and UNISDR (2013), Desinventar Profile, El Salvador; available from: http://www.desinventar.net/DesInventar/profiletab.jsp?countrycode=slv (accessed 21 May 2013).

Schilling, S. P. (1998), LAHARZ: GIS programs for automated mapping of lahar-inundation hazard zones, Open-File Report 98-638, Vancouver, WA.

Schilling, S. P. (2014), Laharz_py–GIS tools for automated mapping of lahar inundation hazard zones, US Geological Survey Open-File Report 2014-1073, 10.3133/ofr20141073.

Schmidt-Thomé, M. (1975), The geology in the San Salvador area (El Salvador, Central America), a basis for city development and planning, *Geol. Jahrb.*, *13*, 207–228.

Scott, K. M., J. W. Vallance, J. W., and P. T. Pringle (1995), Sedimentology, behavior, and hazards of debris flows at Mount Rainier, Washington: US Geological Survey Professional Paper 1547; available from: http://pubs.er.usgs.gov/publication/pp1547 (accessed 16 June 2015).

SNET (2008), Informe técnico sobre el deslizamiento en la parte alta de El Picacho, Volcán de San Salvador y acciones para instalación de Sistema de Alerta Temprana, Reporte de Servicio Nacional de Estudios Territoriales, San Salvador, El Salvador; available from: http://www.snet.gob.sv/Geologia/DeslavePicacho.pdf (accessed 16 June 2015).

SNET (2009), Informe De Los Flujos De Escombros (Deslaves) En Las Ciudades De Verapaz, Guadalupe Y Alrededores De Tepetitan, Reporte de Servicio Nacional de Estudios Territoriales, San Salvador, El Salvador; available from: http://www.snet.gob.sv/googlemaps/inundaciones/pdf/informe340.pdf (accessed 16 June 2015)

Soille, P. (2004), Optimal removal of spurious pits in grid digital elevation models, *Water Resour. Res.*, *40*(12); doi:10.1029/2004WR003060.

Stevens, N. F., V. Manville, and D. W. Heron (2002), The sensitivity of a volcanic flow model to digital elevation model accuracy: experiments with digitized map contours and interferometric SAR at Ruapehu and Taranaki volcanoes, New Zealand, *J. Volcanol. Geotherm. Res.*, *119*, 89–105; doi:10.1016/S0377-0273(02)00307-4.

Tachikawa, T., M. Hato, M. Kaku, and A. Iwasaki (2011), Characteristics of ASTER GDEM version 2, International Geoscience and Remote Sensing Symposium (IGARSS), 3657–3660.

Van Westen, C. J., T. W. J. van Asch, and R. Soeters (2006), Landslide hazard and risk zonation–why is it still so difficult?, *Bull. Eng. Geol. Environ.*, *65*(2), 167–184, doi:10.1007/s10064-005-0023-0.

Weber, H., G. Wiesemann, H. Lorenz, H. Schmdit-Thomé (1978), Mapa Geológico de la República de El Salvador/América Central,. Bundesanstalt Geowissenschaften und Rohstoffe, Hannover, Germany.

Zhang, W. H., and D. R. Montgomery (1994), Digital elevation model grid size, landscape representation and hydrologic simulations, *Water Resour. Res.*, *30*(4), 1019–1028.

Zhang J. X., K. T. Chang, and J. Q. Wu (2008), Effects of DEM resolution and source on soil erosion modelling: A case study using the WEPP model, *Int. J. Geogr. Inf. Sci.*, *22*(8), 925–942.

12

Evaluating the Performance of FLO2D for Simulating Past Lahar Events at the Most Active Mexican Volcanoes: Popocatépetl and Volcán de Colima

Lizeth Caballero,[1] Lucia Capra,[2] and Rosario Vázquez[2]

ABSTRACT

Two lahar episodes that occurred at Colima and Popocatépetl volcanoes (México) are modeled using the FLO2D code, with the scope to define the parameters that control flow simulation and their reliability. The Lahar Patrio at Volcán de Colima was used to evaluate the model performance related to the influence of input hydrograph shape and Manning-*n* value in the absence of real data. The Popocatépetl 2001 lahar was used to evaluate the model response to lahars with different rheologic behaviors. Input parameters for lahar modeling like hydrograph, Manning-*n* values, and rheologic coefficients were derived from geophone data, channel geometry, and textural characteristics of the deposits. To validate the Lahar Patrio simulation, three parameters were used: the percent length ratio, the fitness function, and flow depths. This approach shows that a simplified hydrograph can be a good approximation for lahar simulation. On the contrary, Manning-*n* coefficient has a stronger influence on simulation results. The greatest uncertainty in lahar modeling with FLO2D is the selection of rheologic coefficients since very little changes in their magnitude can affect the simulation results. When a proper assessment of these input parameter uncertainties are performed, results obtained with FLO2D can be very useful for lahar hazard evaluation.

12.1. INTRODUCTION

Lahars are processes that can occur under many different conditions in volcanic environments (both syneruptive and posteruptive; *Manville et al.* [2009]). During the last century they provoked thousands of deaths and several damages (i.e., the 1985 Nevado del Ruiz eruption, *Naranjo et al.* [1986]), even at inactive volcanoes (i.e., the 1988 Hurricane Mitch at Casita volcano, Nicaragua; *Scott et al.* [2005]), examples that pose in evidence the need to improve our understanding of such phenomena and enhance the hazard evaluation process incorporating all available tools for this purpose.

[1] *Facultad de Ciencias/Instituto de Geología, UNAM, Mexico City, Mexico*

[2] *Centro de Geociencias, UNAM, Querétaro, Mexico*

Lahars are mixtures of water and sediment with variable sediment-water content from hyperconcentrated flow (20%–60% vol; *Beverage and Culbertson* [1964]; *Vallance and Scott* [1997]) to debris flow (60%–90% vol; *Vallance and Scott* [1997]). Some lahars show a continuous variation between these two end-members through bulking and debulking processes during transport [*Scott*, 1988], while others maintain their sediment concentration relatively constant. Sediment concentration variation in lahars can induce complex dynamic behaviors. These dynamic characteristics make lahars very complex systems for numerical modeling.

Several modeling approaches have been proposed to assess lahar-related hazards. The semiempirical model LAHARZ [*Schilling*, 1998] has been largely used since it represents a simple tool to obtain a rapid objective hazard delineation based only on flow volume and inundation area.

Natural Hazard Uncertainty Assessment: Modeling and Decision Support, Geophysical Monograph 223, First Edition.
Edited by Karin Riley, Peter Webley, and Matthew Thompson.

In addition to the LAHARZ model, other GIS-based semiempirical models have been proposed [i.e., *Huggel et al.*, 2008]. Besides, several approaches have attempted to propose more accurate numerical models to define lahar transport and flood-prone areas. Different constitutive equations and theoretical approximations are used (Unsteady, open-channel, *Macedonio and Pareschi* [1992]; Two-Phase Titan model, *Córdoba et al.* [2010]; FLO2D model, *O'Brian et al.* [1993]). These models were successfully validated by reproducing observed lahar inundation areas, flow depths, and velocities [i.e., *Worni et al.*, 2012; *Caballero and Capra*, 2014]. The advantage of these models is the possibility to take into account physical and dynamic parameters of the flows, including rheological properties of the fluid and the sediment concentration of the mixture. Nevertheless, one of the issues during the application of these models is to define precisely the input parameters related with fluid properties and sediment concentration. This problem is emphasized in past lahar events whose dynamic characteristics are inferred only by textural characteristics of its deposits. This fact poses a great uncertainty when one wants to define sediment concentrations, rheological properties of the fluid, or a more accurate reconstruction of transport history of a lahar event.

The aim of our work is to define the reliability in determining input parameters for lahar simulation with FLO2D in the absence of direct data. This code was selected since it has been successfully used to delineate lahar inundation zones at Nevado del Huila [*Worni et al.*, 2012] and Popocatépetl volcano [*Caballero and Capra*, 2014]. For this purpose, we use two examples of well-constrained lahar events based on visual images, geophones data, and textural characteristic of their deposits gathered a few days after their emplacement [i.e., *Capra et al.*, 2004; *Vazquez et al.*, 2014]. The first example is the Lahar Patrio that occurred on 15 September 2012 at Volcán de Colima. The second is the 2001 lahar that occurred at Popocatépetl volcano on 22 January 2001 during an explosive episode [*Caballero and Capra*, 2014]. These lahars were selected because they have different triggering mechanisms (rain-triggered and massive remobilization by glacier melting of a pumice flow deposit, respectively), different water content, and rheologic properties that allow us to better understand if FLO2D is an adequate approach to replicate lahars with different dynamic behaviors. Additionally, direct data from real-time observation of these events represent a great opportunity to better understand numerical modeling limitations in poorly monitored volcanoes or when simulation of historic events is needed. This is of primary importance since numerical modeling is now increasingly used for volcanic hazard evaluation.

12.1.1. Geologic Framework of Volcán de Colima

Volcán de Colima is an andesitic stratovolcano (3840 m above sea level) located at the western section of the Trans Mexican Volcanic Belt (TMVB). It is part of the Colima Volcanic Complex, a volcanic chain oriented north to south (Fig. 12.1b). Continuous magmatic activity of Volcán de Colima has deposited large volumes of unconsolidated pyroclastic material on its flanks [*Capra et al.*, 2014]. This material is easily remobilized during

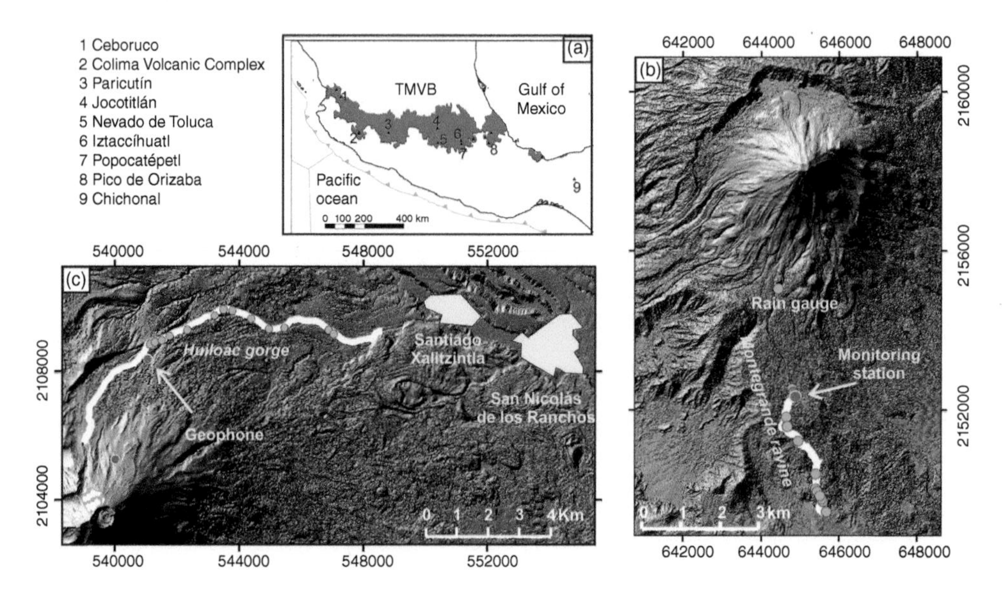

Figure 12.1 (a) Localization of Colima Volcanic Complex and Popocatépetl volcano in the Trans Mexican Volcanic Belt; (b) localization map and digital elevation model of Colima volcano and Montegrande ravine; (c) map of Popocatépetl volcano and Huiloac gorge.

rainy seasons and extreme rainfall events and creates very active gullies where rapid runoff induces lahars [*Capra et al.*, 2014].

One of the most active ravines is called Montegrande. It is located on the southern-central sector of Volcán de Colima (Fig. 12.1b). In 2011, a monitoring station was installed at this ravine. It consists of a 12 m high tower with a directional antenna, a video camera, a rain gauge coupled to a soil moisture sensor, and a 10 Hz geophone. Data from the monitoring station is transmitted in real time to RESCO facilities (the seismological network of Colima University). A three-component Guralp CMG-6TD broadband seismometer is also installed at 500 m northeast of the monitoring site. A second 10 Hz geophone is recording ground vibration 3 km downflow, at the end of the Montegrande ravine. These instruments allow the systematic study of lahar dynamics and its deposits and, more important, the direct application of such data on lahar hazard analysis.

12.1.2. The Lahar Patrio, Volcán de Colima

Lahar Patrio was a rainfall triggered event that occurred on 15 September 2012 along the Montegrande ravine (Fig. 12.1b). Its characteristics were described by *Vázquez et al.* [2014] by coupling seismic data, still images, and field data collected 2 days after the event. The Lahar Patrio event lasted between 40 min and 1 hr, and consisted of three main pulses. The deposit shows three main units. The basal layer is massive, supported by a sandy matrix with dispersed clasts of andesite and scarce pumice. It shows a bimodal granulometric distribution and a thickness of up to 40 cm. The intermediate unit is up to 10 cm in thickness, massive at its base, and laminated at the top. It is composed mainly of sand and has a unimodal granulometric distribution. The upper layer is 20–30 cm in thickness, massive and matrix-supported showing a bimodal granulometric distribution. Based on the textural characteristics of these deposits, the first and the third pulses had a sediment concentration in the lower range of debris flows (approximately 50% vol), whereas the second one, was a hyperconcentrated flow. Maximum flow depths reached are up to 1.7–2 m near to the monitoring station and mean velocity was estimated about 3 m/sec [*Vázquez et al.*, 2014].

12.1.3. The 2001 Lahar, Popocatépetl Volcano

Popocatépetl volcano is a stratovolcano located in the central sector of the Transmexican Volcanic Belt (Fig. 12.1a). It is the southernmost volcano of the Sierra Nevada volcanic chain. Together with the Volcán de Colima, they are the most active volcanoes in Mexico. On 22 January 2001, an intense explosion generated an eruptive column that rose up to 8 km above the crater. Its subsequent collapse produced pyroclastic flows that eroded and melted part of the summit glacier triggering a lahar. The 2001 lahar descended 15 km through Huiloac gorge almost reaching Santiago Xalitzintla town (Fig. 12.1c). Textural and hydraulic characteristics of this lahar were studied by *Capra et al.* [2004] and *Muñoz-Salinas et al.* [2007]. The 2001 lahar deposit is massive, up to 1.9 m in thickness, matrix supported, composed of subrounded clasts of pumice and scarce andesite of up to 70 cm in diameter. The deposit is fine rich (up to 17% of silt and clay) and shows a bimodal granulometric distribution. This lahar was already simulated using the FLO2D model [*Caballero and Capra*, 2014]), but because of its high fine sediment content is here compared with the Lahar Patrio to better understand the role that the rheological characteristics of the flow influence the results of numerical simulations.

12.2. METHODOLOGY

The FLO2D model [*O'Brien et al.*, 1993] routes floods over natural channels solving the full dynamic wave equation. It calculates the velocity in eight directions, channel roughness, the slope between two adjacent cells, and the wetted perimeter, that is the cross-sectional distance along which the streambed and stream banks are in contact with water [*Mackey et al.*, 1998]. For lahar modeling, a quadratic shear model is applied based on five stress terms: the Mohr-Coulomb shear, cohesive yield stress, viscous shear stress, turbulent shear stress, and the dispersive shear stress. The most important input parameters are an accurate digital elevation model (DEM), an input hydrograph, the Manning-*n* coefficient, and the rheologic parameters for yield strength and viscosity [*Worni et al.*, 2012]. The Lahar Patrio and the 2001 lahar simulations were performed on a DEM of 10 m of horizontal resolution obtained by interpolating 10 m spaced contour lines, provided by the Instituto Nacional de Estadística y Geografía (INEGI).

12.2.1. Determination of FLO2D Input Parameters for the Lahar Patrio

12.2.1.1. Hydrograph Reconstruction

The Lahar Patrio hydrograph was constructed based on the geophone data from the monitoring station at Montegrande ravine (Fig. 12.2a). The constructed hydrograph requires three variables: time, discharge, and sediment concentration. Peak discharge at the monitoring station was calculated by *Vázquez et al.* [2014] of around 48 m^3/sec based on inundation limit along topographic profiles and flow velocity. Flow peak discharge was associated with the highest peak from the geophone signal

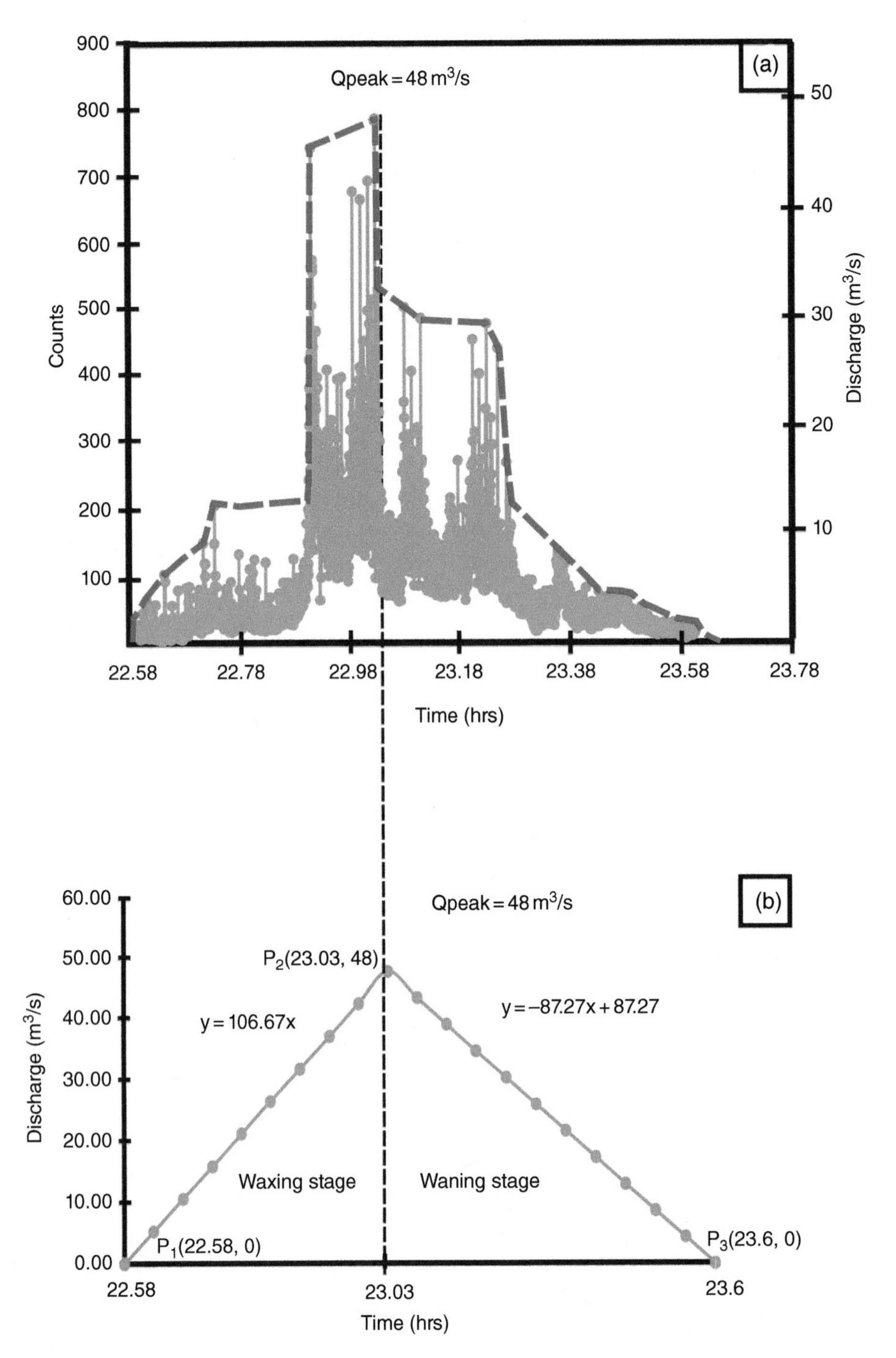

Figure 12.2 (a) Geophone raw data (blue lines and dots). Red dotted line represents the enveloping outline of the geophone signal used as the hydrograph. Peak discharge from *Vazquez et al.* [2014] data. (b) Hypothetical hydrograph constructed.

that corresponds to 786 counts. This peak also corresponds with the arrival of the main front and most energetic pulse. Sediment concentration was assigned based on textural characteristics of the deposits, granulometry, and video camera recordings. The first pulse had a sediment content in the lower limit of debris flows that would correspond to a sediment concentration of 50% vol according to FLO2D user's manual [*O'Brien*, 2001]. The same sediment concentration was used for the third pulse as it shows similar textural characteristics as the first one (i.e., granulometric bimodality; *Vázquez et al.* [2014]). In contrast, the second pulse corresponded to a hyperconcentrated flow deposit, for which a sediment concentration of 30% vol was used. The Lahar Patrio was preceded by a muddy stream flow registered in the geophone with less than 100 counts. Signals registered with less than 100 counts were assumed to be a normal stream flow with sediment concentration under 20% vol. The parts of the hydrograph between the normal streamflow and the lahar event were assigned with a sediment concentration of 20%–30% vol.

The geophone signal is highly unsteady and shows abundant peaks. These peaks represent the passage of the highly energetic lahar main front and the body. FLO2D has no ability to simulate fast flow variations in the range of seconds to minutes. Based on that, the hydrograph was constructed by suppressing the peaks observed in the geophone signal (Fig. 12.2a).

Most volcanoes do not have instrumental equipment to determine high accuracy hydrographs. Historic events at well-monitored volcanoes also lack this kind of data so simulations of these events face the same problem. Determination of hydrograph shape has been developed by *Fairchild* [1987], *Carusso and Pareschi* [1993], and *Iverson et al.* [1998]. All of these authors suggest a simple geometry, like a triangular hydrograph, as the most suitable shape. The two sides of the triangle would represent waxing and waning stages of the flow. Therefore, a triangular hydrograph was developed to observe differences in lahar simulation results between the hydrograph based on geophone data and the simplified hydrograph (Fig. 12.2b). Data used for this hydrograph were based on estimations of lahar duration (circa 1 hr) and peak discharge (48 m^3/s) of *Vázquez et al.* [2014]. For the rising limb of the hydrograph (waxing stage) a linear equation was derived for the slope-intercept form. Recession limb or waning stage was determined from a linear equation of the point-slope form (Fig. 12.2b).

12.2.1.2. Manning Coefficients

Manning-*n* values were calculated based on two methods. The first one was developed by US Geological Survey (USGS) and takes into account only channel characteristics [*Phillips and Tadayon*, 2006]. The second method was developed for high sediment concentration flows and is based on flow depth and channel properties [*Xu and Feng*, 1979]. To calculate Manning-*n* values with USGS empirical tables, the following equation is used [*Phillips and Tadayon*, 2006]:

$$n = (n_0 + n_1 + n_2 + n_3 + n_4)m = \sum n^* m$$

The first step is to set an initial *n*-value based on natural channel features. Substrate characteristics of the Montegrande ravine indicate a predominance of coarse sand and gravel up to 64 mm. Initial values suggested for these type of channels range between 0.028 and 0.035 so the mean value of this interval was used ($n_0 = 0.0315$). Further adjustments to the initial value consider the degree of channel irregularity ($n_1 = 0.008$), transversal section variation ($n_2 = 0.003$), percentage of natural obstructions ($n_3 = 0.025$), and type and amount of vegetation ($n_4 = 0.006$). All of these traits enhance channel roughness so values suggested are added to n_0 ($\Sigma n = 0.0735$). Finally, the degree of channel meandering (m), in this case 1.1 for minor meandering, is multiplied by the total sum, giving a final value n = 0.08.

To observe the differences in the results of lahar simulation relative to different Manning-*n* coefficient, an alternative value was estimated using a *Xu and Feng* [1979] table. Values assigned for irregular natural channels with substrate grain size varying from 0.3 to 0.5 m and flow depths between 1.0 and 2.0 m, vary between 0.125 and 0.167. The mean value of 0.15 was used as an alternative Manning-*n* value.

In the case of Montegrande ravine, we use a unique Manning-*n* value for each lahar simulation since the main channel features like substrate characteristics, channel irregularities, transversal section area, the degree of obstructions, and vegetation are fairly constant across the studied area.

12.2.1.3. Rheologic Coefficients

FLO2D uses rheologic coefficients named α and β that relate yield stress and viscosity with sediment concentration. These values were determined experimentally [*O'Brien and Julien,* 1988], based on rheologic properties of fluid matrix (sediment fraction composed by silt and clay). *O'Brien and Julien* [1988] report granulometric characteristics of the samples used for derivation of the empirical relations. Hence, the values of α and β were chosen based on *O'Brien and Julien* [1988] data for the sample with similar granulometric composition. In order to assess the influence due to variations in rheologic coefficients on lahar simulation outputs, including flow depth and inundation area, the results of *Caballero and Capra* [2014] for the 2001 lahar are used for comparison.

12.3. RESULTS AND DISCUSSION

12.3.1. Quantitative Validation of Simulated Flow

Results of the Lahar Patrio simulation were compared with data reported by *Vázquez et al.* [2014] from monitoring station and field data. Three parameters were used for a quantitative evaluation (Fig. 12.3): percent length ratio (PLR), fitness function (e_1), and flow depth. The first two parameters were used previously by *Proietti et al.* [2009]

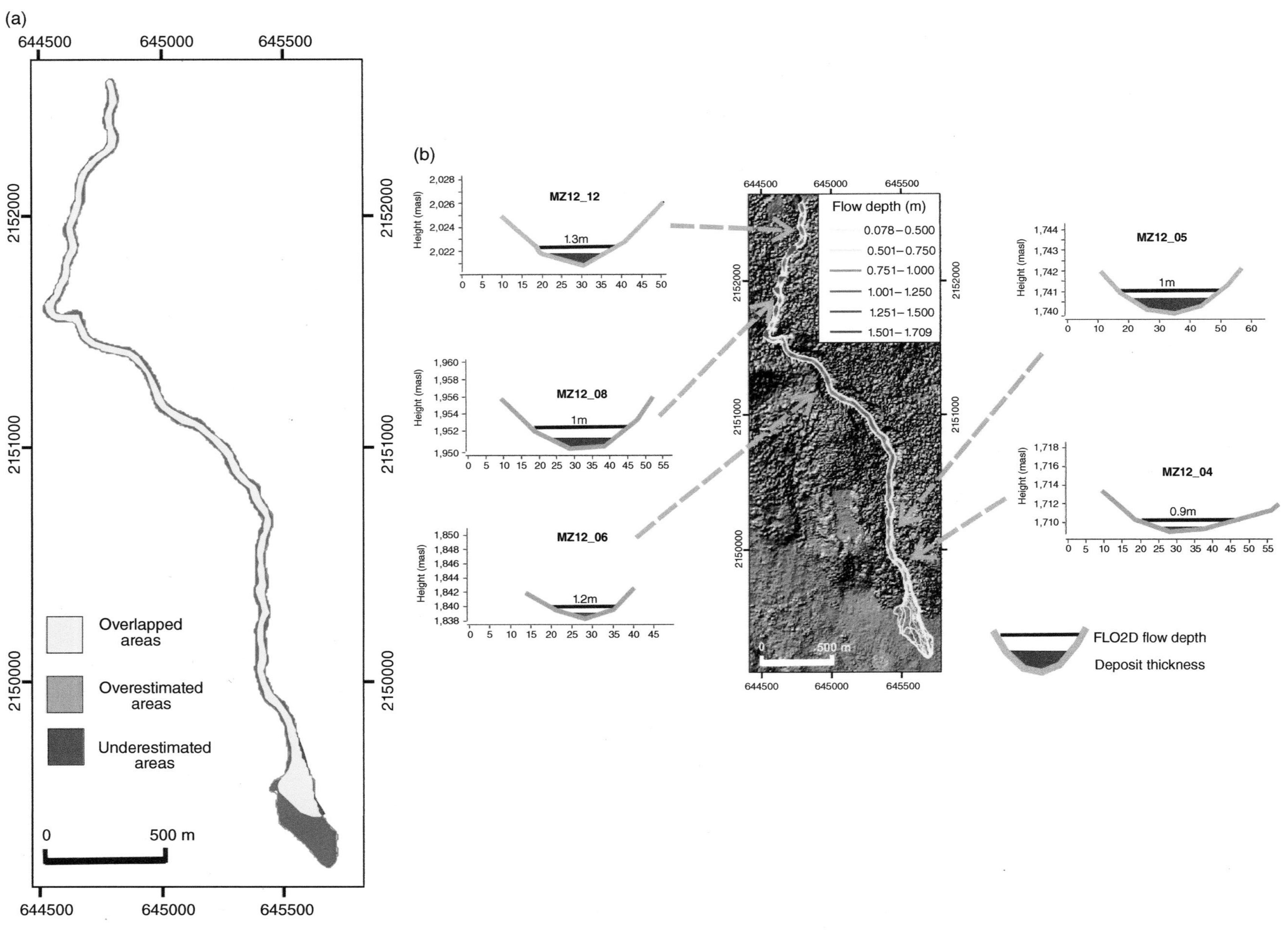

Figure 12.3 Results of Lahar Patrio simulation and comparison with the observed flow: (a) Differences between observed (purple plus yellow areas) and simulated lahar; (b) transversal sections comparing measured deposits and flow depths estimated by FLO2D at the same location.

to evaluate lava flow simulations. The percent length ratio (PLR) compares flow runout and is defined as:

$$PLR = \frac{L_{sim}}{L_{obs}} \times 100$$

where L_{sim} is the simulated length and L_{obs} is the observed length.

The second parameter, a fitness function named e_1, was derived by *Spataro et al.* [2004] and compares the lateral spreading. It is defined as:

$$e_1 = \sqrt{\frac{m(R \cap S)}{m(R \cup S)}}$$

where $R \cap S$ is the overlapping area between the real flow and the simulation. $R \cup S$ is the sum of the underestimated, overlapping, and overestimated areas between the observed event and the simulations. A value of $e_1 = 1$ is a perfect match between the real event and the simulation and $e_1 = 0$ represents a complete disjoint between them.

The first parameter measured PLR, gave 105% indicating that the simulated lahar is 5% larger than the real flow. For the second parameter (e_1), the value obtained was 0.78. Figure 12.3a shows the overlapped, overestimated, and underestimated areas. It could be seen that the greatest difference between them is in the distal region where the flow changes from a confined region to an open channel where the flow spreads laterally.

Finally, to evaluate flow depth, deposit thickness from field data and flow depths obtained during the simulation were compared at the same location (Fig. 12.3b). Considering a maximum sediment concentration of 0.5, flow depth has to be at least twice thicker than the deposit. According to *Vázquez et al.* [2014], the flow reached between 1.7 and 2.0 m in depth at the monitoring station. Maximum flow depth obtained from the simulation at this site (section M12_12) is 1.3 m but 25 m upstream this point it reaches 1.7 m. At the medial zone (sections M12_08 and M12_06), transversal cross sections show that flow depths are twice the deposit thickness so FLO2D makes a good approximation of this portion of the flow. At the distal zone, textural characteristics described by *Vázquez et al.* [2014] indicate a more diluted flow. This behavior is well reproduced at section M12_04 but not at section M12_05, where the relation of flow depth and deposit thickness is pointing to a sediment concentration of 65% vol, possibly due to local morphological conditions not well reproduced in the DEM here used.

FLO2D has the ability to calculate flow velocity along the flow (Fig. 12.4). Flow mean velocity of the Lahar Patrio was estimated at about 3 m/sec. Maximum velocities

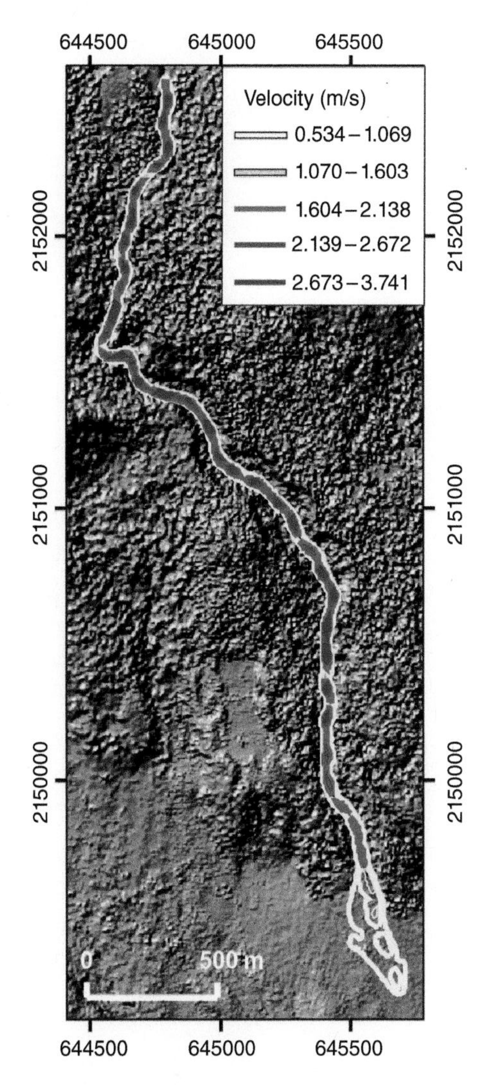

Figure 12.4 Results of flow velocity calculations for lahar simulation of Lahar Patrio.

calculated are between 2.6 and 3.7 m/sec in the central portion of the channel. Lateral and distal parts of the flow had calculated velocities under 1 m/sec. Based on the results about lahar inundation zone, flow depth, and flow velocity, we consider this flow simulation a good approximation of the Lahar Patrio.

12.3.2. Observed Versus Theoretical Hydrograph

To observe numerical modeling accuracy when the input parameters are not well defined, we use the Lahar Patrio to evaluate the influence of the input hydrograph shape and the Manning-*n* value.

Lahar Patrio simulation obtained from the hypothetical hydrograph (Fig. 12.2b) threw almost the same results in flow distribution, flow depth, and flow velocity as the hydrograph based on the geophone data (Fig. 12.5 a,b).

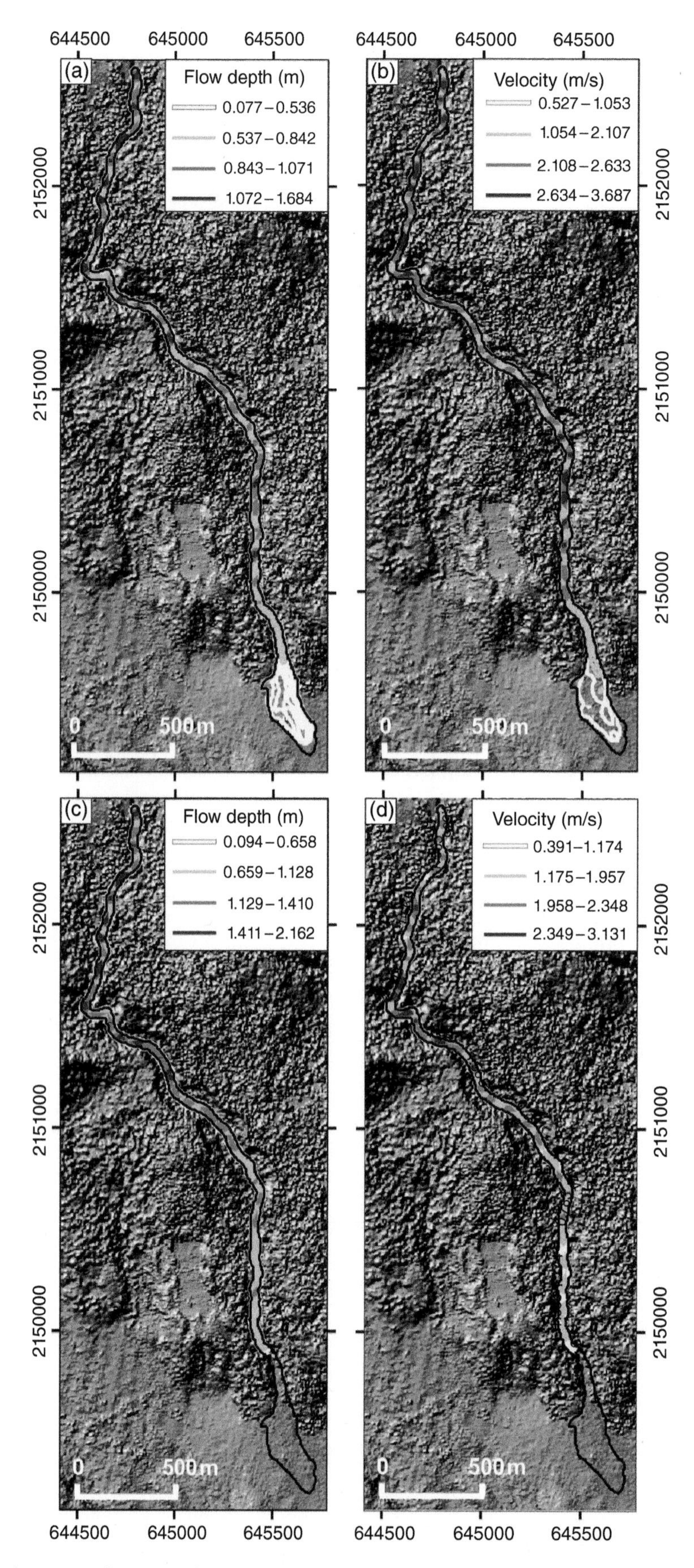

Figure 12.5 Simulation results using alternative input parameters: (a, b) Results of flow depth and flow velocity for the hypothetical hydrograph; (c, d) Simulation results with varying Manning-*n* value for flow depth and flow velocity. Black line represents aerial distribution of lahar simulation using original input parameters.

Therefore, we could state that no major effects on lahar distribution and flow depth is visible when using simple geometries to reconstruct the hydrographs. Nevertheless, this uncertainty could be enhanced if the peak discharge values used are not the same order of magnitude as the real peak discharges.

12.3.3. Sensibility of the Manning-*n* Value

In relation to Manning-*n* value obtained by *Xu and Feng* [1979], results displayed differences in lahar distribution, flow depth, and velocity. Maximum runout distance was underestimated by 750 m (Fig. 12.5c). If we consider that the Lahar Patrio traveled approximately 3.9 km from the monitoring site, distal reach is underestimated by almost 20%. Mean velocity between monitoring station and the distal geophone was calculated about 3 m/sec. FLO2D calculated values from 2.3 m/sec at the monitoring station and 1.9 m/sec at the distal geophone. This implies an underestimation of 23% and 37%, respectively (Fig. 12.5d).

12.3.4. Rheologic Coefficients

Simulation results of the 2001 lahar of Popocatépetl volcano [*Caballero and Capra*, 2014] allowed us to discuss the uncertainty derived from rheologic coefficients. *Caballero and Capra* [2014] simulated alternative scenarios where variations in fine content led to differences in yield strength and viscosity of lahars. Their alternative scenarios were named AS1 (more diluted lahar with lower yield strength and viscosity), AS2 (medium yield strength and viscosity), and AS3 (highest yield strength and viscosity). Their results point to important differences related to aerial distribution (Fig. 12.6a), flow thickness, and flow velocity. Depending on the rheologic coefficient

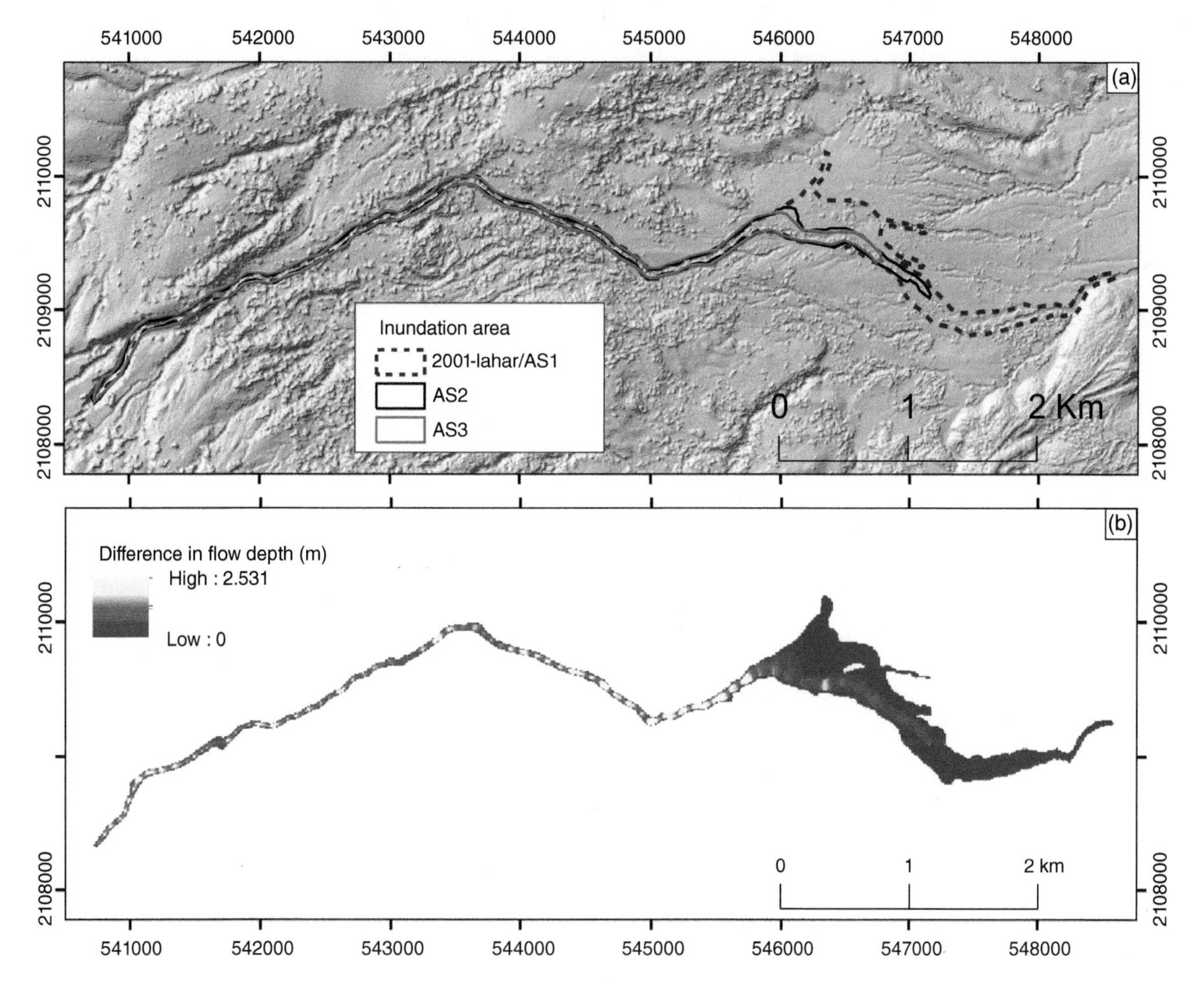

Figure 12.6 Simulation results using different rheologic coefficients for Popocatépetl volcano [modified from *Caballero and Capra*, 2014]: (a) Areas of inundation, (b) differences in flow depths.

values used, flow depths can vary as much as 2.5 m (Fig. 12.6b) and lahar velocities by 3 m/sec. The generation of these alternative scenarios as in *Caballero and Capra* [2014] is not appropriate for Volcán de Colima, since textural analysis of rainfall triggered lahars at Volcán de Colima show low fine content, usually less than 5 wt % of silt and barren of clay. Clay-sized particles in lahars are responsible for the developing of high yield strength and viscosity. Therefore, more viscous lahars are not possible scenarios at this volcano [*Capra et al.*, 2010; *Vázquez et al.*, 2014]. Based on the results obtained, it is clear that the FLO2D code is an appropriated tool to both simulate fine-depleted lahars dominated by particle-particle interaction, as fine-rich lahar dominated by viscous forces. Nevertheless, numerical simulations always include large uncertainties related to input parameters (i.e., Manning-*n* value, sediment concentration, and rheology) that should be properly assessed before using these kinds of tools for lahar hazard evaluation.

12.4. CONCLUSIONS

FLO2D is a numerical code widely used to simulate debris and hyperconcentrated flows [i.e., *Worni et al.*, 2012; *Caballero and Capra*, 2014]. Among the most important factors affecting simulations are input hydrograph, Manning-*n* value, and rheologic coefficients, along with the DEM resolution, which represents an important factor for simulation of gravitational granular flows [*Stevens et al.* 2003; *Huggel et al.*, 2008; *Capra et al.*, 2011]. Some of these parameters like the input hydrograph and the sediment concentration can be estimated accurately on well-monitored volcanoes like Volcán de Colima and Popocatépetl. However, in most cases they had to be determined by field, empirical, and/or previously published data. Results here obtained indicate that a simplified hydrograph can be a suitable approximation for lahar simulation, especially to reproduce historic events. However, an inappropriate selection of Manning-*n* coefficient can lead to an underestimation of lahar runout of about 20%.

The greatest uncertainty in lahar modeling with FLO2D is the selection of rheologic coefficients. The parameters used by *O'Brien and Julien* [1993] were obtained by a regression of yield stress and viscosity taking into consideration only silt and clay. However, other studies show that by measuring more complete granulometric spectrum, yield strength and dynamic viscosities can vary exponentially [*Coussot and Piau*, 1995; *Caballero et al.*, 2014]. The accurate measure of these parameters is of great importance since very little changes in their magnitude can affect the results of FLO2D simulations as shown by *Worni et al.* [2012] and *Caballero and Capra* [2014].

One of the main issues about modeling lahars is that they exhibit highly unsteady behavior, especially regarding their hydraulic and rheologic properties [*Carrivick et al.*, 2009]. Currently, there is no simulation code suitable for hazard assessment that can fully reproduce this dynamic behavior. Therefore, uncertainties related to input parameters must be taken into account when numerical simulations are intended to be used for lahar hazard evaluation.

ACKNOWLEDGMENTS

We thank Sylvain Charbonnier and two anonymous reviewers whose comments helped to improve this paper. This work was supported by CONACyT projects 99486 and 220786.

REFERENCES

Beverage, J. P., and J. K. Culbertson (1964), Hyperconcentrations of suspended sediment, *J. Hydraul. Div.*, *90*(6), 117–128.

Caballero, L., and L. Capra (2014), The use of FLO2D numerical code in lahar hazard evaluation at Popocatépetl volcano: A 2001 lahar scenario, *Nat. Hazards Earth Syst. Sci.*, *14*(12), 3345–3355; doi:10.5194/nhessd-2-4581-2014.

Caballero, L., D. Sarocchi, E. Soto, and L. Borselli (2014), Rheological changes induced by clast fragmentation in debris flows, *J. Geophys. Res.: Earth Surface*, *119*, 1800–1817; doi: 10.1002/2013JF002942.

Capra, L., L. Borselli, N. Varley, J. C. Gavilanes-Ruiz, G. Norini, D. Sarocchi, L. Caballero, and A. Cortes (2010), Rainfall-triggered lahars at Volcán de Colima, Mexico: Surface hydro-repellency as initiation process, *J. Volcanol. Geotherm. Res.*, *189*, 105–107; doi:10.1016/j.jvolgeores.2009.10.014..

Capra, L., J. C. Gavilanes-Ruiz, N. Varley, and L. Borselli (2014), Origin, behavior and hazard of rain-triggered lahars at Volcán de Colima, in *Volcán de Colima: Managing the Threat*, edited by N. Varley and J. C. Komorowski, Springer.

Capra, L., M. A. Poblete, and R. Alvarado (2004), The 1997 and 2001 lahars of Popocatépetl volcano (Central Mexico): Textural and sedimentological constraints on their origin and hazards, *J. Volcanol. Geotherm. Res.*, *131*, 351–369; doi: 10.1016/S0377-0273(03)00413-X.

Capra, L., V. Manea, M. Manea, and G. Norini (2011), The importance of digital elevation model resolution on granular flow simulations: a test case for Colima volcano using TITAN2D computational routine, *Nat. Hazards*, *59*, 665–680; doi: 10.1007/s11069-011-9788-6.

Carrivick, J. L., V. Manville, and S. J. Cronin (2009), A fluid dynamics approach to modelling the 18th March 2007 lahar at Mt. Ruapehu, New Zealand, *Bull. Volcanol.*, *71*(2), 153–169; doi: 10.1007/s00445-008-0213-2.

Caruso, P., and M. Pareschi (1993), Estimation of lahar and lahar-runout flow hydrograph on natural beds, *Environ. Geol.*, *22*(2), 141–152; doi: 10.1007/BF00789326.

Córdoba, G., M. F. Sheridan, and B. Pitman (2010), A two-phase, depth-averaged model for geophysical mass flows in the TITAN code framework, Paper presented at CMG-IUGG, Pisa, Italy.

Coussot, P., and J. M. Piau (1995), The effects of an addition of force-free particles on the rheological properties of fine suspensions, *Can. Geotech. J.*, *32*, 263–270; doi: 10.1139/t95-028.

Fairchild, L. H. (1987), The importance of lahar initiation processes, *Rev. Eng. Geol.*, *7*, 51–62; doi: 10.1130/REG7-p51.

Huggel, C., D. Schneider, P. J. Miranda, H. Delgado Granados, and A. Kääb (2008), Evaluation of ASTER and SRTM DEM data for lahar modeling: a case study on lahars from Popocatépetl Volcano, Mexico, *J. Volcanol. Geotherm. Res.*, *170*(1), 99–110, doi:10.1016/j.jvolgeores.2007.09.005.

Iverson, R. M., S. P. Schilling, and J. W. Vallance (1998), Objective delineation of lahar hazard zones, *Geol. Soc. Am. Bull*, *110*, 972–984, doi: 10.1130/0016-7606(1998)110<0972:ODOLIH>2.3.CO;2.

Macedonio, G., and M. Pareschi (1992), Numerical simulation of some lahars from Mount St. Helens, *J. Volcanol. Geotherm. Res.*, *54*(1), 65–80; doi:10.1016/0377-0273(92)90115-T.

Mackey, P. C., P. M. Barlow, and K. G. Ries (1998), Relations between discharge and wetted perimeter and other hydraulic-geometry characteristics at selected streamflow-gaging stations in Massachusetts, Water-Resources Investigations Report 98-4094, USGS, Marlborough, Massachusetts.

Manville, V., K. Németh, and K. Kano (2009), Source to sink: A review of three decades of progress in the understanding of volcaniclastic processes, deposits, and hazards, *Sed. Geol.*, *220*(3), 136–161; doi: 10.1016/j.sedgeo.2009.04.022.

Muñoz-Salinas, E., V. C. Manea, D. Palacios, and M. Castillo-Rodriguez (2007), Estimation of lahar flow velocity on Popocatépetl volcano (Mexico), *Geomorphology*, *92*, 91–99; doi:10.1016/j.geomorph.2007.02.011.

Naranjo, J. A., H. Sigurdsson, S. N. Carey, and W. Fritz (1986), Eruption of the Nevado del Ruiz volcano, Colombia, on 13 November 1985: Tephra fall and lahars, *Science*, *233*, 961–963; doi: 10.1126/science.233.4767.961.

O'Brien, J. (2001), FLO2D users manual, Nutrioso Arizona.

O'Brien, J., and P. Julien (1988), Laboratory analysis of mudflow properties, *J. Hydraul. Eng.*, *114*(8), 877–887; doi: 10.1061/(ASCE)0733-9429(1988)114:8(877).

O'Brien, J., P. Julien, and W. Fullerton (1993), Two-dimensional water flood and mudflow simulation, *J. Hydraul. Eng.*, *119*(2), 244–261; doi: 10.1061/(ASCE)0733-9429(1993)119:2(244).

Phillips, J. V., and S. Tadayon (2006), Selection of Manning's roughness coefficient for natural and constructed vegetated and non-vegetated channels, and vegetation maintenance plan guidelines for vegetated channels in Central Arizona, US Department of the Interior, US Geological Survey.

Proietti, C., M. Coltelli, M. Marsella, and E. Fujita (2009), A quantitative approach for evaluating lava flow simulation reliability: LavaSIM code applied to the 2001 Etna eruption, *Geochem. Geophys. Geosyst.*, *10*(9);doi:10.1029/2009GC002426.

Schilling, S. (1998), *LAHARZ: GIS programs for automated mapping of lahar-inundation hazard zones, Open-File Report 98–638*, US Geological Survey, Vancouver, WA.

Scott, K. M. (1988), Origin, behavior, and sedimentology of lahars and lahar-runout flows in the Toutle-Cowlitz river system, USGS Professional Paper, 1447-A, 72.

Scott, K. M., J. W. Vallance, N. Kerle, J. L. Macías, W. Strauch, and G. Devoli (2005), Catastrophic, precipitation-triggered lahar at Casita volcano, Nicaragua: ocurrence, bulking and transformation, *Earth Surf. Process. Landforms*, *30*(1), 59–79; doi: 10.1002/esp.1127.

Spataro, W., D. D'Ambrosio, R. Rongo, and G. A. Trunfio (2004), An evolutionary approach for modelling lava flows through cellular automata, *Cellular Automata*, 725–734, Springer; doi: 10.1007/978-3-540-30479-1_75.

Stevens, N., V. Manville, and D. Heron (2003), The sensitivity of a volcanic flow model to digital elevation model accuracy: experiments with digitised map contours and interferometric SAR at Ruapehu and Taranaki volcanoes, New Zealand, *J. Volcanol. Geotherm. Res.*, *119*(1), 89–105, doi:10.1016/S0377-0273(02)00307-4.

Vallance, J. W., and K. M. Scott (1997), The Osceola mudflow from Mount Rainier: sedimentology and hazard implications of a huge clay-rich debris flow, *Geol. Soc. Am. Bull.*, *109*(2), 143–163; doi: 10.1130/0016-606(1997)109<0143:TOMFMR>2.3.CO.

Vázquez, R., L. Capra, L. Caballero, R. Arámbula-Mendoza, and G. Reyes-Dávila (2014), The anatomy of a lahar: Deciphering the 15 September 2012 lahar at Volcán de Colima, Mexico, *J. Volcanol. Geotherm. Res.*, *272*, 126–136; doi:10.1016/j.jvolgeores.2013.11.013.

Worni, R., C. Huggel, M. Stoffel, and B. Pulgarín (2012), Challenges of modeling current very large lahars at Nevado del Huila Volcano, Colombia, *Bull. Volcanol.*, *74*, 309–324; doi: 10.1007/s00445-011-0522-8.

Xu, D., and Q. Feng (1979), Table of roughness of debris flow channel, Abstract of papers of first National Symposium on Debris Flow.

Part III
Biophysical and Climatic Hazards

Karin Riley

Editor-in-Chief

13

An Uncertainty Analysis of Wildfire Modeling

Karin Riley[1] **and Matthew Thompson**[2]

ABSTRACT

Before fire models can be understood, evaluated, and effectively applied to support decision making, model-based uncertainties must be analyzed. In this chapter, we identify and classify sources of uncertainty using an established analytical framework, and summarize results graphically in an uncertainty matrix. Our analysis facilitates characterization of the underlying nature of each source of uncertainty (inherent system variability versus limited knowledge), the location where it manifests within the modeling process (inputs, parameters, model structure, etc.), and its magnitude or level (on a continuum from complete determinism to total ignorance). We adapt this framework to the wildfire context by identifying different planning horizons facing fire managers (near-, mid-, and long-term) as well as modeling domains that correspond to major factors influencing fire activity (fire behavior, ignitions, landscape, weather, and management). Our results offer a high-level synthesis that ideally can provide a sound informational basis for evaluating current modeling efforts and that can guide more in-depth analyses in the future. Key findings include: (1) uncertainties compound and magnify as the planning horizon lengthens; and (2) while many uncertainties are due to variability, gaps in basic fire-spread theory present a major source of knowledge uncertainty.

13.1. INTRODUCTION

13.1.1. Wildfire Modeling and Uncertainty Analysis

Land and fire managers rely on wildfire modeling techniques to better understand potential wildfire activity and evaluate alternative risk management strategies. A wide range of wildfire models exists with varying inputs, structures, outputs, and intended uses [*Sullivan*, 2009a, 2009b, 2009c; *Papadopolous and Pavlidou*, 2011; *Thompson and Calkin*, 2011]. The ability of these models to support efficient and effective risk management is determined to a large degree by how well model-based uncertainties are understood and communicated [*Marcot et al.*, 2012].

Analyzing model-based uncertainties is a process of identifying, classifying, and evaluating sources of uncertainty and their influence on model outputs [*Thompson and Warmink*, Chapter 2, this volume]. Identification and classification, our principal foci here, are systematic and iterative steps that articulate and characterize essential attributes of uncertainties. These steps are a prerequisite for subsequent evaluation of salient sources of uncertainty, which can range from qualitative expert review to computationally demanding quantitative techniques [*Refsgaard et al.*, 2007; *Jimenez et al.*, 2008; *Duff et al.*, 2013]. Uncertainty analysis provides important information for modelers and analysts, aiding selection of appropriate data and modeling techniques, and guiding model calibration and validation efforts. It is crucial that

[1] *Numerical Terradynamic Simulation Group, College of Forestry and Conservation, University of Montana, Missoula, Montana, USA*

[2] *Rocky Mountain Research Station, US Forest Service, Missoula, Montana, USA*

Natural Hazard Uncertainty Assessment: Modeling and Decision Support, Geophysical Monograph 223, First Edition.
Edited by Karin Riley, Peter Webley, and Matthew Thompson.

managers understand model-based uncertainties as well, to establish confidence in results, and to determine the value of investing in additional data collection, research, or more extensive modeling efforts. Adoption of a systematic, rigorous, and consistent process for analyzing uncertainties facilitates the communication of important features of uncertainties faced within modeling and decision contexts.

In this chapter, we focus on uncertainties related to the intersection of wildfire modeling and wildfire management. Our primary objectives are (1) to illustrate application of uncertainty analysis to wildfire modeling, and (2) to introduce a conceptual framework that researchers and managers can apply to guide future modeling and decision support efforts. Ideally, increased adoption of uncertainty analysis principles will lead to targeted and efficient investments in gathering additional information, improved communication between modelers and managers, and more informed decision-making processes [*Walker et al.*, 2003; *Warmink et al.*, 2010].

To begin, we briefly review conceptual models of wildfire activity to help set the stage for our uncertainty analysis framework. Next, we introduce uncertainty analysis techniques in more detail, focusing on the identification and classification of sources of uncertainty, and describe how we tailored our analysis to the wildfire modeling context. We present results that identify where uncertainties manifest in modeling processes, and how uncertainties vary as planning horizons change. Last, we discuss implications of our findings and offer concluding thoughts.

13.1.2. Conceptual Models of Wildfire Activity

The dynamics of wildfire activity and management result from a coupling of human and natural systems with complex feedback loops that operate across a broad spectrum of spatial and temporal scales [*Liu et al.*, 2007; *Spies et al.*, 2014]. (Note that "wildland fire" is a broader term referring to both wildfires and prescribed (i.e., intentionally ignited) fires; although much of this chapter is likely applicable to the prescribed fire context, our focus is on wildfires (i.e., unplanned ignitions)). The management of a single wildfire incident evolves over the course of hours to weeks, with fire sizes ranging from less than a hectare to over a million hectares. Fire growth is dictated by varying weather patterns, landscape conditions, and human responses. Across larger landscapes and longer time horizons, uncertainty about the timing and location of ignitions, along with the weather conditions driving fire behavior, leads to reliance on risk-based characterization of wildfire variables to help support management decisions. At even larger spatiotemporal scales, long-term strategic planning necessarily considers broader drivers of the human-natural system, including changes in wildfire policy, land use and vegetation dynamics, and climate change. These uncertainties pose great challenges to land and fire managers, who are increasingly being asked to account for the effects of climate changes and human-natural system dynamics in their land management plans (for example, US Forest Service direction at http://www.fs.fed.us/emc/nepa/climate_change/includes/cc_land_mgmt_plan _rev_012010.pdf).

To provide background, we review a simple conceptual model of the major factors that influence the number, extent, and intensity of wildfires in the natural environment, in the absence of human management (Fig. 13.1). The frequency and location of ignitions (both human-caused and lightning-caused) determines the number of wildfires. Depending upon the location, ignition frequency can be driven predominately by human activity, climatic and weather patterns that cause lightning, or some combination of the two. Recent weather conditions influence both fuel moisture (via recent precipitation, relative humidity, and temperature) and rate of spread (via wind and fuel moisture). Because weather influences fuel moisture and rate of spread, it is a primary driver of fire intensity as well as fire extent. Landscape conditions

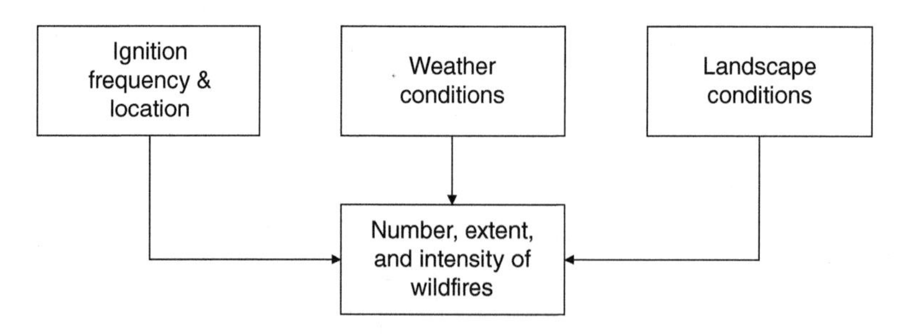

Figure 13.1 A simplified model of factors driving the number, extent, and intensity of wildland fires in the natural environment. (Note that, for example, weather is a source of ignitions, and that ignitions are affected by the landscape as well, since an ignition must land on a receptive fuel in order to be viable. For the sake of simplicity, however, these relationships are not shown.)

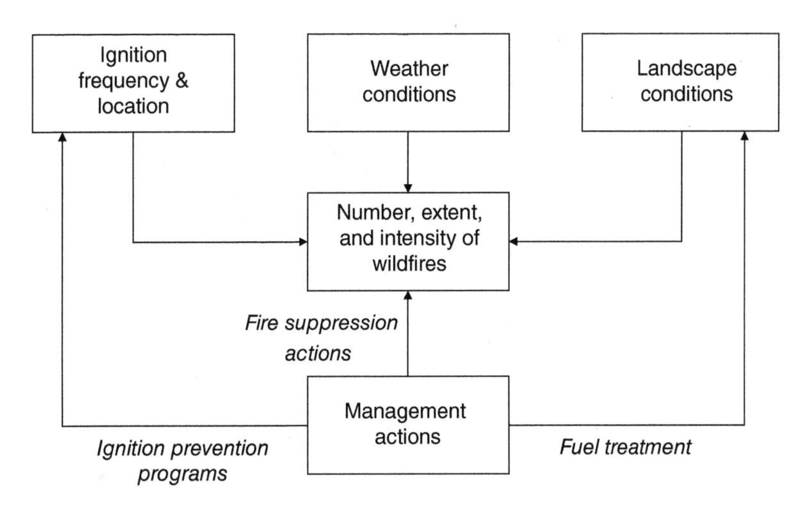

Figure 13.2 A simplified model of factors driving the number, extent, and intensity of wildland fires and how management actions interact with wildfire in a coupled human-natural system.

relate to topography and the composition, structure, and continuity of fuels (flammable vegetation), which influence the intensity of wildfires in concert with weather.

Next, we expand upon this model to illustrate the effects of management actions (Fig. 13.2). Wildfire incident response entails the strategic and tactical deployment of firefighting resources that generally aim to restrict fire growth in order to minimize loss of highly valued resources while balancing safety and cost concerns. Wildfire management options extend beyond incident response to the midterm planning horizon, and include implementing prevention programs to reduce human-caused ignitions and manipulating vegetation and fuel characteristics to reduce extreme fire behavior [*Agee and Skinner*, 2005; *Prestemon et al.*, 2013]. Fuel treatment programs include a range of approaches to remove flammable vegetation via mechanical means as well as through burning from either prescribed fire or managed natural fire.

Last, we incorporate a longer temporal horizon (Fig. 13.3). Policies establishing the emphasis and direction of management actions have and may again change in response to wildfire activity and associated consequences. Landscape conditions are influenced through time by a combination of succession, land use changes resulting from management actions or human development, and natural disturbances like wildfire. Climate can influence all three primary drivers of wildfire activity (for example, changed storm activity leading to changed ignition patterns [*Romps et al.*, 2014]; changed weather patterns; changes in vegetation and fuel composition across the landscape) and these can have a feedback effect on climate (for instance, where increased evapotranspiration due to climate change prevents a burned forest from regenerating, resulting in a net carbon flux). Management actions outside of the wildfire context, such as climate change mitigation activities, are omitted from the figure, but could play a role.

13.2. METHODS

13.2.1. Identifying and Classifying Uncertainties

We consider three primary dimensions of uncertainty in our analysis of wildfire modeling: (1) the underlying cause or *nature* of the uncertainty; (2) where in the modeling or decision process the uncertainty manifests itself, that is, the *location* of the uncertainty; and (3) where along the continuum of total determinism to total ignorance the uncertainty resides, that is, the *level* of the uncertainty [*Walker et al.*, 2003; *Kwakkel et al.*, 2010]. We rely mainly on the work of *Ascough II et al.* [2008], *Warmink et al.* [2010], and *Skinner et al.* [2014] for our characterization of the three dimensions of uncertainty, which we tailor to the wildfire modeling context for our analytical purposes (see Section 13.2.2). Figure 13.4 summarizes the three dimensions, each of which is described in more detail below.

For the nature of uncertainty, we consider two main categories: (1) *knowledge* (also known as epistemic), which refers to limitations of understanding and is considered reducible (in the sense that additional research can increase knowledge and reduce uncertainty); and (2) *variability* (aleatory), which refers to the inherent variation in natural and human systems and is considered irreducible. We consider five main locations of uncertainty: (1) *context*, meaning the assumptions and choices outside of the model boundaries that underlie the modeling process; (2) *inputs*, referring to data for a specific model run; (3) *model structure*, meaning the relationships between

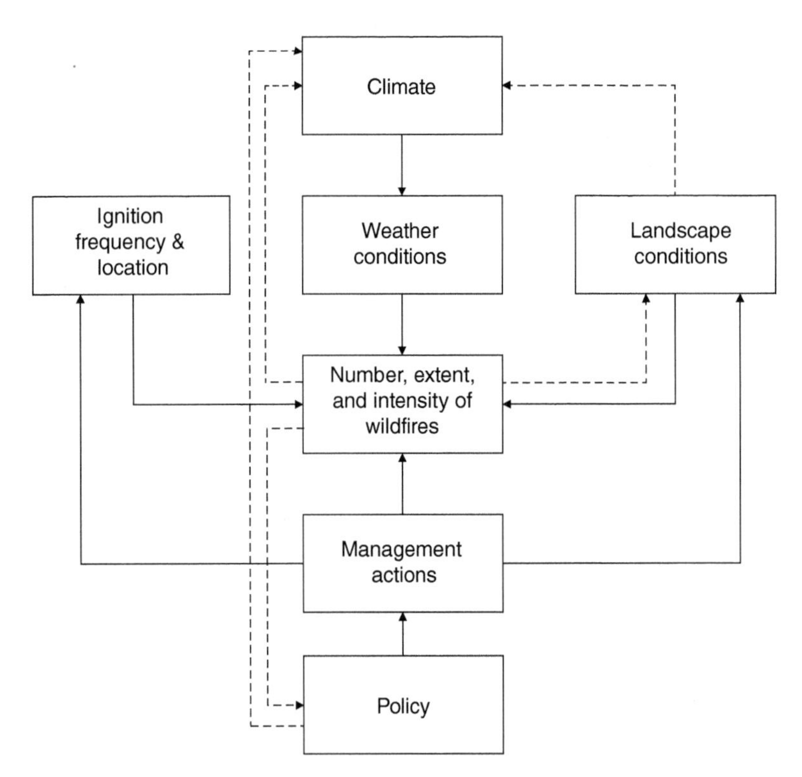

Figure 13.3 A simplified model of factors driving the number, extent, and intensity of wildland fires and how management actions interact with wildfire in a coupled human-natural system, including temporal dynamics. Feedback loops are presented with dashed arrows.

variables and the underlying system; (4) *model technical*, referring to the technical and numerical aspects related to algorithmic and software implementation; and (5) *parameters*, which refers to values invariant within a chosen model context. The third dimension of uncertainty, the level, reflects a spectrum from total determinism (which as it is fully known isn't required in an uncertainty analysis) to total ignorance (which as it is completely unknown cannot be included in an uncertainty analysis) (Fig. 13.4). Between these two endpoints, we consider three levels of uncertainty, in order of increasing uncertainty: (1) *statistical*, in which uncertainties can be quantified probabilistically or quantitatively; (2) *scenario*, in which outcomes or possibilities can be identified but not their likelihood; and (3) *recognized ignorance*, in which factors are known to be a source of uncertainty, but different possibilities or their likelihoods cannot be identified.

13.2.2. Wildfire-Specific Considerations

To organize our uncertainty analysis, we identify three planning horizons facing fire managers: (1) near-term incident response (days to weeks), (2) midterm planning (1–10 yr), and (3) long-term planning (10–50 yr). Note that the relevant spatial scale of interest expands as the temporal horizon increases. Uncertainties for each planning horizon are summarized in Table 13.1. We further organize our analysis according to primary modeling domains, that is, key functions, processes, or actions that drive wildfire activity (see Figs. 13.1–13.3). Specifically, we identify five modeling domains: (1) fire behavior, (2) ignitions, (3) weather, (4) landscape conditions, and (5) management activities. Using this organizational framework (planning horizon and modeling domain), we then identify and classify sources of uncertainty according to the three dimensions listed above (nature, location, and level). We present results in an uncertainty matrix, which is a graphical or tabular summarization of uncertainty analysis findings [*Walker et al.*, 2003; *Warmink et al.*, 2010; *Thompson and Warmink*, Chapter 2, this volume].

To populate our uncertainty matrix, we leverage our own experience applying fire models to support hazard and risk assessment with peer-reviewed literature relating to fire modeling, decision support, and uncertainty analysis. We focus principally on simulation models that implement fire spread in a geospatial context to explicitly model fire growth across a landscape, and that often provide probabilistic outputs for risk-based applications [e.g., *Atkinson et al.*, 2010; *Thompson et al.*, 2011; *Calkin et al.*, 2011; *Salis et al.*, 2013; *Scott et al.*, 2012a; *Scott et al.*, 2012b; *Ager et al.*, 2013; *Han and Braun*, 2013; *Thompson et al.*, 2015; *Thompson et al.*, Chapter 4, this

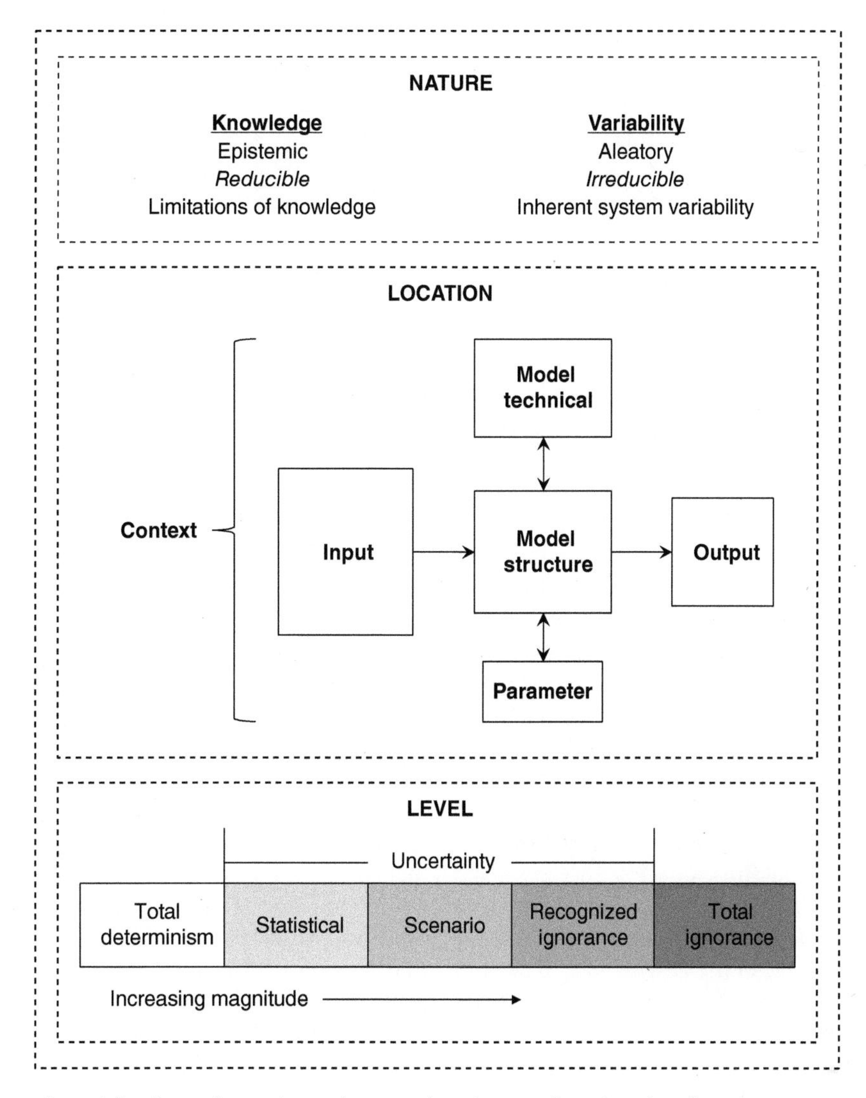

Figure 13.4 Representation of the three dimensions of uncertainty (nature, location, level).

Table 13.1 Factors Influencing Fire Extent and Intensity Across Planning Horizons, in Terms of Uncertain Information

Planning horizon	Ignitions	Weather	Landscape	Management
Wildfire incident response	Observed	Forecasts and historical patterns	Static landscape	Suppression tactics provided by incident commander
Midterm (1–10 yr)	Historical patterns	Historical patterns	Static landscape	Historical patterns of suppression effectiveness; policy scenarios for suppression and fuel management provided by land manager
Long-term (10–50 yr)	Scenarios for changes in patterns due to climate change and land use change	Climate scenarios	Scenarios for biome migration, land use change, management, and disturbance (including no-analog fuel conditions)	Scenarios for policy change in suppression, fuel management, and land use

Note: Fire behavior was not included in this table since it presents a source of uncertainty that is constant across all three planning horizons (the reader is referred to Section 13.3.1 for more information).

volume]. We further limit our focus to operational fire-spread models that have been applied to support analysis and planning efforts, most of which are empirical or quasi-empirical (i.e., a statistical approach based on a physical model; *Sullivan* [2009c]). We draw primarily from our experience with operational fire-modeling systems used in federal wildfire management in the United States [*Finney*, 2004; 2006; *Finney et al.*, 2011a; *Finney et al.*, 2011b], although we expect our analytical framework and findings to be broadly applicable.

In creating the uncertainty matrix, we balance breadth and depth, attempting to address all major influencing factors while recognizing that it is possible to drill down into any given modeling domain or source of uncertainty more comprehensively. Many fire models consist of a collection of equations and submodels; for example, a spatial fire model might employ a one-dimensional surface fire spread equation, a transition to crown fire equation, a one-dimensional crown fire propagation equation, a spatial implementation of fire spread in two dimensions across a heterogeneous landscape, a statistical weather generation algorithm, and a fire suppression module [e.g., *Finney et al.*, 2011b]. Examining the uncertainties in each of the equations and submodels exhaustively would have produced a rather cumbersome and likely unintelligible analysis, so instead we evaluate uncertainties from the perspective of the geospatial fire models themselves (each of which is a collection of submodels and equations). In some cases, separate modeling systems produce inputs for the fire models, for example, in the case of fuel treatment optimization models that provide alternative representations of landscape conditions [e.g., *Ager et al.*, 2013], or in the case of wind models that provide spatially varying wind fields [e.g., *Forthofer et al.*, 2009]. Similarly, we do not examine the uncertainties in these separate modeling systems, but because they can produce landscape or wind data to be used as inputs in the fire models, we consider uncertainty within these models to be a source of input uncertainty from the standpoint of fire modeling. Our results can be interpreted, therefore, as a high-level synthesis that can provide a sound informational basis for evaluating current modeling efforts, and that can guide more in-depth analyses in the future.

13.3. RESULTS

A list of salient uncertainties in fire modeling is presented in the uncertainty matrix, broken down according to planning horizon and primary modeling domain (Table 13.2). Uncertainty in predicting fire behavior is common across planning horizons, with similar factors influencing incident, midterm, and long-term planning. These uncertainties are identified, briefly described, and classified in Section 13.3.1. Sources of uncertainty in the other four domains vary across planning horizons. We discuss how considerations and questions vary, from the perspective of the modeler, across the incident (Section 13.3.2), midterm (Section 13.3.3), and long-term (Section 13.3.4) planning horizons. Building from the uncertainty matrix, we align the primary modeling domain with the five possible locations of uncertainty in the modeling process, following *Warmink et al.* [2010] (Figs. 13.5–13.7).

13.3.1. Uncertainties Common Across Planning Horizons: Fire Behavior

Uncertainty in fire-behavior modeling entails both theoretical [e.g., *Finney et al.*, 2012] and applied [e.g., *Jimenez et al.*, 2008] concerns. Notably, the mechanisms producing fire spread are not yet known despite modeling used extensively for the past few decades [e.g., *Rothermel*, 1972]. Recent work demonstrates that radiation is insufficient to cause ignition in fine fuels, with direct flame contact produced by buoyancy-driven instabilities being a more likely mechanism for fire spread [*Cohen and Finney*, 2014; *Finney et al.*, 2015]. Within the buoyancy dynamics, there is a certain amount of stochasticity in intermittent flame contacts, which would introduce variability uncertainty into particle ignition. It may be a number of years before new physically-based models are ready for operational implementation in the incident, midterm, and long-term planning horizons. We classify the current state of the science regarding fire physics as a source of recognized ignorance and model context uncertainty, since at this time researchers are not sure which physical processes are operating and need to be included in models.

We deem fire behavior to have a location of "model context," since the current lack of physical understanding affects the assumptions and choices underlying the construction of the model (Table 13.2). Once improved physical models are available, they will undoubtedly have sources of input, structure, technical and parameter uncertainty, but at this stage, such uncertainties cannot be listed. Because that is the case, every other source of uncertainty identified in this chapter could be considered to fall into the realm of recognized ignorance and model context uncertainty. Rather than discussing fire modeling in that sense, which would seem to have limited utility, the remainder of the uncertainty analysis regards uncertainties in fire occurrence and behavior as they are currently implemented in fire models and thus used to support decision making.

As they are currently constructed, we classify rate-of-spread and intensity equations as a source of model structure uncertainty, rooted in the mathematical relations between variables. There is a degree of uncertainty

Table 13.2 An Uncertainty Matrix Identifying and Classifying Uncertainties in Fire Modeling

Management and planning context	Uncertainty source		Nature		Location					Level		
			Knowledge	Variability	Context	Input	Model structure	Model technical	Parameter	Statistical	Scenario	Recognized ignorance
Across Contexts												
	Fire Behavior (Physical Basis)											
		Convective and radiative transfer mechanisms	x		x							x
		Rate-of-spread and intensity equations	x		x							x
	Fire Behavior (As Currently Implemented)											
		Rate-of-spread and intensity equations	x				x				x	
		Empirical model coefficients	x						x	x		
		Fire behavior fuel model parameters	x						x	x		
		Fire spread algorithms (2D)	x					x		x		
Wildfire Incident Response												
	Weather											
		Wind speed and direction forecasts		x		x				x		
		Temperature and relative humidity forecasts		x		x				x		
	Landscape											
		Vegetation type and configuration	x			x					x	
		Surface and canopy fuel models	x			x					x	
		Discretized landscape representation	x					x			x	
		Fuel moisture	x			x					x	
	Management											
		Tactics		x	x						x	

(Continued)

Table 13.2 (Continued)

Management and planning context	Uncertainty source		Nature		Location					Level		
			Knowledge	Variability	Context	Input	Model structure	Model technical	Parameter	Statistical	Scenario	Recognized ignorance
		Productive capacity and effectiveness of firefighting resources	x				x				x	
Midterm (1–10 yr)												
	Ignitions (natural and human)											
		Timing		x		x				x		
		Location		x		x				x		
		Frequency/number		x		x				x		
	Weather											
		Wind speed and direction distributions		x		x				x		
		Temperature and relative humidity distributions		x		x				x		
	Landscape											
		Changes in vegetation type and configuration										
		Local, due to disturbance, management, and land use change		x		x					x	
		Systematic, due to succession and regeneration		x		x					x	
		Changes in surface and canopy fuel models										
		Local, due to disturbance, management, and land use change		x		x					x	

		Systematic, due to succession and regeneration		x	x			x	
	Management								
		Response capacity and initial attack success	x		x			x	
		Ignition prevention effectiveness	x		x		x		
		Hazardous fuel reduction effectiveness	x		x			x	
Long-term (10–50 yr)									
	Fire behavior (physical basis)								
		Changes in relationship between fuel characteristics and fire spread	x			x			x
		Changes in relationship between fuel characteristics and ignition probability	x		x				x
	Ignitions								
		Changes in lightning patterns		x	x				x
		Changes in development patterns (human-caused ignitions)		x	x			x	
	Weather								
		Changes in precipitation regimes		x	x			x	
		Changes in wind patterns		x	x			x	
		Changes in temperature and relative humidity distributions		x	x			x	
		Changes in season length		x	x			x	
	Landscape								
		Biome migration		x	x			x	

(*Continued*)

Table 13.2 (Continued)

Management and planning context	Uncertainty source		Nature		Location					Level		
			Knowledge	Variability	Context	Input	Model structure	Model technical	Parameter	Statistical	Scenario	Recognized ignorance
		No-analog vegetation conditions		x		x						x
		Changes in vegetation type, cover, and height (systematic, due to changes in climate)		x		x					x	
		Changes in surface and canopy fuel models (systematic, due to changes in climate)		x		x					x	
		Changes in land use		x		x					x	
	Management											
		Wildfire policy		x		x					x	

Note: Across the incident, midterm, and long-term planning horizons, sources of uncertainty are listed in the uncertainty matrix according to the fire behavior, ignition, weather, landscape, and management domains. Sources of uncertainty from the incident planning horizon are also present at the midterm planning horizon, and sources of uncertainty from the midterm are also present at the long-term planning horizon, so only additional sources of uncertainty are listed as the temporal scope increases.

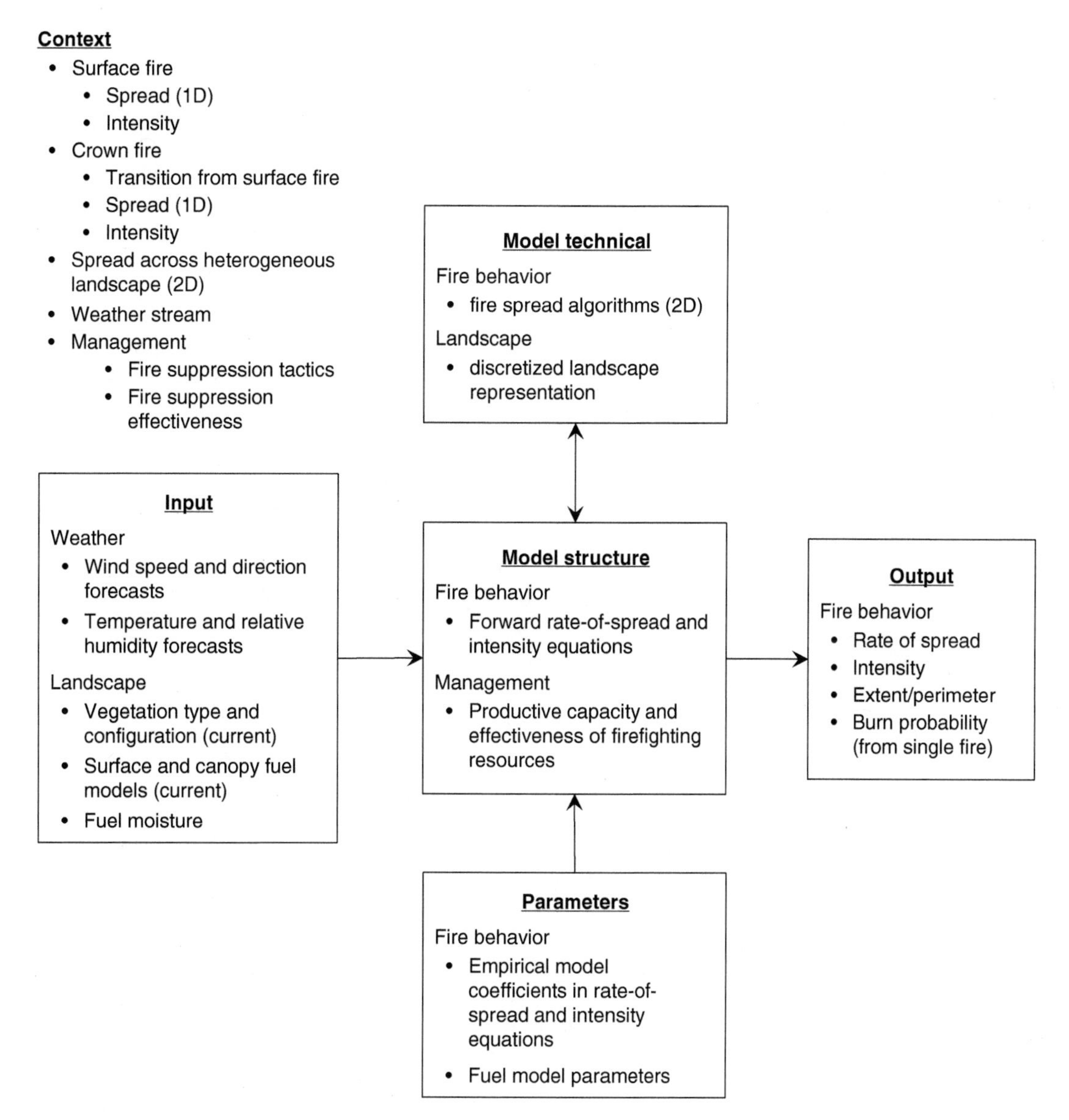

Figure 13.5 Location of uncertainties in wildfire model architecture at the near-term (incident) planning horizon [format after *Warmink et al.*, 2010].

in the factors chosen to be included in the rate of spread models, resulting in model structure uncertainty [*Wilson*, 1990]. Current fire spread models used commonly in the United States [e.g., *Rothermel*, 1972] are based on empirically derived coefficients and provide reasonable estimates of fire spread regardless of the mechanism. The values of the coefficients themselves are a source of parametric uncertainty, and are highly dependent on the source data with which the statistical relationships were built [*Liu et al.*, 2015; *Sullivan*, 2009b]. This uncertainty can be particularly relevant for prediction outside the scope of the initial experiments. Due to factors not included in surface and crown fire spread models (including subpixel variability in fuels, empirical tuning of coefficients that "ought" to be physically related to fire spread, and fire spotting), however, the results are widely considered to be accurate within a factor of two or three [*Albini*, 1976; *Wilson*, 1990]. This source of uncertainty is sometimes handled by modelers testing different rate-of-spread parameters based on expert judgment, producing a set of scenarios.

Two separate classes of models exist for predicting surface fire and crown fire. The *Rothermel* [1972] rate of spread model accounts for the spread of surface fire only, across litter, grass, and shrubs. Fire behavior fuel models developed to integrate with the *Rothermel* [1972] spread model have a standardized set of parameters defining fuel characteristics such as loading, bulk density, and

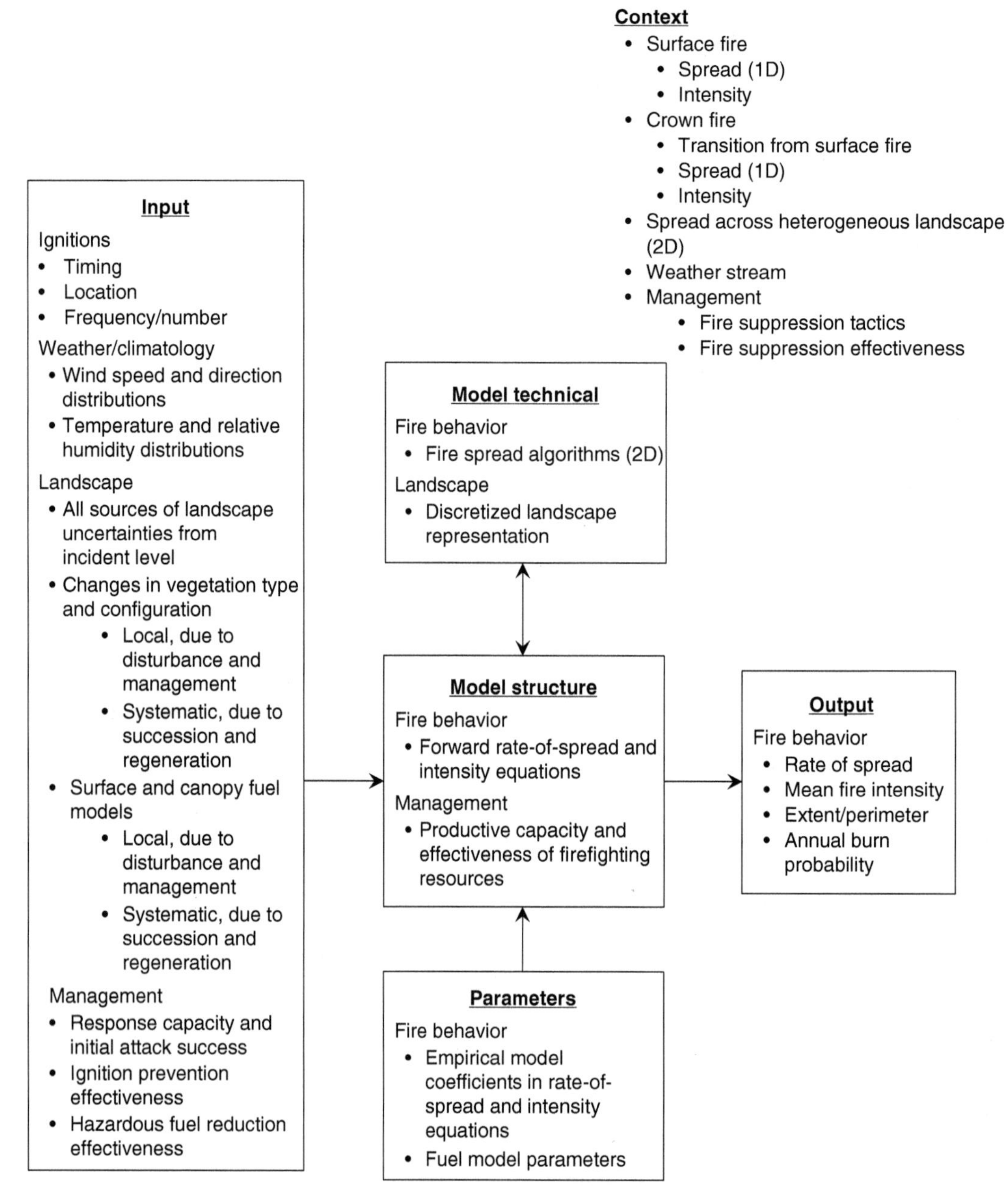

Figure 13.6 Location of uncertainties in wildfire model architecture in the midterm (1–10 yr) planning horizon [format after *Warmink et al.*, 2010]. While the sources of model technical, structural, and context uncertainties remain the same compared to the incident context (Fig. 13.5), additional uncertainties in model inputs are in play and different model outputs are available.

particle size [*Scott and Burgan*, 2005], and the degree to which these values reflect actual physical conditions represents a source of parametric uncertainty in fire behavior modeling. Several models are commonly used to simulate the transition of fire from surface to crown fire and the subsequent spread of crown fire [*Rothermel*, 1991; *Van Wagner*, 1977; *Finney*, 2004; *Scott and Reinhardt*, 2001]. These models have been criticized for underrepresenting or overrepresenting the initiation and spread of crown fire [e.g., *Alexander and Cruz*, 2013]; however, comparison of empirical measurements of the active crown fire rate of spread with modeled estimates suggests that the error rates are low enough that in practical terms the models can be utilized to support decision making during fires [*Cruz et al.*, 2005].

As noted above, fire-spread models make a number of simplifying assumptions, for example, that fuels are homogeneous, which we classify as a source of model

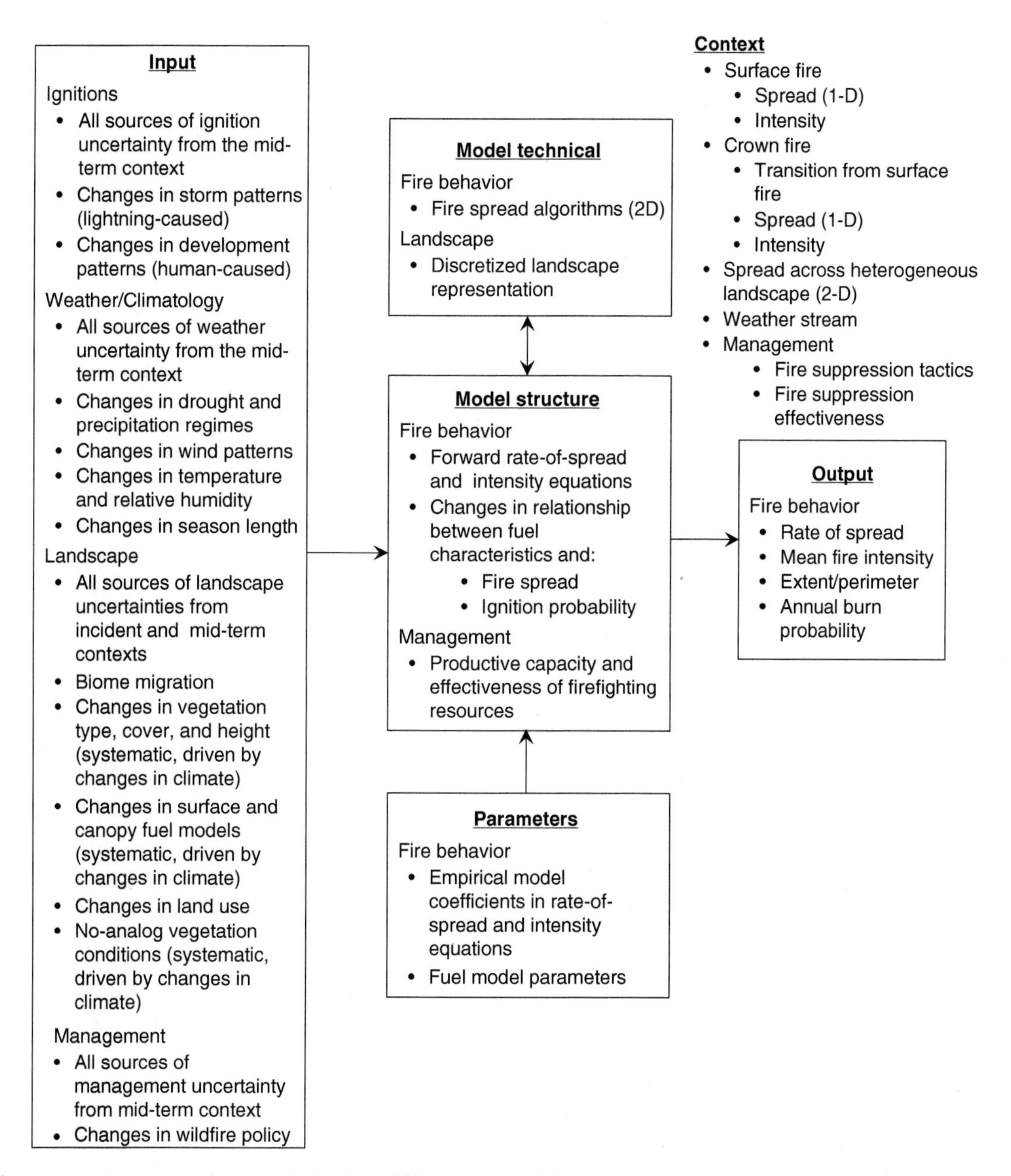

Figure 13.7 Location of uncertainties in wildfire model architecture in the long-term (10–50 yr) context [format after *Warmink et al.*, 2010]. Additional sources of input uncertainty have come into play in four of the five domains (ignitions, weather, landscape, and management). Uncertainties in long-term planning inputs are of greater magnitude, including shifts in vegetation type and composition, changes in wildfire policy, and possible no-analog fuel conditions.

structure uncertainty. In addition, fire-spread algorithms may then be implemented spatially in a model, where the paths of fastest spread across heterogeneous landscape can be quite complex [i.e., *Finney*, 2002], a source of model technical uncertainty. Field measurements of fire-spread rate and historical fire sizes serve as a check on model outputs in the face of these uncertainties regarding fire behavior and spread [e.g., *Cruz and Alexander*, 2013; *Finney et al.*, 2011b]. Model inputs are often adjusted by analysts in order to calibrate model outputs such as fire size to better match historical data.

13.3.2. Near-Term Modeling and Uncertainty: The Wildfire Incident (1–30 Days)

In the context of a given wildfire incident, one of the domains of uncertainty can be eliminated: the timing and location of the ignition is more or less known (subject to measurement error). Weather presents a form of variability uncertainty, however, in the incident planning horizon (typically a week or less), fire behavior analysts often utilize forecasts of temperature, precipitation, relative humidity, and wind. Where weather forecasts are not available or reliable, statistical predictions can be made based on historical weather records [*Finney et al.*, 2011a]. Multiple statistical weather predictions can be combined by some models to produce an ensemble simulation that outputs spatially resolved estimates of the probability of burning [e.g., *Finney et al.*, 2011a].

The landscape is rendered for use in common fire models based upon measurements and expert judgment, subject to knowledge uncertainty (since landscape configuration theoretically can be known). A static landscape reflecting current topographic and vegetation conditions is commonly used as a basic input for modeling [*Ryan and Opperman*, 2013; *Scott et al.*, 2013], with the landscape itself being a stylized model of reality. By this, we mean that the landscape must be discretized into homogenous spatial units for the model, rendering the landscape layers into a map of polygons or grids of a certain resolution (a source of model technical uncertainty). In addition, vegetation conditions are typically discretized; for example, surface fuel conditions are often represented using a discrete set of fire-behavior fuel models [*Scott and Burga*, 2005]. The degree to which assignment of these fire-behavior fuel models reflects on-the-ground fuel conditions is a source of uncertainty. Fuel moisture conditions, which are important in rate of spread calculations, vary across the landscape based on aspect and elevation. Some fire models use fuel moisture conditioning to geospatially estimate this variability [*Nelson*, 2000] based on recent weather conditions, while others assign constant fuel moisture values for the entire landscape. In either case, fuel moisture values are discretized to a pixel or assumed to be homogenous for the landscape as a whole. Uncertainties in fuel moisture and landscape vegetation values (e.g., crown bulk density and crown base height) can be quantified statistically at small scales (i.e., forest plot). However, at the landscape level, probabilities of various vegetation configurations relevant to fire models are not currently known, so we assigned a scenario level of uncertainty to the landscape representation at landscape scales. In the future, remote sensing techniques may make statistical quantification of uncertainties in some landscape variables possible at larger scales.

When fire suppression is practiced for multiple decades, the effects are thought to create positive feedbacks with negative consequences in certain ecosystems (in other words, fire suppression may lead to a buildup of vegetation that makes future fires more severe, more difficult to control, and larger), a phenomenon sometimes termed "the wildfire paradox" [*Arno and Brown*, 1991; *Calkin et al.*, 2014; *Calkin et al.*, 2015]. However, how suppression activities affect the growth and evolution of individual incidents is poorly understood. This uncertainty is particularly salient for the management of fires that escape initial containment efforts, especially under extreme weather conditions leading to intense fire behavior inherently resistant to control. While the modeling of initial containment typically assumes a single function for suppression resources, to build fire line, and compares the rate of fire line production to the rate of fire spread [*Fried and Fried*, 1996], large fire containment efforts are far more dynamic and complex [*Thompson*, 2013], with a very limited empirical basis to characterize effectiveness or efficiency [*Finney et al.*, 2009; *Holmes and Calkin*, 2013; *Thompson et al.*, 2013b; *Calkin et al.*, 2014b]. It is important to note that although large escaped fires typically account for only a small fraction of total ignitions, they also account for the majority of area burned and suppression expenditures [*Calkin et al.*, 2005; *Short*, 2013]. Productive capacity and effectiveness of firefighting resources theoretically could be known (knowledge uncertainty) and form a source of model structure uncertainty, but tactics are subject to human variability and the factors that need to be included in models are not clear (model context uncertainty).

Under some circumstances, fire behavior analysts may work with incident personnel to incorporate potential barriers to future fire spread such as known or expected fire line, burnouts, or natural barriers into model runs. However, existing operational models don't have the capacity to directly model fire line construction activities based on the amount and type of firefighting resources present [*Calkin et al.*, 2011]. Further, these models are unable to account for the broader suite of suppression tactics such as aerial retardant delivery. Reliance on expert judgment and intuition is therefore common.

The location of each uncertainty in incident-level wildfire modeling is shown in Figure 13.5. Within the incident planning horizon, sources of uncertainty are mostly located in model inputs, with the remainder being more or less equally divided between model technical, model structure, parameters, and model context. However, this doesn't mean that the input uncertainties necessarily outweigh the other locations of uncertainty in magnitude, as the respective levels of each source of uncertainty also come into play (Table 13.2).

13.3.3. The Midterm Planning Horizon: Modeling and Uncertainty in the 1–10 Year Time Frame

As the temporal dimension of fire modeling expands from the incident to the midterm (next 1–10 years), sources of uncertainty are compounded in four of the five domains (ignitions, weather, landscape, and management; Table 13.2, Fig. 13.6). In this time frame, the location and number of ignitions cannot be known, as it will be based on natural variability in weather (for example, storm tracks, temperature, and precipitation patterns) as well as variability in human behavior resulting in human-caused ignitions. Historical ignition patterns often form the basis for statistical estimates of future ignition likelihood in the modeling environment [*Andrews et al.*, 2003; *Syphard et al.*, 2008], although the accuracy of these estimates can be degraded by incomplete or inadequate fire history records [*Short*, 2013].

Weather inputs for models in the midterm planning horizon are often based on historical distributions of wind speed and direction, temperature, relative humidity, and precipitation [*Finney et al.*, 2011b]. An array of weather station records and more recently interpolated gridded weather products [e.g., *Abatzoglou*, 2011] are available to analysts. Relative to the incident planning horizon where the sequence of weather is known for the previous days (and can be used as a starting point for modeled weather) and forecasts are generally available, these midterm estimates entail a greater degree of uncertainty, being based on statistical distributions and time-series analysis of historical weather. Broader climatic factors such as interannual and decadal oscillations, including the El Niño Southern Oscillation, Atlantic Multidecadal Oscillation, and Pacific Decadal Oscillation, also exert an influence on drought and weather patterns at this temporal scale. The effects of such oscillations vary spatially, and neither their onset nor influence on fire behavior via weather can be modeled reliably at this point in time, adding to uncertainty in weather variability in the midterm planning horizon.

The uncertainties regarding the accuracy of the depicted landscape at the incident level are still present, and will increase over time as disturbances, plant growth, and succession affect the landscape. In addition, landscape changes are likely to occur due to management, including fuel reduction and timber harvest. Both the rate and location of these changes are uncertain due to factors such as market volatility and litigation. Land-use changes, including expansion of the wildland-urban interface, are also likely during this time frame. In contrast to the landscape at the incident planning horizon, which theoretically can be known, the landscape during the midterm planning horizon is subject to variability uncertainty. This uncertainty could be addressed in the modeling realm by using different landscape scenarios as inputs to fire models.

Similarly, sources of uncertainty in management present at the incident planning horizon still exist, and are augmented by additional factors. Historical patterns of suppression effectiveness often form the basis for model structure in this planning horizon [*Finney et al.*, 2009]. The effectiveness of fire suppression during both initial and extended attack is not currently well understood, but can be modeled statistically. The effectiveness of other management actions such as fire prevention and fuel management can be inferred empirically or through simulation modeling (sources of knowledge uncertainty, with statistical and scenario level, respectively) [*Ager et al.*, 2010; *Prestemon et al.*, 2013; *Thompson et al.*, 2013a]. Additional sources of uncertainty affect fire management at this scale. Burn probability models often simulate only "large" fires, since small fires contribute little to landscape-scale burn probability [e.g., *Finney et al.*, 2011b]. The number of "large" fires is affected by initial attack success and response capacity across potentially simultaneous wildfires, as well as the effectiveness of prevention programs, forming a source of input uncertainty. Rate of spread and intensity may be affected by fuel treatment programs via their effect on fuel model assignment, producing additional input uncertainty. At this planning horizon, most management factors can be considered a source of input uncertainty, but can be modeled using a statistical or scenario approach.

In summary, additional sources of input uncertainty appear when moving from the incident to midterm planning horizon, while the sources of uncertainty in the model technical, model structure, parameter, and model context locations are common across the two planning horizons (Fig. 13.6). Additional model outputs (including mean fire intensity and annual burn probability) are typically available as well.

13.3.4. The Long-Term Planning Horizon: Modeling and Uncertainty During the 10–50 Yr Time Frame

At the long-term planning horizon, many of the uncertainties present in the incident and midterm planning horizons increase in magnitude, and several new sources of uncertainty appear, with the result of moving the level of uncertainty toward recognized ignorance in some cases. Because shifts in vegetation composition and climate may produce nonequilibrium and no-analog conditions, unprecedented changes in fire behavior are possible. For example, changes in the relationship between fuel characteristics (such as fuel moisture) and ignition probability may occur, since these probabilities vary across different ecosystems. In addition, changes in the relationship between fuel conditions (such as fuel moisture) and fire spread may occur, as different pairings between vegetation and climate manifest. These possible changes fall into the realm of recognized ignorance, and present a major challenge for fire modeling in this time frame.

Within the ignitions domain, the number, frequency, and timing of ignitions are likely to change as storm tracks are altered due to climate change [*Romps et al.*, 2014]. However, the mechanisms are driven by natural variability and are not well understood, resulting in a level of recognized ignorance. In addition, human development may change the number and location of ignitions, for example, by increasing the presence of humans in previously remote areas and by making some previously burnable areas nonburnable through irrigated agriculture or paving. The patterns are unpredictable, resulting in a scenario level of uncertainty in model inputs for human ignitions.

As the climate changes, so will the short-term weather that drives large wildfires. Among these predicted changes are alterations in drought and rainfall regimes; changes in wind patterns, temperature, and relative humidity; lightning activity; and fire season length [*Kirtman et al.*, 2013; *Westerling et al.*, 2006]. These projected changes vary spatially, with some areas predicted to become wetter and others predicted to become drier, for example. Interactions among these changing weather factors and how they affect the availability of fuels to burn are complex [*Loehman et al.*, 2014]. For example, if precipitation and temperature during fire season both increase, are fires likely to become more or less frequent? Further complicating the issue, outputs from General Circulation Models (GCMs) indicate a range of possible future climate predictions based on both different carbon emissions trajectories and differences in model architecture [*Kirtman et al.*, 2013]. Changes in global, national, and local policies may affect carbon emissions trajectories and the resulting degree of climate change. In any case, because changes in climate are expected to alter weather probability distributions, uncertainty in weather inputs increases to scenario level, where emissions pathways and the outputs of GCMs present possible scenarios for model inputs.

In addition, changes in local and national policies could result in systematic land-use change. Changes in land management policy may result in differences in harvest rates, fuel treatment rates, and how disturbances such as wildfires are managed, causing widespread changes in landscape conditions. These changes have a nature of variability, and produce sources of input uncertainty.

At this timescale, landscape conditions will be affected by complex feedbacks driven by climate change. These changes could include biome migration and type conversions (e.g., from forest to grassland) driven by changes in climate [*Loehman et al.*, 2014]. No-analog vegetation conditions may result from changes in climate, meaning that the set of current fuel models may need to be augmented or modified in ways that aren't clear yet, resulting in a source of recognized ignorance. Climate change will likely affect the rate and magnitude of disturbances including wildfire and bark beetle infestations, which may systematically impact the landscape. While alterations in the landscape are likely to be mainly local in scale in the midterm planning horizon, they are likely to be widespread and systematic during the long-term planning horizon. Since stochastic natural variability cannot be predicted, different landscape scenarios might be utilized as model inputs.

Because it is not possible to predict the exact trajectory of climate and because possible no-analog weather conditions may occur, as well as natural and human variability inherent in the system, definitive predictions of changes in fire behavior and occurrence are currently not feasible to predict during this time frame [*Kirtman et al.*, 2013]. However, scenario planning based on the outputs of GCMs can be used as a tool for the long-term planning horizon.

In summary, when moving from the midterm to the long-term planning horizon, more sources of input uncertainty come into play in four of the five domains (ignitions, weather, landscape, and management; Table 13.2; Fig. 13.7). Uncertainties in long-term planning inputs are of greater magnitude, including systematic shifts in vegetation type and composition, changes in wildfire policy, and possible no-analog fuel conditions.

13.4. DISCUSSION

Globally, a wide variety of approaches to fire modeling exist, ranging from rapid simulation to inform incident decision making to computationally intensive fluid dynamic models aimed more at improving physical understanding of fire propagation [*Linn*, 1997; *McGrattan*, 2010]. In this chapter, we largely abstracted from specific modeling approaches to focus on uncertainties surrounding the major drivers of wildfire activity and their role in modeling to support wildfire management. That multiple uncertainties are identified doesn't mean models aren't useful, and in fact models will likely grow in importance moving forward, due to a number of factors including: (1) increasing human development in some parts of the world with commensurate increases in values at risk, and (2) climate change likely to cause widespread changes in fire dynamics as well as fires in some areas that haven't previously experienced them.

Our analysis revealed that that the current state of knowledge about fire physics places modeling efforts in the realm of recognized ignorance, pending improved understanding of fire physics [*Finney et al.*, 2012]. However, multiple existing fire models have empirically tuned coefficients over the past few decades, resulting in model outputs that replicate fire spread and intensity reasonably enough for use across planning horizons. Future research may reduce knowledge gaps as a better understanding of the physical process is gained, and it is possible that in the future, natural variability may be the dominant source of uncertainty due to turbulence and buoyancy dynamics.

Weather variables presented a form of variability uncertainty across all planning contexts. Ignitions are known at the incident planning horizon, but fall into the category of variability uncertainty at the midterm and long-term planning horizons, where they are driven by weather as well as human behavior. Landscape variables are a source of knowledge uncertainty at the incident planning horizon (where they theoretically could be measured and known), but become a source of variability uncertainty at midterm and long-term planning horizons (where they are driven by changes due to natural disturbance, management, succession, and climate change). Many management variables could theoretically be known (for example, success rates from fire prevention programs), while others (such as choice of suppression tactics during an incident) fall into the category of variability uncertainty.

At the near-term incident response horizon, sources of uncertainty are fairly evenly divided in their location across model structure, model technical, and input uncertainty. However, additional sources of input uncertainty come into play at the midterm and long-term planning horizons, so that location of uncertainty is dominated by inputs at the long-term planning horizon.

At the incident and midterm planning horizons, ignitions and weather tended to be possible to quantify with a statistical level of uncertainty, but landscape and management factors tended to have a scenario level of uncertainty across all planning horizons. The proportion of factors with a scenario rather than statistical level increased in the long-term context, reflecting the increasing degree of uncertainty as the scope of the planning horizon expands. On the whole, while some sources of uncertainty could be represented with a statistical level of uncertainty (for example, distributions of temperature and precipitation from weather station records), most factors could only be represented with a scenario level of uncertainty (including landscape vegetation variables), reflecting, for instance, the inability to quantify spatiotemporal landscape patterns, or to place error bounds on measurements of input variables such as vegetation characteristics at the landscape scale. Recognized ignorance is rare in our classification, relating only to conceptual understanding of basic fire physics across all planning horizons, and several uncertainties in the long-term planning horizon, specifically changes in the relationship between fuel characteristics and ignition probability and fire spread, changes in storm patterns, and no-analog vegetation conditions. Admittedly, the level and location dimensions were difficult to ascertain, and are in some sense subjective judgments given our experience and understanding of contemporary model application. Future research may reduce the level of some of these uncertainties from recognized ignorance to scenario, or from scenario to statistical.

One important finding of this uncertainty analysis is that uncertainty increases as the spatial and temporal scale of the fire modeling analysis increases. Figure 13.8

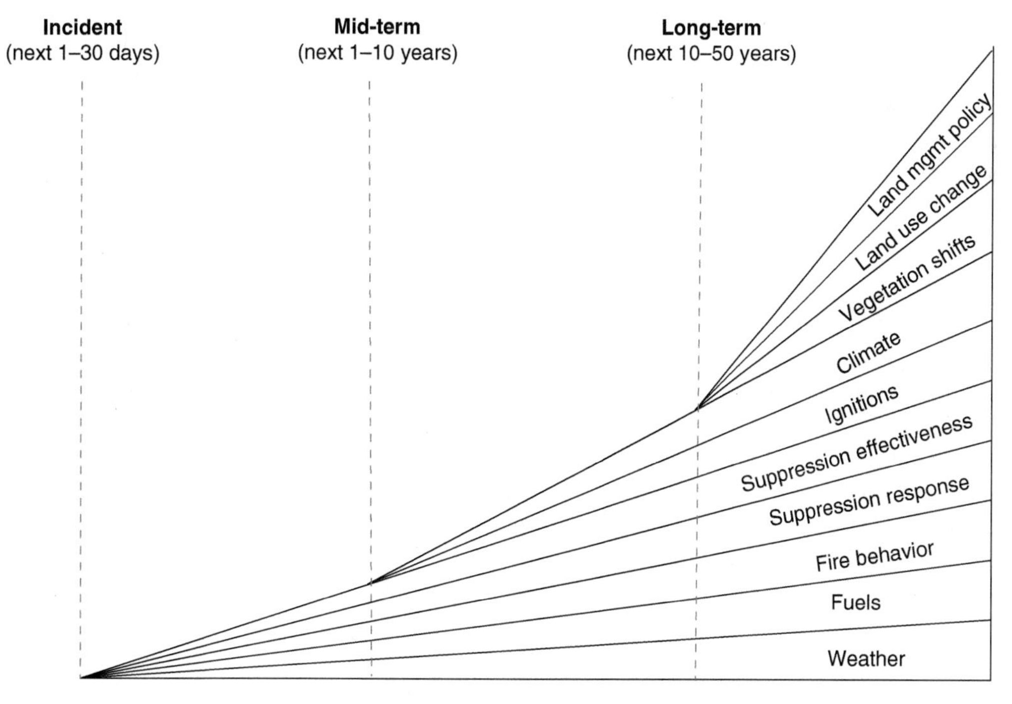

Figure 13.8 Compounding uncertainty across planning levels. As modeling frameworks move from shorter to longer-term planning contexts, additional sources of uncertainty come into play, and existing sources of uncertainty grow in magnitude.

illustrates how uncertainty compounds across the three planning horizons. Many of the sources of uncertainty remain nominally the same across planning contexts, but expand in magnitude as the time frame grows longer. For example, during an incident, weather during the next few days is uncertain but can be modeled using weather forecasts with better accuracy than can be derived from using climatological averages. The uncertainty in weather grows at the landscape assessment level, but can still be modeled using weather records and a statistical approach. Moving to the long-term planning horizon, climate itself is uncertain, producing a higher magnitude of uncertainty in predicting the weather events that produce the ignition and spread of wildland fire. In addition, moving from the incident to midterm to long-term planning context, additional sources of uncertainty come into play, for example, changes in land use. Awareness of how uncertainty increases with spatial and temporal scale can help researchers to encompass uncertainty meaningfully into study designs.

Although uncertainty increased with the time frame of the planning horizon in this analysis, this is not always the case: more uncertainty can be present in the short term than in the long term. For example, this pattern was seen in projections of polar bear populations, based on arctic sea ice extent under various scenarios for atmospheric greenhouse gas concentrations. Due to dynamics of regional warming in these scenarios, far lower certainty existed during the first 2–3 decades of the analysis, but by the end of the 21st century, projections converged on high certainty of low sea ice coverage and commensurately low polar bear populations [*Amstrup et al.*, 2008]. Whether uncertainty increases or decreases with time, the conclusion is that planning across both short- and long-term time frames is important, and enables management decisions in the short term, which can set the stage for ecosystem health and resiliency in the longer term in light of uncertainty.

13.5. CONCLUSIONS

This work systematically identifies and classifies model-based uncertainties faced in wildfire modeling, and represents an expansion of uncertainty analysis as applied to wildfire risk management. We organize presentation of our uncertainty matrix around real-world planning horizons (incident, midterm, and long-term), and primary modeling domains (fire physics, ignitions, weather, landscape, and management), which allows for an enhanced identification of salient uncertainties. Within this framework, we delineate and distinguish commonalities and differences in sources of uncertainty and their respective natures, locations, and levels. An important finding of this study is that while some sources of uncertainty are common across all planning horizons, more sources of uncertainty appear while others grow in magnitude (i.e., change in the level dimension) as the scale of the planning horizon increases. The result is compounding uncertainty and, more important, a need to rethink whether modeling approaches applied in one planning horizon are appropriate for use in other planning horizons.

The presence of all these sources of uncertainty need not deter analysts and need not undermine confidence in model predictions. Explicit recognition and analysis of uncertainties can increase the confidence of managers in model predictions, improve the modeling process, improve study design, and enhance communication across modelers, analysts, decision makers, and stakeholders. The framework developed here can provide a powerful tool for future analyses of wildfire activity, and we hope will help organize critical thinking to ensure the right questions are being asked, the right models are being used for the right reasons, and model outputs are properly understood in the context of model-based uncertainty analysis.

ACKNOWLEDGMENTS

We are grateful to Mark Finney for providing comments on an early version of the manuscript, and to three anonymous reviewers for their helpful suggestions.

REFERENCES

Abatzoglou, J. T. (2011), Development of gridded surface meteorological data for ecological applications and modelling, *Int. J. Climatol.*, *33*(1), 121–131.

Agee, J. K., and C. N. Skinner (2005), Basic principles of forest fuel reduction treatments, *For. Ecol. Man.*, *211*, 83–96.

Ager, A. A., M. Buonopane, A. Reger, and M. A. Finney (2013), Wildfire exposure analysis on the national forests in the Pacific Northwest, USA, *Risk Anal.*, *33*(6), 1000–1020.

Ager, A. A., N. M. Vaillant, and A. McMahan (2013). Restoration of fire in managed forests: a model to prioritize landscapes and analyze tradeoffs, *Ecosphere*, 4:art29, 10.1890/ES13-00007.1.

Ager, A. A., N. M. Vaillant, and M. A. Finney (2010), A comparison of landscape fuel treatment strategies to mitigate wildland fire risk in the urban interface and preserve old forest structure, *For. Ecol. Man.*, *259*(8), 1556–1570.

Albini, F. A. (1976), Estimating wildfire behavior and effects, USDA Forest Service General Technical Report INT-30, USDA Forest Service Intermountain Forest and Range Experiment Station, Ogden, Utah.

Alexander, M. E., and M. G. Cruz (2013), Are the applications of wildland fire behaviour models getting ahead of their evaluation again?, *Environ. Mod. Soft.*, *41*, 65–71.

Amstrup, S. C., B. G. Marcot, and D. C. Douglas (2008), A Bayesian network modeling approach to forecasting the 21st century worldwide status of polar bears, 213–268, in *Arctic*

Sea Ice Decline: Observations, Projections, Mechanisms, and Implications, edited by E. T. DeWeaver, C. M. Bitz, and L.-B. Tremblay, Geophysical Monograph 180, American Geophysical Union, Washington, DC.

Andrews, P. L., D. O. Loftsgaarden, and L. S. Bradshaw (2003), Evaluation of fire danger rating indexes using logistic regression and percentile analysis, *Int. J. Wildland Fire*, *12*, 213–226.

Arno, S. F., and J. K. Brown (1991), Overcoming the paradox in managing wildland fire, *West. Wildlands*, *17*(1), 40–46.

Ascough, J. C., II, H. R. Maier, J. K. Ravalico, and M. W. Strudley (2008), Future research challenges for incorporation of uncertainty in environmental and ecological decision-making, *Ecol. Mod.*, *219*(3–4), 383–399.

Atkinson, D., M. Chladil, V. Janssen, and A. Lucieer (2010), Implementation of quantitative bushfire risk analysis in a GIS environment, *Int. J. Wildland Fire*, *19*(5), 649–658.

Calkin, D. E., C. S. Stonesifer, M. P. Thompson, and C. W. McHugh (2014b), Large airtanker use and outcomes in suppressing wildland fires in the United States, *Int. J. Wildland Fire*, *23*(2), 259–271.

Calkin, D. E., J. D. Cohen, M. A. Finney, and M. P. Thompson (2014a), How risk management can prevent future wildfire disasters in the wildland-urban interface, *Proc. Nat. Acad. Sci.*, *111*(2), 746–751.

Calkin, D. E., K. M. Gebert, J. G. Jones, and R. P. Neilson (2005), Forest Service large fire area burned and suppression expenditure trends, 1970–2002, *J. Forestry*, *103*(4), 179–183.

Calkin, D. E., M. P. Thompson, and M. A. Finney (2015), Negative consequences of positive feedbacks in US wildfire management, *For. Ecosys.*, *2*(9), http://dx.doi:10.1186/s40663-015-0033-8.

Calkin, D. E., M. P. Thompson, M. A. Finney, and K. D. Hyde (2011), A real-time risk assessment tool supporting wildland fire decision making, *J. Forestry*, *109*(5), 274–280.

Cohen, J. D., and M. A. Finney (2014), Fine fuel particle heating during experimental laboratory fires, Paper presented at VII International Conference on Forest Fire Research, Associacao para o Desenvolvimento da Aerodinamica Industrial, Coimbra, Portugal.

Cruz, M. G., and M. E. Alexander (2013), Uncertainty associated with model predictions of surface and crown fire rates of spread, *Environ. Mod. Soft.*, *47*, 16–28.

Cruz, M. G, M. E. Alexander, and R. H. Wakimoto (2005), Development and testing of models for predicting crown fire rate of spread in conifer forest stands, *Can. J. For. Res.*, *35*, 1626–1639.

Duff, T. J., D. M. Chong, and K. G. Tolhurst. (2013), Quantifying spatio-temporal differences between fire shapes: Estimating fire travel paths for the improvement of dynamic spread models, *Environ. Mod. Soft.*, *46*, 33–43.

Finney, M., I. C. Grenfell, and C. W. McHugh (2009), Modeling containment of large wildfires using generalized linear mixed-model analysis, *For. Sci.*, *55*(3), 249–255.

Finney, M. A. (2002), Fire growth using minimum travel time methods, *Can. J. For. Res.*, *32*(8), 1420–1424.

Finney, M. A. (2004), FARSITE: Fire area simulator, model development and evaluation, USDA Forest Service Research Paper RMRS-RP-4, Revised, USDA Forest Service Rocky Mountain Research Station, Missoula, Montana.

Finney, M. A. (2006), FlamMap 3.0, USDA Forest Service, Rocky Mountain Research Station, Fire Sciences Laboratory, Missoula, Montana, Rep., 213–220, US Department of Agriculture, Forest Service, Rocky Mountain Research Station, Portland, Oregon.

Finney, M. A., C. W. McHugh, I. C. Grenfell, K. L. Riley, and K. C. Short (2011b), A simulation of probabilistic wildfire risk components for the continental United States, *Stochas. Environ. Res. Risk Assess.*, *25*(7), 973–1000.

Finney, M. A., I. C. Grenfell, C. W. McHugh, R. C. Seli, D. Tretheway, R. D. Stratton, and S. Brittain (2011a), A method for ensemble wildland fire simulation, *Environ. Mod. Assess.*, *16*(2), 153–167.

Finney, M. A., J. D. Cohen, J. M. Forthofer, S. S. McAllister, M. J. Gollner, D. J. Gorham, K. Saito, N. K. Akafuah, B. A. Adam, and J. D. English (2015), Role of buoyant flame dynamics in wildfire spread, *Proc. Nat. Acad. Sci.*, www.pnas.org/cgi/doi/10.1073/pnas.1504498112.

Finney, M. A., J. D. Cohen, S. S. McAllister, and W. M. Jolly (2012), On the need for a theory of wildland fire spread, *Int. J. Wildland Fire*, *22*(1), 25–36.

Forthofer, J., K. Shannon, B. Butler (2009), Simulating diurnally driven slope winds with WindNinja, in Proceedings of 8th Symposium on Fire and Forest Meteorological Society, American Meteorological Society, Kalispell, Montana, USA, 13–15 October 2009.

Fried, J. S., and B. D. Fried (1996), Simulating wildfire containment with realistic tactics, *For. Sci.*, *42*(3), 267–281.

Han, L., and W. John Braun (2013), Dionysus: A stochastic fire growth scenario generator, *Environmetrics*, *25*(6), 431–442.

Holmes, T. P., and D. E. Calkin (2013), Econometric analysis of fire suppression production functions for large wildland fires, *Int. J. Wildland Fire*, *22*(2), 246–255, http://dx.doi.org/10.1071/WF11098.

Jimenez, E., M. Y. Hussaini, and S. Goodrick (2008), Quantifying parametric uncertainty in the Rothermel model, *Int. J. Wildland Fire*, *17*(5), 638–649.

Kirtman, B., S. B. Power, J. A. Adedoyin, G. J. Boer, R. Bojariu, I. Camilloni, F. J. Doblas-Reyes, A. M. Fiore, M. Kimoto, G. A. Meehl, M. Prather, A. Sarr, C. Schär, R. Sutton, G. J. van Oldenborgh, G. Vecchi, and H. J. Wang (2013), Near-term climate change: Projections and predictability, 953–1028, in *Climate Change 2013: The Physical Science Basis. Contribution of Working Group I to the Fifth Assessment Report of the Intergovernmental Panel on Climate Change*, edited by T. F. Stocker, D. Qin, G.-K. Plattner, M. Tignor, S. K. Allen, J. Boschung, A. Nauels, Y. Xia, V. Bex and P.M. Midgley, Cambridge University Press, Cambridge, UK, and New York, USA.

Kwakkel, J. H., W. E. Walker, and V. A. W. J. Marchau, (2010), Classifying and communicating uncertainties in model-based policy analysis, *Int. J. Tech. Pol. Man.*, *10*(4), 299–315.

Linn, R. R. (1997), Transport model for prediction of wildfire behavior, Los Alamos National Laboratory, Scientific Report LA 13334-T, Los Alamos, New Mexico.

Liu, J., T. Dietz, S. R. Carpenter, M. Alberti, C. Folke, E. Moran, A. N. Pell, P. Deadman, T. Kratz, J. Lubchenco, E. Ostrom, Z. Ouyang, W. Provencher, C. L. Redman, S. H.

Schneider, and W. W. Taylor (2007), Complexity of coupled human and natural systems, *Science*, *317*, 1513–1516.

Liu, Y., E. Jimenez, M. Y. Hussaini, G. ökten, and S. Goodrick. (2015), Parametrick uncertainty quantification in the Rothermel model with randomized quasi-Monte Carlo methods, *Int. J. Wildland Fire*, *24*(3), 307–316.

Loehman, R. A., E. Reinhardt, and K. L. Riley (2014), Wildland fire emissions, carbon, and climate: Seeing the forest and the trees–a cross-scale assessment of wildfire and carbon dynamics in fire-prone, forested ecosystems, *For. Ecol. Man.*, *317*, 9–19.

Marcot, B. G., M. P. Thompson, M. C. Runge, F. R. Thompson, S. McNulty, D. Cleaves, M. Tomosy, L. A. Fisher, and A. Bliss. (2012) Recent advances in applying decision science to managing national forests, *For. Ecol. Man.*, *285*, 123–132.

McGrattan, K., S. Hostikka, J. Floyd, H. Baum, R. G. Rehm, W. E. Mell, and R. McDermott. (2010), *Fire Dynamics Simulator (Version 5) Technical Reference Guide: Mathematical Model, 1*, 108, Nist Special Publication 1018-5, Washington, DC.

Nelson, R. M. (2000). Prediction of diurnal change in 10-h fuel stick moisture content, *Can. J. For. Res.*, *30*, 1071–1087.

Papadopoulos, G. D., and F.-N. Pavlidou (2011), A comparative review on wildfire simulators, *IEEE Sys. J.*, *5*(2), 233–243.

Prestemon, J. P., T. J. Hawbaker, M. Bowden, J. Carpenter, M. T. Brooks, K. L. Abt, R. Sutphen, and S. Scranton (2013), Wildfire ignitions: A review of the science and recommendations for empirical modeling, USDA Forest Service General Technology Report SRS-GTR-171, USDA Forest Service Southern Research Station, Asheville, North Carolina.

Refsgaard, J. C., J. P. van der Sluijs, A. L. Højberg, and P. A. Vanrolleghem (2007), Uncertainty in the environmental modelling process: A framework and guidance, *Env. Mod. Soft.*, *22*(11), 1543–1556.

Romps, D. M., J. T. Seeley, D. Vollaro, and J. Molinari (2014), Projected increase in lightning strikes in the United States due to global warming, *Science*, *346*(6211), 851–854.

Rothermel, R. C. (1972), A mathematical model for predicting fire spread in wildland fuels, USDA Forest Service Research Paper INT-115, USDA Forest Service Intermountain Forest and Range Experiment Station, Ogden, Utah.

Rothermel, R. C. (1991), Predicting behavior and size of crown fires in the northern Rocky Mountains, USDA Forest Service Intermountain Research Station, Research Paper, INT-438.

Ryan, K. C., and T. S. Opperman (2013), LANDFIRE: A national vegetation/fuels data base for use in fuels treatment, restoration, and suppression planning, *For. Ecol. Man.*, *294*, 208–216.

Salis, M., A. A. Ager, B. Arca, M. A. Finney, V. Bacciu, P. Duce, and D. Spano (2013), Assessing exposure of human and ecological values to wildfire in Sardinia, Italy, *Int. J. Wildland Fire*, *22*(4), 549–565.

Scott, J., D. Helmbrecht, M. P. Thompson, D. E. Calkin, and K. Marcille (2012b), Probabilistic assessment of wildfire hazard and municipal watershed exposure, *Nat. Hazards*, *64*(1), 707–728.

Scott, J., D. Helmbrecht, S. Parks, and C. Miller (2012a), Quantifying the threat of unsuppressed wildfires reaching the adjacent wildland-urban interface on the Bridger-Teton National Forest, Wyoming, USA, *Fire Ecol.*, *8*(2), 125–142.

Scott, J. H., and E. D. Reinhardt (2001), Assessing crown fire potential by linking models of surface and crown fire behavior, USDA Forest Service Research Paper RMRS-RP-29, USDA Forest Service Rocky Mountain Research Station, Fort Collins, Colorado.

Scott, J. H., and R. E. Burgan (2005), Standard fire behavior fuel models: A comprehensive set for use with Rothermel's surface fire spread model, The Bark Beetles, Fuels, and Fire Bibliography, 66.

Scott, J. H., M. P. Thompson, and D. E. Calkin (2013), A wildfire risk assessment framework for land and resource management, USDA Forest Service General Technical Report RMRS-GTR-315, USDA Forest Service Rocky Mountain Research Station.

Short, K. (2013), A spatial database of wildfires in the United States, 1992–2011, *Earth System Science Data Discussions*, *6*(2), 297–366.

Skinner, D. J., S. A. Rocks, and S. J. Pollard (2014), A review of uncertainty in environmental risk: characterising potential natures, locations and levels, *J. Risk Res.*, *17*(2), 195–219.

Spies, T. A., E. M. White, J. D. Kline, A. P. Fischer, A. Ager, J. Bailey, J. Bolte, J. Koch, E. Platt, and C. S. Olsen (2014), Examining fire-prone forest landscapes as coupled human and natural systems, *Ecol. Soc.*, *19*(3), 9.

Sullivan, A. L. (2009a), Wildland surface fire spread modelling, 1990–2007, 1: Physical and quasi-physical models, *Int. J. Wildland Fire*, *18*, 349–368.

Sullivan, A. L. (2009b), Wildland surface fire spread modelling, 1990–2007, 2: Empirical and quasi-empirical models, *Int. J. Wildland Fire*, *18*, 369–386.

Sullivan, A. L. (2009c), Wildland surface fire spread modelling, 1990–2007, 3: Simulation and mathematical analogue models, *Int. J. Wildland Fire*, *18*, 387–403.

Syphard, A. D., V. C. Radeloff, N. S. Keuler, R. S. Taylor, T. J. Hawbaker, S. I. Stewart, and M. K. Clayton (2008), Predicting spatial patterns of fire on a southern California landscape, *Int. J. Wildland Fire*, *17*(5), 602–613.

Thompson, M. P. (2013), Modeling wildfire incident complexity dynamics, *PloS one*, *8*(5), e63297.

Thompson, M. P., and D. E. Calkin (2011), Uncertainty and risk in wildland fire management: A review, *J. Environ. Man.*, *92*(8), 1895–1909.

Thompson, M. P., D. E. Calkin, J. Herynk, C. W. McHugh, and K. C. Short (2013b), Airtankers and wildfire management in the US Forest Service: examining data availability and exploring usage and cost trends, *Int. J. Wildland Fire*, *22*(2), 223–233.

Thompson, M. P., D. E. Calkin, M. A. Finney, A. A. Ager, and J. W. Gilbertson-Day (2011), Integrated national-scale assessment of wildfire risk to human and ecological values, *Stochas. Environ. Res. Risk Assess.*, *25*(6), 761–780.

Thompson, M. P., J. R. Haas, J. W. Gilbertson-Day, J. H. Scott, P. Langowski, E. Bowne, and D. E. Calkin (2015),

Development and application of a geospatial wildfire exposure and risk calculation tool, *Environ. Mod. Soft.*, *63*, 61–72.

Thompson, M. P., N. M. Vaillant, J. R. Haas, K. M. Gebert, and K. D. Stockmann (2013a), Quantifying the potential impacts of fuel treatments on wildfire suppression costs, *J. For.*, *111*(1), 49–58.

Van Wagner, C. E. (1977), Conditions for the start and spread of crown fire, *Can. J. For. Res.*, *7*, 23–34.

Walker, W. E., Harremoës, J., Rotmans, J. P. van der Sluijs, M. B. A. van Asselt, P. Janssen, and M. P. Krayer von Krauss (2003), Defining uncertainty: A conceptual basis for uncertainty management in model-based decision support, *Integrated Assessment*, *4*(1), 5–17.

Warmink, J., J. Janssen, M. J. Booij, and M. S. Krol (2010), Identification and classification of uncertainties in the application of environmental models, *Environ. Mod. Soft.*, *25*(12), 1518–1527.

Westerling, A. L., H. G. Hidalgo, D. R. Cayan, and T. W. Swetnam (2006), Warming and earlier spring increase western U.S. forest wildfire activity, *Science*, *313*(5789), 940–943.

Wilson, R. A. (1990), Reexamination of Rothermel's fire spread equations in no-wind and no-slope conditions, USDA Forest Service Research Paper INT-434, USDA Forest Service Intermountain Research Station, Ogden, Utah.

14

Fire and Smoke Remote Sensing and Modeling Uncertainties: Case Studies in Northern Sub-Saharan Africa

Charles Ichoku,[1] Luke T. Ellison,[2] Yun Yue,[3] Jun Wang,[3,4] and Johannes W. Kaiser[5]

ABSTRACT

Significant uncertainties are incurred in deriving various quantities related to biomass burning from satellite measurements at different scales, and, in general, the coarser the resolution of observation the larger the uncertainty. WRF-Chem model simulations of smoke over the northern sub-Saharan African (NSSA) region for January–February 2010, using fire energetics and emissions research version 1.0 (FEERv1) aerosol emissions derived from MODIS measurements of fire radiative power (FRP) and aerosol optical depth (AOD), resulted in a severe model underestimation of AOD compared with satellite retrievals. Such uncertainties are attributable to three major factors: limitations in the spatial and temporal resolutions of the satellite observations used to quantify emissions, modeling parameters and assumptions, and the unique geographic characteristics of NSSA. It is recommended that field campaigns involving synergistic coordination of ground-based, airborne, and satellite measurements with modeling be conducted in major and complex biomass burning regions such as the NSSA, and that significant improvements in the spatial and temporal resolutions of observation systems needed to reduce uncertainties in biomass burning characterization be seriously considered in future satellite missions.

14.1. INTRODUCTION

Wildfires and other types of open biomass burning represent one of the most ubiquitous disturbances to vegetated land ecosystems globally [e.g., *Andreae*, 1991; *Ichoku et al.*, 2008a, 2012]. These vegetation fires are either ignited by natural processes such as lightning or by human action such as arson, accident, prescribed (controlled) burning for land management, or societal cultural practices as applicable to game hunting, slash-and-burn agriculture, and other forms of land clearing. Whatever the nature or purpose of ignition, depending on circumstances, such open fires can easily become hazardous to life and property. The hazardous effects of fires are not limited to the destructive effects of the associated flame and heat [e.g., *Cohen*, 2010], but also extend to the potential adverse impacts of the emitted smoke on air quality and human health both near and far [e.g., *Colarco et al.*, 2004; *Wang et al.*, 2006; *Wiedinmyer et al.*, 2006; *Henderson et al.*, 2011], as well as those of the postburn land surface processes that may include erosion, landslides, mud deposits, and pollution of water resources by soot and other residues [e.g., *Moody et al.*, 2013].

Determination of the areas and quantities of biomass consumed by fires, and their resulting emissions and impacts, can be done at local to global scales, depending on the targeted application(s) and the available tools and resources [e.g., *Michalek et al.*, 2000; *Ichoku and Kaufman*, 2005; *Roberts et al.*, 2005; *van der Werf et al.*, 2006, 2010; *de Groot et al.*, 2007; *Pouliot et al.*, 2008; *Schultz et al.*, 2008; *Vermote et al.*, 2009; *Giglio et al.*, 2010; *Roy et al.*,

[1]Climate and Radiation Laboratory, NASA Goddard Space Flight Center, Greenbelt, Maryland, USA

[2]Climate and Radiation Laboratory, NASA Goddard Space Flight Center, Greenbelt, Maryland, and Science Systems and Applications Inc., Lanham, Maryland, USA

[3]Department of Earth and Atmospheric Sciences, University of Nebraska, Lincoln, Nebraska, USA

[4]Now at Center for Global and Regional Environmental Research, and Department of Chemical and Biochemical Engineering, University of Iowa, Iowa City, Iowa, USA

[5]Max Planck Institute for Chemistry, Mainz, Germany

Natural Hazard Uncertainty Assessment: Modeling and Decision Support, Geophysical Monograph 223, First Edition.
Edited by Karin Riley, Peter Webley, and Matthew Thompson.

2010; *French et al.*, 2011; *Kaiser et al.*, 2012; *Miettinen et al.*, 2013; *Peterson and Wang*, 2013; *Peterson et al.*, 2013; *Ichoku and Ellison*, 2014; *Schroeder et al.*, 2014a]. Irrespective of the approach or scale, such exercises are generally associated with a wide range of uncertainties, which are partly because of the dynamic and intractable nature of biomass burning processes, and partly due to the imperfections in the measurement approaches and modeling assumptions used. Measurement methods may be ground based, airborne, or satellite based. Ground-based methods are typically used for localized measurements with high precision over a short time period, whereas satellite methods can be applied regionally or globally for an extended time period albeit with a lower accuracy and precision. Based on the analysis of burned areas retrieved from multiple satellite sensors during 1997–2008, it was estimated that between 330 and 430 Mha were burned annually globally, of which ~250 Mha (i.e., ~70%) was estimated to have burned each year on the continent of Africa alone [*Giglio et al.*, 2010]. These numbers were used, within the Global Fire Emissions Database version 3 (GFED3) framework, to estimate that the global annual carbon emissions from open biomass burning for 1997–2009 was in the range of 1.6 Pg C yr^{-1} to 2.8 Pg C yr^{-1}, with an annual average of 2.0 Pg C yr^{-1}, of which Africa alone contributes ~52% [*van der Werf et al.*, 2010]. Although the emission uncertainties associated with such satellite-based global estimates are large, they can be even larger at regional scales. For instance, *Zhang et al.* [2014] found a factor of 12 difference when comparing seven satellite-derived fire emissions inventories for February 2010 in the northern sub-Saharan African (NSSA) region. Therefore, although the current paper will examine these uncertainties from a global perspective, case studies will be mainly based on data from the NSSA region, which comprise mostly savanna fires [e.g., *Gatebe et al.*, 2014].

Some of the main uncertainties in quantifying biomass burning parameters stem from a variety of factors, including the difficulty in addressing the following questions: (1) Where and when exactly does a fire occur? (2) What are the mass loadings and conditions of the biomass fuel? (3) What is the fire intensity and/or size? (4) What are the relative proportions of the fire phases (flaming, smoldering, and glowing) per unit area and how does this distribution vary in space and time? (5) How long does a given fire burn, and how does it affect (and is it affected by) environmental conditions? (6) How far does a given fire spread, when is an area considered burned, and what is the total burned area when the fire ends? (7) How much smoke is emitted per unit time from a given fire? (8) How high is the plume injected and how far is it spreading? (9) What are the important constituents of the smoke and what are their respective concentrations? (10) How do smoke constituents interact with one another and with other atmospheric constituents to change and/or form new ones over time? (11) How do the characteristics of different fires in similar ecosystems differ? (12) What are the fire diurnal cycle and the seasonal burn pattern in a given area or region?

These questions are not an exhaustive list of the essential questions concerning the quantification of biomass burning characteristics and emission constituents. Yet no single measurement or modeling approach can address any of them to the required accuracy at various spatial and temporal scales. For instance, although ground-based and airborne systems can be used for limited active fire measurements at high temporal frequency over an extended part of a day, only portions of the fire or smoke can be observed at any given time. Conversely, satellite measurements can cover much larger regions or even the entire globe, but only for a smaller set of parameters at a much reduced spatial resolution and/or temporal frequency, depending on whether the satellite is geostationary or polar orbiting. Ideally, the ability to address most of the above questions to an acceptable level of accuracy should involve proper synergy between the different (ground-based, airborne, and satellite) measurement approaches and appropriate modeling systems [e.g., *Schroeder et al.*, 2014a].

This study addresses uncertainties related to the satellite approach, which has become more and more widely used for fire characterization and emissions estimation at local to global scales. It is recognized that satellite observation systems are numerous and varied, thereby offering a similarly diverse range of capabilities for remote sensing of fires and smoke. However, the pyrolysis and emissions processes of biomass burning are extremely dynamic and continuous, and cannot be adequately followed by satellite observations, which can only provide highly discretized and sparse (both spatially and temporally) sampling of such processes. This gap in observation resulting from the intrinsic sampling intervals of different satellite systems represents a significant fundamental uncertainty in biomass burning characterization. Furthermore, even at the satellite sampling times, errors of omission or commission do occur, imposing another layer of uncertainty. These uncertainties related to non-observation of existing fires or false alarms on nonexistent fires, typically quantified in terms of errors of omission and commission, respectively, have been quite extensively investigated in the literature [e.g., *Ichoku et al.*, 2003; *Li et al.*, 2003; *Morisette et al.*, 2005; *Csiszar et al.*, 2006, 2014; *Schroeder et al.*, 2008; *Freeborn et al.*, 2014]. Therefore, this study focuses on uncertainties of measured parameters of actually observed fires, burned areas, and smoke constituents.

The objective of this study is to investigate uncertainties associated with the satellite characterization of biomass burning, as they relate to the derived geophysical products such as smoke constituents and their applications. These uncertainties will be examined in the context of the 12 basic relevant questions outlined above. To anchor this study to contemporary reality, the analysis will be limited to satellite observation systems that are currently (or have been recently) operational, and known to provide data products that are related to biomass burning (Table 14.1). Then, we will explore how the observation uncertainties can propagate when used in deriving smoke emissions as well as in regional modeling. The conclusions will include an outlook on the potential for integration of available airborne and ground-based measurements to improve results.

14.2. METHODS

Satellite measurements related to biomass burning may be categorized into five groups of parameters, namely: active fires, burned surfaces, smoke plume dispositions, aerosol distribution and particle properties, and trace gas concentrations [*Ichoku et al.*, 2012]. Whereas parameters of "active fires" (i.e., fire location, fire temperature and area, and fire radiative power [FRP]) and those of "burned surfaces" (i.e., burned area and burn severity proxy indices such as the differenced normalized burn ratio [dNBR]) are uniquely retrievable from satellite measurements within the limitations of remote sensing uncertainties [e.g., *Roy et al.*, 2006, 2008; *French et al.*, 2008; *Roy and Boschetti*, 2009; *Freeborn et al.*, 2011; *Randerson et al.*, 2012; *Hyer et al.*, 2013; *Miettinen et al.*, 2013; *Mouillot et al.*, 2014], direct satellite retrieval of smoke constituents is somewhat more ambiguous because they are often mixed with similar particulate and gaseous constituents from nonfire sources [e.g., *Deeter et al.*, 2003; *Kaufman et al.*, 2005]. Therefore, at regional to global scales, the most frequent use of satellite active-fire and burned-area products is for the estimation of smoke emissions, which are subsequently applied to various uses, including air quality and climate modeling [e.g., *Heald et al.*, 2003; *Kasischke and Bruhwiler*, 2003; *Kukavskaya et al.*, 2013].

The amount of a specific carbonaceous aerosol or trace gas species emitted as a smoke constituent is traditionally derived as follows [e.g., *Lavoué et al.*, 2000; *Andreae and Merlet*, 2001]:

$$M_x = EF_x * M_{biomass} \qquad (14.1)$$

where M_x is the mass of the emitted smoke constituent x, EF_x is its emission factor, and $M_{biomass}$ is the mass of the dry biomass burned. $M_{biomass}$ can be estimated as follows [*Seiler and Crutzen*, 1980]:

$$M_{biomass} = A \times B \times \alpha \times \beta \qquad (14.2)$$

where A is the burned area, B is the biomass density, α is the fraction of aboveground biomass, and β is the fraction consumed or combustion completeness.

Typically, EF_x is derived from laboratory or field experimentation, whereas A, B, α, and β are derived through satellite or airborne remote sensing, though they can be based on hybrid approaches. Although most current global and regional models employ emissions derived on the basis of equations (14.1) and (14.2), there are numerous uncertainties associated with this approach, particularly with regard to the accuracy of determination of the constituent parameters: EF_x A, B, α, and β, as well as the error propagation that results when they are combined [e.g., *French et al.*, 2004].

In an effort to alleviate the complexity imposed by requiring the solution of equation (14.2) as a prerequisite to solving equation (14.1), *Ichoku and Kaufman* [2005] established a similar relationship to equation (14.1), in which EF_x is replaced with C_e^x, which is designated as the emission coefficient (for any given smoke constituent x), and $M_{biomass}$ is replaced with either fire radiative energy (FRE) or its release rate R_{fre} (i.e., FRP). Thus,

$$M_x = C_e^x \cdot FRE \quad \text{or} \quad R_x = C_e^x \cdot R_{fre} \qquad (14.3)$$

where R_x is the rate of emission of species x (expressed in kg/s) since R_{fre} is the FRE release rate expressed in MJ/s, or MW. C_e^x is therefore expressed in kg/MJ. The validity of the relationship in equation (14.3) has been verified in a laboratory experiment, where satellite measurements of fire energetics and smoke were replicated by burning small biomass fuel samples in a burn chamber equipped with a giant smoke stack upon which the relevant instruments were set up, and the retrieved FRP and AOD were used to derive C_e for smoke aerosols [*Ichoku et al.*, 2008b].

Based on equation (14.3), a new emissions dataset, known as the fire energetics and emissions research version 1.0 (FEERv1), has been developed from Terra- and Aqua-MODIS measurements of FRP and AOD [*Ichoku and Ellison*, 2014]. FEER.v1 is composed of a global gridded C_e^x dataset at 1° × 1° grid spatial resolution for smoke aerosols and a number of other important constituents. These gridded C_e^x values for smoke aerosols were applied to equation (14.3) together with FRE data obtained through time integration of MODIS FRP measurements that have been gridded at 0.5° × 0.5° resolution within the Global Fire Assimilation System [GFASv1.0; *Kaiser et al.*, 2012]. The resulting daily emissions of smoke aerosols are then utilized as input into the Weather Research and Forecasting coupled with

Table 14.1 Selected Current or Recent Satellite Sensors Providing Observations of Fires and Smoke That Are Relevant to This Study

Satellie/sensor name	Description	Spatial resolution	Period	Reference (e.g.)
Quickbird	High-resolution satellite-borne sensors	2–4 m	2001–present	*Barbosa et al.* [2014]
Ikonos	High-resolution satellite-borne sensors	4 m	1999–present	*Barbosa et al.* [2014]
Landsat	Satellite series carrying the Thematic Mapper (TM) or enhanced TM (ETM) sensors	30 m	1979–present	*Chander et al.* [2009]
ASTER	Advanced Spaceborne Thermal Emission and Reflection Radiometer	15 m, 30 m, 90 m	1999–present	*Abrams,* [2000]
MODIS	Moderate-resolution Imaging Spectro-radiometer aboard Terra and Aqua	0.25 km, 0.5 km, 1 km	1999–present 2002–present	*Xiong et al.* [2009]
VIIRS	Visible–Infrared Imaging Radiometer Suite on the Suomi National Polar-Orbiting Partnership (NPP) satellite	375 m, 750 m	2011–present	*Hillger et al.* [2013]
BIRD	Bi-spectral Infra-Red Detection on a German Space Agency (DLR) small satellite	185 m, 370 m	2001–2004	*Lorentz et al.* [2013]
TET-1	Technologie Entwicklungstraeger on a German Space Agency (DLR) small satellite	42.4 m, 356 m	2012–present	*Lorentz et al.* [2013]
CALIOP	Cloud-Aerosol Lidar with Orthogonal Polarization on the Cloud-Aerosol Lidar and Infrared Pathfinder Satellite Observations (CALIPSO) satellite	5 km	2006–present	*Winker et al.* [2009]
MISR	Multi-angle Imaging Spectro-Radiometer aboard Terra	0.275 km, 1.1 km	1999–present	*Diner et al.* [2005]
AVHRR	Advanced Very High Resolution Radiometer	1 km	1978–present	*Ichoku et al.* [2003]
SPOT-VGT	The VEGETATION sensor aboard the European SPOT-4 satellite	1 km	1998–present	*Tansey et al.* [2004]
GOES	Sensors aboard the Geostationary Operational Environmental Satellite series	4 km	1994–present	*Zhang and Kondragunta* [2008]
SEVIRI	Spinning Enhanced Visible and Infrared Imager aboard the European Meteosat	3 km	2004–present	*Roberts and Wooster* [2008]
MOPITT	Measurements of Pollution in the Troposphere aboard the Terra satellite	22 km	1999–present	*Deeter et al.* [2003]
AIRS	Atmospheric Infrared Sounder aboard the Aqua satellite	90 km	2002–present	*Warner et al.* [2007]
TES	Tropospheric Emission Spectrometer aboard the Aura satellite	5×8 km	2004–present	*Luo et al.* [2007]
SCIA	SCIAMACHY on the European ENVISAT	30×120 km	2003–present	*Buchwitz et al.* [2006]
GOSAT	Greenhouse Gases Observing Satellite	10.5 km	2009–present	*Yokota et al.* [2009]

Note: The cited reference for each is just an example and not necessarily the official reference.

Chemistry (WRF-Chem) regional model for simulation of biomass burning aerosol emissions and dispersion over the NSSA region [*Zhang et al.*, 2014]. The WRF-Chem AOD simulations are compared against MODIS-derived AOD for January and February 2010.

14.3. RESULTS AND DISCUSSION

Uncertainties associated with satellite measurements can vary widely because fires occur in different ecosystems at various scales under a diversity of conditions. Table 14.2 provides a summary of uncertainties associated with some of the satellite-based measurements of fire- and smoke-related variables, as obtained from literature, classified according to the 12 essential questions identified in the introduction, and expressed under different ranges of sensor spatial resolutions (very high: 0.001–0.01 km, high: 0.01–0.1 km, medium: 0.1–1 km, coarse: 1–10 km, and very coarse: 10–100 km), for ease of reference. These sensor-resolution classifications were determined based on a reasonable assessment of typical contemporary satellite instruments used for regional-global remote sensing. The reported uncertainty value ranges represent rough averages (not actual arithmetic means) estimated from the variety of values and plots published in the respective cited references. Overall, it is noticeable that uncertainties not only differ by variable but also by resolution, generally getting worse the coarser the resolution, as can be observed in cases represented in at least two spatial resolution categories. From the partial distribution of values in Table 14.2, it is obvious that most of the variables related to active fires and burned areas are observed at medium to coarse resolutions, whereas those associated with smoke are observed at coarse to very coarse resolutions. Analysis of global fire distributions has shown that lower FRP fires (which can be either relatively small hot fires or cooler fires of various sizes) occur much more frequently than larger ones in virtually all regions of the world [e.g., *Ichoku et al.*, 2008a]. Thus, most fire-related variables are observed at resolutions that are much coarser than their scale of occurrence, thereby contributing to the uncertainty. Also, because of the temporally discrete nature of satellite observations, time-dependent fire and emissions characteristics such as fire duration, smoke emission rates, and transformations are not directly retrieved, though when the fires are large enough to be observed from geostationary satellites, it may be possible to determine fire duration. Otherwise, such time-dependent phenomena are typically derived through postproduction modeling that incorporates additional parameters from other sources.

One of the outcomes of the survey in Table 14.2 is that all satellite retrievals are subject to significantly large uncertainties (underestimation and overestimation). However, at each scale, fire radiative power (FRP) appears to be more prone to underestimation relative to higher resolutions [*Wooster et al.*, 2003; *Roberts and Wooster*, 2008]. Burned area (BA) also appears to have a greater tendency toward underestimation [e.g., *Roy and Boschetti*, 2009]. This is probably because of the relatively coarse resolutions at which they are observed, causing nondetection of smaller or less intense fires and smaller burned areas [e.g., *Wang et al.*, 2009; *Tsela et al.*, 2014]. Since FRP and BA are the satellite-retrieved variables that are most commonly used for emissions estimates as in equations (14.2) and (14.3), the implications of their uncertainties for emissions require evaluation. Part of the reason why FRP and BA can be severely underestimated is because of the imaging geometry constraints of most satellite sensors, whereby pixels become large, fewer, and sometimes overlap away from nadir, resulting in lower total FRP, as illustrated in Figure 14.1. Similarly, BA has the tendency toward underestimation, whether it is derived using a change detection approach [e.g., *Roy et al.*, 2008] or estimated from the active fire-pixel counts [e.g., *Giglio et al.*, 2009]. A global assessment of the overall effect of this phenomenon based on a long record (2003–2009) of MODIS active fire observations in relation to scan angles is illustrated in Figure 14.2. By comparing fires observed at a single pixel at different off-nadir scan angles (starting from 25° up to the MODIS maximum of 55°) to the corresponding nadir pixel counts for the same fire, it has been found that a single fire pixel observed by MODIS at 55° off nadir can be equivalent to up to 16 fire pixels observed at nadir. In terms of FRP, although the value can be doubled at 55° off nadir, it becomes less than 30% when evaluated per km^2, which amounts to a net underestimation, since there are considerably fewer observations off nadir than at nadir.

To evaluate the uncertainty of aerosol emission estimates on model simulations, FEERv1 aerosol emissions were implemented in WRF-Chem over the NSSA region. Recent results of comparisons between FEERv1 aerosol emissions against other major emissions inventories in this region show that FEERv1 emissions are higher (by up to a factor of two) than many of the commonly used global fire emissions inventories that are based on bottom-up approaches [*Ichoku and Ellison*, 2014; *Zhang et al.*, 2014]. Those bottom-up emissions inventories are typically used with enhancement factors in model simulations of smoke aerosols to match observed atmospheric aerosol distributions [e.g., *Kaiser et al.*, 2012]. However, even when provided with uniform emissions, different models also have intrinsic characteristics that can significantly affect the uncertainty of simulations of smoke aerosol processes, transport, and impacts [e.g., *Textor et al.*, 2007]. The quantitative evaluation performed in this study involves deriving aerosol optical depth (AOD)

Table 14.2 The Uncertainty Ranges of Satellite-Derived Fire and Smoke Variables

				Uncertainty levels*					
				Spatial resolution	Very high (0.001–0.01 km)	High (0.01–0.1 km)	Medium (0.1–1 km)	Coarse (1–10 km)	Very coarse (10–100 km)
Item no.	Essential questions	Satellite retrieved variable	Symbol	Satellite sensors**	QuickBird, Ikonos	Landsat, ASTER, SPOT, (Lidar/SAR)[#]	MODIS[a] VIIRS, BIRD, TET-1, CALIP	MODIS[a] MISR, AVHRR, SPOT-VGT, GOES, SEVIRI	MOPITT, AIRS, TES, SCIA, GOSAT
1	Fire location	Fire location	FL[b]			~0.15 km	~0.75 km	~5 km	
2	Fuel load and conditions	Biomass	BM[c]		±50%	±50%			
3	Fire size/intensity	Fire Area	FA[d]				65–250%	±50%	
		Fire Temp	FT[e]				±30%	±100 K	
		Fire Radiative Power	FRP[f]				±30%	±50%	
4	Fire characteristics (flaming/ smoldering)	Flaming ratio	FSR[g]				40%–140%		
5	Fire duration	N/A							
6	Burned area	Burned area	BA[h]			±10%	±20%	±30%	
		Burn Severity	BS[i]			±70%			
7	Smoke emission rate	N/A							
8	Plume injection height	Plume top height	PTH[j]					±0.5 km	
		Plume vertical profile	PVP[k]				±7%		
9	Major smoke constituents	Aerosol Optical Depth	AOD[l]					±0.15	
		Carbon Monoxide	CO[m]						±50%
		Carbon Dioxide	CO_2[n]						97%–102%
		Methane	CH_4[o]						96%–102%

10	Smoke transformation	N/A
11	Fire behavior	N/A
12	Fire diurnal/seasonal cycles	N/A

* Uncertainty levels are expressed in measurement units or percentages. In the case of the latter, the measured value is 100%, such that the range shows average uncertainty range.

** These are only selected currently or recently orbiting satellite sensors that are relevant to this study, and the details and relevant reference for each are given in Table 14.1.

[#] This represents a variety of spaceborne and airborne Lidar and synthetic aperture radar (SAR) systems that is used to estimate biomass (Period: variable, Res: 5–60 km) [e.g., *Montesano et al.*, 2014].

[a] MODIS currently offers spatial resolutions of: 0.5 km for BA, 1 km for FL and FRP, and 3–10 km for AOD.

[b] Location uncertainty depends on spatial resolution and observation geometry. These are expressed here as approximately half the typical average nominal pixel size in each resolution group. See *Zhukov et al.* [2006], *Hyer and Reid* [2009], *Csiszar et al.* [2014], *Schroeder et al.* [2014a,b].

[c] Biomass (BM) is used as a generic designation for fuel load. Airborne radar shows potential for spaceborne radar. See *Brandis and Jacobson* [2003], *Jin et al.* [2012].

[d] Uncertainty range in % about the mean estimates. See *Lorenz et al.* [2013], *Peterson et al.* [2013], *Peterson and Wang* [2013], *Giglio and Kendall* [2001].

[e] Uncertainty range in % about the mean estimates or in actual temperature (K) values. See *Lorenz et al.* [2013], *Giglio and Kendall* [2001].

[f] FRP is retrieved from satellite, but what is really needed in subpixel fire intensity. See *Kaufman et al.* [1998], *Wooster* [2003], *Zhukov et al.* [2006], *Roberts and Wooster* [2008, 2014], *Peterson et al.* [2013], *Peterson and Wang* [2013].

[g] See *Lorenz et al.* [2013].

[h] See *Loboda et al.* [2007], *Giglio et al.* [2009], *Roy and Boschetti* [2009], *Tsela et al.* [2010, 2014], *Stroppiana et al.* [2012], *Padilla et al.* [2014].

[i] BS is typically expressed in the form of differenced Normalized Burn Ratio (dNBR) or Relative dNBR (RdNBR); See *Epting et al.* [2005], *Miller and Thode* [2007].

[j] See *Scollo et al* [2012].

[k] Because of its curtain character, CALIOP seldom scans near smoke plume source, which is where it is most needed to characterize plume injection. See *Winker et al.* [2009, 2013], *Kacenelenbogen et al.* [2014].

[l] AOD is retrieved from satellite, but what is really needed is particulate matter (PM) concentrations in smoke. Typical range of AOD values is 0–5 (unitless). See *Petrenko and Ichoku* [2013].

[m] See *Kasibhatla et al.* [2002], *Kopacz* [2010].

[n] What is actually evaluated is dry air column-averaged mole fractions of CO_2 (XCO_2). See *Schneising et al.* [2008], *Morino et al.* [2011], *Reuter et al.* [2011].

[o] See *Schneising et al.* [2009], *Morino et al.* [2011].

N/A = parameters that are currently not directly retrieved from satellite measurements.

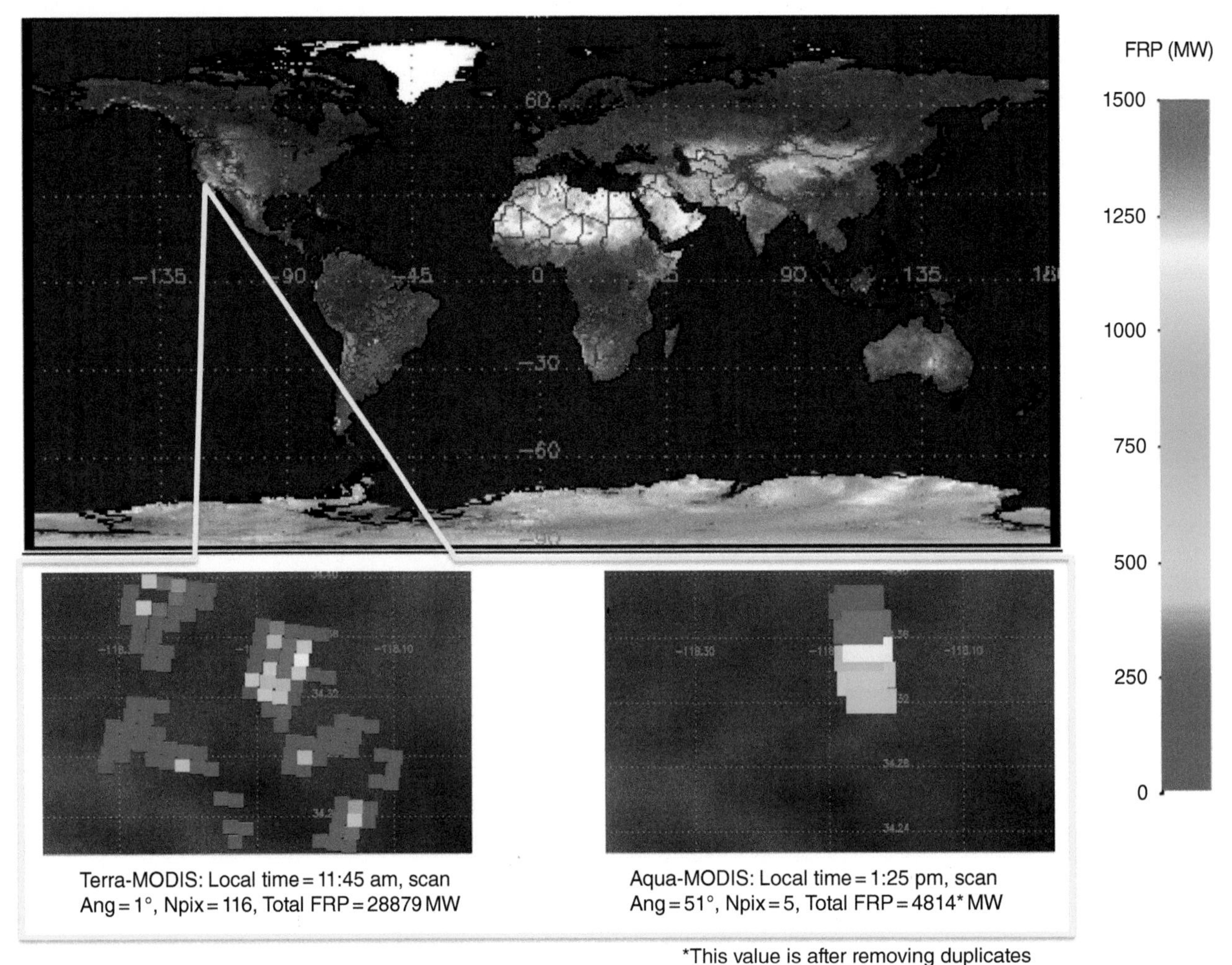

Figure 14.1 Effect of scan angle on MODIS observation of the Station Fire in Pasadena, California, on 30 August 2009. Fire detections near nadir (bottom left) show pixels to be almost square shaped at 1 × 1 km resolution, whereas near scan edge (bottom right) pixels are much fewer, individually stretched almost up to 4 × 2.5 km resolution, duplicated, and overlapping one another, and total FRP is underestimated.

from WRF-Chem based on FEERv1 emissions and comparing this with direct AOD retrievals from MODIS. This is done for January and February 2010, which is the typical peak of the burning season in NSSA. Incidentally, significant dust emissions also occur in this region during this season, as indicated by very heavy aerosol loading that appears prominently in dark red colors in Figure 14.3b, which represents a simple combination of both Terra- and Aqua-MODIS Collection 5 (C5) AOD retrievals from the Dark Target, Deep Blue, and Ocean algorithms. Although the current MODIS Collection 6 (C6) AOD product has a combined version [e.g., *Levy et al.*, 2013], C5 is used for the current comparison to avoid an attempt to characterize additional discrepancy due to version differences, as the FEERv1 emissions were based on C5. Since WRF-Chem simulations did not include dust emissions, to avoid (or at least limit) dust influence in the satellite AOD samples, it was decided that these comparisons would be most realistic at areas that are not in the normal seasonal dust trajectory. Four areas were selected for the MODIS/WRF-Chem AOD comparisons and labeled according to the main country or region covered, namely: Senegal, Gabon, Central Africa, and Southern Sudan (Figs. 14.3b and 14.3c). Terra- and Aqua-MODIS C5 AOD are in general good agreement overall, but WRF-Chem AOD simulations are very low (Fig. 14.3d), in spite of the fact that the FEERv1 emissions upon which they are based are higher than those of most other existing smoke emissions inventories. This AOD underestimation may be due to a combination of multiple factors, one of which may be emissions underestimation, while others may include WRF-Chem model variables and parameters as well as assumptions and process treatment algorithms. Also, although the main areas

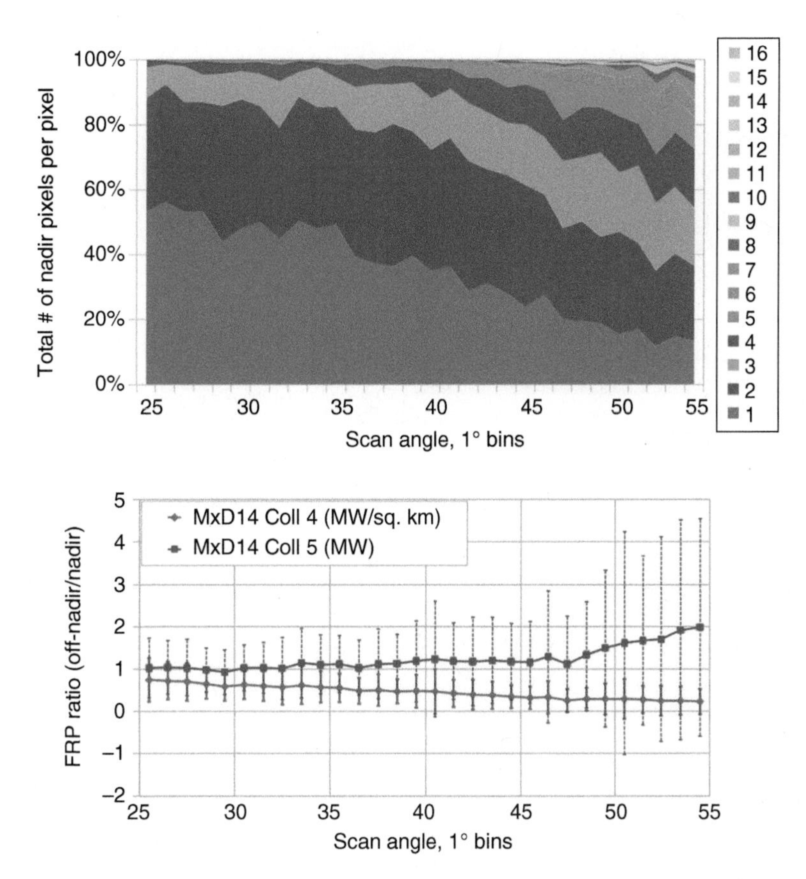

Figure 14.2 Analysis of the effect of scan angle on collocated MODIS fire observation from Terra and Aqua, within 20 min of each other globally for 2003–2009. This collocation is only possible within the high latitudes (>55°N) where there is significant overlap of MODIS swaths between the two satellites. MxD14 represents the official MODIS fire products from Terra (MOD14) and Aqua (MYD14). There were 11,295 pairs of Terra/Aqua fire observations. (Top) Relative percentages of off-nadir single pixel fire detection from one satellite and corresponding number of near-nadir pixels of the same fire from the other satellite. (Bottom) Ratios of the FRP value of single off-nadir pixels to total FRP value of the corresponding near-nadir fire pixels, expressed both in terms of FRP (as in MODIS collection 5) and FRP per unit area (as in MODIS collection 4). The point values are the means of such ratios for bins of 1° off-nadir observations starting at 25° scan angle, whereas error bars are the corresponding standard deviations of the FRP ratios.

of dust loading have been avoided in these comparisons, there could still be some residual dust or even cloud contamination in the MODIS-retrieved AOD. Because of various types of AOD retrieval constraints and MODIS swath coverage limitations, typical maps of AOD contain significant data gaps, such that the boxes for the regions of interest are seldom completely filled, as exemplified by Figure 14.3b, unlike Figure 14.3c, which shows complete coverage offered by the model. In Figure 14.3d, different circle symbol sizes on the Terra and Aqua curves depict the degree of coverage of sample areas for the MODIS AOD curves. The plots show that WRF-Chem AOD tends to agree better when the sample boxes have higher coverage by MODIS retrievals, as the root mean square error (RMSE) values denote in Table 14.3. Based on these results, it can be inferred that WRF-Chem regional modeling of smoke aerosols over NSSA using the FEERv1 satellite-based emissions estimates produce a net underestimation of AOD relative to satellite AOD, with the discrepancy becoming larger as the gap in satellite AOD coverage increases.

14.4. CONCLUSIONS

Satellite fire observation is relied upon for many applications. However, significant uncertainty is incurred in the satellite retrieval or estimation of biomass burning quantities, such as active fire location, area, temperature, radiative power, burned area and burn severity, plume injection and profile, and smoke constituents including aerosols and trace gases. Typically, the uncertainties tend to increase as the spatial and temporal resolutions of the

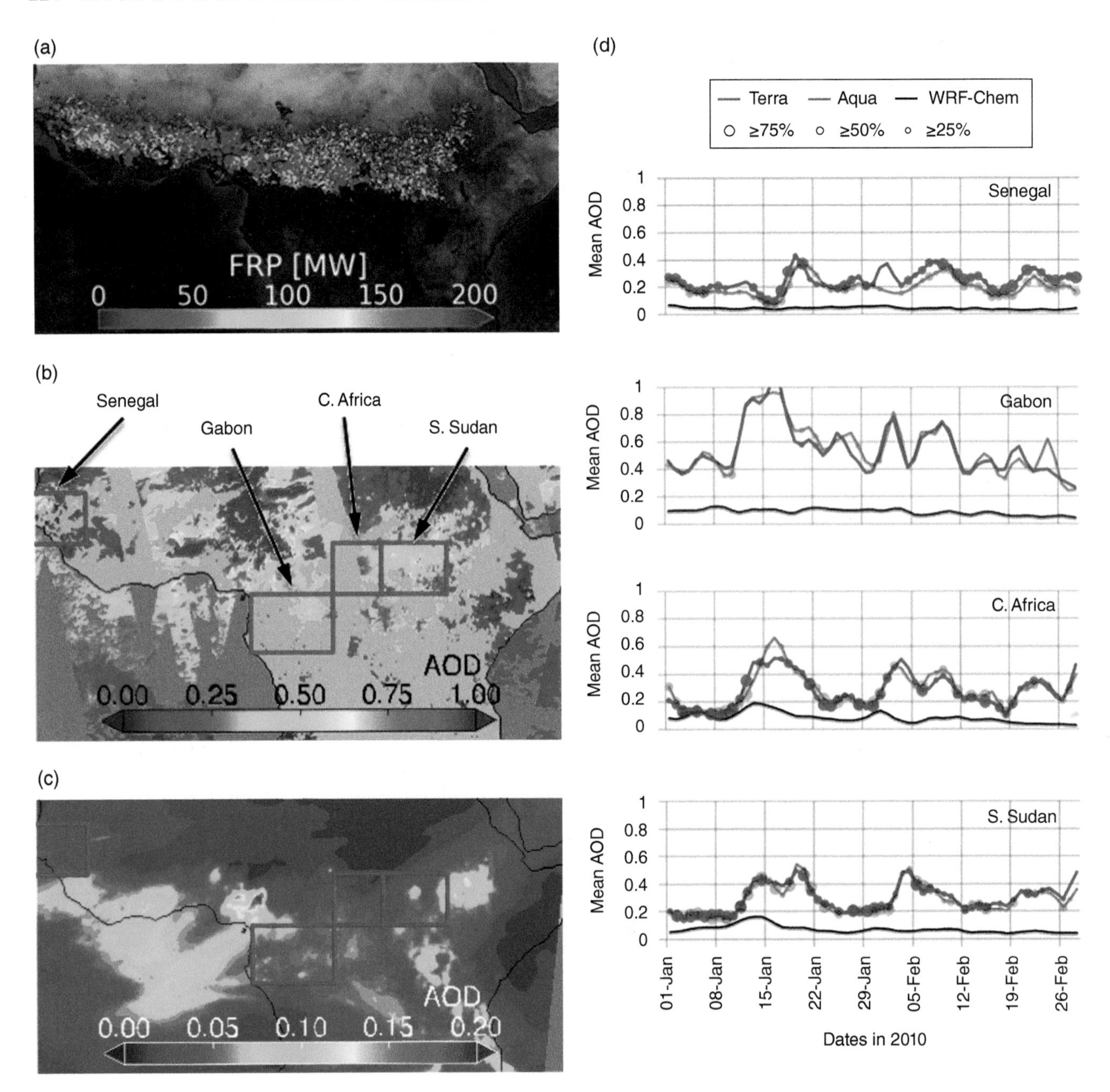

Figure 14.3 Evaluation of uncertainty in aerosol optical depth (AOD) generated from WRF-Chem model based on FEERv1 aerosol emissions, by comparison to satellite-observed AOD over northern sub-Saharan Africa (NSSA) during January–February 2010: (a) Fire locations and associated FRP values from MODIS on Terra and Aqua; (b) composited Terra and Aqua MODIS mean AOD for 5 February 2010, showing boxes where AOD comparisons are made; (c) WRF-Chem simulation of only smoke aerosol AOD for 5 February 2010, also showing the sampling box locations. AOD values increase from blue to red. Notice the difference in AOD value ranges as indicated by the color scales between (b) and (c). Boxed areas are selected to avoid the main dust trajectory (as indicated by the dark-red thick aerosol plume in [b]), such that the sampled AOD may be mainly smoke aerosols; (d) daily MODIS average AOD at Terra and Aqua overpass times (colored curves) and corresponding WRF-Chem simulations (black curves) for 12–1 p.m. local time, which coincides approximately with the average of the local times of Terra and Aqua overpasses. The size of the circles on the satellite-AOD curves indicate the extent of spatial coverage of the satellite retrievals within the sample boxes, as gaps do occur due to cloud or other factors that can cause AOD retrieval to fail (as seen in [b]).

Table 14.3 Root Mean Square Error (RMSE) Values Between WRF-Chem AOD Simulations and MODIS AOD Retrievals for Terra and Aqua According to Bins of 25% Coverage of MODIS AOD Retrievals Over Each Sample Box Area Shown in Figure 14.3b.

	Senegal		Gabon		C. Africa		S. Sudan	
Box coverage (cov.)	Terra	*Aqua*	*Terra*	*Aqua*	*Terra*	*Aqua*	*Terra*	*Aqua*
75% < = cov. <100%	– (0)	0.22 (16)	– (0)	– (0)	0.13 (20)	0.13 (13)	0.20 (20)	0.20 (12)
50% < = cov. <75%	0.17 (27)	0.20 (20)	0.27 (1)	– (0)	0.24 (14)	0.20 (13)	0.20 (17)	0.20 (19)
25% < = cov. <50%	0.18 (25)	0.25 (17)	0.39 (11)	0.47 (2)	0.29 (13)	0.25 (21)	0.29 (14)	0.25 (15)
0% < = cov. <25%	0.13 (7)	0.24 (6)	0.53 (47)	0.49 (56)	0.31 (11)	0.29 (12)	0.27 (8)	0.33 (13)

Note: The numbers in parentheses represent the sample size (i.e., the number of days in January–February 2010 falling within the respective coverage bins for each case).

satellite observations decrease. Incidentally, most of these biomass-burning quantities are currently observed at suboptimal spatial and temporal resolutions. For instance, the current operational systems, such as MODIS and VIIRS, that provide the most commonly used active fire products, observe these fires at nominal 1000 m and dual (375 m and 750 m) spatial resolutions, respectively, even though most open fires exist at much smaller scales. As a result, most of these fires are omitted and the FRP for those that are observed are mostly underestimated. In the same way, burned areas are underestimated. Since FRP and burned areas are used mostly to estimate smoke emissions, these also become underestimated and are propagated into modeling simulations of smoke distributions from fires.

Although such uncertainties affect fire measurements and modeling everywhere, the northern sub-Saharan African (NSSA) region has been used as a case study to evaluate the effect of emissions uncertainty on aerosol estimates for this study. This is fitting, given that NSSA contributes 20%–25% of global biomass burning, and together with southern sub-Saharan Africa (SSSA) make up > 50% of the annual global biomass burning. Nevertheless, NSSA biomass burning has been one of the least investigated by means of ground-based or airborne measurement techniques, and therefore potentially harbors the largest uncertainty, as estimates of its biomass burning parameters are based mainly on satellite observations and other proxy information. Overall, it is found that FEERv1 emissions, which are based on a top-down approach from MODIS measurements of FRP and AOD, when used in regional smoke modeling with the WRF-Chem model can underestimate AOD relative to MODIS by 0.13 to 0.27 RMSE in AOD when MODIS has AOD retrievals in 50% or more of the area of interest. Paradoxically, a similar comparison of MODIS C5 AOD against simulated AOD from the Goddard Chemistry Aerosol Radiation and Transport (GOCART) global model using emissions based on satellite BA products through a variety of bottom-up approaches show a severe overestimation in the NSSA region [*Petrenko et al.*, 2012]. This is even more surprising because those bottom-up emissions based on BA had been shown to produce lower smoke emissions than FEERv1, which is based on a top-down approach using FRP measurements [*Ichoku and Ellison*, 2014]. This type of obvious discrepancy causes a general confusion regarding which of the following three areas could be the main source of the uncertainty: emissions, model, or geographic region.

Uncertainties in the quantification of fire output, particularly smoke, by satellite and modeling can be affected by a variety of factors, including: satellite measurement characteristics, parameter retrieval algorithms, contamination of desired variables by other undesired targets such as clouds, model assumptions and resolution, and the surface and atmospheric characteristics of the geographic region of study. There is need for a well-coordinated, comprehensive, and robust strategy to address such uncertainty. Based on the results of the current study and those cited here, the following three recommendations become appropriate: (1) Conduct integrated field experiments combining ground-based, airborne, and satellite measurements and linking them to modeling in a synergistic way [e.g., *Schroeder et al.*, 2014a] to better characterize biomass burning energetics and emissions in a coherent manner. (2) Conduct such integrated field studies in the NSSA region, which contributes 20%–25% of global biomass burning emissions and even a larger proportion of atmospheric dust loading within the same season, making remote-sensing discrimination of dust and smoke almost impossible over land, and thus far investigated mainly over ocean [e.g., *Kaufman et al.*, 2005; *Guo et al.*, 2013]. (3) Design future fire-related satellite missions with specific attention toward significantly improving the spatial, temporal, spectral, and radiometric resolutions of sensors to maximize the retrieval of the various variables related to fires and smoke, as listed in Table 14.2, in order to optimally address their associated essential questions.

ACKNOWLEDGMENTS

This research was funded by NASA under its Research Opportunities in Space and Earth Sciences (ROSES), 2009 and 2013 Interdisciplinary Studies (IDS) Program (Dr. Jack Kaye, Earth Science Research Director), through the Radiation Sciences Program Managed by Dr. Hal Maring.

REFERENCES

Abrams, M. (2000), The Advanced Spaceborne Thermal Emission and Reflection Radiometer (ASTER): Data products for the high spatial resolution imager on NASA's Terra platform, *Int. J. Remote Sens.*, *21*(5), 847–859.

Andreae, M. O. (1991), Biomass burning: Its history, use, and distribution and its impact on environmental quality and global climate, 3–21, in *Global Biomass Burning: Atmospheric, Climatic, and Biospheric Implications*, edited by J. S. Levine, MIT Press, Cambridge, Massachusetts.

Andreae, M. O., and P. Merlet, (2001), Emission of trace gases and aerosols from biomass burning, *Glob. Biogeochem. Cycles*, *15*, 955–966.

Barbosa, J. M., E. N. Broadbent, and M. D. Bitencourt (2014), Remote sensing of aboveground biomass in tropical secondary forests: A review, *Int. J. For. Res.*, 2014.

Brandis, K., and C. Jacobson (2003), Estimation of vegetative fuel loads using Landsat TM imagery in New South Wales, Australia, *Int. J. Wildland Fire*, *12*(2), 185–194.

Buchwitz, M., R. De Beek, S. Noël, J. P. Burrows, H. Bovensmann, O. Schneising, I. Khlystova, M. Bruns, H. Bremer, P. Bergamaschi, S. Komer, and M. Heimann (2006), Atmospheric carbon gases retrieved from SCIAMACHY by WFM-DOAS: Version 0.5 CO and CH 4 and impact of calibration improvements on CO 2 retrieval, *Atmos. Chem. Phys.*, *6*(9), 2727–2751.

Chander, G., B. L. Markham, and D. L. Helder (2009), Summary of current radiometric calibration coefficients for Landsat MSS, TM, ETM+, and EO-1 ALI sensors, *Remote Sens. Environ.*, *113*(5), 893–903.

Cohen, J. (2010), The wildland-urban interface fire problem, *Fremontia*, *16*.

Colarco, P. R., M. R. Schoeberl, B. G. Doddridge, L. T. Marufu, O. Torres, and E. J. Welton (2004), Transport of smoke from Canadian forest fires to the surface near Washington, D.C.: Injection height, entrainment, and optical properties, *J. Geophys. Res.*, *109*, D06203; doi:10.1029/2003JD004248.

Csiszar, I., J. Morisette, and L. Giglio (2006), Validation of active fire detection from moderate-resolution satellite sensors: The MODIS example in Northern Eurasia, *IEEE Trans. Geosci. Remote Sens.*, 1757–1764; doi:10.1109/TGRS.2006.875941.

Csiszar, I., W. Schroeder, L. Giglio, E. Ellicott, K. P. Vadrevu, C. O. Justice, and B. Wind (2014), Active fires from the Suomi NPP Visible Infrared Imaging Radiometer Suite: Product status and first evaluation results, *J. Geophys. Res. Atmos.*, *119*, 803–816; doi:10.1002/2013JD020453.

Deeter, M. N., L. K. Emmons, G. L. Francis, D. P. Edwards, J. C. Gille, J. X. Warner, B. Khattatov, D. Ziskin, J.-F. Lamarque, S.-P. Ho, V. Yudin, J.-L. Attie, D. Packman, J. Chen, D. Mao, and J. R. Drummond (2003), Operational carbon monoxide retrieval algorithm and selected results for the MOPITT instrument, *J. Geophys. Res. Atmos.*, *108*(D14), 1984–2012; doi: 10.1029/2002JD003186.

de Groot, W. J., R. Landry, W. A. Kurz, K. R. Anderson, P. Englefield, R. H. Fraser, R. J. Hall E. Banfield, D. A. Raymond, V. Decker, T. J. Lynham, and J. M. Pritchard (2007), Estimating direct carbon emissions from Canadian wildland fires, *Int. J. Wildland Fire*, *16*, 593–606; doi:10.1071/WF06150.

Diner, D. J., B. H. Braswell, R. Davies, N. Gobron, J. Hu, Y. Jin, R. A. Kahn, Y. Knyazikhin, N. Loeb, J.-P. Muller, A. W. Nolin, B. Pinty, C. B. Schaaf, G. Seiz, and J. Stroeve (2005), The value of multiangle measurements for retrieving structurally and radiatively consistent properties of clouds, aerosols, and surfaces, *Rem. Sens. Environ.*, *97*, 495–518.

Epting J., D. Verbyla, and B. Sorbel (2005), Evaluation of remotely sensed indices for assessing burn severity in interior Alaska using Landsat TM and ETM+, *Remote Sens. Environ.*, *96*(3–4), 328–339.

Freeborn, P. H., M. J. Wooster, and G. Roberts (2011), Addressing the spatiotemporal sampling design of MODIS to provide estimates of the fire radiative energy emitted from Africa, *Remote Sens. Environ.*, *115*(2), 475–489; doi:10.1016/j.rse.2010.09.017.

Freeborn, P. H., M. J. Wooster, G. Roberts, and W. Xu (2014), Evaluating the SEVIRI fire thermal anomaly detection algorithm across the Central African Republic using the MODIS active fire product, *Remote Sens.*, *6*, 1890–1917.

French, N. H. F., E. S. Kasischke, R. J. Hall, K. A. Murphy, D. L. Verbyla, E. E. Hoy, and J. L. Allen (2008), Using Landsat data to assess fire and burn severity in the North American boreal forest region: An overview and summary of results, *Int. J. Wildland Fire*, *17*(4), 443–462; doi:10.1071/WF08007.

French, N. H. F., et al. (2011), Model comparisons for estimating carbon emissions from North American wildland fire, *J. Geophys. Res.*, *116*, G00K05; doi:10.1029/2010JG001469.

French, N. H. F., P. Goovaerts, and E. S. Kasischke (2004), Uncertainty in estimating carbon emissions from boreal forest fires, *J. Geophys. Res.*, *109*, D14S08; doi:10.1029/2003JD003635.

Gatebe, C. K., C. M. Ichoku, R. Poudyal, M. O. Román, and E. Wilcox (2014), Surface albedo darkening from wildfires in northern sub-Saharan Africa, *Environ. Res. Lett.*, *9*(6), 065003.

Giglio, L., and J. D. Kendall (2001), Application of the Dozier retrieval to wildfire characterization: A sensitivity analysis, *Remote Sens. Environ.*, *77*(1), 34–49. doi:10.1016/S0034-4257(01)00192-4.

Giglio, L., J. T. Randerson, G. R. Van der Werf, P. S. Kasibhatla, G. J. Collatz, D. C. Morton, and R. S. DeFries (2010), Assessing variability and long-term trends in burned area by merging multiple satellite fire products, *Biogeosciences*, *7*(3), 1171–1186.

Giglio, L., T. Loboda, D. P. Roy, B. Quayle, and C. O. Justice (2009), An active-fire based burned area mapping algorithm for the MODIS sensor, *Remote Sens. Environ.*, *113*(2), 408–420. doi:10.1016/j.rse.2008.10.006.

Guo, Y., B. Tian, R. A. Kahn, O. Kalashnikova, S. Wong, and D. E. Waliser (2013), Tropical Atlantic dust and smoke aerosol variations related to the Madden-Julian Oscillation in MODIS and MISR observations, *J. Geophys. Res. Atmos.*, *118*, 4947–4963; doi:10.1002/jgrd.50409.

Heald, C. L., D. J. Jacob, P. I. Palmer, M. J. Evans, G. W. Sachse, H. B. Singh, and D. R. Blake (2003), Biomass burning emission inventory with daily resolution: Application to aircraft observations of Asian outflow, *J. Geophys. Res.*, *108*(D21), 8811; doi:10.1029/2002JD003082.

Henderson, S. B., M. Brauer, Y. C. MacNab, and S. M. Kennedy (2011), Three measures of forest fire smoke exposure and their associations with respiratory and cardiovascular health outcomes in a population-based cohort, *Environ. Health Pers.*, *119*(9), 1266.

Hillger, D., T. Kopp, T. Lee, D. Lindsey, C. Seaman, S. Miller, J. Solbrig, S. Kidder, S. Bachmeier, T. Jasmin, and T. Rink (2013), First-Light Imagery from Suomi NPP VIIRS, *Bull. Amer. Meteor. Soc.*, *94*, 1019–1029. doi: 10.1175/BAMS-D-12-00097.1.

Hyer, E. J., and J. S. Reid (2009), Baseline uncertainties in biomass burning emission models resulting from spatial error in satellite active fire location data, *Geophys. Res. Lett.*, *36*, L05802; doi:10.1029/2008GL036767.

Hyer, E. J., J. S. Reid, E. M. Prins, J. P. Hoffman, C. C. Schmidt, J. I. Miettinen, and L. Giglio (2013), Patterns of fire activity over Indonesia and Malaysia from polar and geostationary satellite observations, *Atmos. Res.*, *122*, 504–519.

Ichoku, C., and L. Ellison (2014), Global top-down smoke-aerosol emissions estimation using satellite fire radiative power measurements, *Atmos. Chem. Phys.*, *14*(13), 6643–6667; doi:10.5194/acp-14-6643-2014.

Ichoku, C., and Y. J. Kaufman (2005), A method to derive smoke emission rates from MODIS fire radiative energy measurements, *IEEE Trans. Geosci. Remote Sens.*, *43*(11), 2636–2649.

Ichoku, C., J. V. Martins, Y. J. Kaufman, M. J. Wooster, P. H. Freeborn, W. M. Hao, S. Baker, C. A. Ryan, and B. L. Nordgren (2008b), Laboratory investigation of fire radiative energy and smoke aerosol emissions, *J. Geophys. Res.*, *113*, D14S09; doi:10.1029/2007JD009659.

Ichoku, C., L. Giglio, M. J. Wooster, and L. A. Remer (2008a), Global characterization of biomass-burning patterns using satellite measurements of Fire Radiative Energy, *Remote Sens. Environ.*, *112*, 2950–2962.

Ichoku, C., R. Kahn, and M. Chin (2012), Satellite contributions to the quantitative characterization of biomass burning for climate modeling, *Atmos. Res.*, *111*, 1–28.

Ichoku, C., Y. J. Kaufman, L. Giglio, Z. Li, R. H. Fraser, J.-Z. Jin, and W. M. Park (2003), Comparative analysis of daytime fire detection algorithms using AVHRR data for the 1995 fire season in Canada: Perspective for MODIS, *Int. J. Remote Sens.*, *24*(8), 1669–1690.

Jin, S., and S.-C. Chen (2012) Application of QuickBird imagery in fuel load estimation in the Daxinganling region, China, *Int. J. Wildland Fire*, *21*(5), 583–590.

Kacenelenbogen, M., J. Redemann, M. A. Vaughan, A. H. Omar, P. B. Russell, S. Burton, R. R. Rogers, R. A. Ferrare, and C. A. Hostetler (2014), An evaluation of CALIOP/CALIPSO's aerosol-above-cloud detection and retrieval capability over North America, *J. Geophys. Res. Atmos.*, *119*, 230–244; doi:10.1002/2013JD020178.

Kaiser, J. W., et al. (2012), Biomass burning emissions estimated with a global fire assimilation system based on observations of fire radiative power, *Biogeosciences*, *9*, 527–554; doi:10.5194/bg-9-527-2012.

Kasibhatla, P., A. Arellano, J. A. Logan, P. I. Palmer, and P. Novelli (2002), Top-down estimate of a large source of atmospheric carbon monoxide associated with fuel combustion in Asia, *Geophys. Res. Lett.*, *29*(19), 6–1.

Kasischke, E. S., and L. M. Bruhwiler (2003), Emissions of carbon dioxide, carbon monoxide and methane from boreal forest fires in 1998, *J. Geophys. Res.*, *107*, 8146; doi:10.1029/2001JD000461.

Kaufman, Y. J., C. O. Justice, L. P. Flynn, J. D. Kendall, E. M. Prins, L. Giglio, D. E. Ward, W. P. Menzel, and A. W. Setzer (1998), Potential global fire monitoring from EOS-MODIS, *J. Geophys. Res. Atmos.*, *103*(D24), 32215–32238.

Kaufman, Y. J., O. Boucher, D. Tanré, M. Chin, L. A. Remer, and T. Takemura (2005), Aerosol anthropogenic component estimated from satellite data, *Geophys. Res. Lett.*, *32*, L17804; doi:10.1029/2005GL023125.

Kopacz, M., D. J. Jacob, J. A. Fisher, J. A. Logan, L. Zhang, I. A. Megretskaia, R. M. Yantosca, K. Singh, D. K. Henze, J. P. Burrows, M. Buchwitz, I. Khlystova, W. W. McMillan, J. C. Gille, D. P. Edwards, A. Eldering, V. Thouret, and P. Nedelec (2010), Global estimates of CO sources with high resolution by adjoint inversion of multiple satellite datasets (MOPITT, AIRS, SCIAMACHY, TES), *Atmos. Chem. Phys.*, *10*(3), 855–876; doi:10.5194/acp-10-855-2010.

Kukavskaya, E. A., A. J. Soja, A. P. Petkov, E. I. Ponomarev, G. A. Ivanova, and S. G. Conard (2013), Fire emissions estimates in Siberia: evaluation of uncertainties in area burned, land cover, and fuel consumption, *Can. J. For. Res.*, *43*(5), 493–506; doi:10.1139/cjfr-2012-0367.

Lavoué, D., C. Liousse, H. Cachier, B. J. Stocks, and J. G. Goldammer (2000), Modeling of carbonaceous particles emitted by boreal and temperate wildfires at northern latitudes, *J. Geophys. Res.*, *105*(D22), 26,871–26,890; doi: 10.1029/2000JD900180.

Levy, R. C., S. Mattoo, L. A. Munchak, L. A. Remer, A. M. Sayer, and N. C. Hsu (2013), The Collection 6 MODIS aerosol products over land and ocean, *Atmos. Meas. Tech. Discuss.*, *6*, 159–259.

Li, Z., R. Fraser, J. Jin, A. A. Abuelgasim, I. Csiszar, P. Gong, R. Pu, and W. M. Hao (2003), Evaluation of algorithms for fire detection and mapping across North America from satellite, *J. Geophys. Res.*, *108*(D2), 4076; doi:10.1029/2001JD001377.

Loboda, T., K. J. O'Neal, and I. Csiszar (2007), Regionally adaptable dNBR-based algorithm for burned area mapping from MODIS data, *Remote Sens. Environ.*, *109*(4), 429–442. doi:10.1016/j.rse.2007.01.017.

Lorenz, E. (2013), Thermal remote sensing with small satellites: BIRD, TET and the next generation BIROS, 149–176, in *Thermal Infrared Remote Sensing*, Springer Netherlands.

Luo, M., C. P. Rinsland, C. D. Rodgers, J. A. Logan, H. Worden, S. Kulawik, A. Eldering, A. Goldman, M. W. Shephard, M. Gunson, and M. Lampel (2007), Comparison of carbon monoxide measurements by TES and MOPITT: Influence of a priori data and instrument characteristics on nadir atmospheric species retrievals, *J. Geophys. Res.*, *112*, D09303; doi:10.1029/2006JD007663.

Michalek, J. L., N. H. F. French, E. S. Kasischke, R. D. Johnson, and J. E. Colwell (2000), Using Landsat TM data to estimate carbon release from burned biomass in an Alaskan spruce complex, *Int. J. Remote Sens.*, *21*, 323–338.

Miettinen, J., E. Hyer, A. S. Chia, L. K. Kwoh, S. C. and Liew (2013), Detection of vegetation fires and burnt areas by remote sensing in insular Southeast Asian conditions: current status of knowledge and future challenges, *Int. J. Remote Sens.*, *34*(12), 4344–4366.

Miller, J. D., and A. E. Thode (2007), Quantifying burn severity in a heterogeneous landscape with a relative version of the delta Normalized Burn Ratio (dNBR), *Remote Sens. Environ.*, *109*(1), 66–80; doi:10.1016/j.rse.2006.12.006.

Montesano, P. M., R. F. Nelson, R. O. Dubayah, G. Sun, Bruce D. Cook, K. J. R. Ranson, E. Næsset, and V. Kharuk (2014), The uncertainty of biomass estimates from LiDAR and SAR across a boreal forest structure gradient, *Remote Sens. Environ.*, *154*, 398–407.

Moody, J. A., R. A. Shakesby, P. R. Robichaud, S. H. Cannon, D. A. Martin (2013), Current research issues related to post-wildfire runoff and erosion processes, *Earth Sci. Rev.*, *122*, 10–37.

Morino, I., O. Uchino, M. Inoue, Y. Yoshida, T. Yokota, P. O. Wennberg, G. C. Toon, D. Wunch, C. M. Roehl, J. Notholt, T. Warneke, J. Messerschmidt, D. W. T. Griffith, N. M. Deutscher, V. Sherlock, B. Connor, J. Robinson, R. Sussmann, and M. Rettinger (2011), Preliminary validation of column-averaged volume mixing ratios of carbon dioxide and methane retrieved from GOSAT short-wavelength infrared spectra, *Atmos. Meas. Tech.*, *4*, 1061–1076; doi:10.5194/amt-4-1061-2011.

Morisette, J. T., L. Giglio, I. Csiszar, and C. O. Justice (2005), Validation of MODIS active fire detection product over Southern Africa using ASTER data, *Int. J. Remote Sens.*, *26*(19), 4239–4264.

Mouillot, F., M. G. Schultz, C. Yue, P. Cadule, K. Tansey, P. Ciais, and E. Chuvieco (2014), Ten years of global burned area products from spaceborne remote sensing, a review: Analysis of user needs and recommendations for future developments, *Int. J. Appl. Earth Obs. Geoinf.*, *26*, 64–79.

Padilla, M., S. V. Stehman, and E. Chuvieco (2014), Validation of the 2008 MODIS-MCD45 global burned area product using stratified random sampling, *Remote Sens. Environ.*, *144*, 187–196; doi:10.1016/j.rse.2014.01.008.

Peterson, D., and J. Wang (2013), A Sub-pixel-based calculate of fire radiative power from MODIS observations: 2. Sensitivity analysis and potential fire weather application, *Remote Sens. Environ.*, *129*, 231–249.

Peterson, D., J. Wang, C. Ichoku, E. Hyer, and V. Ambrosia (2013), A sub-pixel-based calculation of fire radiative power from MODIS observations: 1. Algorithm development and initial assessment, *Remote Sens. Environ.*, *129*, 262–279; doi:10.1016/j.rse.2012.10.036.

Petrenko, M., and C. Ichoku (2013), Coherent uncertainty analysis of aerosol measurements from multiple satellite sensors, *Atmos. Chem. Phys.*, *13*, 6777–6805; doi:10.5194/acp-13-6777-2013.

Petrenko, M., R. Kahn, M. Chin, A. Soja, T. Kucsera, and Harshvardhan (2012), The use of satellite-measured aerosol optical depth to constrain biomass burning emissions source strength in the global model GOCART, *J. Geophys. Res.*, *117*, D18212; doi:10.1029/2012JD017870.

Pouliot, G., T. G. Pace, B. Roy, T. Pierce, and D. Mobley (2008), Development of a biomass burning emissions inventory by combining satellite and ground-based information, *J. Appl. Remote Sens.*, *2*(1), 021501–021501.

Randerson, J. T., Y. Chen, G. R. van der Werf, B. M. Rogers, and D. C. Morton (2012), Global burned area and biomass burning emissions from small fires, *J. Geophys. Res.*, *117*, G04012; doi:10.1029/2012JG002128.

Reuter, M., et al. (2011), Retrieval of atmospheric CO2 with enhanced accuracy and precision from SCIAMACHY: Validation with FTS measurements and comparison with model results, *J. Geophys. Res.*, *116*, D04301; doi:10.1029/2010JD015047.

Roberts, G., and M. J. Wooster (2014), Development of a multi-temporal Kalman filter approach to geostationary active fire detection and fire radiative power (FRP) estimation, *Remote Sens. Environ.*, *152*, 392–412.

Roberts, G., M. J. Wooster, G. L. W. Perry, N. Drake, L.-M. Rebelo, and F. Dipotso (2005), Retrieval of biomass combustion rates and totals from fire radiative power observations: Application to southern Africa using geostationary SEVIRI imagery, *J. Geophys. Res.*, *110*, D21111; doi:10.1029/2005JD006018.

Roberts, G. J., and M. J. Wooster (2008), Fire detection and fire characterization over Africa using Meteosat SEVIRI, *IEEE Trans. Geosci. Remote Sens.*, *46*(4), 1200–1218; doi:10.1109/TGRS.2008.915751.

Roy, D. P., and L. Boschetti (2009), Southern Africa validation of the MODIS, L3JRC and GLOBCARBON burned area products, *IEEE Trans. Geosci. Remote Sens.*, *47*(4), 1032–1044; doi:10.1109/TGRS.2008.2009000.

Roy, D. P., L. Boschetti, and S. N. Trigg (2006), Remote sensing of fire severity: assessing the performance of the normalized burn ratio, *Geosci. Remote Sens. Lett., IEEE*, *3*(1), 112–116.

Roy, D. P., L. Boschetti, C. O. Justice, and J. Ju, (2008), The collection 5 MODIS burned area product: Global evaluation by comparison with the MODIS active fire product, *Remote Sens. Environ.*, *112*(9), 3690–3707.

Roy, D. P., L. Boschetti, S. W. Maier, and A. M. S. Smith (2010), Field estimation of ash and char colour-lightness using a standard grey scale, *Int. J. Wildland Fire*, *19*(6), 698–704.

Schneising, O., M. Buchwitz, J. P. Burrows, H. Bovensmann, M. Reuter, J. Notholt, R. Macatangay, and T. Warneke (2008),

Three years of greenhouse gas column-averaged dry air mole fractions retrieved from satellite, Part 1: Carbon dioxide, *Atmos. Chem. Phys.*, *8* (14), 3827–3853.

Schneising, O., M. Buchwitz, J. P. Burrows, H. Bovensmann, P. Bergamaschi, and W. Peters (2009), Three years of greenhouse gas column-averaged dry air mole fractions retrieved from satellite, Part 2: Methane, *Atmos. Chem. Phys.*, *9*, 443–465; doi:10.5194/acp-9-443-2009.

Schroeder, W., E. Ellicott, C. Ichoku, L. Ellison, M. B. Dickinson, R. D. Ottmar, C. Clements, D. Hall, V. Ambrosia, and R. Kremens (2014a), Integrated active fire retrievals and biomass burning emissions using complementary near-coincident ground, airborne and spaceborne sensor data, *Remote Sens. Environ.*, *140*, 719–730.

Schroeder, W., E. Prins, L. Giglio, I. Csiszar, C. Schmidt, J. Morisette, and D. Morton (2008), Validation of GOES and MODIS active fire detection products using ASTER and ETM plus data, *Remote Sens. Environ.*, *112*, 2711–2726.

Schroeder, W., P. Oliva, L. Giglio, and I. A. Csiszar (2014b), The New VIIRS 375m active fire detection data product: Algorithm description and initial assessment, *Remote Sens. Environ.*, *143*, 85–96.

Schultz, M. G., A. Heil, J. J. Hoelzemann, A. Spessa, K. Thonicke, J. G. Goldammer, A. C. Held, J. M. C. Pereira, and M. van het Bolscher (2008), Global wildland fire emissions from 1960 to 2000, *Glob. Biogeochem. Cycles*, *22*, GB2002; doi:10.1029/2007GB003031.

Scollo, S., R. A. Kahn, D. L. Nelson, M. Coltelli, D. J. Diner, M. J. Garay, and V. J. Realmuto (2012), MISR observations of Etna volcanic plumes, *J. Geophys. Res.*, *117*, D06210; doi:10.1029/2011JD016625.

Seiler W., and P. J. Crutzen (1980), Estimates of gross and net fluxes of carbon between the biosphere and the atmosphere from biomass burning, *Clim. Change*, *2*, 207– 248.

Stroppiana, D., G. Bordogna, P. Carrara, M. Boschetti, L. Boschetti, and P. A. Brivio (2012), A method for extracting burned areas from Landsat TM/ETM+ images by soft aggregation of multiple Spectral Indices and a region growing algorithm, *ISPRS J. Photogramm. Remote Sens.*, *69*, 88–102.

Tansey, K., J.-M. Grégoire, D. Stroppiana, A. Sousa, J. Silva, J. Pereira, L. Boschetti, et al. (2004),Vegetation burning in the year 2000: Global burned area estimates from SPOT VEGETATION data, *J. Geophys. Res. Atmos.*, *109*(D14), 1984–2012

Textor, C., M. Schulz, S. Guibert, S. Kinne, Y. Balkanski, S. Bauer, T. Berntsen, et al. (2007), The effect of harmonized emissions on aerosol properties in global models, an AeroCom experiment, *Atmos. Chem. Phys.*, *7*(17), 4489–4501.

Tsela, P., K. Wessels, J. Botai, S. Archibald, D. Swanepoel, K. Steenkamp, and P. Frost (2014), Validation of the two standard MODIS satellite burned-area products and an empirically-derived merged product in South Africa, *Remote Sens.*, *6*(2), 1275–1293; doi:10.3390/rs6021275.

Tsela, P. L., P. van Helden, P. Frost, K. Wessels, and S. Archibald (2010), Validation of the MODIS burned-area products across different biomes in South Africa, 3652–3655, in 2010 *IEEE International Geoscience and Remote Sensing Symposium, IEEE*; doi:10.1109/IGARSS.2010.5650253.

van der Werf, G. R., J. T. Randerson, L. Giglio, G. J. Collatz, M. Mu, P. S. Kasibhatla, D. C. Morton, R. S. DeFries, Y. van Jin, and T. T. van Leeuwen (2010), Global fire emissions and the contribution of deforestation, savanna, forest, agricultural, and peat fires (1997–2009), *Atmos. Chem. Phys.*, *10*(23), 11707–11735.

van der Werf, G. R., J. T. Randerson, L. Giglio, G. J. Collatz, P. S. Kasibhatla, and A. F. Arellano Jr. (2006), Interannual variability in global biomass burning emissions from 1997 to 2004, *Atmos. Chem. Phys.*, *6*, 3423–3441; doi:10.5194/acp-6-3523-2006.

Vermote, E., E. Ellicott, O. Dubovik, T. Lapyonok, M. Chin, L. Giglio, and G. J. Roberts (2009), An approach to estimate global biomass burning emissions of organic and black carbon from MODIS fire radiative power, *J. Geophys. Res.*, *114*, D18205; doi:10.1029/2008JD011188.

Wang, J., S. A. Christopher, U. S. Nair, J. S. Reid, E. M. Prins, J. Szykman, and J. L. Hand (2006), Mesoscale modeling of Central American smoke transport to the United States: 1. "Top-down" assessment of emission strength and diurnal variation impacts, *J. Geophys. Res.*, *111*, D05S17; doi:10.1029/2005JD006416.

Wang, W., Y. Liu, J. J. Qu, and X. Hao (2009), Analysis of the moderate resolution imaging spectroradiometer contextual algorithm for small fire detection, *J. Appl. Remote Sens.*, *3*(1), 031502–031502.

Warner, J., M. M. Comer, C. D. Barnet, W. W. McMillan, W. Wolf, E. Maddy, and G. Sachse (2007), A comparison of satellite tropospheric carbon monoxide measurements from AIRS and MOPITT during INTEX-A, *J. Geophys. Res.*, *112*, D12S17; doi:10.1029/2006JD007925.

Wiedinmyer, C., B. Quayle, C. Geron, A. Belote, D. McKenzie, X. Zhang, S. O'Neill, and K. K. Wynne (2006), Estimating emissions from fires in North America for air quality modeling, *Atmos. Environ.*, *40*, 3419–3432; doi:10.1016/j.atmosenv.2006.02.010.

Winker, D. M., J. L. Tackett, B. J. Getzewich, Z. Liu, M. A. Vaughan, and R. R. Rogers (2013), The global 3-D distribution of tropospheric aerosols as characterized by CALIOP, *Atmos. Chem. Phys.*, *13*, 3345–3361; doi:10.5194/acp-13-3345-2013.

Winker, D. M., M. A. Vaughan, A. Omar, Y. Hu, K. A. Powell, Z. Liu, W. H. Hunt, and S. A. Young (2009), Overview of the CALIPSO mission and CALIOP data processing algorithms, *J. Atmos. Oceanic Technol.*, *26*, 2310–2323; doi:10.1175/2009JTECHA1281.1.

Wooster, M. J. (2003), Fire radiative energy for quantitative study of biomass burning: derivation from the BIRD experimental satellite and comparison to MODIS fire products, *Remote Sens. Environ.*, *86*(1), 83–107; doi:10.1016/S0034-4257(03)00070-1.

Xiong, X., K. Chiang, J. Sun, W. L. Barnes, B. Guenther, and V. V. Salomonson (2009), NASA EOS Terra and Aqua MODIS on-orbit performance, *Adv. Space Res.*, *43*(3), 413–422.

Yokota, T., Y. Yoshida, N. Eguchi, Y. Ota, T. Tanaka, H. Watanabe, and S. Maksyutov (2009), Global concentrations of CO2 and CH4 retrieved from GOSAT: First preliminary results, *Sola*, *5*(0), 160–163.

Zhang, F., J. Wang, C. Ichoku, E. J. Hyer, Z. Yang, C. Ge, S. Su, X. Zhang, S. Kondragunta, J. W. Kaiser, C. Wiedinmyer, and A. da Silva (2014), Sensitivity of mesoscale modeling of smoke direct radiative effect to the emission inventory: a case study in northern sub-Saharan African region, *Environ. Res. Lett.*, *9*(7), 075002.

Zhang, X., and S. Kondragunta (2008), Temporal and spatial variability in biomass burned areas across the USA derived from the GOES fire product, *Remote. Sens. Environ.*, *112*, 2886–2897.

Zhukov, B., E. Lorenz, D. Oertel, M. Wooster, and G. Roberts (2006), Spaceborne detection and characterization of fires during the Bispectral Infrared Detection (BIRD) experimental satellite mission (2001–2004), *Remote Sens. Environ.*, *100*, 29–51.

15

Uncertainty and Complexity Tradeoffs When Integrating Fire Spread with Hydroecological Projections

Maureen C. Kennedy[1] and Donald McKenzie[2]

ABSTRACT

The discipline of hydroecology integrates biological and hydrological dynamics for the purpose of understanding and projecting watershed processes. Fire is a disturbance that both shapes and responds to ecosystem dynamics and impacts the provisioning of ecosystem services, yet few hydroecological models integrate dynamically with fire. Such integration requires using hydroecological projections as inputs to a fire model, which then predicts fire spread. It is important to match the complexity of the fire-spread model with data uncertainty of the hydroecological model, thereby minimizing total model uncertainty. We present the preliminary development and uncertainty assessment of a stochastic model of fire spread, with falsifiable components that make it possible to characterize uncertainty in model structure and parameter estimates. We compare simulated fire to the spatial pattern of observed spread for a wildfire in Washington state (USA) and discover issues in parameter identifiability. The parameter identifiability issues illuminate a deficiency in how the spread and hydroecological models are integrated. By developing a model of fire spread of complexity appropriate to the uncertainty in projected data we can identify model inadequacies, thereby reducing overall uncertainty in integrating fire with projections of watershed dynamics in a changing climate.

15.1. INTRODUCTION

The basis of the discipline of hydroecology is the understanding that hydrological processes are both impacted by, and in turn impact, biological dynamics [*Hannah et al.*, 2004; *Wood et al.*, 2007]. Disturbances such as fire also have feedbacks with both biological and hydrological processes [*DeBano et al.*, 1998], yet hydroecological research is lacking an integration with disturbance regimes [*Hannah et al.*, 2007]. Since future fires will likely become more extensive and severe in many regions [*Flannigan et al.*, 2009; *Littell et al.*, 2010; *Stavros et al.*, 2014] and fires represent a possibly increasing risk to natural resources, property, and ecosystem services such as water quality [*Hurteau et al.*, 2014; *Rocca et al.*, 2014], future projections of watershed dynamics and disturbance regimes would benefit from integrating fire spread and hydroecological models. Such an integration should not ignore the uncertainties introduced by the linking of models.

Sources of uncertainty in ecosystem models fall into three broad categories: data uncertainty, model-structure uncertainty, and natural variability including stochastic events [*O'Neill and Gardner*, 1979; *Beck*, 1987; *Turley and Ford*, 2009]. Data uncertainty and model-structure uncertainty can both result in uncertainty in parameter estimation and cumulative error from these sources of uncertainty can be nonlinear; for example, a 10% error in parameter estimation can propagate to an order of magnitude greater error in prediction [*O'Neill et al.*, 1980]. We adapt from *Hanna* [1988] the concept that total model

[1] *University of Washington, School of Interdisciplinary Arts and Sciences, Division of Sciences and Mathematics, Tacoma, Washington, USA*

[2] *Pacific Wildland Fire Sciences Laboratory, Pacific Northwest Research Station, US Forest Service, Seattle, Washington, USA*

Natural Hazard Uncertainty Assessment: Modeling and Decision Support, Geophysical Monograph 223, First Edition.
Edited by Karin Riley, Peter Webley, and Matthew Thompson.

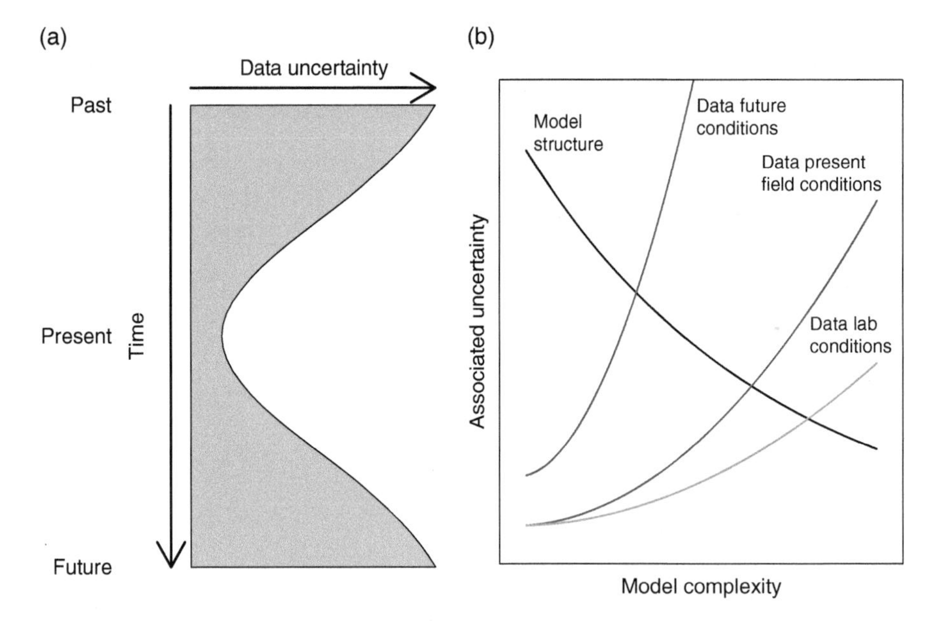

Figure 15.1 (a) Uncertainty associated with data is at a minimum in the present or near present time, with uncertainty in data increasing into the past and future timeframes. (b) If processes are correctly represented then model structure uncertainty is expected to decrease with model complexity. In the laboratory the associated increase in data uncertainty is flat, if the model is applied to data directly measured in the field for current conditions then the data uncertainty is higher but still relatively low and can accommodate a fairly complex model. If the data are from future projected conditions then the data uncertainty is expected to increase steeply with increasing model complexity. Figure adapted from the concepts presented in *Hanna* [1988] and visualized by the *US Environmental Protection Agency* [2009].

uncertainty is a combination of the reduction in model structure uncertainty with increasing model complexity (assuming the processes are represented correctly) and an increase in data uncertainty with increasing model complexity (Fig. 15.1), on top of a baseline of presumably irreducible uncertainty due to natural variability. The shape of the tradeoff between model complexity and data uncertainty determines the model complexity at which minimum uncertainty is obtained, and changes with the context of a model application. This conceptual framework underpins many well-known statistical methods for model selection such as the Akaike information criterion (AIC), which balances model fit with model complexity.

The ability to falsify component hypotheses is at the core of scientific inquiry, and science in the form of model development and use is not exempt from this requirement. The tendency is for model complexity to increase, especially as computational facilities improve [*Beck,* 1987], with the implicit assumption that this will produce more accurate and precise outcomes. This tendency occurs without compensating for the associated increase in data uncertainty or undertaking appropriate evaluation [*Alexander and Cruz,* 2013]. A large, complex model comprises many interacting component hypotheses. If there is a mismatch between model outputs and observations, such a large model makes it difficult to trace the mismatch to the component of the model that is the source of the model deficiency. In that sense it may not be possible to falsify and evaluate a more complex model effectively [*Beck,* 1987], because individual component hypotheses cannot be identified. Under these conditions, modelers may identify a mismatch between observations and model output, but cannot use that observation to improve model performance and reduce model uncertainty.

It is important to match the level of complexity of a given model to the scope of the application and the extent and quality of input data [*Jackson et al.*, 2000; *McKenzie and Perera*, 2014]. For example, a fire-spread model that simulates fires for the present, the recent past, or the near future [e.g., *Finney,* 2004] has access to better input data (presumably observed) than a fire-spread model that simulates fires farther into the future, with inputs that are projections from other models (Fig. 15.1a). For projecting future fire, the tradeoff in data uncertainty with model complexity is steeper because of the considerable uncertainty in projections of regional climate, weather and fuels that are not observed directly (Fig. 15.1b). Given this uncertainty in projections, integrating a hydroecological model with a fire-spread model that requires detailed data to predict fire perimeters gives false precision to estimates of fire spread.

An individual fire perimeter is the result of numerous complex relationships among fuels, climate, weather, and topography, and a faithful prediction of an individual fire perimeter requires high-quality data of intricate detail and fine scale (akin to the current laboratory or field conditions curves in Fig. 15.1b) well outside of the capabilities of current hydroecological models. To apply a model of such detail using insufficient data inputs produces predictions with false precision because the presumed certainty in the simulated fire perimeter does not adequately reflect the true uncertainty stemming from the uncertainty in the data inputs. Therefore, a relatively simpler model is appropriate for integration with hydroecological projections. Such a model represents adequately the way spread is expected to respond to the underlying template of topography, fuels, and moisture, but not necessarily to replicate fire perimeters, which requires a more complex model structure.

We are developing a stochastic model of fire spread (WMFire_beta) at a level of aggregation and process detail that is commensurate with the inputs provided by RHESSys, a watershed level hydroecological model [*Tague and Band*, 2004]. The structure of WMFire_beta readily facilitates quantifying uncertainties in both model structure and data, with respect to fire spread in general and in the integration of the hydroecological and fire-spread models. As a predictor of the pattern of fire spread for individual fires, and of characteristics of fire regimes when integrated with a hydroecological model, WMFire_beta can be compared to observations and modified if it cannot replicate the expected patterns in data. We describe here a process of model evaluation through which we are able to assess parameter uncertainty, and thereby diagnose inadequacies in model structure and integration with the hydroecological model, improving model structure and reducing overall model uncertainty.

15.2. METHODS

15.2.1. Regional HydroEcological Simulation System

The regional hydroecologic simulation system (RHESSys) is a GIS-based hydroecological model of water-carbon dynamics including water, carbon, and nutrient cycling (described in detail by *Tague and Band* [2004]). RHESSys is organized into hierarchical layers that aggregate different processes spatially. The largest spatial layer is the basin, which aggregates net fluxes of carbon, water, and nitrogen over the study area. The basin is composed of hillslopes, which aggregate drainage into a single stream reach. Hillslopes contain zones, which aggregate climate data, and zones contain patches, the smallest horizontal spatial layer in RHESSys. Patches are divided into vertical canopy strata that are defined by the model user, each with a set height and vegetation type. These strata correspond to vegetation layers (e.g., understory and overstory), which modify incoming radiation, precipitation, and wind. Multiple vegetation types can be assigned to strata of a given height, accommodating mixed vegetation. A single litter layer is applied to the entire patch, with inputs received from the canopy strata. The inputs provided by RHESSys to the fire-spread model are quantified at the patch level.

15.2.2. WMFire_Beta Structure

WMFire_beta is an expansion of the exogenously constrained dynamic percolation (ECDP) model developed to replicate patterns of past fire regimes as recorded in a spatiotemporal record of fire-scar data [*Kennedy and McKenzie*, 2010; *McKenzie and Kennedy*, 2012]. On a raster grid, a single ignition point is located either randomly or by the user. Fire spread from the ignition point to each of its four orthogonal neighbors is tested independently against some probability of spread (p_s). If spread is successful, then the next iteration the neighbors of the newly burned cell are tested for spread. If spread is not successful, then it is not tested again from the source cell to the corresponding neighbor. In ECDP the probability of spread is assigned uniformly across the landscape, where changes in the probability of spread result in varied fire size distributions and fire shapes [*Kennedy and McKenzie*, 2010; *McKenzie and Kennedy*, 2012]. A uniform assignment of spread probability aggregates dynamic interactions of topography, vegetation, climate, and weather into a single static value. This static aggregation is inadequate to understand how changing climate and watershed processes predicted by the hydroecological model may modify patterns of fire spread and fire regimes. To that end, the p_s parameter should be disaggregated to calculate dynamically the effects of fuel load ($p_s(l)$) and moisture ($p_s(m)$) of the neighbor cell, the wind direction relative to the direction of spread ($p_s(w)$), and the topographic slope relative to the direction of spread ($p_s(S)$; Table 15.1). In WMFire_beta, the overall probability of spread is the product of the component probabilities:

$$p_s(l,m,w,S)=p_s(l)\,p_s(m)\,p_s(w)\,p_s(S) \quad (15.1)$$

In broad terms, it is understood that fire spread increases with fuel load, with decreasing fuel moisture, and in the wind direction and upslope (Fig. 15.2). This mathematical expression can be interpreted as the accumulation of barriers to fire spread: if there is no barrier to spread, then all of the component probabilities are 1 and spread will occur. If any one of the components absolutely prevents spread (e.g., no fuel in the pixel), then

Table 15.1 WMFire_Beta Parameter Values, Interpretations, and Search Ranges

Parameter	Interpretation	Range (S1)	Range (S2)
k_{1_slope}	Value of $p_s(S)$ at zero slope	(0.05,1)	(0.75,1)[a]
k_{2_slope}	Steepness of the increasing (decreasing) relationship between $p_s(S)$ and positive (negative) slope, a symmetric relationship	(0,3)	(0,3)
k_{1_wind}	Defines the maximum value $p_s(w)$ takes in the wind direction at the maximum wind speed, and if > 0.5 the flanking direction at which $p_s(w)$ falls below one	(0.75,2)	(0.75,2)
k_{2_wind}	Probability of spread directly against the wind direction (backing fire)	(0,0.5)	(0,0.5)
k_{1_load}	Steepness of the increasing relationship between $p_s(l)$ and fuel load	(0.05,3)	(0.05,5)[a]
k_{2_load}	Value of fuel load at which $p_s(l) = 0.5$	(0,2)	(0,2)
$k_{1\_moisture}$	Steepness of the decreasing relationship between $p_s(m)$ and fuel moisture	(0.05,3)	(0.05,3)
$k_{2\_moisture}$	Value of fuel moisture at which $p_s(m) = 0.5$	(15,30)	(15,30)

[a] Search range modified between S1 (500 replicate simulations to estimate a p-value for the observed statistic) and S2 (a single simulated realization to compare to acceptable limits for each statistic).

the component probability and the overall probability of spread are zero.

15.2.2.1. Fuels

For both fuel load (l) and fuel moisture (m), we chose a sigmoid function that takes values between zero and one with two parameters (Table 15.1):

$$p_s(l) = \frac{1}{1+e^{-k_{1\_load}(l-k_{2\_load})}}; \quad (15.2)$$

$$p_s(m) = 1 - \frac{1}{1+e^{-k_{1\_moisture}(m-k_{2\_moisture})}} \quad (15.3)$$

k_{1*} defines the shape of the curve (its steepness), k_{2*} defines where along the x-axis the function crosses a value of 0.5 (Fig. 15.2a, 15.2b), where 0.495 is estimated to be a percolation threshold for a dynamic percolation model that spreads fire on a raster grid [*Kennedy and McKenzie*, 2010]. The percolation threshold is the landscape mean probability of spread below which fires cannot span the landscape.

15.2.2.2. Wind

For wind (w), we adapted from *Weisberg et al.* [2008] a trigonometric function:

$$p_s(w) = k_{1\_wind} * v_{relative}\left(1+cos(\gamma-\omega)\right) + k_{2\_wind} \quad (15.4)$$

where $v_{relative} = 1$ if the wind speed (v) is ≥ the maximum wind speed (v_{max}) for a given landscape, v/v_{max} otherwise, ω is the wind direction (rad), γ is the orientation of the neighbor pixel relative to the pixel spreading fire (rad), and k_{2_wind} is the intercept, or the probability of spread against the direction of the wind (Fig. 15.2c; Table 15.1). Note that this function mathematically can take values > 1.0, in which case $p_s(w)$ is set to 1.

15.2.2.3. Topography

For topographic slope (S), we borrowed from LANDSUM [*Keane et al.*, 2002] the functional form:

$$p_s(S) = k_{1_{slope}} e^{Ik_{2_{slope}}S^2} \quad (15.5)$$

where k_{1_slope} gives the value of p_s at zero slope, k_{2_slope} defines the steepness of the curve, and I = 1 if S > 0, –1 otherwise (Fig. 15.2d; Table 15.1). Note that this function mathematically can take values > 1.0, in which case $p_s(S)$ is set to 1.

We first evaluate WMFire_beta for whether the model can adequately represent the relationship between fire spread and the underlying template of topography, fuels, and weather. For this first test, we compare simulated spatial patterns of fire spread to an observed fire event.

15.2.3. Study Site

To evaluate WMFire_beta, we use the initial progression of the Tripod complex fire (first 5 days of burning), which burned through the Okanogan-Wenatchee National Forest in north-central Washington state (Fig. 15.3). This initial progression was described and evaluated by *Prichard and Kennedy* [2014], and we chose this progression because of our previous empirical work in the area [*Prichard and Kennedy*, 2012; *Prichard and Kennedy*, 2014]. We used maps of fuel models using the Scott and Burgan 40-model classification [*Scott and Burgan*, 2005] provided by the LANDFIRE project (available online; http://landfire.cr.usgs.gov/viewer) to extract expected loads of 100 hr and finer woody fuels across the fire area. We assume that this litter classification is commensurate with the patch-level litter layer simulated by RHESSys.

We then used remotely sensed vegetation layers and a 30 m resolution digital elevation model (DEM) for the

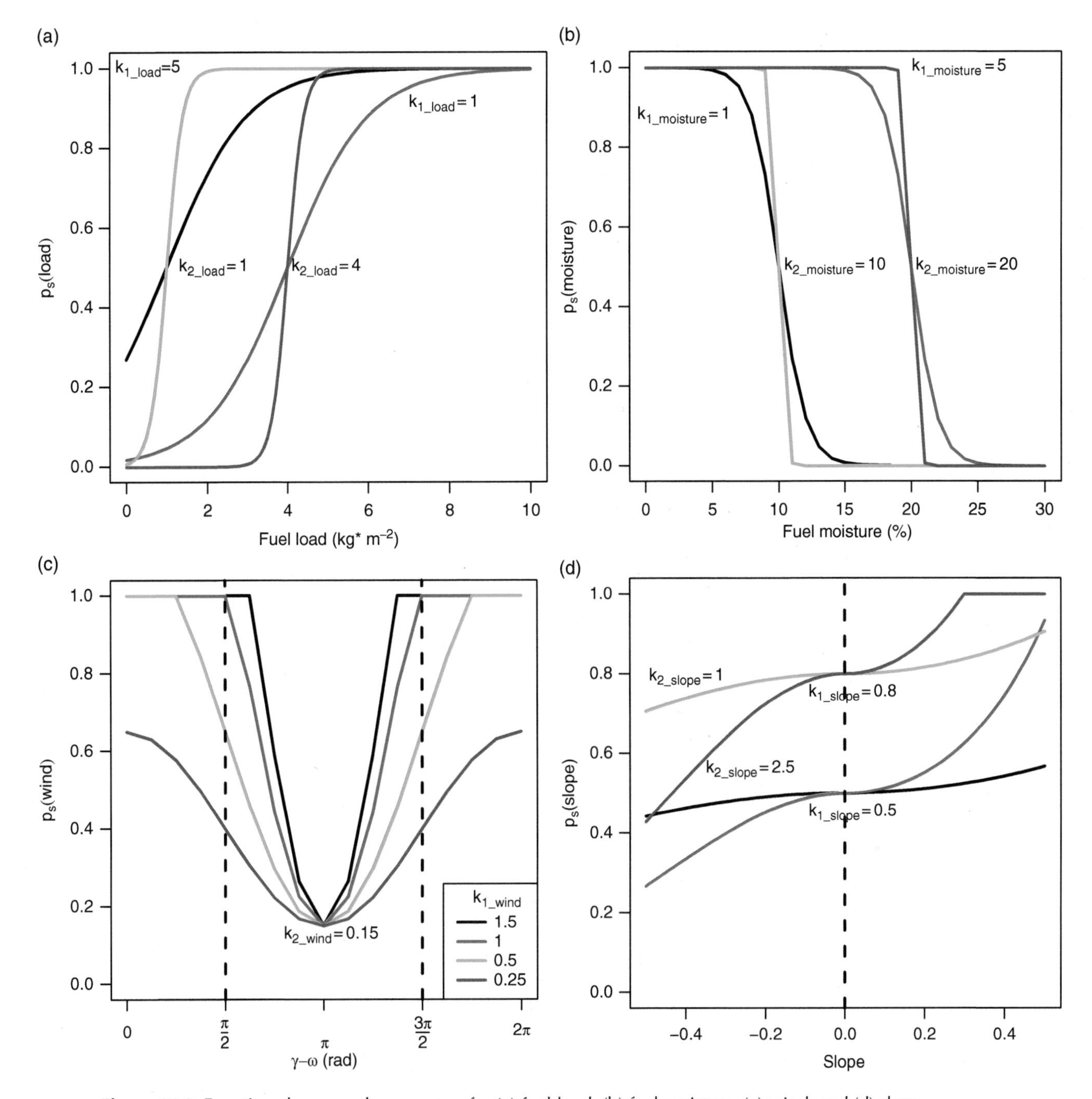

Figure 15.2 Function shapes and parameters for (a) fuel load, (b) fuel moisture, (c) wind, and (d) slope.

area to perform an uncalibrated RHESSys run to project litter fuel moisture across the extent of the fire. We chose a wind direction from weather data on those first 5 days of spread [*Prichard and Kennedy*, 2014] and an ignition point by visually locating an ignition pixel near the center of the first day of fire progression. The fire severity map obtained from monitoring trends in burn severity [MTBS; *Eidenshenk et al.*, 2007] was used to define the fire perimeter (Fig. 15.3, see *Prichard and Kennedy* [2014]) and to calculate spatial statistics for comparison of simulated to observed spread. For the purposes of this analysis, the low-severity burn category was assumed to be unburned. This reduces the number of burned pixels within this section of the fire, and restricts the representation of fire spread to those sections with substantial modifications to the burned area within the perimeter.

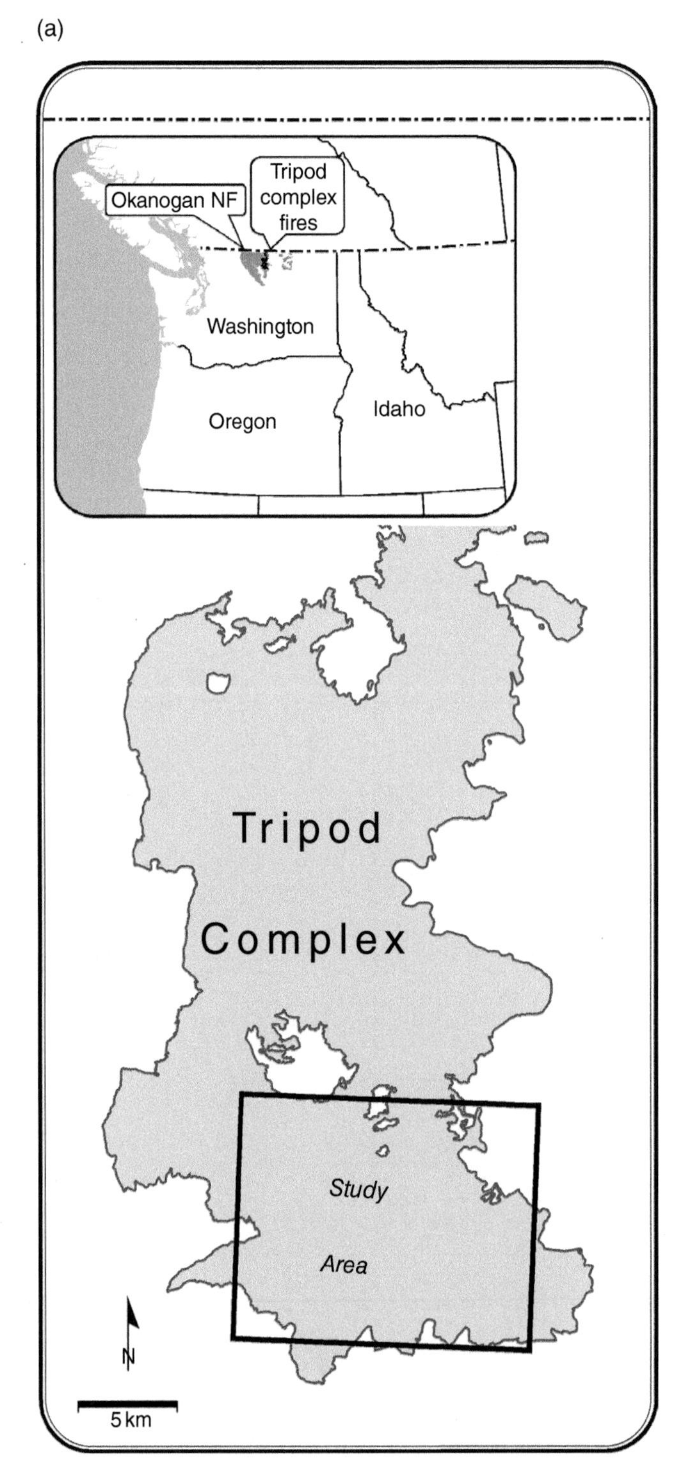

Figure 15.3 (a) Tripod fire study area; (b) Tripod fire perimeter within the study area, overlaid on the digital elevation model; (c) example simulated perimeter from search 1; (d) example simulated perimeter from search 2.

15.2.4. Evaluation Criteria: Spatial Statistics of Fire Spread

WMFire_beta is not designed to predict fire perimeters, rather it is expected that the model capture the major aggregate patterns that characterize how a fire spreads under a given set of circumstances. For the spatial pattern of fire spread, we seek statistics that are able to differentiate complex patterns of fire spread from relatively simple ones, understanding that for example more complex topography may map to complex spatial patterns of spread [*Kennedy and McKenzie*, 2010; *McKenzie and*

(b)

(c)

Figure 15.3 (Continued)

Kennedy, 2012]. Fractal dimension is a measure of the complexity of the fire pattern and is calculated as [*McGarigal et al.*, 2002]:

$$fd = \frac{2ln(0.25p)}{ln(a)} \quad (15.6)$$

where p is the length of the fire perimeter (interior and exterior; m), and a is the burned area (m^2). This measure is 1 for a completely compact square fire and increases with increasing complexity of the burned area.

Lacunarity is a measure of the unburned space within the fire perimeter and is estimated using a gliding box method [*Allain and Cloitre*, 1991; *Plotnick et al.*, 1993].

Figure 15.3 (Continued)

A square box with each side a given number of pixels (r) is overlain on the grid at the top corner and the number of burned pixels in the box is counted. The box is moved one pixel over and the number of burned pixels is again counted, and so on until the entire fire is traversed. A frequency table of the number of burned pixels (S) is used to quantify a probability distribution of the number of burned pixels (Q[S,r]). Lacunarity is calculated as the ratio of the second moment of the probability distribution to the square of the first moment:

$$lac = \frac{\Sigma S^2 Q(S,r)}{\left(\Sigma SQ(S,r)\right)^2}. \tag{15.7}$$

If for a given mean the variance increases, then lacunarity increases.

15.2.5. Model Evaluation: Parameter Search

We conducted two different parameter searches to compare simulated to observed patterns with the goal to identify parameter combinations able to produce patterns of spread within reasonable limits of those observed. The first parameter search (S1) follows the method of *McKenzie and Kennedy* [2012] to create a distribution of criteria values for a given parameter combination and then compare the observed statistic to the distribution. This is accomplished for the stochastic model by generating n_{sim} (500) independent replicate fires for a given parameter combination (where we simulated a total of 1500 parameter combinations using the evolutionary algorithm described below, for a total of 750,000 simulated fires), then calculating the p-value for the observed statistic (θ_{obs}) within the distribution of simulated statistics (θ_j, j = 1,…, n_{sim}). The two-tailed p-value is calculated as:

$$\hat{p} = 2 * \frac{\sum_j I}{n_{sim}+1}, \tag{15.8}$$

where I tallies the number of simulated statistics more extreme than the observed. If the observed statistic is > the median simulated value (in the right-hand tail), then I = 1 if $\theta_{obs} < \theta_j$, 0 otherwise. Alternatively, if the observed statistic is < the median simulated value (in the left-hand tail), then I = 1 if $\theta_{obs} > \theta_j$, 0 otherwise. If the p-value is sufficiently large, that indicates that the observed statistic cannot be distinguished from the distribution of simulated statistics (i.e., that the observed value is a typical model realization and that parameter set cannot be rejected). This sets the burden of inference on Type II error, the probability of failing to reject when you should reject. To that end, we set the threshold p-value for accepting the parameter set to 0.2. A parameter set that accomplishes a p-value > 0.2 for both fd and lac is retained.

For the second parameter search (S2), we ask whether any realization of the model for a given parameter set is capable of achieving an individual value for both fd and lac simultaneously within reasonable bounds of the observed statistic. This is an adaptation of the concept of behavioral models [*Hornberger and Cosby*, 1985] and the binary criteria utilized by *Reynolds and Ford* [1999]. We set limits of acceptability around each statistic and accept only parameter combinations for which an individual realization accomplishes values of the statistic within the acceptable limit for both statistics simultaneously. The acceptable limits were determined by approximating the expected error for each statistic determined by replicate WMFire_beta simulations. We generated simulated distributions of both lacunarity and fractal dimensions by performing replicate simulations of a single parameter vector. We calculated the standard deviation of each statistic among the replicate simulations and used the standard deviation as an approximation of the expected error. For this search, we simulated 250,000 parameter combinations, resulting in 250,000 simulated fires.

For both searches, we use the evolutionary algorithm developed by *Reynolds and Ford* [1999] for multicriteria model assessment [*Gupta et al.*, 1998]. The algorithm returns distributions of parameter values able to satisfy both criteria simultaneously, or if no such parameter combination exists, the algorithm returns the set of parameter combinations for which no further improvement among the multiple criteria is possible [*Reynolds and Ford,* 1999; *Kennedy and Ford,* 2011].

Parameter identifiability is the problem of finding a unique parameter set that unambiguously provides the best model performance [*Beck,* 1987]. A lack of identifiability is a symptom of uncertainty in model structure, uncertainty in data, or both. The diagnosis of parameter identifiability guides improvements to model structure, with an overall reduction in model uncertainty [*Reynolds and Ford,* 1999; *Turley and Ford,* 2009]. It is assumed that for a low-uncertainty, identifiable parameter, the distribution of values able to accomplish both statistics simultaneously would occupy a narrow range with an obvious mode. Therefore, the model is first assessed for whether it can match both criteria, then for whether parameters able to match both criteria are identifiable.

15.3. RESULTS

15.3.1. Observed Fire-spread Patterns

The observed fractal dimension is 1.27 and the observed lacunarity is 1.36 for the first 5 days of progression for the Tripod fire. We used replicate simulations of WMFire_beta for several parameter combinations to provide estimates of the standard deviation of each statistic and found the standard deviations of both to be consistently below 0.1. We decided to round the standard deviation of both to 0.1 to provide a preliminary screening of those parameter values able to produce an individual fire within reason of the observed patterns. This gave a range of the statistic (1.17, 1.37) for fd and (1.26, 1.46) for lac.

15.3.2. Parameter Identifiability

For both searches, WMFire_beta was able to achieve both the fractal dimension and the lacunarity criteria simultaneously for multiple parameter combinations. See Figure 15.3c and d for example simulated fires. Based on preliminary analysis of the first search, the search range for some of the parameters was modified for the second search (one run per parameter set). The distribution of parameter values able to replicate both the fractal dimension and the lacunarity of the observed fires shows varying identifiability among the parameters (Fig. 15.4).

The parameter k_{1_slope} defines the probability of spread, with no other barriers, on a flat slope (Table 15.1). In S1, this parameter has a narrow range of values near 1 that match both criteria, whereas in S2, this parameter occupies a broader range and lacks identifiability (Fig. 15.4a). In S2, k_{1_slope} has a bimodal distribution with modes near 0.75 and >0.95. The parameter k_{2_slope} defines the steepness of the increase with increasing positive slope and decrease with increasing negative slope (Table 15.1), and both searches show a very broad range and lack of identifiability for this parameter. The underlying distribution for k_{2_slope} for both searches is bimodal, with obvious modes at values <0.5 and at values >2 (Fig. 15.4b).

The parameter k_{1_wind} determines the deviation of the wind direction from the spread direction at which the probability of spread associated with wind falls from 1 (Table 15.1), and both searches show broad ranges in parameter values able to replicate both criteria. S1 has an obvious mode near 0.8 with a tail to the right, and S2 shows bimodality in k_{1_wind} with modes at values near 0.8 and >1.8 (Fig. 15.4c). The parameter k_{2_wind} defines the probability of spread directly against the wind direction (Table 15.1) and this parameter shows values <0.2 are able to replicate both criteria, with a right-skewed distribution for both searches (Fig. 15.4d).

The parameter k_{1_load} defines the steepness of the curve with increasing load (Table 15.1), and in S1, parameter values able to replicate both criteria simultaneously are distributed near the upper edge of the search range (Fig. 15.4e), whereas, for S2, the parameter values are distributed broadly across the expanded search range, although the main portion of the distribution is located above a value of 2. The parameter k_{2_load} is the value of fuel load at which the curve crosses a probability of spread of 0.5 (Table 15.1), and for both searches this shows a narrower range of values able to replicate both criteria (Fig. 15.4f), with values distributed lower for S2 than S1.

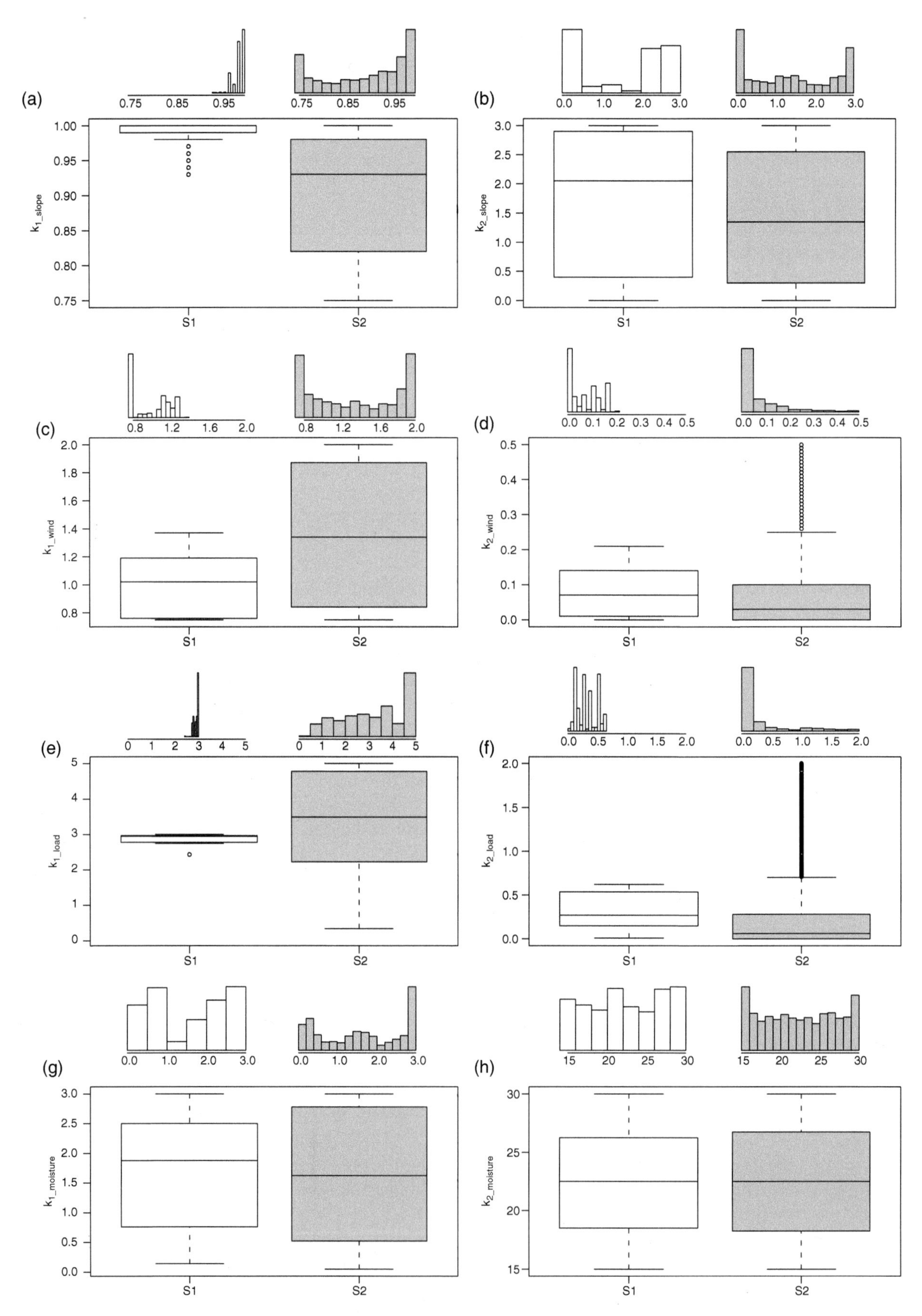

Figure 15.4 Distribution of parameter values not rejected for each search. A broad distribution with no obvious modes indicates an identifiability problem for that parameter.

The two parameters associated with fuel moisture (Table 15.1) both lack identifiability for both searches, occupying nearly uniformly the full search range for both (Fig. 15.4g–h).

15.4. DISCUSSION

15.4.1. Sources of Parameter Identifiability

The relative simplicity of WMFire_beta enables us to trace the sources of the problems with parameter identifiability that were illuminated by this model analysis. Our ability to trace the source of those problems means we have been able to falsify a component hypothesis represented by the model structure. The problems with parameter identifiability arise from the lack of sensitivity in model outputs to changes in parameter values. There are two main sources of this lack of sensitivity: the mathematical structure of the model, and the translation between the hydroecological and fire-spread model.

The mathematical structure of WMFire_beta decreases sensitivity of model predictions to increasing values of several of the parameters. For example, we can calculate the value of fuel load at which $p_s(l)$ crosses 0.9 and evaluate how that threshold fuel load changes with k_{1_load}. We find that the value of load at which the probability crosses 0.9 decreases rapidly with increasing parameter values up to $k_{1\_load} = 2$, from a value above 5 kg*m^{-2} to one near 2 kg*m^{-2} (Fig. 15.5). The threshold value for fuel load then levels off with increasing k_{1_load}, with much smaller changes with increasing parameter value. This lack of sensitivity does not present a model inadequacy or parameter uncertainty, rather it is an artifact of the mathematical model structure (equations [15.2], [15.3]).). In this case, the parameter is identifiable in the sense that k_{1_load} should be of a value sufficiently high to ensure that the probability of spread associated with fuel load crosses 0.9 at a value near 2 kg*m^{-2}.

In contrast to k_{1_load}, the distributions of the parameters associated with moisture ($k_{1\_moisture}$, $k_{2\_moisture}$) encompass the full range of values at which the mathematical structure indicates that the model should be highly sensitive to changes in them (Fig. 15.4g,h; Fig. 15.5b). For example, values of $k_{1\_moisture}$ returned by the parameter search span the range between zero and three, which as shown by Figure 15.5b should produce different patterns of fire spread. The origin of this issue with identifiability seems to be an inadequate translation between hydroecological and fire-spread models, rather than a mathematical artifact. Our ability to identify this source of model uncertainty, and thereby falsify a model component, is facilitated by the relative simplicity of our model formulation.

In our first model formulation, we chose litter moisture content to represent the overall moisture of fuels for fire spread, but it appears that this is inadequate. Litter is a quickly drying fuel layer, which is predicted to have near zero variability across the landscape under the weather during the Tripod fire. Therefore, predicted spread is insensitive to the moisture parameters because moisture itself is not found to vary across the landscape. The upper range for $k_{2\_moisture}$ in both searches is well above the maximum value predicted for litter moisture across the Tripod landscape, so all moisture parameter combinations predict probability of spread near 1 regardless of the location of the pixel. Although the lack of variability in litter moisture is a reasonable prediction, this illuminates that litter moisture as represented by equation (15.3) is an inadequate representation of moisture to inform fire spread across the landscape. The model is unable to map the interaction of spread and moisture on the landscape if there is no variability in moisture, and this result implies that litter moisture is not an important factor in fire spread. This process of model analysis enables us to falsify the moisture component of the model structure, and leads us to develop a new representation in order to improve the model and reduce model uncertainty.

There are other issues found in the distributions of parameter values, including the bimodal distribution of the parameters associated with topographic slope (Fig. 15.4a,b). It is common in models for parameters to have dependencies or to accommodate each other in order for model predictions to match data [*Beck,* 1987]. Given the inadequately represented relationship between moisture and fire spread in WMFire_beta, it may be that the remaining parameters are compensating and resulting in these anomalous distributions. The inadequate representation of litter moisture for fire spread must be corrected in the model structure before other identifiability issues are investigated. One possibility is to calculate the water balance deficit, the difference between the potential evapotranspiration and the actual evapotranspiration [*Stephenson,* 1998], which is a representation of moisture condition that is related to fire regimes [*Littell and Gwozdz,* 2011].

15.4.2. Uncertainty Assessment and Understanding Hazards

The two parameter searches represent distinct interpretations of the fire event. For S1, in order to calculate the simulated null distribution, it was de facto necessary to simulate n_{sim} realizations with fires of sufficient size to calculate the spatial statistic. We thereby established a rule wherein if a certain number of realizations failed to spread a fire of sufficient size before 500 acceptable fires are simulated, then that parameter set is rejected.

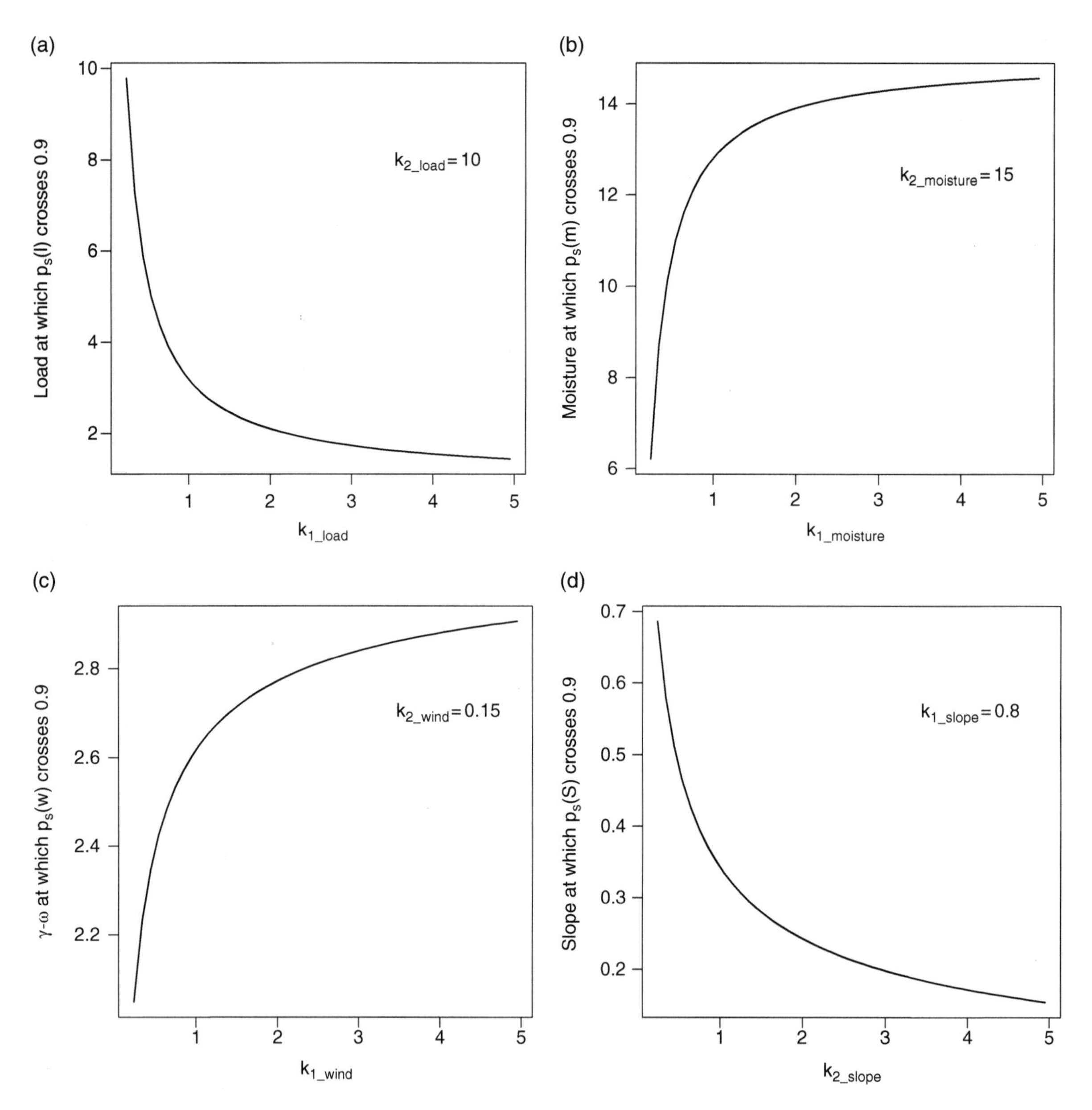

Figure 15.5 Sensitivity of WMFire_beta equations to the shape parameters for each component used to calculate the probability of fire spread is indicated by a steep relationship between the parameter value and the response, in this case threshold values at which the component spread probability crosses 0.9. If the threshold value changes rapidly with increasing parameter value then the model is judged sensitive, whereas if the threshold value shows little change with increasing parameter value then sensitivity is low. Sensitivity to the shape parameter decreases (becomes less steep) as the parameter value increases.

This implicitly creates a third criterion that is accepted only when the majority of the realizations from a given parameter set create a large extreme fire as observed. This is interpreted to mean that on that day, given that ignition, and under those conditions, the Tripod fire was an inevitable outcome rather than a rare stochastic event. In contrast, S2 allows for the possibility that the observed fire event was one such possible outcome even under those conditions, and accepts as adequate a parameter set that can replicate that outcome (within error) as one possibility among many more. This has the implicit assumption that even under the conditions of the Tripod fire, most ignitions are not expected to cause an extreme fire event.

It is important to resolve the differences between the two searches when projecting future fire regimes and effects. For example, if parameters from S1 are chosen but the interpretation that this fire was an inevitable event is incorrect, then subsequent projections could be biased to more extreme fire regimes. Conversely, if the interpretation that this fire was one possibility among other stochastic realizations is incorrect, then parameters from S2 would bias projections to less extreme fire regimes. Therefore, deciding between S1 and S2 parameters is an important step in identifying an adequate model structure and reducing model uncertainty. There is also a risk in comparing model outputs to a single pattern of spread that the model will be overcalibrated to that one fire event and lack the desired robustness to a variety of conditions.

The addition of another criterion can aid in remedying problems of model identification [*Reynolds and Ford,* 1999]. WMFire_beta can be integrated with RHESSys to test their combined ability to replicate expected past fire regimes for watersheds in different vegetation types and expected fire regimes. We will quantify criteria for expected fire regimes of watersheds currently calibrated for RHESSys runs. If an S1 parameter set best replicates statistics that described the expected fire regime (e.g., fire return interval, seasonality), then that is evidence in favor of the interpretation that extreme fires are inevitable under extreme conditions. If an S2 parameter set best replicates the observed fire regime, then that is evidence in favor of extreme fires being rare events even under extreme conditions. Model evaluation thereby reduces uncertainty both in model projections and in our understanding of those natural hazards that are being modeled. If a unique model can be identified to satisfy both the pattern of spread for an individual fire and the fire regimes of diverse watersheds, we will have identified a model of sufficient complexity and robustness to be used in the prediction of fire in future hydroecological projections.

15.4.3. Conclusions

A projection of the impacts of climate change on hydrological processes in the context of disturbance regimes is complicated by multiple sources of uncertainty. There is uncertainty in the underlying model structure required to integrate hydroecological projections with fire spread and effects. The data input into the fire-spread model are fuels and moistures projected by the hydrological model. As a projection of future values, the hydrological model outputs have their own associated uncertainties, and such uncertainty does not justify implementing a complex model of spread intended to predict fire perimeters with precision (Fig. 15.1). Yet the fire-spread model must be sufficiently complex to respond dynamically to changes in fuels and weather. We therefore require a fire-spread model that is sufficiently complex to be of use, but not too complex relative to the uncertainty in model inputs.

Fire in the future under a changing climate and integrated with hydrological projections is best evaluated by changing patterns of fire spread and properties of fire regimes, rather than an attempt to predict the perimeters of individual fires. A hydroecological model such as RHESSys does not produce outputs of sufficient detail and fine scale to be credibly integrated with fire-spread models designed to replicate fire perimeters. A relatively simple fire-spread model with falsifiable components makes it possible to understand uncertainty in model structure and projections, where the relationship between model components and model performance is tractable both computationally and cognitively.

ACKNOWLEDGMENTS

This research is a product of the Western Mountain Initiative. Christina Tague and Janet Choate at the University of California, Santa Barbara, provided RHESSys runs for moisture estimates. Robert Norheim provided the GIS maps for this manuscript. David Ford provided invaluable comments on an earlier version of this manuscript.

REFERENCES

Alexander, M. E., and M. G. Cruz (2013), Are the applications of wildland fire behaviour models getting ahead of their evaluation again? *Environ. Mod. Soft.*, *41*, 65–71; doi:10.1016/j.envsoft.2012.11.001.

Allain, C., and M. Cloitre (1991), Characterizing the lacunarity of random and deterministic fractal sets, *Phys. Rev. A.*, *44*(6), 3552–3558.

Beck, M. B. (1987), Water quality modeling: A review of the analysis of uncertainty, *Water Resour. Res.*, *23*(8), 1393–1442; doi:10.1029/WR023i008p01393.

DeBano, L. F., D. G. Neary, and P. F. Folliott (1998), *Fire Effects on Ecosystems*, John Wiley and Sons, Inc., New York.

Eidenshink, J., B. Schwind, K. Brewer, and Z. Zhu (2007), A project for monitoring trends in burn severity, *Fire Ecol.*, *3*(1), 3–21.

Finney, M. A. (2004), FARSITE: Fire Area Simulator--Model development and evaluation, USDA, Rocky Mountain Research Station Research Paper RMRS-RP-4, Revised.

Flannigan, M. D., M. A. Krawchuk, W. J. de Groot, B. M. Wotton, and L. M. Gowman (2009), Implications of changing climate for global wildland fire, *Int. J. Wildland Fire*, *18*, 483–507.

Gupta, H., S. Sorooshian, and P. Yapo (1998), Toward improved calibration of hydrologic models: Multiple and noncommensurable measures of information, *Water Resour. Res.*, *34*(4), 751–763.

Hanha, S. R. (1988), Air quality model evaluation and uncertainty, *JAPCA J. Air Waste Man. Assoc.*, *38*(4), 406–412; doi:10.1080/08940630.1988.10466390.

Hannah, D. M., J. P. Sadler, and P. J. Wood (2007), Hydroecology and ecohydrology: Challenges and future prospects, 421–429, in *Hydroecology and Ecohydrology: Past, present, and Future*, edited by P. J. Wood, D. M. Hannah, and J. P. Sadler, John Wiley and Sons, West Sussex, England.

Hannah, D. M., P. J. Wood, and J. P. Sadler (2004), Ecohydrology and hydroecology: A new paradigm? *Hydrolog. Processes*, *18*(17), 3439–3445; doi:10.1002/hyp.5761.

Hornberger, G., and B. Cosby (1985), Selection of parameter values in environmental models using sparse data: A case study, *Appl. Math. Comput.*, *17*(4), 335–355; doi:10.1016/0096-3003(85)90040-2.

Hurteau, M. D., J. B. Bradford, P. Z. Fulé, A. H. Taylor, and K. L. Martin (2014), Climate change, fire management, and ecological services in the southwestern US, *For. Ecol. Man.*, *327*, 280–289; doi:10.1016/j.foreco.2013.08.007.

Jackson, L. J., A. S. Trebitz, and K. L. Cottingham (2000), An introduction to the practice of ecological modeling, *BioScience*, *50*(8), 694–706.

Keane, R. E., R. A. Parsons, and P. F. Hessburg (2002). Estimating historical range and variation of landscape patch dynamics: limitations of the simulation approach, *Ecol. Mod.*, *151*(1), 29–49.

Kennedy, M. C., and D. McKenzie (2010), Using a stochastic model and cross-scale analysis to evaluate controls on historical low-severity fire regimes, *Landscape Ecol.*, *25*(10), 1561–1573; doi:10.1007/s10980-010-9527-5.

Kennedy, M. C., and E. D. Ford (2011), Using multicriteria analysis of simulation models to understand complex biological systems, *BioScience*, *61*(12), 994–1004; doi:10.1525/bio.2011.61.12.9.

Littell, J. S., E. E. Oneil, D. McKenzie, J. A. Hicke, J. A. Lutz, R. A. Norheim, and M. M. Elsner (2010), Forest ecosystems, disturbance, and climatic change in Washington State, USA, *Clim. Change*, *102*(1–2), 129–158; doi:10.1007/s10584-010-9858-x.

Littell, J. S., R. B. Gwozdz (2011), Climatic water balance and regional fire years in the Pacific Northwest, USA: Linking regional climate and fire at landscape scales, in *The Landscape Ecology of Fire, Ecological Studies 213*, edited by D. McKenzie, C. Miller, D. Falk, Springer, New York.

McGarigal, K., S. A. Cushman, M. C. Neel, and E. Ene. (2002), FRAGSTATS: Spatial pattern analysis program for categorical maps, Version 3.3, University of Massachusetts, Amherst, Massachusetts, USA, http://www.umass.edu/landeco/research/fragstats/fragstats.html.

McKenzie, D., and M. C. Kennedy (2012), Power laws reveal phase transitions in landscape controls of fire regimes, *Nature Communications*, *3*, 726; doi:10.1038/ncomms1731.

McKenzie, D., A. Perera (2014), Modeling wildfire regimes in forest landscapes: Abstracting a complex reality, in *Modeling and Mapping Forest Landscape Patterns*, in press, edited by A. H. Perera, B. R. Sturdevant, and L. J. Buse, Springer, New York.

O'Neill, R.V., and R. H. Gardner (1979), Sources of uncertainty in ecological models, 447–463, in *Methodology in Systems Modelling and Simulation*, edited by B. P. Zeilgler, M. S. Elzas, G. J. Klir, I. Ö. Tuncer, North-Holland Publishing Company, Amsterdam.

O'Neill R. V., R. H. Gardner, and J. B. Mankin (1980), Analysis of parameter error in a nonlinear model, *Ecol. Mod.*, *8*, 297–311.

Plotnick, R. E., R. H. Gardner, and R. V. O'Neil (1993), Lacunarity indices as measures of landscape texture, *Landscape Ecol.*, *8*(3), 201–211.

Prichard, S. J., and M. C. Kennedy (2012), Fuel treatment effects on tree mortality following wildfire in dry mixed conifer forests, Washington State, USA, *Int. J. Wildland Fire*, *21*, 1004–1013, 10.1071/WF11121.

Prichard, S. J., and M. C. Kennedy (2014), Fuel treatments and landform modify landscape patterns of burn severity in an extreme fire event, *Ecol. Appl.*, *24*(3), 571–90.

Reynolds, J. H., and E. D. Ford (1999), Multi-criteria assessment of ecological process models, *Ecology*, *80*(2), 538–553.

Rocca, M. E., P. M. Brown, L. H. MacDonald, and C. M. Carrico (2014), Climate change impacts on fire regimes and key ecosystem services in Rocky Mountain forests, *For. Ecol. Manag.*, *327*, 290–305; doi:10.1016/j.foreco.2014.04.005.

Scott, J. H., and R. E. Burgan (2005), Standard fire behavior fuel models: a comprehensive set for use with Rothermel's surface fire spread model, Gen. Tech. Rep. RMRS-GTR-153, Fort Collins, Colorado, U.S. Department of Agriculture, Forest Service, Rocky Mountain Research Station.

Stavros, E. N., J. Abatzoglou, N. A. Larkin, and D. McKenzie (2014), Regional projections of the likelihood of very large wildland fires under a changing climate in the contiguous Western United States, *Clim. Change*, *126*(3), 455–468; doi 10.1007/s10584-014-1229-6.

Stephenson, N. L. (1998), Actual evapotranspiration and deficit: Biologically meaningful correlates of vegetation distribution across spatial scales, *J. Biogeogr.*, *25*, 855–870.

Tague, C. L., and L. E. Band (2004), RHESSys: Regional Hydro-Ecologic Simulation System, An object- oriented approach to spatially distributed modeling of carbon, water, and nutrient cycling, *Earth Interac.*, *8*, paper no. 19, 1–42.

Turley, M. C., and E. D. Ford (2009), Definition and calculation of uncertainty in ecological process models, *Ecol. Mod.*, *220*(17), 1968–1983; doi:10.1016/j.ecolmodel.2009.04.046.

US Environmental Protection Agency, office of the Science Advisor (2009), Guidance on the development, evaluation, and application of environmental models, EPA/100/K-09/003| March2009, accessed 4 November 2014 at http://epa.gov/crem/library/cred_guidance_0309.pdf.

Weisberg, P. J., D. Ko, C. Py, and J. M. Bauer (2008), Modeling fire and landform influences on the distribution of old-growth pinyon-juniper woodland, *Landscape Ecol.*, *23*, 931–943; doi:10.1007/s10980-008-9249-0.

Wood P. J., D. M. Hanna, and J. P. Sadler (2007), Ecohydrology and hydroecology: An introduction, 1–6, in *Hydroecology and Ecohydrology: Past, Present, and Future*, edited by P. J. Wood, D. M. Hannah, and J. P. Sadler, John Wiley and Sons, West Sussex, England.

16

Uncertainty Quantification and Propagation for Projections of Extremes in Monthly Area Burned Under Climate Change: A Case Study in the Coastal Plain of Georgia, USA

Adam J. Terando,[1] Brian Reich,[2] Krishna Pacifici,[3] Jennifer Costanza,[4] Alexa McKerrow,[5] and Jaime A. Collazo[6]

ABSTRACT

Human-caused climate change is predicted to affect the frequency of hazard-linked extremes. Unusually large wildfires, individually or expressed as area burned over time, are a type of extreme event that is constrained by climate and can be a hazard to society but also an important ecological disturbance. Here we project changes in the frequency of extreme monthly area burned by wildfires for the end of the 21st century for a wildfire-prone region in the southeast United States. Predicting changes in area burned is complicated by the large and varied uncertainties in how the climate will change and in the models used to predict those changes. We characterize and quantify multiple sources of uncertainty and propagate the expanded prediction intervals of future area burned. We find nontrivial probabilities for an increasing number of extreme wildfire months for the period 2070–2099 (95% projection interval of 5 fewer to 28 more extreme fire months for a high fossil-fuel emissions scenario), resulting from the warmer climate, but also due to the inherent uncertainty when dealing with extreme events. Our approach illustrates that while accounting for multiple sources of uncertainty in global change science problems is a difficult task, it will be necessary in order to properly assess the risk of increased exposure to these society-relevant events.

[1]*US Geological Survey, Southeast Climate Science Center, Raleigh, North Carolina, and Department of Applied Ecology, North Carolina State University, Raleigh, North Carolina, USA*

[2]*Department of Statistics, North Carolina State University, Raleigh, North Carolina, USA*

[3]*Department of Forestry and Environmental Resources, Program in Fisheries, Wildlife, and Conservation Biology, North Carolina State University, Raleigh, North Carolina, USA*

[4]*Department of Forestry and Environmental Resources, North Carolina State University, Raleigh, North Carolina, USA*

[5]*Department of Applied Ecology, North Carolina State University, Raleigh, North Carolina, and Core Science Analytics, Synthesis & Libraries, US Geological Survey, Raleigh, North Carolina, USA*

[6]*Department of Applied Ecology, North Carolina State University, Raleigh, North Carolina, and US Geological Survey, North Carolina Cooperative Fish and Wildlife Research Unit, North Carolina State University, Raleigh, North Carolina, USA*

16.1. INTRODUCTION

Human-caused global warming (that is, *anthropogenic* climate change, or ACC) has the potential to increase exposure to hazards by increasing the frequency or severity of damaging extreme events [*Walsh et al.*, 2014]. Extreme wildfires present one of the most visible and concerning examples of a hazard that could be affected by ACC because of the potential for large-scale destruction and disruption of human-dominated landscapes. And yet, although wildfires can be a hazard, they are also a vital part of ecosystem disturbance and nutrient cycling dynamics [*Certini*, 2005]. As such, efforts to promote ecologically beneficial fires, which can also reduce the risk of damaging fires and promote system resilience, will suffer if changing climatic conditions increase the risk of

Natural Hazard Uncertainty Assessment: Modeling and Decision Support, Geophysical Monograph 223, First Edition.
Edited by Karin Riley, Peter Webley, and Matthew Thompson.

large uncontrollable fires [*Ryan et al.*, 2013]. Furthermore, in areas with large unfragmented forest ecosystems, increases in wildfire activity or the area burned by wildfires could also impact the global carbon cycle, further exacerbating ACC through the rapid release of carbon from forests [*Liu et al.*, 2014].

Recent studies have shown that some regions are currently experiencing more frequent large wildfires due to more conducive climatic conditions linked to ACC [*Dennison et al.*, 2014]. Given the changes already observed, and the strong expectation that global warming will accelerate in the future, researchers have developed models to predict how future ACC could impact different aspects of wildfire regimes, such as the frequency of large wildfires [*Girardin and Mudelsee*, 2008], the severity of wildfires as characterized by burn rates [*Fried et al.*, 2004], and the total area burned [*Litschert et al.*, 2012]. The potential for future changes in extreme wildfire occurrence or as expressed as the total area burned is an example of an indirect consequence of ACC. Unlike direct consequences of ACC or changes to the climate system itself (e.g., changes in climate phenomena such as temperature, precipitation, and hurricanes) that can be modeled solely through the use of climate models, indirect consequences require additional models to link the climate forcing to the ecological response. And with this additional layer of modeling comes additional uncertainty associated with the predicted consequences of ACC because of our imperfect knowledge of the system and the future.

Uncertainties in the context of global change have been extensively explored and debated within the climate science literature, particularly with regard to how best to characterize and quantify uncertainty given our inability to (1) construct controlled experiments on the climate system, (2) make deterministic predictions for this highly nonlinear system beyond 2–3 weeks [*Lorenz*, 1963], or (3) resolve many key processes in the climate system [*Palmer et al.*, 2005]. Generally, in climate change prediction problems (and likely the broader class of global change prediction problems), three forms of uncertainty are identified: internal variability, scenario uncertainty, and structural uncertainty [*Hawkins and Sutton*, 2009]. Internal or "natural" variability refers to the unforced variability that arises in highly nonlinear complex systems such as the Earth's climate system. Scenario uncertainty in the context of ACC refers to uncertainty in future anthropogenic emissions of greenhouse gases that lead to a perturbed climate response. While multiple models may be used to evaluate the same or alternate scenarios, by definition no formal likelihood or probability structure is attached to each independent vision of the future. Finally, structural uncertainty, or model uncertainty, occurs when competing models exist to describe the unknown state of nature relative to the parameter of interest, represented as θ [*Draper*, 1995]. Each general circulation model (GCM) that is a member of a larger ensemble of models (such as those found in the Coupled Model Intercomparison Project[s]) represents a particular set of assumptions about how nature (i.e., the climate system) operates which can be used to make predictions about θ. We also note that a fourth type of uncertainty, parametric uncertainty, is also present and represents uncertainty about parameters within or common to the individual models that affect the predictive uncertainty about θ [cf. *Sriver et al.*, 2012]. Depending on the problem and context, this form of uncertainty is sometimes subsumed under the more general category of structural or model uncertainty.

With these forms of uncertainty defined, the question becomes how to characterize and quantify each so that predictions are most useful to decision makers. Typically, usefulness in this context is defined as providing predictions about the future state of θ that maximize sharpness in a probabilistic sense (subject to calibration) [*Gneiting et al.*, 2007] while also avoiding underdispersion, which leads to overconfident predictions [*Draper*, 1995; *Raftery et al.*, 2005]. Note that this formulation focuses on probabilistic predictions, suggesting that a Bayesian approach (or at least partially Bayesian when considering calibration [*Draper*, 2013]) is appropriate to quantify uncertainty.

Various methods have been employed to account for these sources of uncertainty, particularly as it relates to structural uncertainty in climate models. Sometimes referred to as "kernel dressing," these methods transform point predictions into distributional functions that allow for a formalization of the degree of belief that is assigned to each model's prediction [*Brocker and Smith*, 2008]. Bayesian Model Averaging (BMA) [*Raftery et al.*, 1997], an increasingly popular way to quantify uncertainty across competing models, has been employed in fields such as meteorology [*Raftery et al.*, 2005], climate change science [*Terando et al.*, 2012], and ecology [*Wintle et al.*, 2003] where forecasts or predictions normally must be developed for nonlinear, unbounded systems characterized by partial observability, partial controllability, or both. However, far fewer attempts have been made to *combine* and *propagate* the independent forms of uncertainty in global change predictions where models of climatic responses to fossil fuel emissions are linked to indirect ACC consequences such as changes in large wildfires. Fully propagating uncertainty from ACC scenarios to climate models to ecological models is certainly a nontrivial task. But if predictions are to be of use to decision makers that must adapt or respond to ACC impacts on ecosystems, then attempts to more fully characterize and quantify both climate model and ecological model uncertainty will be necessary.

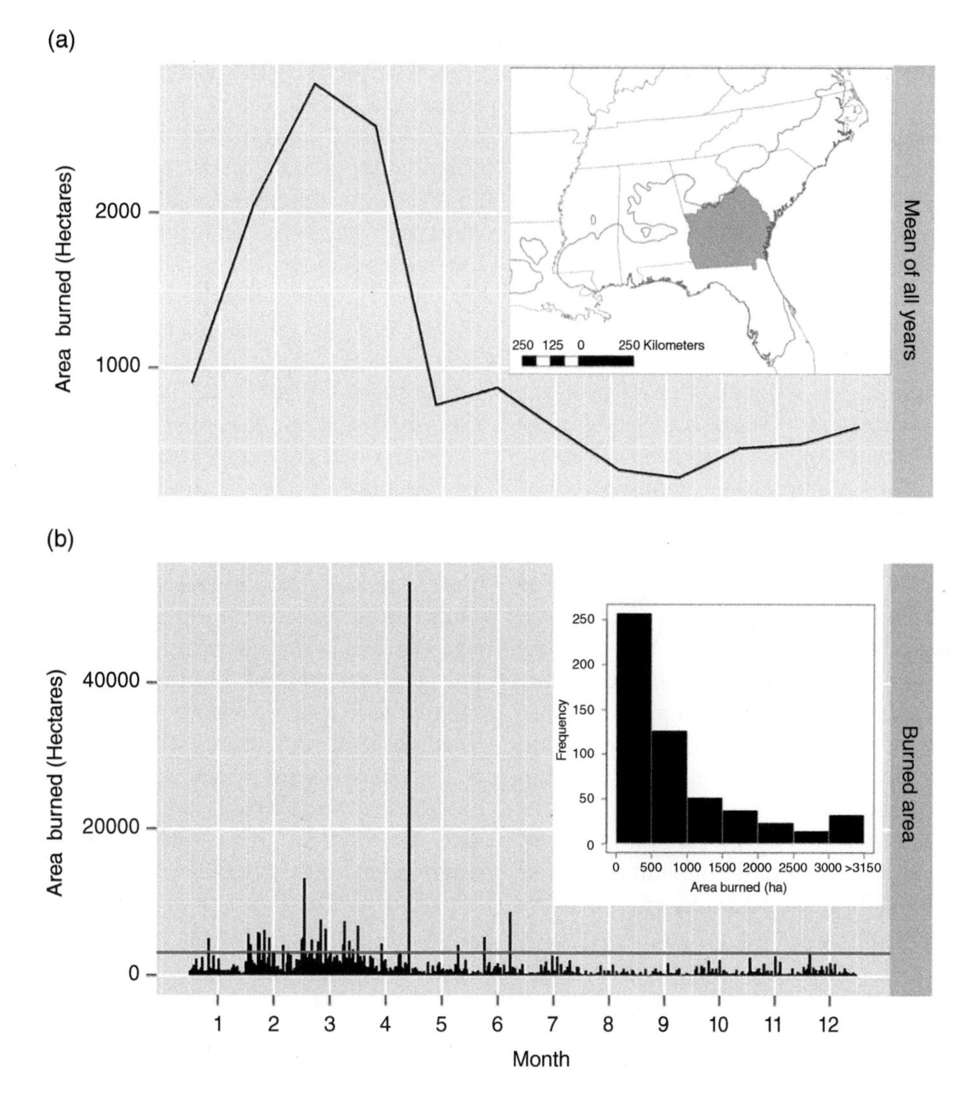

Figure 16.1 Wildfire characteristics in the study region: (a) Mean monthly area burned over the period 1966–2010; (b) monthly area burned over the period of record (1966–2010) and the 96th sample quantile (horizontal red line, equal to 4047 ha or ~10,000 acres). Also shown is the study area in the context of the historic range of the longleaf pine (inset in [a]) and the histogram of observations in the inset in (b).

Here we provide an example of fully propagated structural uncertainty in a global change prediction problem, in which θ represents *months with an unusually large (i.e., extreme) amount of hectares burned by wildfires* (hereafter denoted as EBA for extreme burned area) across the coastal plain region of Georgia, USA (Fig. 16.1). The model is developed in the context of the longleaf pine ecosystem, a critically endangered and highly altered pyrophytic ecosystem in the southeastern United States. In this system, the response of θ to ACC is highly uncertain but requires that decision makers develop conservation plans that account for a changing climate. We use BMA to quantify structural uncertainty among competing ecological models that relate climatic conditions to EBA months, as well as among the actual climate models. Internal climate variability is estimated and incorporated into the BMA when deriving model-specific weights. Scenario uncertainty is assessed by examining the results in light of multiple fossil fuel emissions trajectories. Our work provides a template for global change studies that must account for and propagate uncertainty about the effects of ACC in a way that informs natural resource decision makers.

16.2. DATA

16.2.1. Study Area

We focus on a 9 million ha region of the coastal plain of Georgia, USA (Fig. 16.1). The mix of land uses, the occurrence of large natural areas, as well as the recent occurrence of several extreme fires make the region an ideal study area for our research. The region is dominated by agricultural and forest land uses, with rural communities distributed throughout. Several public landholdings with large natural areas are present, including the US Fish and Wildlife Service's Okefenokee National Wildlife Refuge. Historically, wildfires across the southeast coastal plain were very common and although individual fires could cover vast tracts of land (>1000 km^2), burn intensities were low and crown fires rare [*Frost*, 2006; *Peet*, 2006]. For example, savannas dominated by longleaf pine (*Pinus palustris*) once covered millions of hectares in the southeast United States and burned as frequently as every 1–3 yr [*Brown and Smith*, 2000]. When frequently burned, this ecosystem can harbor among the highest levels of plant diversity in the world [*Peet and Allard*, 1993]. However, because of recent fire suppression and forest conversion, the ecosystem is now critically endangered [*Noss et al.*, 1995]. This fire suppression occurs directly through firefighting activities, but also indirectly through fragmentation of the landscape, which increases fire breaks and decreases the ability of any single ignition to spread across the landscape [*Frost*, 1993]. This active and passive fire suppression has resulted in much less frequent fire return intervals compared to pre-European settlement conditions. *Costanza et al.* [2015], for example, examined wildfire records for the coastal plain regions of Alabama, Georgia, and Florida and found wildfire return intervals that ranged from ~40 yr for surface fires in early succession longleaf pine forests to over 100 yr in degraded sites. Although wildfire return intervals are now much longer resulting in less area burned across the landscape, the peak fire season still occurs in the spring (Fig. 16.1) when large amounts of fine dead fuels (such as grasses and pine needles) from the previous winter are present and when humidity levels are low.

Even though wildfires are a natural part of all ecosystems in the coastal plain, large, high-intensity fires still pose unique risks to ecosystems, human health, and property, and are costly to manage. For example, within the past decade in this study area, the Okefenokee Refuge has experienced two extreme fire events. The Big Turnaround Complex Fire in 2007 was the most costly fire ever in the US Fish and Wildlife Refuge System [*US Fish and Wildlife Service*, 2007], burning over 388,000 acres. Only 4 yr later, the Honey Prairie Fire started in the swamp and burned for over 11 mo, involving over 1400 personnel [*US Fish and Wildlife Service*, 2012]. These wildfires impacted ecological communities [*Beganyi and Batzer*, 2011] and emitted aerosols that can affect human health [*Bhoi et al.*, 2009]. Some fires on the southeast coastal plain can also be intense ground fires that consume large amounts of fuel contained in peat soils [*Poulter et al.*, 2006]. These fires can lead to high carbon emissions, while also retarding peat development and vegetation recovery following the fire [*Poulter et al.*, 2006; *Richardson*, 2012]. Given the impacts of extreme events such as these, the difficulty of managing them, and the potential for climate change to alter wildfire regimes, it is important to quantify the uncertainty associated with wildfires in this region in order to better plan and manage the risks they pose.

16.2.2. Wildfire Observations

Monthly wildfire records were obtained for the period 1966–2010 for counties in the study area (Fig. 16.1d; *Chan*, Georgia Forestry Commission, unpublished data). We aggregated the monthly area burned across all counties into a single value for the entire study area for each month from 1966 to 2010. The wildfire data are based on state agency records of reported wildfires. Thus, some biases may exist in the data; for example, small fires may have been underreported. However, large fires are likely captured in these data, making them well suited for our purpose of modeling extremes. Furthermore, while other studies have reported bias in wildfire records over time due to changes in reporting methods through time [*Littell et al.*, 2009], because the data for Georgia were collected and distributed by a single agency, this bias should be reduced. A bias is apparent for the years prior to 1966, wherein the time series of monthly burned area is less variable (i.e., a "smoother" time series). This likely marks the point at which reporting methods changed, leading to more accurate and precise surveys of wildfire size. The combination of better reporting of wildfires, continual data collection by a single agency, and a stationary (log-transformed) time series from 1966 to 2010, indicates that this dataset is appropriate to use to evaluate the relationship between climate and EBA months.

16.2.3. Climate Data and Models

Climate data are taken from [*Maurer et al.*, 2002], a 12 km resolution gridded dataset of daily meteorological observations from 1950 to 2010. After calculating derived daily climate variables such as diurnal temperature range (see Section 16.3.1), we average the daily values over space and time to obtain monthly time series of these variables for the study region for the same period as the wildfire observations (1966–2010). We use bias-corrected statistically downscaled climate model output for our ensemble of climate projections [*Stoner et al.*, 2013]. The downscaled models are derived from GCM output produced as part of the Third Coupled Model Intercomparison Project [CMIP3; *Meehl et al.*, 2007]. The downscaled models are bias corrected to the [*Maurer et al.*, 2002] dataset, have the same spatial and temporal resolution, and have output available for the period 1960–2099. The composition of the GCM ensemble varies by emission scenario, with 8, 6, and 11 models available for the B1, A1b, and A2 scenarios, respectively (see Table 16.3 in Section 16.4.2).

16.3. METHODS

16.3.1. Structural Uncertainty in the Ecological Model

We build a statistical model to predict extremely large values of the number of hectares burned per month (EBA) Y_t as a function of covariates, X_t Our use of monthly burned area as our measure of interest is based on issues of practicality and utility. First, while the ignition and spread of individual fire events will respond strongly to short-term fluctuations in variables such as wind speed and relative humidity (affecting fuel moisture) or lightning (affecting ignition probability), average monthly climate variables will still reflect the broad-scale climatic conditions that are necessary for extreme wildfires to occur. Second, few long-term wildfire databases exist for the Southeast that encompass a large sample of EBA months. Using the monthly wildfire database available from the Georgia Forestry Commission (see Section 16.2.2) with a period of record that stretches back to the 1960s allows for a more robust estimation of the parameters used to predict extremes. Finally, we argue that projecting changes to EBA months is consistent with the scale or specificity of planning conducted by managers with regard to climatic changes occurring over multiple decades in the future.

The covariates X_t include an intercept (I), the mean monthly diurnal temperature range (T_{dn}), the mean monthly maximum daily temperature (T_{mx}), the detrended with seasonal cycle removed maximum daily temperature (T_{anom}), and the 3 mo (short-term) and 24 mo (long-term) standardized precipitation index (SPI) values (SPI_3, SPI_{24}) [*Guttman*, 1998]. Diurnal temperature range is used to represent atmospheric moisture content, an important consideration for fuel moisture content. More direct measures of moisture content such as relative humidity were ultimately not able to be included because of the limited set of climate variables available in the statistically downscaled climate model dataset. We do not consider ignition source in this model since our interest lies in predicting the occurrence of EBA months. Regardless of whether the ignition was due to lightning or humans, the occurrence of an EBA month reflects an inability to rapidly extinguish wildfires due to fuel and climatic conditions.

Three seasonal terms (S_1, S_2, S_3) are included based on the amount of chilling hours [*Fishman et al.*, 1987], which we use to signal the onset and termination of the growing season, and therefore the changing ratio between dry combustible fuels and wet noncombustible fuels. This also allows for dynamic seasons whose length can change as the climate warms. Season S_1 corresponds to the high fire season in the late winter and early spring, S_2 coincides with the nadir of the fire season in the summer months when moisture levels are high and available combustible fuels are low, and S_3 is the fall and winter season when more acreage is typically burned than in the S_2 season but extremes in monthly area burned are rare (Fig. 16.1). For simplicity, we refer to the seasons S_1, S_2, S_3 as the spring, summer, and fall fire seasons. We also include interactions between each of these 6 covariates and the three season indictors giving a total of 18 covariates in X_t, to account for differences in fire patterns by season.

To estimate the relationship between climate and EBA months, the model is initially fit with X_t taken to be the observed monthly weather values from the *Maurer et al.* [2002] dataset for the years 1966–2010. Because of the large number of covariates and the small number of EBA months, we consider several statistical models defined by subsets of the 18 variables. We averaged over predictions from the statistical models using Bayesian Model Averaging (BMA). BMA [*Raftery et al.*, 1997] allows for a probabilistic estimation of structural uncertainty between models by producing a weighted predictive distribution according to relative skill of each candidate model (cf. *Wintle et al.* [2003] for a discussion of BMA in ecological applications). The BMA predictive model for observation Y_t is:

$$f(Y_t \mid Y_0) = \sum_{k=1}^{k} f(Y_t \mid M_k) Prob(M_k \mid Y_0), \quad (16.1)$$

where $M_1, \ldots, M_k$ are the k competing models, and Y_o is the training data used to fit the model.

We use a points-above-threshold analysis [*Cole*, 2001] for extreme events (EBA months). Observations are considered extreme if they exceed the threshold T hectares. We use fivefold cross-validation to select a value of T that maximizes model skill based on comparison of Brier scores [*Brier*, 1950] and quantile scores. The covariates are included in both the probability of exceeding the threshold and the severity of exceedances. We use logistic regression to model exceedance probabilities

$$logit\left[Prob(Y_t > T)\right] = X_t^T \beta_1 \quad (16.2)$$

Exceedances of the threshold T are modeled using the generalized Pareto distribution (GPD). The GPD has two parameters: scale $\sigma_t > 0$ and shape ξ. Under the GPD model, the conditional probability of exceeding *y* hectares burned is

$$Prob\left(Y_t > y \middle| Y_t > T\right) = \left[1 + \frac{\xi}{\sigma_t}(y - T)\right]^{-\frac{1}{\xi}} \quad (16.3)$$

The shape parameter controls the tail behavior, with large and positive ξ corresponding to a heavy tail and thus very large extreme values and negative ξ corresponding to a bounded distribution without severe extremes. In our analysis, we allow the scale parameter to vary with covariates, $\log(\sigma_t) = X_t^T \beta_2$. The shape parameter is held constant over time because this parameter is notoriously difficult to estimate. We assume all subsets of covariates are equally likely a priori and that the same subset of covariates appears in the logistic probability of an EBA month and the GPD log scale parameter. The covariates are centered and scaled to have mean zero and variance one, and the elements of β_j have prior normal distributions of $N(0,\sigma^2)$, where the variances σ^2 have inverse-gamma priors of (0.1,0.1). Finally, the GPD shape parameter has a $N(0, 0.25^2)$ prior. We note that regardless of the *decision-dependent* definition of a catastrophic fire (tied to actual economic and environmental losses), we assume that our model-based selection of *T* will be sufficiently low to permit inference on the actual decision-relevant threshold.

16.3.2. Structural and Emissions Uncertainty in the Climate Models

The ensembles of downscaled climate models contain structural uncertainty due to different parameterizations and model resolutions, leading to different predictions given the same climate forcings. We again use BMA to account for this structural uncertainty. *Terando et al.*, [2012], used a bootstrapping procedure to gauge climate model prediction skill for long-term evolving temperature trends. In brief, training datasets of the estimated trend over a 50 yr period (1961–2010) are constructed for the observations and downscaled GCM output. The temporal trends form the basis of the estimated model skill that is used to estimate the BMA weights. The bootstrapping procedure is used to estimate the uncertainty around the trend by sampling 1000 times from the residuals of a best fit statistical model of the trend in the observations and climate model output. Both linear and polynomial models are fit based on AIC selection and after accounting for autocorrelation in the residuals. These resampled residuals are then recombined with the trend estimate and a new trend estimate is produced. The result is 1000 pseudo time series for each observed and modeled climate variable that reflects decadal climate variability and the ability of the downscaled GCMs to reproduce the observed time-evolving trends. This method limits spuriously low or high model weights that can result from using BMA with climate models whose experiments use observed climate forcings, but are not meant to simulate the actual observed weather. It also implicitly takes into account the internal unforced variability of the climate system (see *Terando et al.* [2012] for further details). We extend the univariate model in that work to the multivariate case and evaluate model skill for the downscaled ensemble for maximum temperature, diurnal temperature, 3 mo SPI, 24 mo SPI, and monthly precipitation.

Scenario uncertainty is evaluated in the standard manner by comparing results for multiple fossil fuel emission scenarios. We compare three emission scenarios used in the Fourth Assessment Report of the Intergovernmental Panel on Climate Change (AR4 IPCC) corresponding to a low (B1), medium (A1b), and high (A2) future level of fossil fuel emissions [*Nakicenovic et al.*, 2000]. Based on our structural uncertainty characterization in the joint ecological-climatological model, we substitute the BMA posterior distribution for X_t to make predictions for two 30 yr periods, present (1980–2009) and future (2070–2099), conditional on the chosen emission scenario, to evaluate the potential for ACC to increase the number of EBA months in this region.

16.4. RESULTS

16.4.1. Cross-Validation Results and GPD Fitted Parameters

The Brier scores and quantile scores from the fivefold cross-validation indicated that a threshold *T* of 1465 ha burned provides the best estimate of the GPD parameters for predicting EBA months with 4047 or more hectares burned. This threshold corresponds to the 0.8 sample quantile over the 45 yr observation period. Brier scores were similar for the three quantiles (0.5, 0.8, 0.95) that were tested against six extreme quantiles for prediction (Table 16.1). However the quantile scores were lowest (lower error) for the 0.8 quantile across all EBA thresholds above the prediction 0.8 quantile. Given the larger relative differences between the quantile scores that favored using the 0.8 quantile threshold compared to the Brier scores, we used the 0.8 quantile to fit the GPD parameters. This threshold likely represents a compromise between lower thresholds that increase the number of observations available for fitting the model and higher thresholds that are more representative of the processes that are responsible for EBA months.

The 90% posterior interval for the GPD shape parameter ξ is [0.11, 0.44]. The shape is positive so the distribution of extreme monthly burned area is unbounded; however, the shape parameter is less than 0.5 so the distribution is not extremely heavy-tailed and has finite mean and variance. The posterior probabilities that each covariate in X_t is included in the statistical model of EBA months range from a low of 0.08 (T_{anom}. S_1) to 1 for the three seasonal intercept terms and diurnal temperature in the high fire season (Table 16.2). Consistent with the peak fire season occurring when conditions are dry and combustible fuels are present, there is strong support for inclusion of the spring season diurnal temperature (positive posterior mean values) and three month SPI covariates (negative posterior mean values). Note that the estimated posterior means

Table 16.1 Brier Scores and Quantile Scores for the Ecological Model Based on Fivefold Cross Validation for Three EBA Thresholds Tested Against Six Prediction Thresholds

		Threshold Tested					
Score Criteria	*T*	0.95	0.96	0.97	0.98	0.99	0.995
Brier	0.5	0.040	0.033	0.028	0.018	0.011	0.006
	0.8	0.041	0.034	0.029	0.019	0.011	0.006
	0.95	0.043	0.035	0.029	0.019	0.012	0.006
Quantile	0.5	2215	2036	1807	1484	1015	730
	0.8	1536	1382	1210	1025	809	643
	0.95	2891	2551	2175	1748	1235	911

Note: Lower scores in both cases indicate less error between the model prediction and the verifying observations.

Table 16.2 Estimated Bayesian Model Averaging Probabilities ($[P(\beta)]_j \neq 0$) for the Climate and Seasonal Covariates (X_t) and Best-Fit Values for the β_j Used to Estimate the Probability of Exceeding a Threshold (Hectares Burned Month^{-1}), β_1, and the Magnitude of the Extreme Burned Area, β_2.

		β_1			β_2		
X_t	$P(\beta_j \neq 0)$	μ	σ	90% CI	μ	σ	90% CI
$I.S_1$	1	−0.78	0.31	[−1.31,−0.32]	7.51	0.30	[6.99,7.97]
$T_{dn}.S_1$	1	1.00	0.28	[0.56,1.48]	0.50	0.21	[0.16,0.84]
$T_{mx}.S_1$	0.29	−0.10	0.24	[−0.65,0]	−0.09	0.21	[−0.57,0]
$T_{anom}.S_1$	0.08	0.00	0.05	[0,0]	0.00	0.04	[0,0]
$SPI_3.S_1$	0.69	−0.28	0.26	[−0.72,0]	−0.25	0.21	[−0.59,0]
$SPI_{24}.S_1$	0.1	0.02	0.08	[0,0.19]	0.01	0.05	[0,0.06]
$I.S_2$	1	−3.02	0.74	[−4.37,−2.09]	7.06	0.59	[5.99,7.89]
$T_{dn}.S_2$	0.98	0.98	0.40	[0.33,1.64]	0.36	0.31	[−0.1,0.9]
$T_{mx}.S_2$	0.67	0.68	0.76	[0,2.08]	0.24	0.50	[−0.39,1.2]
$T_{anom}.S_2$	0.38	0.11	0.28	[−0.14,0.73]	0.18	0.32	[0,0.89]
$SPI_3.S_2$	0.46	0.18	0.28	[0,0.78]	0.13	0.25	[−0.1,0.67]
$SPI_{24}.S_2$	0.26	−0.03	0.14	[-0.34,0.08]	0.05	0.17	[−0.03,0.44]
$I.S_3$	1	−2.52	0.62	[−3.65,−1.72]	7.28	0.56	[6.24,8.07]
$T_{dn}.S_3$	0.27	0.10	0.25	[0,0.71]	−0.01	0.21	[−0.42,0.32]
$T_{mx}.S_3$	0.57	−0.39	0.48	[−1.31,0]	−0.14	0.42	[−0.94,0.44]
$T_{anom}.S_3$	0.28	−0.04	0.17	[−0.42,0.1]	−0.08	0.19	[−0.53,0]
$SPI_3.S_3$	0.66	−0.33	0.34	[−0.93,0]	−0.35	0.37	[−1,0]
$SPI_{24}.S_3$	0.27	0.05	0.16	[−0.04,0.41]	−0.09	0.22	[−0.58,0]

Note: The covariates are included in the model estimation of both the probability of exceeding the EBA threshold and the severity of EBA.

for the spring season T_{dn} are the largest in absolute terms amongst all nonintercept covariates. It is interesting that T_{mx} and SPI_3 have the largest probabilities in the fall season, but T_{mx} has negative posterior means, suggesting that while "warm and dry" conditions increase the probability of EBA months in the spring season, "cool and dry" conditions are more important in the fall (but with fewer and less extreme EBA months overall).

16.4.2. GCM BMA Weights

The posterior climate model weights for the full ensemble in the A2 scenario include many low model probabilities (Table 16.3). The median model weight is 0.06, compared with the prior probability of 0.09 for an equally weighted model ensemble (where $n = 11$). Only two models have posterior probabilities of 0.18 or higher (twice the prior probability). Model weights are rebalanced for the other two emission scenarios since fewer GCMs are available, resulting in model weights as high as 0.39 in the case of the A1b scenario.

16.4.3. Projected Extreme Fires

Figure 16.2 shows the histogram of simulations and PDFs of the posterior predictive distributions for the number of months with 4047 ha (10,000 acres) or more burned over a 30 yr period (corresponding to the 0.96 quantile for the observation period). In Figure 16.2a, the histogram of 10,000 simulation results from the cross-validated predictive distribution for the 30 yr period that overlaps with the observations, indicates good agreement between the ecological model and the actual observed number of extreme burn months (vertical dashed line, equal to 12 EBA months). Similarly, the BMA-based joint ecological-climatological model PDF shows that the verifying observation is near the fiftieth percentile of the downscaled ensemble over the same 30 yr period (Fig. 16.2b).

Table 16.3 BMA Model Weights for Each Emissions Scenario for the Empirically Downscaled Members of the CMIP3 GCM Ensemble Based on Observations and Simulations From the 20C3M Experiment

		Model Weights		
n	Model	A2	B1	A1b
1	ccsm3.0	0.17	0.19	
2	cgcm3.1(T47)	0.02	0.03	0.03
3	cgcm3.1(T63)	0.12	0.13	0.19
4	cnrm	0.01	0.01	0.02
5	echam5	0.18	0.20	0.28
6	echo	0.25	0.28	0.39
7	gfdl-cm2.0	0.04	0.04	
8	gfdl-cm2.1	0.11	0.12	
9	hadcm3	0.01		
10	hadgem	0.06		0.09
11	pcm	0.04		

Note: Blank spaces indicate the absence of GCM output for that particular emissions scenario.

For the forward simulations depicting the projected probability of EBA months at the end of the 21st century (2070–2099), "long tails" in the BMA PDFs (solid black lines) are present for all three emissions scenarios, indicating an increased probability for more frequent EBA months (Fig. 16.2c–e). The 95% projection intervals for the B1 (low emissions level), A1b (medium emissions level), and A2 (high emissions level) scenarios are [−5, 18], [−4, 33], and [−6, 28], respectively. The 50th percentile value for the B1 scenario shows 0.7 more EBA months in the future, 0.2 more EBA months for the A1b scenario, and 0.5 fewer months for the A2 scenario. Accordingly, all scenarios indicate that roughly half of the PDF area lies above zero and that the long heavy tails are skewed toward an increase in extreme wildfire months. The apparently nonlinear relationship between the upper bounds of the projection interval and the emission scenario (i.e., a longer tail in the A1b scenario compared to the A2 scenario) is due to the reduced number of GCMs available for the A1b scenario, which causes a higher weight to be placed on one or more models that project more frequent EBA months. When the same set of models is used for the B1, A1b, and A2 emissions scenarios, the 95% intervals are [−5, 21], [−4, 31], and [−5, 30], respectively, resulting in very similar results for the mid- and high-emissions scenarios while still showing smaller increases in EBA months for the low-emissions scenario.

16.5. DISCUSSION

The potential for increases in EBA months is likely due to longer and hotter summer wildfire seasons that increase the probability for large fires relative to historic conditions. Most GCM projections of the diurnal temperature range remain constant or show declines over time, indicating that this variable is not contributing to the projected increases, although it could be contributing to the possibility of decreases in EBA months. Overall, the positive long-tailed PDF projections suggest a trend toward the summer season becoming a more important source for EBA months.

Although large uncertainties are associated with these projections, we would argue that the inclusion of structural uncertainty in both the candidate ecological models and the downscaled GCMs is still a more informative

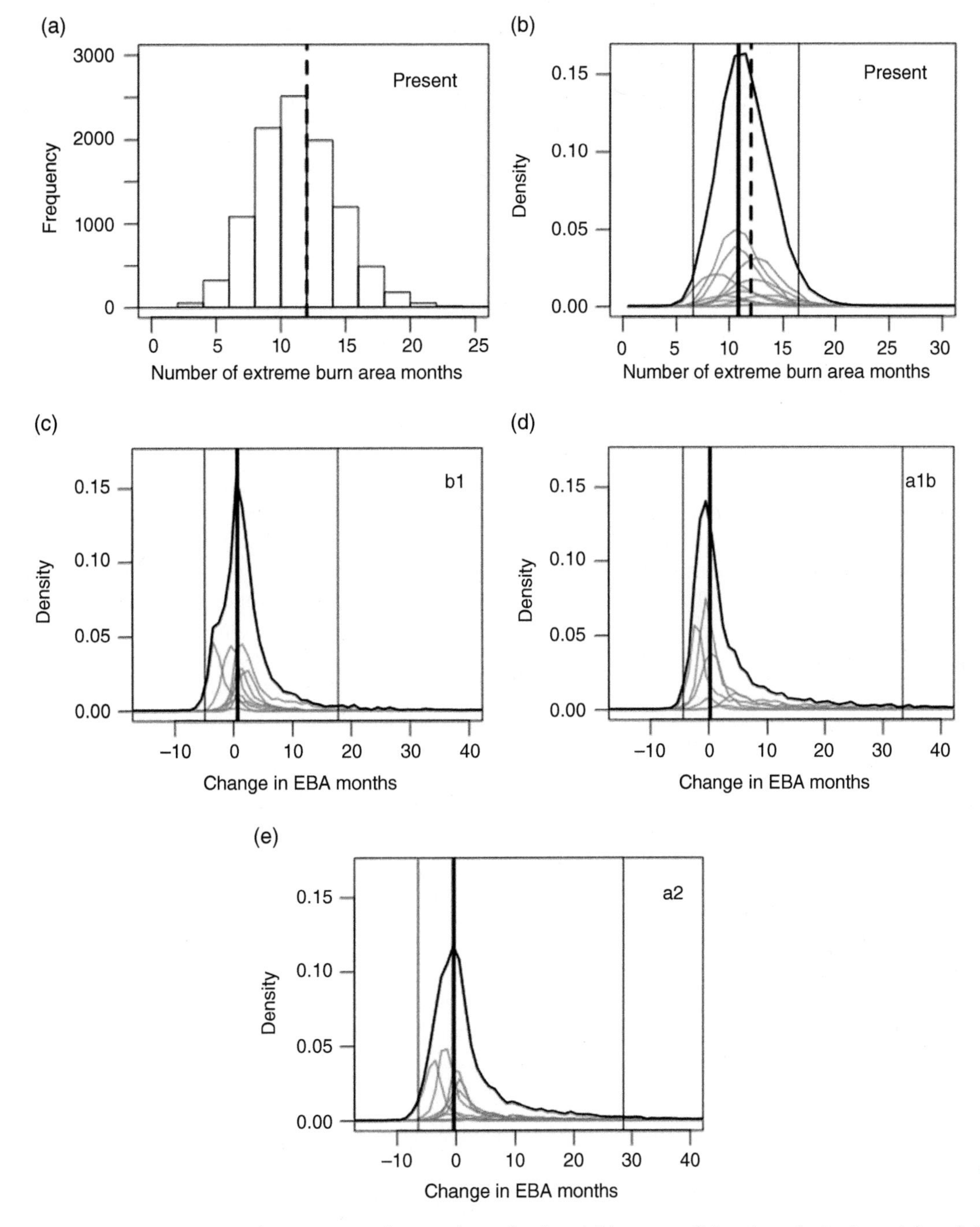

Figure 16.2 (a) Histogram of 10,000 simulations from the five-fold cross-validated ecological model and the verifying observation (vertical dashed line) of the number of EBA months (>4047 ha) for the period 1980–2009. (b) BMA PDF of the predicted number of EBA months for the period 1980–2009 for the joint ecological-climate model. Verifying observation is the dashed vertical line, fiftieth percentile is the solid thick vertical line, thin vertical lines cover the 95% interval, and grey lines are individual GCM PDFs. (c) BMA PDF of projected difference between the number of EBA months in 2070–2099 versus 1980–2009 for the b1 emission scenario. (d) Same as (c) but for the a1b emission scenario. (e) Same as (c) but for the a2 emission scenario.

and faithful depiction of our current knowledge about how ACC could impact EBA months. In contrast, simple point projections would likely result in a truncated uncertainty range that is underdispersive and overconfident [*Terando et al.*, 2012]. The long tails showing nonnegligible probabilities of increasing EBA months indicate that large damages in the future cannot be completely discounted from a risk assessment by a decision maker, even though the entirety of the PDFs spans positive (more EBA months) and negative values (fewer EBA months). Somewhat analogous to the uncertain estimates of the Earth's climate sensitivity to a doubling of CO_2 [*Knutti and Hegerl*, 2008], decision-relevant predictions with PDFs that are characterized by asymmetric tails may warrant special scrutiny of the potential consequences of the more extreme, but still possible, future state of nature.

Predictions about climate change are characterized by deep uncertainty [*Lempert*, 2002]. This is an inherent characteristic of climate change models and represents a major challenge for decision analysts who are trying to appropriately estimate and incorporate uncertainty into decision frameworks, and for decision makers, who ultimately have to choose among alternative actions. We address one such consequence, namely, extreme wildfire months in the southeastern United States. Extreme events represent a unique class of possible outcomes in the advent of climate change. Uncertainty is compounded by the lack of observations that can be used to predict future occurrences, and this low probability of occurrence could lead to an undervaluation of importance in decision problems. Such events may have profound consequences on many human endeavors, including land management decisions, and thus, their inclusion and treatment in decision problems is warranted [*Lempert et al.*, 2002].

Much of the impetus for characterizing and quantifying the different sources of uncertainty is for use in decision making wherein we directly incorporate the different sources of uncertainty into predictive models evaluating the consequences of potential actions [*Walters*, 1986]. Ignoring this uncertainty in decision analysis results in overconfidence not only in the effects of our actions, but also in the robustness of our decisions. Therefore, we subscribe to and support the idea that providing probabilistic measures of events (e.g., EBA months) using all available information and uncertainties is an essential step in decision making.

Incorporating uncertainties derived from climate change models has not been handled explicitly in most land-management decision problems, although several authors have proposed a framework for doing so [cf., *Conroy et al.*, 2011; *Nichols et al.*, 2011], while recognizing two potential challenges. The first is the realization that land managers do not have control over actions (e.g., emission controls) that might affect the course of climate change dynamics, and the second is that quantifying and predicting changes in the climate system is not a trivial matter. We have attempted to address the latter by coupling an estimate of climatic change uncertainty with the *extradisciplinary* structural uncertainty present in ecological models to predict how climate change effects the system dynamics of interest (e.g., wildfire).

Several caveats and limitations apply to our approach used for quantifying uncertainty in global change prediction problems. First, our model projections assume that fuel characteristics remain stationary through time. If fuel loads or fuel types change, even moderately, the projections are less likely to hold. Second, suppression (both direct and indirect through urbanization) could change through time, which would have an attendant effect on the changing EBA probabilities.

As a final caveat, we stress that model weighting continues to be a subject of great debate within the climate impacts community [cf., *Knutti*, 2010]. Prior studies have shown that applying model weights to GCMs can result in overconfident projections [*Terando et al.*, 2012] while equally weighted projections did not. Such a risk would also apply in this study and future work could incorporate projections that also derive projections from equally weighted GCM ensembles. However, as noted in *Knutti* [2010], the line between unweighted and weighted model ensembles is not so clear cut considering that other, implicit forms of model weighting are commonly practiced. For instance, older generations of GCMs are continually cast aside in climate impact studies, which in effect places a weight of zero on these GCMs. In addition, some GCMs may share similar "DNA" [*Masson and Knutti*, 2011], which in effect results in higher weights being placed on models with shared or very similar parameterizations and subroutines.

Our view is that fully using the information contained within the observations in order to construct a model-weighted prediction still has merit if done judiciously and with careful consideration of the trade-offs. From a decision-making perspective, a high utility outcome might involve a weighted prediction that is frequently updated as new data (and new models) are received, thus allowing managers to estimate new expected values for the portfolio of actions available to them. Such situations currently exist in areas such as hurricane forecasts [*Hamill et al.*, 2011] and wildlife harvesting limits [*Nichols et al.*, 2007]. Bringing this type of system to the realm of global change problems is daunting but perhaps necessary for realizing successful climate adaptation strategies. This study continues toward this goal by providing examples of how model weighting could be undertaken in joint ecology-climate modeling problems.

In this chapter, we presented an explicit framework that couples multiple sources of uncertainty from different disciplines into probabilistic global change projections. This coupling could have significant implications for decision makers as it represents a quantitative end point that can be incorporated into a myriad of land management and hazard mitigation decision problems. Our intent was to underscore the potential value of this estimation problem in decision making and stimulate further discussion and research in this area.

Further advances in uncertainty estimation and model weighting will provide an opportunity to improve the quality of uncertainty estimates for optimization of actions or for the development of robust climate adaptation strategies.

ACKNOWLEDGMENTS

We thank the two anonymous reviewers and J. Littell for their comments and insights, which greatly improved this work. Any use of trade, product, or firms' names is for descriptive purposes only and does not imply endorsement by the US government.

REFERENCES

Beganyi, S. R., and D. P. Batzer (2011), Wildfire induced changes in aquatic invertebrate communities and mercury bioaccumulation in the Okefenokee Swamp, *Hydrobiologia*, *669*(1), 237–247; doi:10.1007/s10750-011-0694-4.

Bhoi, S., J. Qu, and S. Dasgupta (2009), Multi-sensor study of aerosols from 2007 Okefenokee forest fire, *J. Appl. Remote Sens.*, *3*(1), 031501; doi:10.1117/1.3078070.

Brier, G. W. (1950), Verification of forecasts expressed in terms of probability, *Monthly Weather Rev.*, *78*(1), 1–3.

Brocker, J., and L. a. Smith (2008), From ensemble forecasts to predictive distribution functions, *Tellus A*, *60*(4), 663–678; doi:10.1111/j.1600-0870.2008.00333.x.

Brown, J. K., and J. K. Smith (2000), *Wildland fire in ecosystems: Effects of fire on flora*, Gen. Tech., edited by J. K. Brown and J. K. Smith, USDA Forest Service, Rocky Mountain Research Station, Ogden, UT.

Certini, G. (2005), Effects of fire on properties of forest soils: A review, *Oecologia*, *143*(1), 1–10; doi:10.1007/s00442-004-1788-8.

Cole, S. (2001), *An Introduction to Statistical Modeling of Extreme Values*, Springer, London.

Conroy, M. J., M. C. Runge, J. D. Nichols, K. W. Stodola, and R. J. Cooper (2011), Conservation in the face of climate change: The roles of alternative models, monitoring, and adaptation in confronting and reducing uncertainty, *Biol. Conserv.*, *144*(4), 1204–1213; doi:10.1016/j.biocon.2010.10.019.

Costanza, J. K., A. J. Terando, A. J. McKerrow, and J. A. Collazo (2015), Modeling climate change, urbanization, and fire effects on *Pinus palustris* ecosystems of the southeastern U.S., *J. Environ. Man.*, *151*, 186–199; doi:10.1016/j.jenvman.2014.12.032.

Dennison, P. E., S. C. Brewer, J. D. Arnold, and M. A. Moritz (2014), Large wildfire trends in the western United States, 1984–2011, *Geophys. Res. Lett.*, *41*(8), 2928–2933; doi:10.1002/2014GL059576.

Draper, D. (1995), Assessment and propagation of model uncertainty, *J. R. Statist. Soc. B*, *57*(1), 45–97.

Draper, D. (2013), Bayesian model specification: Heuristics and examples, Oxford Scholarship, 702, in *Bayesian Theory and Applications*, edited by P. Damien, P. Dellaportas, N. G. Polson, and D. A. Stephens, Oxford University Press, Oxford.

Fishman, S., A. Erez, and G. A. Couvillon (1987), The temperature dependence of dormancy breaking in plants: Mathematical analysis of a two-step model involving a cooperative transition., *J. Theor. Biol.*, *124*(4), 473–483.

Fried, J. S., M. S. Torn, and E. Mills (2004), The impact of climate change on wildfire severity: A regional forecast for northern California, *Clim. Change*, *64*(1–2), 169–191; doi:10.1023/B:CLIM.0000024667.89579.ed.

Frost, C. C. (1993), Four centuries of changing landscape patterns in the longleaf pine ecosystem, edited by S. Hermann, *Proc. Tall Timbers Fire Ecol. Conf.*, *18*, 17–43.

Frost, C. C. (2006), History and future of the longleaf pine ecosystem, 9–48, in *Longleaf Pine Ecosystems: Ecology, Management, and Restoration.*, edited by S. Jose, E. Jokela, and D. Miller, Springer, New York.

Girardin, M. P., and M. Mudelsee (2008), Past and future changes in Canadian boreal wildfire activity, *Ecol. Appl.*, *18*(2), 391–406; doi:10.1890/07-0747.1.

Gneiting, T., F. Balabdaoui, and A. E. Raftery (2007), Probabilistic forecasts, calibration and sharpness, *J. R. Stat. Soc. B*, *69*(2), 243–268.

Guttman, N. B. (1998), Comparing the Palmer Drought Index and the Standardized Precipitation Index, *J. Amer. Water Resour. Assoc.*, *34*(1), 113–121.

Hamill, T. M., J. S. Whitaker, M. Fiorino, and S. G. Benjamin (2011), Global Ensemble predictions of 2009's tropical cyclones initialized with an Ensemble Kalman Filter, *Monthly Weather Rev.*, *139*(2), 668–688; doi:10.1175/2010MWR3456.1.

Hawkins, E., and R. Sutton (2009), The potential to narrow uncertainty in regional climate predictions, *Bull. Amer. Meteor. Soc.*, *90*(8), 1095–1107.

Knutti, R. (2010), The end of model democracy?, *Clim. Change*, *102*(3–4), 395–404; doi:10.1007/s10584-010-9800-2.

Knutti, R., and G. C. Hegerl (2008), The equilibrium sensitivity of the Earth's temperature to radiation changes, *Nat. Geosci.*, *1*(11), 735–743; doi:10.1038/ngeo337.

Lempert, R., S. Popper, and S. Bankes (2002), Confronting surprise, *Soc. Sci. Comput. Rev.*, *20*(4), 420–440; doi:10.1177/089443902237320.

Lempert, R. J. (2002), A new decision sciences for complex systems., *Proc. Nat. Acad. Sci., 99, Suppl 3*, 7309–13; doi:10.1073/pnas.082081699.

Litschert, S. E., T. C. Brown, and D. M. Theobald (2012), Historic and future extent of wildfires in the Southern Rockies ecoregion, USA, *For. Ecol. Man.*, *269*, 124–133; doi:10.1016/j.foreco.2011.12.024.

Littell, J. S., D. McKenzie, D. L. Peterson, and A. L. Westerling (2009), Climate and wildfire area burned in western U.S. ecoprovinces, 1916–2003, *Ecol. Appl.*, *19*(4), 1003–1021; doi:10.1890/07-1183.1.

Liu, Y., S. Goodrick, and W. Heilman (2014), Wildland fire emissions, carbon, and climate: Wildfire-climate interactions, *For. Ecol. Man.*, *317*(SI), 80–96; doi:10.1016/j.foreco.2013.02.020.

Lorenz, E. N. (1963), Deterministic nonperiodic flow, *J. Atmos. Sci.*, *20*(2), 130–141.

Masson, D., and R. Knutti (2011), Climate model genealogy, *Geophys. Res. Lett.*, *38*(8), n/a–n/a; doi:10.1029/2011GL046864.

Maurer, E. P., A. W. Wood, J. C. Adam, D. P. Lettenmaier, and B. Nijssen (2002), A long-term hydrologically based dataset of land surface fluxes and states for the conterminous United States, *J. Climate*, *15*(22), 3237–3251.

Meehl, G. a., C. Covey, K. E. Taylor, T. Delworth, R. J. Stouffer, M. Latif, B. McAvaney, and J. F. B. Mitchell (2007),

The WCRP CMIP3 multimodel dataset: A new era in climate change research, *Bull. Amer. Meteor. Soc.*, *88*(9), 1383–1394; doi:10.1175/BAMS-88-9-1383.

Nakicenovic, N., et al. (2000), *Special report on emissions scenarios: A special report of Working Group III of the Intergovernmental Panel on Climate Change*, edited by N. Nakicenovic and R. Swart, Cambridge University Press, Cambridge.

Nichols, J. D., M. C. Runge, F. A. Johnson, and B. K. Williams (2007), Adaptive harvest management of North American waterfowl populations: a brief history and future prospects, *J. Ornithol.*, *148*(S2), 343–349; doi:10.1007/s10336-007-0256-8.

Nichols, J. D., M. D. Koneff, P. J. Heglund, M. G. Knutson, M. E. Seamans, J. E. Lyons, J. M. Morton, M. T. Jones, G. S. Boomer, and B. K. Williams (2011), Climate change, uncertainty, and natural resource management, *J. Wild. Man.*, *75*(1), 6–18; doi:10.1002/2010-33.

Noss, R., E. LaRoe, and J. Scott (1995), Endangered ecosystems of the United States: a preliminary assessment of loss and degradation, *Biological Report 28*, US Department of the Interior National Biological Service, Washington, D.C.

Palmer, T. N., G. J. Shutts, R. Hagedorn, F. J. Doblas-Reyes, T. Jung, and M. Leutbecher (2005), Representing model uncertainty in weather and climate prediction, *Ann. Rev. Earth Planet. Sci.*, *33*(1), 163–193; doi:10.1146/annurev.earth.33.092203.122552.

Peet, R. K. (2006), Ecological classification of longleaf pine woodlands, 51–94, in *Longleaf Pine Ecosystems: Ecology, Management, and Restoration.*, edited by S. Jose, E. Jokela, and D. Miller, Springer, New York.

Peet, R. K., and D. J. Allard (1993), Longleaf pine vegetation of the Southern Atlantic and Eastern Gulf Coast Regions: A preliminary classification, in *The Longleaf Pine ecosystem: Ecology, restoration and management*, edited by S. M. Hermann, *Proc. Tall Timbers Fire Ecol. Conf. 18*, 45–81.

Poulter, B., N. L. Christensen, and P. N. Halpin (2006), Carbon emissions from a temperate peat fire and its relevance to interannual variability of trace atmospheric greenhouse gases, *J. Geophys. Res.*, *111*(D6), D06301; doi:10.1029/2005JD006455.

Raftery, A. E., D. Madigan, and J. A. Hoeting (1997), Bayesian model averaging for linear regression models, *J. Amer. Stat. Assoc.*, *92*(437), 179–191.

Raftery, A. E., T. Gneiting, F. Balabdaoui, and M. Polakowski (2005), Using Bayesian model averaging to calibrate forecast ensembles, *Monthly Weather Rev.*, *133*(5), 1155–1174; doi:10.1175/MWR2906.1.

Richardson, C. J. (2012), Pocosins evergreen shrub bogs of the Southeast, 189–202, in *Wetland Habitats of North America: Ecology and Conservation Concerns*, edited by D. P. Batzer and A. Baldwin, University of California Press, Berkeley, CA.

Ryan, K. C., E. E. Knapp, and J. M. Varner (2013), Prescribed fire in North American forests and woodlands: history, current practice, and challenges, *Front. Ecol. Environ.*, *11*(s1), e15–e24; doi:10.1890/120329.

Sriver, R. L., N. M. Urban, R. Olson, and K. Keller (2012), Toward a physically plausible upper bound of sea-level rise projections, *Clim. Change*, *115*(3–4), 893–902; doi:10.1007/s10584-012-0610-6.

Stoner, A. M. K., K. Hayhoe, X. Yang, and D. J. Wuebbles (2013), An asynchronous regional regression model for statistical downscaling of daily climate variables, *Int. J. Climatol.*, *33*(11), 2473–2494; doi:10.1002/joc.3603.

Terando, A., K. Keller, and W. E. Easterling (2012), Probabilistic projections of agro-climate indices in North America, *J. Geophys. Res.*, *117*(D8), D08115; doi:10.1029/2012JD017436.

US Fish and Wildlife Service (2007), OkefenokeeFire, *Big Turnaround Fire Biggest Most Expens. FWS*, available from http://www.fws.gov/fire/news/ga/big_turnaround_fire.shtml.

US Fish and Wildlife Service (2012), Honey Prairie Fire Declared Out (912), available from http://www.fws.gov/okefenokee/PDF/honey prairie fire declared out.pdf.

Walsh, J., et al. (2014), Our changing climate, 19–67, in *Climate Change Impacts in the United States: The Third National Climate Assessment*, edited by J. M. Melillo, T. C. Richmond, and G. W. Yohe, US Global Change Research Program.

Walters, C. (1986), *Adaptive Management of Renewable Resources*, Macmillan Pub. Co., New York.

Wintle, B. A., M. A. McCarthy, C. T. Volinsky, and R. P. Kavanagh (2003), The use of Bayesian model averaging to better represent uncertainty in ecological models, *Conserv. Biol.*, *17*(6), 1579–1590; doi:10.1111/j.1523-1739.2003.00614.x.

17

Simulating Vegetation Change, Carbon Cycling, and Fire Over the Western United States Using CMIP5 Climate Projections

D. Bachelet,[1,2] T. Sheehan,[1] K. Ferschweiler,[1] and J. Abatzoglou[3]

ABSTRACT

For the "Integrated Scenarios of Climate, Hydrology, and Vegetation for the Northwest" project, the dynamic global vegetation model MC2 was run at 4 km resolution with a suite of climate futures from the Coupled Model Intercomparison Project (CMIP5) downscaled using the statistical downscaling approach Multivariate Adaptive Constructed Analogs over the western United States. While all climate models project warmer conditions, they differ in their projections of the seasonality and magnitude of rainfall. The vegetation model is sensitive to the water available for plant production and soil organic matter decomposition, but also for fuel buildup and wildfire occurrence. Results show large shifts in vegetation toward warmer types (e.g., temperate to subtropical forest types, warm subtropical grasslands replacing cool temperate grasslands) and an expansion of forest/woodland types enhanced by local increases in annual precipitation and a moderate CO_2 effect on water-use efficiency, important when water availability declines. Complex interactions of climate and disturbance drive the large changes the model is simulating with much geographic patchiness due in part to soil types as well as temporal variability due to changes in rainfall seasonality.

17.1. INTRODUCTION

As a result of both natural and anthropogenic forcings, altered seasonal weather patterns [*Anderson et al.*, 2010; *Nayak et al.*, 2010; *Gillies et al.*, 2012; *Wang et al.*, 2014] and concomitant changes in long-term temperature and precipitation regimes [*Abatzoglou and Barbero,* 2014; *Kam et al.*, 2014] are having a pronounced effect on ecosystems and the services they provide throughout the western United States [*Abatzoglou and Kolden,* 2011; *Brutsaert,* 2011; *Chmura et al.*, 2011; *Jiménez et al.*, 2011; *Munson et al.*, 2011; *Adams et al.*, 2012, *Bowker et al.*, 2012, *Reed et al.*, 2012; *Anderegg et al.*, 2013; *Abatzoglou and Barbero,* 2014; *Sherriff et al.*, 2014]. Such ongoing changes are posing formidable challenges for those who manage the land and its variety of ecosystem services. One critical piece of knowledge required by natural resource managers is the understanding of how the climate, the water cycle, and the vegetation will change in the future and how they will be affected by natural and human-caused disturbances.

Land managers have recently begun considering how to adapt to and mitigate potential climate-related impacts [e.g., *GAO*, 2007; *Furniss et al.*, 2010; *CEQ,* 2010; *Peterson et al.*, 2011]. A variety of regional studies in the western United States have examined potential climate-driven changes in the distribution of vegetation cover types and species [e.g., *Lenihan et al.*, 2008 a,b; *Coops and Waring* 2011; *Rehfeldt et al.*, 2012; *Halofsky et al.*, 2014]. Other studies have focused on the potential effects of regional climate change on fire regime [e.g., *Whitlock et al.*, 2003; *Westerling et al.*, 2006; *Littell et al.*, 2010], insect population dynamics [e.g., Bentz *et al.*, 2010], and forest productivity [e.g., *Latta et al.*, 2009]. These studies have used

[1] Conservation Biology Institute, Corvallis, Oregon, USA
[2] Oregon State University, USA
[3] University of Idaho, Moscow, Idaho, USA

Natural Hazard Uncertainty Assessment: Modeling and Decision Support, Geophysical Monograph 223, First Edition.
Edited by Karin Riley, Peter Webley, and Matthew Thompson.

various modeling approaches and sources of climate change scenarios, including regional climate models (RCMs) and statistically downscaled general circulation model (GCM) projections.

The objectives of this study were to examine how the latest projections of future climate, using 20 climate models from Phase 5 of the Coupled Model Intercomparison Project (CMIP5), would directly affect future vegetation distribution, carbon capture and storage, as well as the fire regimes over the western United States. CMIP5 climate models include for the first time a series of earth system models (ESMs) that incorporate further biological and chemical effects and detailed ecosystem modeling, some even including dynamic vegetation. Feedbacks from these land cover changes could affect climate projections in ways that have not been documented previously [e.g., *Brovkin et al.*, 2013].

We used the MC2 dynamic vegetation model, which has been widely used at multiple scales including National Parks [e.g., *King et al.*, 2013], individual states [e.g. *Lenihan et al.*, 2008a], across the nation [e.g., *Bachelet et al.*, 2008] and the continent as well as globally [*Gonzalez et al.*, 2010], to simulate vegetation dynamics, carbon cycling and wildfire occurrence and effects with downscaled CMIP5 future climates (20 climate models and two greenhouse gas concentration trajectories). The model simulates the complex interactions between climate and vegetation by simulating direct effects on ecosystem processes such as carbon capture or decomposition as well as indirect effects such as changes in fire disturbance. Much of the variability or uncertainty in the results stems from the wide range (both in magnitude and seasonality) of precipitation and evaporative demand projections because they affect directly the competition for water resources between trees and grasses and as a consequence, the buildup and moisture content of fuel loads driving changes in the fire regime.

17.2. METHODS

17.2.1. General Model Description

MC2, the C++ version of MC1 [*Daly et al.*, 2000; *Bachelet et al.*, 2001; *King et al.*, 2013], is a dynamic global vegetation model (DGVM) that includes three modules that simulate biogeography, biogeochemistry, and wildfire interactions. The model runs at a monthly time step over a grid of a specified spatial resolution. Each grid cell is simulated independently (time before space), with no cell-to-cell communication. The model simulates potential vegetation that would occur without direct intervention by humans but indirect effects such as increasing greenhouse gas concentrations, grazing, and fire suppression can be included. For this project, neither grazing nor fire suppression was explicitly simulated.

In the biogeography module, woody lifeforms are distinguished by phenology (evergreen vs. deciduous) and leaf shape (needle leaf vs. broadleaf). An environmental gradient algorithm based on coldest month temperature and growing season precipitation predicts the relative dominance of lifeforms (shrub, tree) based on the observed distribution of lifeform mixtures along temperature and precipitation gradients. The relative dominance of C3 and C4 grasses is simulated by comparing the potential production of pure C3 and pure C4 grass stands based on an algorithm using summer temperature, assuming C4 grasses are better adapted to warmer conditions. Thresholds of maximum monthly woody and herbaceous carbon pool values are used to distinguish forest, savanna, shrubland, and grassland classes. There are 36 vegetation types possible, 14 within the temperate zone. We have aggregated these types into 8 broader categories to facilitate visualization over the western United States. The climate data used in the biogeography model have been smoothed over 15 yr to account for the inherent inertia of long-lived woody types to short-term climate variability.

The model always simulates competition for light, water, and nutrients between woody lifeforms and herbaceous lifeforms, including grasses, forbs, and sedges. It does not simulate individual species. The biogeochemistry module is a modified version of the CENTURY model [*Metherell et al.*, 1993] that simulates the cycling of carbon and nitrogen among plant parts, multiple classes of litter, and soil organic matter pools. It calculates ecosystem carbon fluxes such as net primary production, soil respiration including decomposition, and fire emissions. This module also simulates actual and potential evapotranspiration (AET and PET) and soil-water content in multiple soil layers. Woody and herbaceous production is limited by suboptimal temperature, water availability, and atmospheric CO_2 that differ for woody vs. herbaceous lifeforms [*Bachelet et al.*, 2001]. Woody leaf and herbaceous moisture contents are calculated as functions of the ratio of available water to PET, and are interpreted as live fuel moisture contents affecting fire behavior.

Fire occurrence is simulated as discrete events driven by pseudodaily weather, fuel amount, and moisture [*Lenihan et al.*, 1998, 2008a,b]. At each time step, a fuel buildup and moisture index is calculated and compared to a threshold that indicates the presence of sufficient dry fuels to sustain a fire. The model assumes ignition sources are always available. The direct effects of fire include mortality, consumption of aboveground carbon, and volatilization of nitrogen stocks.

17.2.2. Detailed Description of the Fire Module

The fire module includes a set of mechanistic fire behavior and effects functions [*Rothermel*, 1972; *Peterson and Ryan*, 1986; *van Wagner*, 1993]. Live and dead fuel loads in 1 hr, 10 hr, 100 hr, and 1000 hr fuel classes are estimated from the carbon pool sizes produced by the biogeochemistry module. Allometric functions based on lifeform characteristics from the biogeography module are used to calculate total height, crown base height, and bark thickness for an average-sized tree as a function of carbon pool sizes. The fractions of tree carbon pools killed by fire are functions of simulated fire intensity, crown position, and bark thickness, with complete mortality occurring in the case of crown fires.

The moisture content of the different fuel classes and the potential fire behavior are calculated each day using pseudodaily data generated from simple interpolations of monthly climate inputs between midmonth values [*Lenihan et al.*, 1998]. Potential fire behavior (including rate of spread) is calculated each day based on daily interpolated fuel loads, their moisture contents, and weather. Potential fire behavior is modulated by vegetation type (provided by the biogeography model), which affects fuel properties and realized wind speeds (higher for grassland than forest).

Actual fire is simulated whenever the calculated rate of spread is greater than zero and user-specified thresholds are exceeded for the fine fuel moisture code (FFMC) and the buildup index (BUI) of the Canadian fire weather index system. These two indices are inverse functions of fine fuel and coarser fuel moisture contents, respectively, as specified by *van Wagner and Pickett* [1985].

Only one fire is simulated per year per cell on the first day when all thresholds are exceeded. As the daily inputs to the fire module are generated in a deterministic fashion, there is no stochastic variation in fire behavior or other outputs among multiple runs with the same inputs and initial conditions. Fires do not extend beyond one time step.

17.2.3. Model Inputs

The MC2 model requires inputs of soil depth, texture and bulk density. Soils data from *Kern* [1995; 2000] were obtained from R. Drapek (USFS PNW) who reprojected the original 1 km data to a 4 km grid using the "majority" rule such that the soil-related value that occupies the majority of the area of a particular grid-cell gets assigned to the entirety of that grid-cell.

Climate inputs to the vegetation model include monthly precipitation, mean vapor pressure or dew-point temperature, and mean daily maximum and minimum temperatures averaged monthly. Historical climate data (1895–2010) were acquired from the PRISM (Parameter-elevation Regressions on Independent Slopes Model) group at Oregon State University [*Daly et al.*, 2008] at 2.5 arc-minute grid resolution and reprojected to a 4 km grid.

Twenty future climate time series (2010–2100) were acquired from the WCRP (World Climate Research Programme) CMIP5 (Coupled Model Intercomparison Project Phase 5) multimodel database website (http://cmip-pcmdi.llnl.gov/cmip5/data_getting_started.html) (Table 17.1) for two representative concentration pathways (RCPs) 4.5 and 8.5 (Table 17.2). The performance, or credibility, of those 20 climate models was assessed based on their ability to reproduce the observed twentieth-century climate of the Pacific Northwest United States (PNW) and the surrounding region [*Rupp et al.*, 2013]. They were downscaled to 4 km at the University of Idaho using the Multivariate Adaptive Constructed Analogues (MACA) method [*Abatzoglou and Brown*, 2011] for the western United States (31.02–49.1 N latitude, -124.77 to -103.02 E longitude) and used as input to MC2.

17.2.4. Run Protocol

The vegetation model is run in three distinct phases. First, we use the static biogeography model MAPSS [*Neilson,* 1995] to generate a map of potential vegetation distribution at the beginning of the twentieth century using average climate (12 monthly values for each climate variable used iteratively). This map is used in the biogeochemistry module to calculate initial conditions for carbon and nitrogen pools associated with each vegetation type with their prescribed fire return intervals. This phase ends when the resistant soil carbon pool size changes by less than 1% from one year to the next. Its duration varies across the map from a few decades in the Great Plains grasslands up to 3000 yr in the rain forests of the PNW. Second, during the spinup phase, the model is run iteratively using a detrended historical climate time series (1895–2009) to allow for readjustments of vegetation type and carbon pool sizes in response to interannual variability and prognostic wildfires. Fires cause episodic disequilibrium in the carbon cycle. Decomposition of the plant material killed by fires causes ecosystems to become temporary sources of carbon to the atmosphere. The size of live vegetation pools increases until net primary production again matches heterotrophic respiration. The time to reach this quasi equilibrium varies greatly between individual grid cells. When looking at a large region like the western United States, we assume that the spinup phase ends when the overall net biological production (net ecosystem production minus carbon consumed by wildfire) reaches a stable state near zero. During the third and last phase, the model is run with time series of historical and future climate.

Table 17.1 List of the 20 CMIP5 Climate Models That Were Downscaled and Used in This Project

	GCM (general circulation model) or ESM (Earth System Model)	Origin	Atmosphere resolution Grid-cell size in degree Lat × Lon; L: # vertical levels
1	BCC-CSM1-1	Beijing Climate Center, China Meteorological Administration	2.8×2.8 L26
2	BCC-CSM1-1-M	Beijing Climate Center, China Meteorological Administration	1.12×1.12 L26
3	BNU-ESM	College of global change and earth system science, Beijing Normal University, China	2.8×1.4 L26
4	CanESM2	Canadian Center for Climate Modelling and Analysis (Canada)	2.8×2.8 L35
5	CCSM4	National Center for Atmospheric Research (USA)	1.25×.94 L26
6	CNRM-CM5	Meteo France and CNRS (France)	1.4×1.4 L31
7	CSIRO-MK3-6.0	Commonwealth Scientific and Industrial Research Organization, Queensland Climate Change Center of Excellence (Australia)	1.8×1.8 L18
8	GFDL-ESM2G	National Oceanic and Atmospheric Administration/Goddard Fluid Dynamics Laboratory (USA)	2.5×2.0 L48
9	GFDL-ESM2M	National Oceanic and Atmospheric Administration/Goddard Fluid Dynamics Laboratory (USA)	2.5×2.0 L48
10	HadGEM2-CC	Meteorological Office Hadley Center, (UK)	1.88×1.25 L60
11	HadGEM2-ES	Meteorological Office Hadley Center, (UK)	1.88×1.25 L38
12	INM-CM4	Institute for Numerical Mathematics (Russia)	2.0×1.5 L21
13	IPSL-CM5A-LR	Institut Pierre Simon Laplace (France)	3.75×1.8 L39
14	IPSL-CM5A-MR	Institut Pierre Simon Laplace (France)	2.5×1.25 L39
15	IPSL-CM5B-LR	Institut Pierre Simon Laplace (France)	3.75×1.8 L39
16	MIROC5	Atmosphere and Ocean Research Institute (U. Tokyo), National Institute for Environmental Studies, Japan Agency for Marine-Earth Science and Technology (Japan)	1.4×1.4 L40
17	MIROC-ESM	Atmosphere and Ocean Research Institute (U. Tokyo), National Institute for Environmental Studies, Japan Agency for Marine-Earth Science and Technology (Japan)	2.8×2.8 L80
18	MIROC-ESM-CHEM	Atmosphere and Ocean Research Institute (U. Tokyo), National Institute for Environmental Studies, Japan Agency for Marine-Earth Science and Technology (Japan)	2.8×2.8 L80
19	MRI-CGCM3	Meteorological Research Institute (Japan)	1.1×1.1 L48
20	NorESM1-M	Norwegian Climate Center (Norway)	2.5×1.9 L26

Table 17.2 Brief Summary of the Two RCP (Representative Concentration Pathways) Scenarios Used in This Project With the Associated Climate Change Estimates for the Western United States Averaged Across 20 CMIP5 Climate Models for the Period 2081–2100 Relative to 1986–2005

	Atmospheric CO_2 concentration in 2100 in ppm	Changes in temperature (in °C)	Changes in annual precipitation (in mm)
RCP 8.5	936	4.7–5.8	15.0–60.3
RCP 4.5	538	2.6–3.0	−8.9–29.7

17.3. RESULTS

17.3.1. Background on Climate Projections

The basic trends in the climate time series used for this study can serve as background information to better interpret the vegetation response to change. Cooler temperatures occurred before the 1930s followed by warmer temperatures through the 1950s. An extended period of cooling from the late 1950s to the 1980s was followed by a more stable period that ended around 2000 when temperatures warmed again. These multidecadal patterns have been attributed to natural climate variability and various strengths of existing teleconnections [*McCabe et al.*, 2004; *Wang et al.*, 2009; *Arriaga-Ramirez and Cavazos*, 2010; *Kam et al.*, 2014; *Wang et al.*, 2014]. Starting in 2010, temperatures simulated by the climate models are increasing above their historical range of variability to a greater extent for RCP 8.5 (Fig. 17.1, Table 17.2). Note that the transition between the PRISM data for the historical period (1895–2010) and the ensemble mean of the climate projections can be large. This is particularly evident for maximum temperature and dewpoint temperature despite the fact that the downscaled climate futures were bias corrected with the MACA method [*Abatzoglou and Brown*, 2011].

A period of low precipitation occured around the 1930s (Dust Bowl period) followed by a relatively stable period until the 1980s (Fig. 17.1). Multiyear periods of low annual averages reminiscent of the 1930s alternated with high precipitation years around the 1980s. As soon as the 21st century starts, the ensemble mean annual precipitation projected by the 20 climate models seems to trend upward during the 21st century especially with RCP 8.5. Projected extreme values (low and high) are mostly outside the 20th century range (Fig. 17.1) particularly for RCP 8.5 in the second half of the century with a few very dry years (300 mm) as well as extremely wet years (>700 mm) simulated by a handful of climate models. Climate models have recognized limitations including that of simulating natural climate variability with accuracy but the CMIP5 models have been shown to include significant improvements [*Polade et al.*, 2013]. We can thus assume that the projected precipitation variability during the 21st century includes reasonable representations of the El Niño Southern Oscillation and of the Pacific Decadal Oscillation, which have been influencing both regional precipitation patterns and extreme events in the western United States.

17.3.2. Vegetation Model Projections

17.3.2.1. Change in Vegetation

The MC2 model was calibrated for a previous project [*Bachelet et al.*, 2015] to best approximate the potential vegetation map produced by *Küchler* [1964, 1975]. The model simulates maritime evergreen forests along the coast of the Pacific Northwest and warmer forest types in California, conifer forests in mountainous regions, shrublands and woodlands occupying most of the intermountain west, large desert areas in the southwest, and a matrix of western grasslands extending eastward. Fires maintain the open grasslands and cool temperatures favor conifer dominance. Mild climate inputs along the coast cause the model to include deciduous lifeforms in areas where evergreens dominate such as the Olympics.

For all the climate futures under RCP 8.5, the vegetation model simulates (1) the disappearance of alpine areas (particularly in the Rocky Mountains) by the end of the 21st century (Table 17.3) as forests move uphill, (2) the "woodification" of western states including both shrub invasion of grassland areas and shifts from woodlands to forests (Fig. 17.2), (3) shifts from cool vegetation types to warmer types including shifts from evergreen to mixed forests especially along the Pacific coast (Fig. 17.2), and shifts from C3 to C4 grasslands (not shown here). Regional warming (see Fig. 17.3), increased annual average precipitation (see Fig. 17.3), and enhanced water-use efficiency due to elevated CO_2 particularly in the second half of 21st century are all contributing factors responsible for these changes. However, it is difficult to separate and quantify the relative contributions of each of these factors since complex interactions between them vary both spatially and temporally. Warming is clearly the direct cause of the

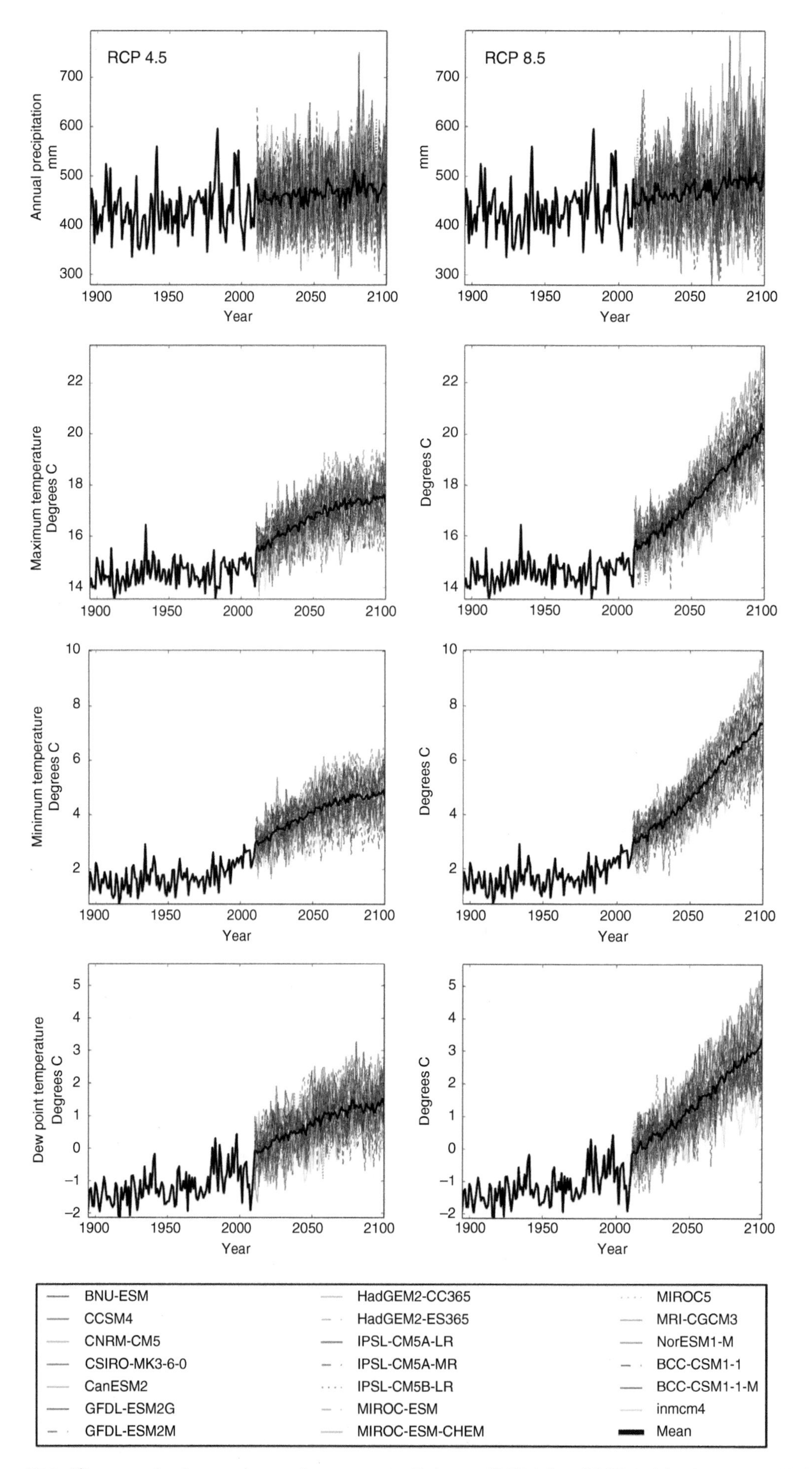

Figure 17.1 Climate projections under two Representative Pathways (RCP 4.5 and RCP 8.5) for the western United States. For the historical period (1895–2010), the thick black line corresponds to PRISM data. From 2011 forward, each climate model projection has been drawn and the thick black line corresponds to the ensemble mean.

Table 17.3 Changes in Areal Extent (in %) of the Vegetation Types from 1971–2000 to 2071–2100 Using Projected Climates with Two Greenhouse Gas Concentration Trajectories

A. RCP 8.5

	Alpine tundra	Conifer forest	Mixed forest	Deciduous forest	Wood land	Shrub land	Grass land	Arid land
CCSM4	−100	−6	149	58	−24	−1	−2	−57
CNRM-CM5	−100	14	170	624	43	1	−12	−99
CSIRO-Mk3-6-0	−100	−14	149	144	−20	−3	4	−72
CanESM2	−100	−2	195	1453	162	−7	−4	−94
HadGEM2-ES	−100	−14	148	−27	12	1	−2	−21
HadGEM2-CC	−100	−20	148	−84	−41	5	0	−44
IPSL-CM5A-MR	−100	−4	155	−92	−40	−9	2	−33
MIROC5	−99	5	133	261	−22	−30	11	16
NorESM1-M	−100	−4	136	250	−26	−13	3	5
BCC-CSM1-1-M	−99	−2	144	60	−6	3	−5	−85

B. RCP 4.5

	Alpine tundra	Conifer forest	Mixed forest	Deciduous forest	Wood land	Shrub land	Grass land	Arid land
CCSM4	−99	2	100	70	62	21	−39	−27
CNRM-CM5	−98	15	127	499	103	21	−50	−51
CSIRO-Mk3-6-0	−99	−3	100	95	87	20	−35	−39
CanESM2	−100	8	146	865	158	14	−42	−61
HadGEM2-ES	−99	2	91	216	105	23	−43	−33
HadGEM2-CC	−100	0	98	21	98	28	−46	−66
IPSL-CM5A-MR	−99	6	92	−95	90	18	−35	−73
MIROC5	−99	7	59	262	92	19	−42	31
NorESM1-M	−99	8	89	111	90	14	−35	−13
BCC-CSM1-1-M	−98	0	85	46	53	27	−42	−35

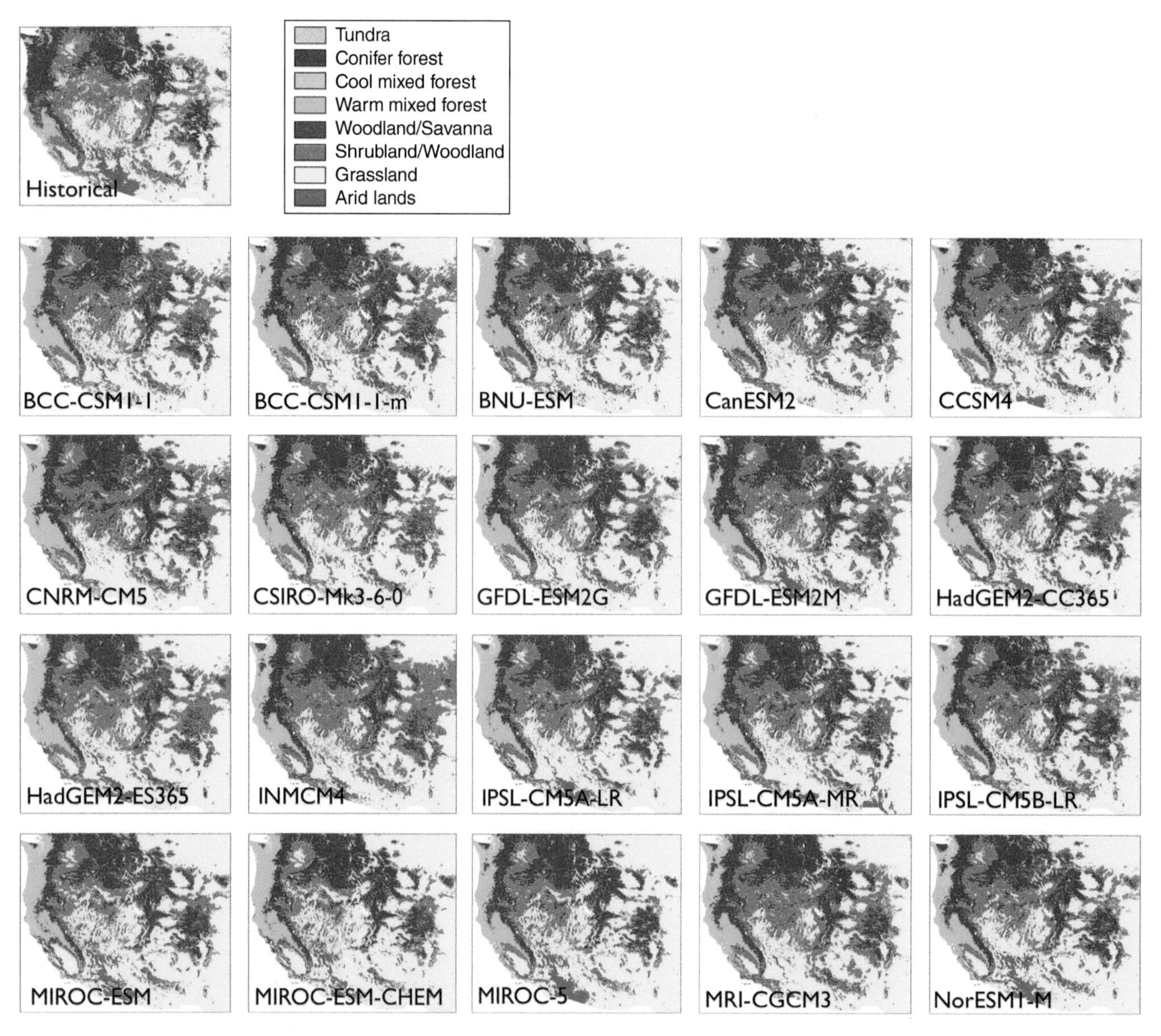

Figure 17.2 Simulated vegetation distribution (mode for 1971–2000) and projected vegetation distribution for the end of the 21st century (mode for 2071–2100) by 20 climate models from CMIP5 under RCP8.5.

disappearance of alpine areas as forests move uphill and shift to warmer vegetation types. The expansion of the woody lifeforms is due in part to enhanced water-use efficiency but also to the increase in available soil water from increased rainfall that allows deep-water recharge accessible to tree roots in the model. We ran the model with no CO_2 effect (not shown here) and did not see much change in historical and early 21st century vegetation dynamics. The effect of CO_2 becomes important when its concentration increases more significantly after 2060.

The extent of arid lands (defined in the model by extremely low average biomass, or NPP) is variable across all climate futures (Fig. 17.2), depending on how each climate model projected seasonal precipitation and the strength of the Arizona monsoon. Wetter future winters, such as with CanESM2 (wettest future among the 20 climate models with highest summer rainfall and second highest winter precipitation), CNRM-CM5 (highest winter precipitation and third highest summer rainfall among the 20 climate models), and MRI-CGCM3 (third highest winter precipitation), drive the disappearance of arid lands and their replacement by grasslands. Increased winter precipitation (over at least 15 yr) allows C4 grasses to establish and grow, producing enough biomass to have grid-cells previously classified as arid lands exceed the biomass threshold and switch to a grassland type.

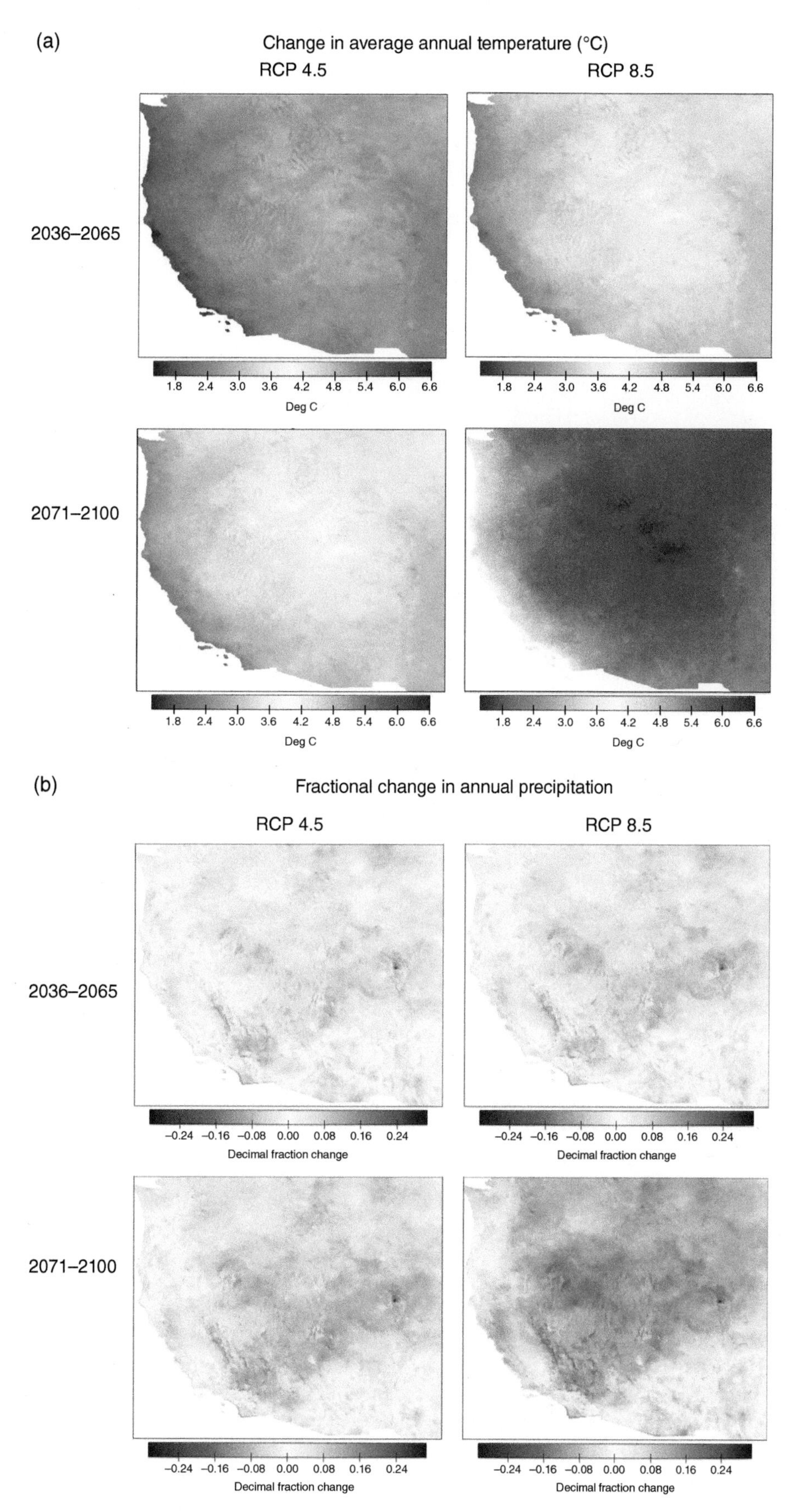

Figure 17.3 Changes between mid-21st century (2036–2065) or the end of the 21st century (2071–2100) and the historical baseline (1971–2000) for (a) average annual temperature and (b) average annual precipitation under RCP 4.5 and 8.5 across 20 CMIP5 climate models.

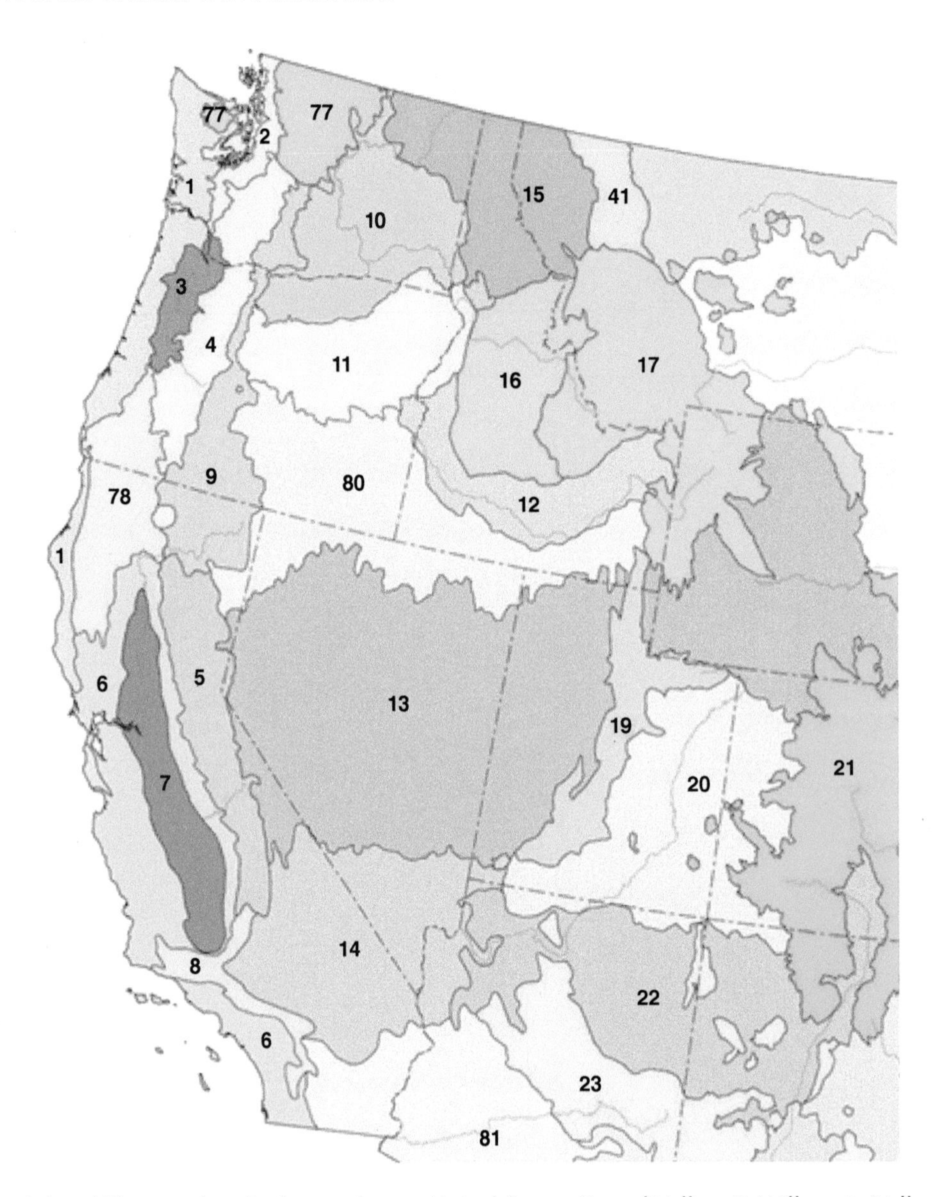

Figure 17.4 Level III ecoregions in the contiguous United States. Central Valley: 7; Willamette Valley: 3; Sierra Nevada: 7, and Cascades: 4.

17.3.2.2. Carbon Stocks and Fluxes

17.3.2.2.1. Temporal Variability in Carbon Sources and Sinks

Simulated soil carbon stocks in the western United States increased throughout the 20th century while live vegetation stocks varied but remained fairly level (Fig. 17.5). During the first half of the 21st century, large wildfires (illustrated by a rise in the maximum fraction of area burned in Fig. 17.5) cause a decrease in the live vegetation carbon pool (illustrated by sharp episodic drops in live C stocks in Fig. 17.5). Under RCP 4.5, the fraction of area burned simulated under most climate futures can be episodically larger than under the warmer RCP 8.5 scenario because milder climate conditions allow for fuel buildup under reduced drought stress and subsequently greater fire effects (Fig. 17.5). Note that under wetter or cooler (less evaporative demand) climate futures projected by some climate models, live vegetation carbon simply increases.

Because the 21st century is projected to be wetter on average across all climate futures and because elevated CO_2 concentration is enhancing water-use efficiency in the second half of the century, the woody expansion the model is simulating in the intermountain West (Fig. 17.2) contributes to increasing live vegetation carbon pools under warmer conditions. In other areas, once the simulated vegetation has switched (often after disturbance such as fire) to a type better adapted to the new climate

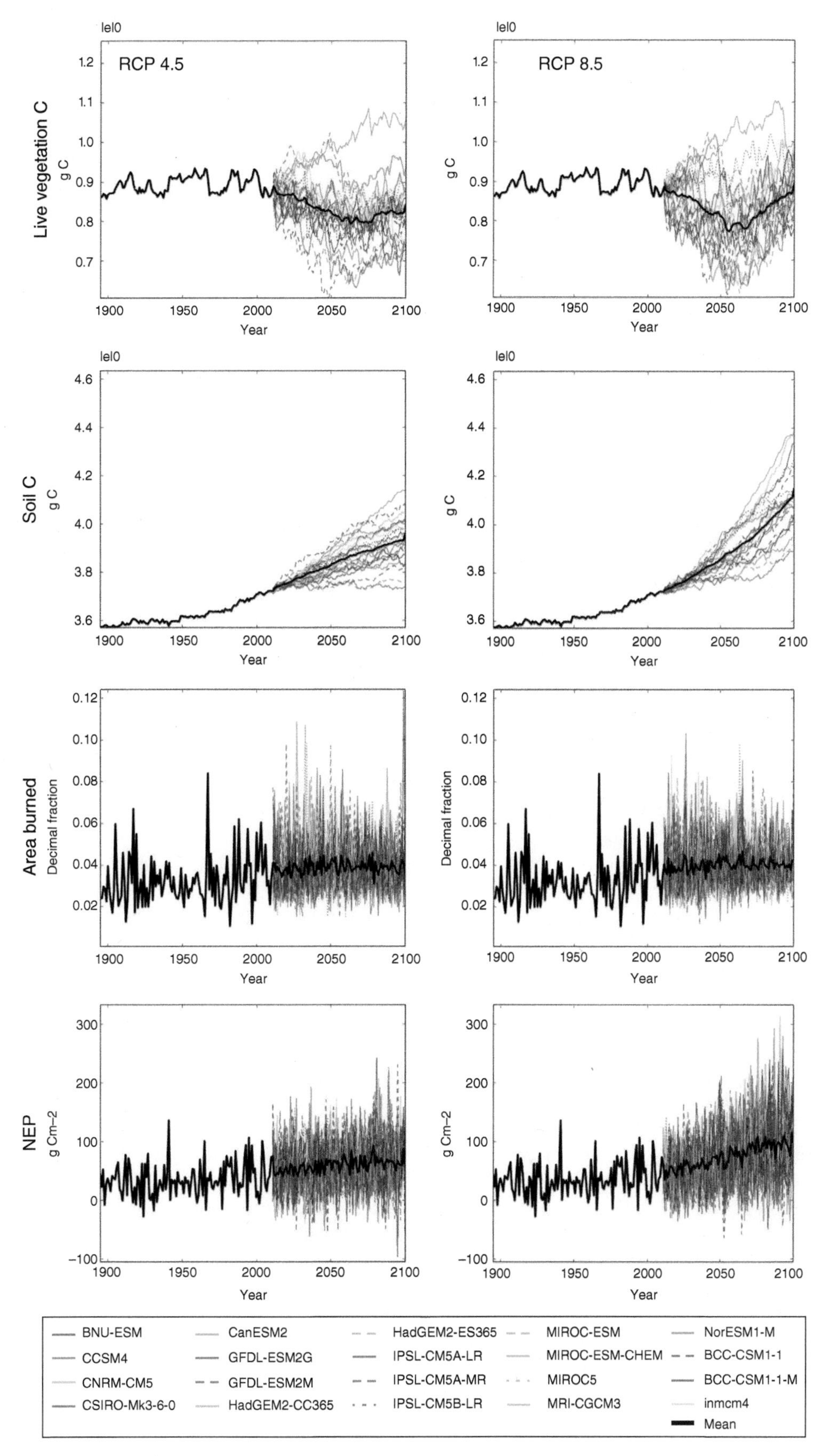

Figure 17.5 Carbon stocks and fluxes dynamics as well as fraction of area burned by wildfires over the western United States from 1895 to 2100 using climate projections from 20 CMIP5 climate models under RCP 4.5 and 8.5. From 2011 forward, each climate model-driven projection has been drawn and the thick black line corresponds to the ensemble mean. NEP stands for net ecosystem production (net primary production minus heterotrophic respiration).

conditions, live vegetation carbon stocks can also increase. It is important to note that while the ensemble average shows an overall increase during the second half of the century, some of the outcomes from particular climate futures show declines in carbon stocks well below the historical average during the first half of the century and remaining so by 2100, especially under RCP 4.5.

The ensemble mean for net ecosystem production (the balance between net primary production and soil respiration) remains fairly stable throughout the 21st century under the RCP 4.5 but increases on average under RCP 8.5 after 2050 with much year-to-year variability, well beyond that of the 20th century after 2050.

17.3.2.2.2. Geographic Variability in Carbon Sources and Sinks

While we have shown the temporal variability of carbon fluxes, there are also regional differences in the strength of the carbon sources and sinks across western states. Carbon density decreases by midcentury in areas where it is simulated to be the highest under baseline (1971–2000) conditions (Fig. 17.6). The western part of the Pacific Northwest region as well as Rocky Mountains show declines in carbon density due to an increase in wildfires where fuels are not limiting but drier conditions increase fire risk. Patterns of carbon losses can be compared to patterns of climate change over the area of study (see Fig. 17.3) but simple correlations do not explain the whole story. The following section illustrates the complex interactions that link climate and disturbance regimes to drive these carbon dynamics along the West Coast.

17.3.2.2.3. Regional Dynamics: The Role of Fire

Using level-III Ecoregional boundaries (Fig. 17.4), we compared changes in carbon stocks in the Northwest (Willamette Valley in Oregon) where warming is projected to be moderate and precipitation still abundant to those in the Southwest (Central Valley in California) where drought conditions are already limiting production. The range of conditions simulated by climate models in the Central Valley with RCP 8.5 produce a variety of vegetation responses with + or - 50% change in the size of the live vegetation pool and possible increases of ~20% in the area burned by wildfire (Fig. 17.7). Such is not the case in the Willamette Valley where under all warmer climate futures, wildfires, inexistent during the 20th century, cause large carbon losses early in the 21st century accompanied by a rapid shift to vegetation types more adapted to warmer and drier conditions. The projected decrease in precipitation (Fig. 17.3) allows fuel moisture to decrease sufficiently to meet fire occurrence thresholds. However, the new vegetation starts accumulating carbon under all climate futures soon after the 2050s under the RCP 8.5 such that it almost doubles the live carbon stocks in the last 40 yr of the century. Concurrent with the loss of two-thirds of the live vegetation carbon before 2050, the model simulates a sharp increase in litter inputs (fire-caused mortality) to the soil carbon pool that lasts for over a decade but is quickly followed by a sustained decrease due in part to more fires eliminating litter (Fig. 17.7). Moreover, warmer soil conditions are likely sustaining high rates of decomposition and slowing the stabilization of the soil carbon pool.

We compared the dynamics in these low-elevation valleys to those in nearby mountains. In the Sierra Nevada of California, the model simulates an increase in total ecosystem carbon due to the uphill expansion of woodlands and forests with an overall decrease in the area burned by wildfires although with much year-to-year variability (Fig. 17.8). Such changes are driven mostly by an increase in precipitation at high elevation (see Fig. 17.3). In the Cascades of Oregon, where, unlike in the Sierras, precipitation is projected to decrease, large fires cause declines in live biomass with much carbon lost to fire emissions. Some stabilization occurs after 2050. Soil carbon increases at first, due to fire-caused mortality and increased litter inputs, but it stabilizes by mid-21st century (Fig. 17.8).

In summary, at lower elevations where currently mild conditions are projected to be replaced by much warmer and drier ones, fires (or other disturbances that would eliminate current vegetation) could cause very rapid shifts to new vegetation types better adapted to the new climate, with significant carbon losses followed by some recovery in the second half of the century. The level of agreement between the responses to climate futures in such areas as the Willamette Valley is amazingly tight and, if model agreement can be taken as a low level of uncertainty, this might be a dire projection for an area dominated by urban and agricultural areas. At high elevation, however, the change in climate is not as fast and impacts develop later in the 21st century when fire limits carbon sequestration potential.

In areas where warm and dry conditions already exist, like in California, climate change does not transform the landscape as dramatically as in the Northwest especially at high elevations where climate models project an increase in precipitation, which could reduce the fire impacts without eliminating the risk of extreme fire years.

17.3.3. Uncertainty

The uncertainty associated with precipitation projections has been discussed abundantly elsewhere [e.g., *Meehl et al.*, 2007] but it is important to emphasize how large its effects on simulated impacts can be, particularly when simulating fire.

While all climate models simulate warmer conditions, they differ in their projections of both seasonality and

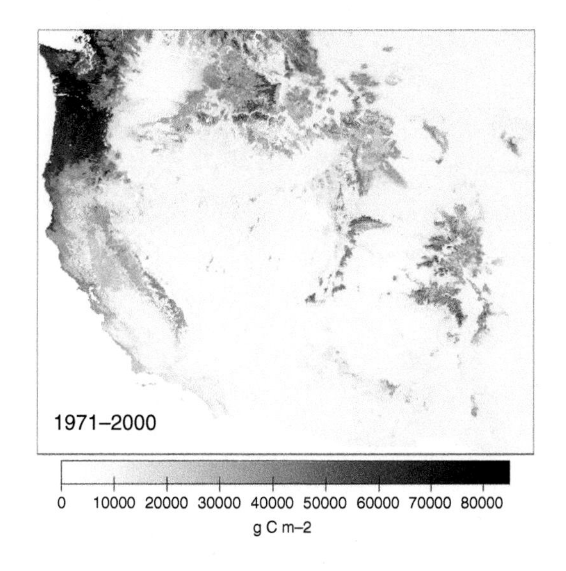

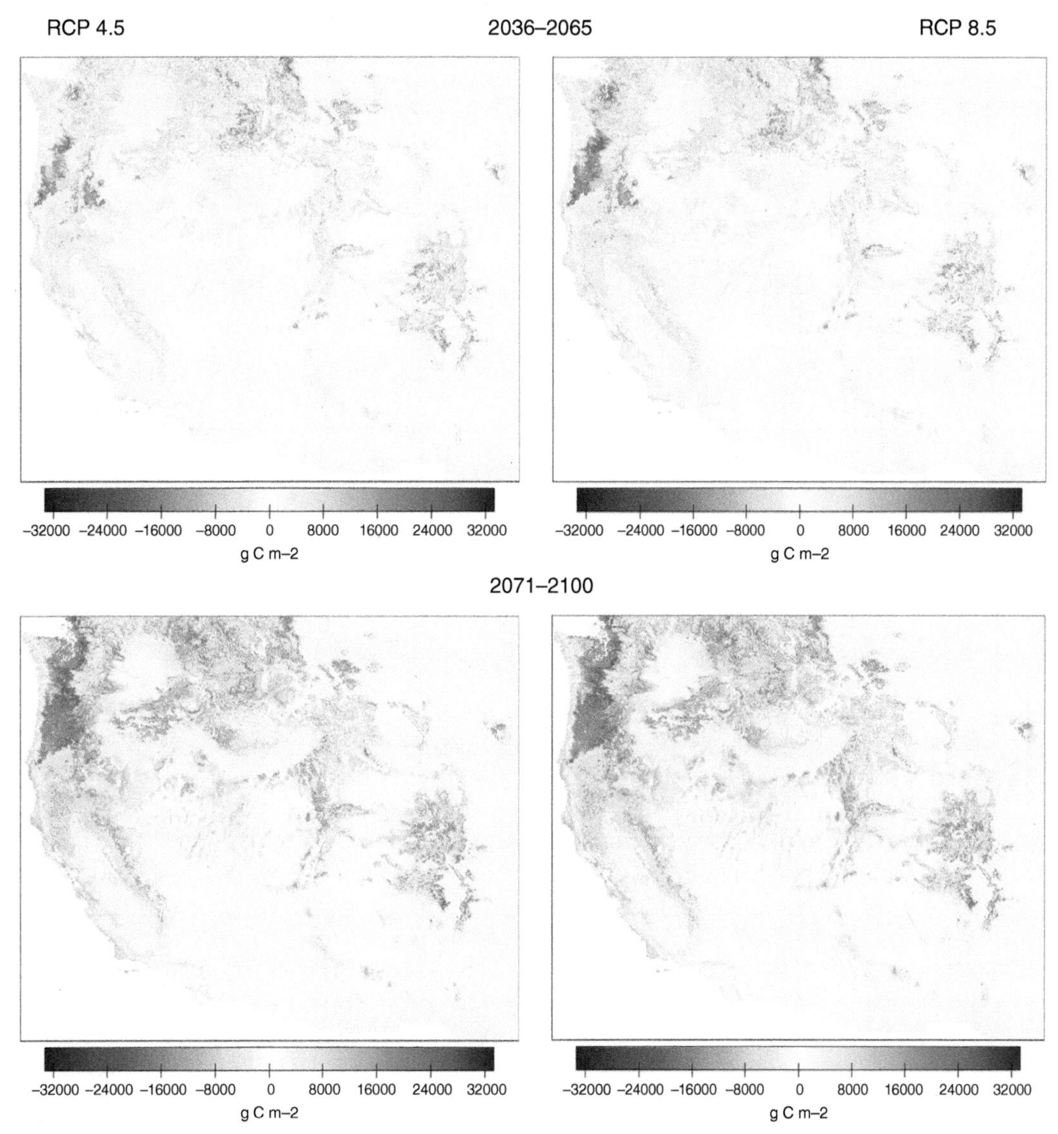

Figure 17.6 Ecosystem carbon density simulated for baseline conditions (1971–2000) and simulated change under future conditions for the middle and end of the 21st century relative to baseline.

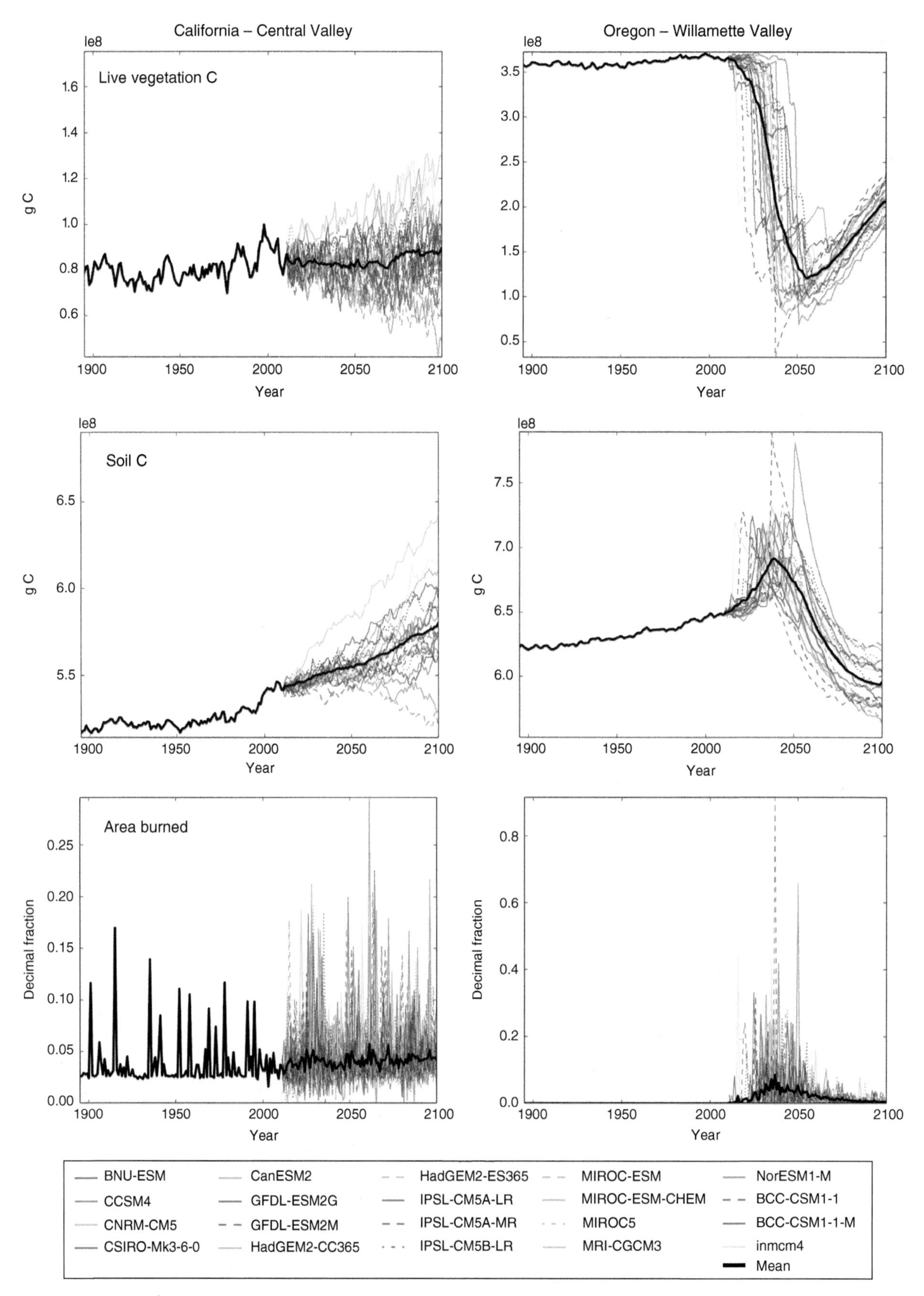

Figure 17.7 Temporal dynamics of the live vegetation carbon pool, soil and litter carbon pools, and area burned for the Willamette Valley of Oregon and the central valley of California as simulated by the MC2 dynamic vegetation model. From 2011 forward, results are shown for 20 of the CMIP5 climate futures under RCP 8.5 and the thick black line corresponds to the ensemble mean.

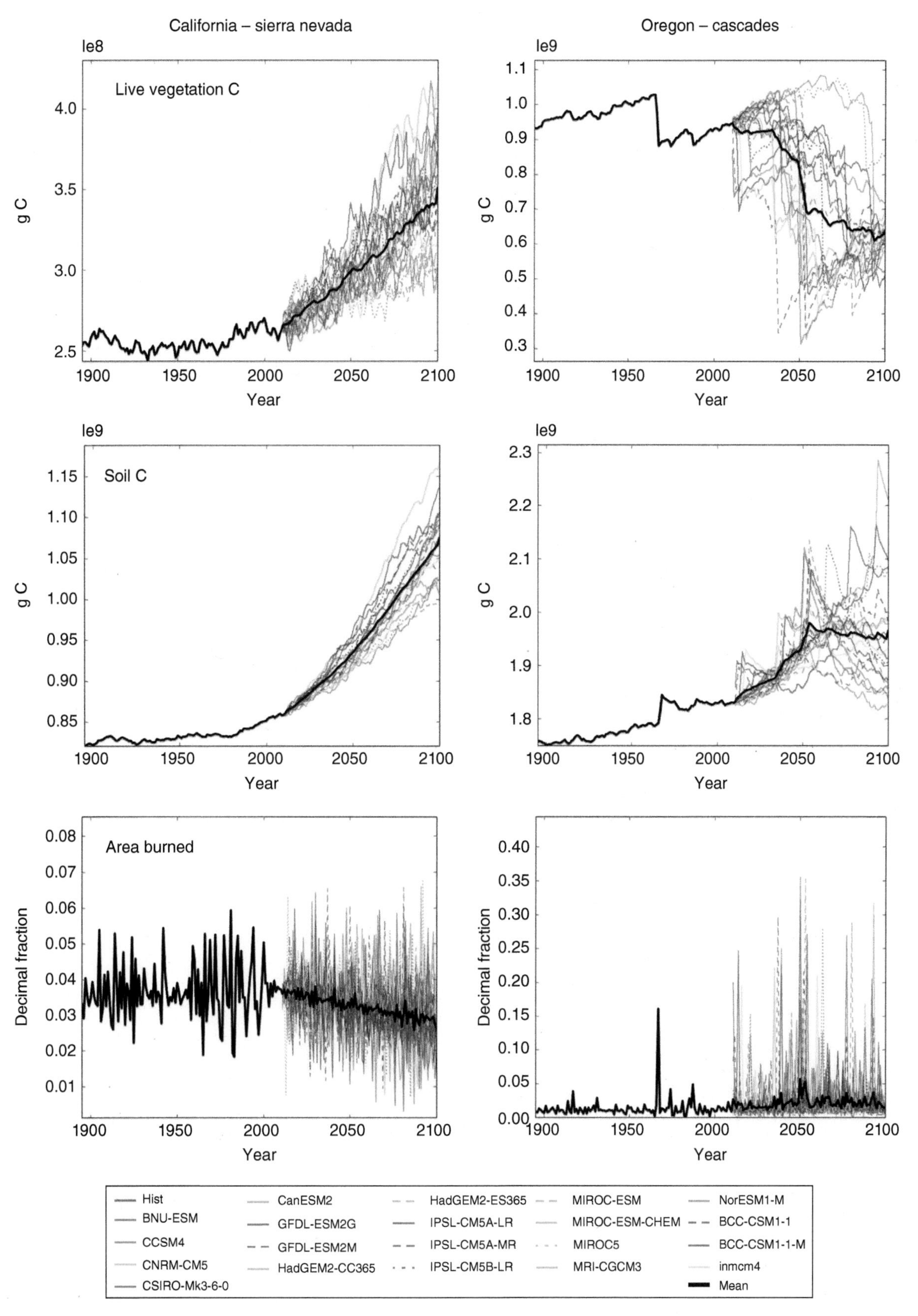

Figure 17.8 Temporal dynamics of the live vegetation carbon pool, soil and litter carbon pools, and area burned for the Sierra Nevada of California and the Cascades of Oregon as simulated by the MC2 dynamic vegetation model. From 2011 forward, results are shown for 20 of the CMIP5 climate futures under RCP 8.5 and the thick black line corresponds to the ensemble mean.

magnitude of rainfall, which has a direct effect on fire behavior. Wet years allow increased production and thus litter accumulation (i.e., fuel buildup) in normally dry fuel-limited areas, increasing fire risk in the following drier years. Dry years in areas when fuels are not limiting become likely candidates for extensive fires. The MC2 fire model is sensitive to the amount of water available for fuel buildup and moisture, but also for plant production and turnover of soil organic matter. The complex interactions between climate, vegetation growth, and disturbance drive the large changes the model is simulating with much geographic patchiness due to soil types (the key to soil-water availability) as well as temporal variability due to changes in rainfall seasonality. Accurate soil properties are essential to accurately project soil-water availability and seasonal drought stress [*Peterman et al.*, 2014]. Furthermore, aside from temperature and precipitation, projections of vapor pressure deficit are crucial to draw an accurate portrait of environmental conditions.

The MC2 vegetation model simulates potential vegetation cover resulting only from a legacy of past climatic conditions. Consequently, we used PRISM (climate) data for the period 1895–2010 rather than simulated climate. We found that the transition between late 20th century conditions from PRISM and early 21st century downscaled climate projections was abrupt (particularly with regard to dew-point temperature, Fig. 17.5). Such fast transition to warmer conditions may have caused the fire model to respond quickly and simulate early 21st century catastrophic fires causing swift transitions to warmer vegetation cover types where the discrepancy was greatest. This abrupt shift to warmer and drier conditions at the beginning of the 21st century has probably caused an overestimation of the fire risk, or at least its urgency, in areas where the vegetation cover suddenly became maladapted to the climate (e.g., Willamette Valley where the potential vegetation cover switches from evergreen dominated to a mixed vegetation type after wildfire).

Furthermore, because our model uses monthly values and does not limit fire ignitions, it likely overestimates fire risk early in the year by allowing ignition as soon as weather conditions derived from monthly climate variables become dry and fuel is not limiting. The importance of scattered rainy days that may reduce fire risk early in the year may be missed since daily precipitation interpolations in the fire model are stochastic and projections of rainfall seasonality uncertain.

High atmospheric concentrations of CO_2 may be associated with increased water-use efficiency and thus could significantly ameliorate impacts of increasing vapor pressure deficits and reduce the incidence of summer soil drought [*Keenan et al.*, 2013]. The simulation of increased water-use efficiency with higher levels of atmospheric CO_2 allowed woody lifeforms not only to survive under drier conditions but also to expand into drier areas traditionally dominated by herbaceous lifeforms. This effect is moderate in MC2 but continues to be the subject of scientific enquiry as data are scarce to document the response of many species, especially mature trees, to increases in atmospheric CO_2 concentrations [*Battipaglia et al.*, 2013]. Using a higher or lower CO_2 effect could modulate the response of the simulated vegetation to the CMIP5 climate drivers.

The model MC2 has other limitations. It was designed to work at coarse scale and consequently misses ecosystem processes that are important at fine scale and often require cell-to-cell communication, which is missing in the model. For example, the routing of water through an entire watershed is done a posteriori rather than through lateral flow. The model calibration for such a large domain as the entire western United States requires compromises that can be eliminated when the model is run for limited study areas [e.g., *King et al.*, 2013]. Moreover, while the model includes fire, the driving disturbance that contributed to creating the western landscapes we see today, it does not include other disturbances such as insect or pest outbreaks that have significantly affected western ecosystems in the last 20 yr [e.g., *Robinet and Roques*, 2010; *Kasischke et al.*, 2013]. Furthermore, it was designed to simulate land without human intervention and thus our runs did not include land use. We have since revised the model code to include the direct human impacts on vegetation distribution.

17.4. DISCUSSION AND CONCLUSIONS

There is much temporal and regional variability in the results from the MC2 vegetation model mostly caused by the wide range of climate futures provided by the CMIP5 climate models. While they all agree on a warming trend, their projections of future precipitation magnitude and seasonality vary widely. Because MC2 includes a dynamic fire model that is very responsive to fuel load and fuel moisture, its results span a wide range of possibilities for the region.

However, we found much agreement between results from all climate futures for the following:

1. Large shifts in potential vegetation toward warmer types (e.g., temperate to subtropical forest types, warm subtropical grasslands replacing cool temperate grasslands) due to a shift in climate zone driven by a general warming trend over all western states.

2. An expansion of forest and woodland types enhanced by a moderate CO_2 effect on water-use efficiency and production when water availability declines. This effect becomes significant when concentrations exceed 550 ppm by midcentury. Projected increases in precipitation in the intermountain West also contribute to woody expansion and fuel buildup.

3. An overall increase in carbon stocks mostly because woody lifeforms extend their range. This increase is modulated by fire events that cause large carbon losses through fire emissions but also stimulate nutrient release and fast regrowth.

4. An increase in the occurrence of fires where fuel loads are not limiting (such as high elevation forests) because warmer drier conditions cause the drying of fuels.

We have run the MC2 model with fire suppression (using *Rogers et al.* [2011] approach) but are not reporting these results here. Results confirm its importance in exacerbating woody expansion, carbon capture as well as fuel buildup. In a future project, we hope to include the effects of land use with CMIP5 futures and report on the full suite of effects of human activities on vegetation distribution and fire regime as we have done with CMIP3 futures [*Bachelet et al.*, 2015].

ACKNOWLEDGMENTS

Funding for this research was provided by the US Department of the Interior via the Northwest Climate Science Center through agreement #G12AC20495 within the framework of the research project entitled "Integrated Scenarios of climate, hydrology and vegetation for the Northwest," P. Mote (Oregon State University) principal investigator.

We acknowledge the World Climate Research Programme's Working Group on Coupled Modelling, which is responsible for CMIP, and we thank the climate modeling groups (listed in Table 17.1 of this paper) for producing and making available their model output. For CMIP, the US Department of Energy's Program for Climate Model Diagnosis and Intercomparison provides coordinating support and led development of software infrastructure in partnership with the Global Organization for Earth System Science Portals.

The authors want to acknowledge R. Nemani (NASA) for allowing free access to the Pleiades NASA supercomputer. We thank D. King (Oregon State University) and B. Baker (CBI) for useful discussions about the model, its strengths and limitations, and for their comments on early versions of the manuscript. And we thank two anonymous reviewers for their helpful comments.

REFERENCES

Abatzoglou, J. T., and C. A. Kolden (2011), Climate change in western US deserts: Potential for increased wildfire and invasive annual grasses, *Rangeland Ecol. Man.*, *64*, 471–478.

Abatzoglou, J. T., and R. Barbero (2014), Observed and projected changes in absolute temperature records across the contiguous United States, *Geophys. Res. Lett.*, *41*, 2014GL061441.

Abatzoglou, J. T., and T. J. Brown (2011), A comparison of statistical downscaling methods suited for wildfire applications, *Int. J. Clim.*; doi:10.1002/joc.2312.

Adams, H. D., C. H. Luce, D. D. Breshears, C. D. Allen, M. Weiler, V. C. Hale, A. M. S. Smith, and T. E. Huxman (2012), Ecohydrological consequences of drought- and infestation-triggered tree die-off: insights and hypotheses, *Ecohydrology*, *5*, 145–159.

Anderegg, W. R. L., J. M. Kane, and L. D. L. Anderegg (2013), Consequences of widespread tree mortality triggered by drought and temperature stress, *Nature Climate Change*, *3*, 30–36.

Anderson, B. T., J. Y. Wang, G. Salvucci, S. Gopal, and S. Islam (2010), Observed trends in summertime precipitation over the southwestern United States, *J. Climate*, *23*, 1937–1944.

Arriaga-Ramirez, S., and T. Cavazos (2010), Regional trends of daily precipitation indices in northwest Mexico and southwest United States, *J. Geophys. Res. Atmos.*, *115*(D14); doi: 10.1029/2009JD013248.

Bachelet, D., R. P. Neilson, J. M. Lenihan, and R. J. Drapek (2001), Climate change effects on vegetation distribution and carbon budget in the United States, *Ecosystems*, *4*, 164–185.

Bachelet, D., J. Lenihan, R. Drapek, and R. Neilson (2008), VEMAP vs VINCERA: A DGVM sensitivity to differences in climate scenarios, *Global and Planetary Change*, *64*(1–2), 38–48.

Bachelet, D., K. Ferschweiler, T. Sheehan, B. Sleeter, and Z. Zhu (2015), Projected carbon stocks in the conterminous US with land use and variable fire regimes, *Glob. Change Biol.*, *21*(12), 4548–4560; doi: 10.1111/gcb.13048.

Battipaglia, G., M. Saurer, P. Cherubini, C. Calfapietra, H. R. McCarthy, R. J. Norby, and M. F. Cotrufo (2013), Elevated CO_2 increases tree-level intrinsic water use efficiency: Insights from carbon and oxygen isotope analyses in tree rings across three forest FACE sites, *New Phytol.*, *197*, 544–554.

Bentz, B. J., J. Regniere, C. J. Fettig, E. M. Hansen, J. L. Hayes, J. A. Hicke, R. G. Kelsey, J. F. Negron, and S. J. Seybold (2010), Climate change and bark beetles of the western United States and Canada: Direct and indirect effects, *Bioscience*, *60*, 602–613.

Bowker, M. A., A. Munoz, T. Martinez, and M. K. Lau (2012), Rare drought-induced mortality of juniper is enhanced by edaphic stressors and influenced by stand density, *J. Arid Environ.*, *76*, 9–16.

Brovkin, V., L. Boysen, V. K. Arora, J. Boisier, P. Cadule, L. Chini, M. Claussen, P. Friedlingstein, V. Gayler, B. van den Hurk, G. Hurtt, C. Jones, E. Kato, N. de Noblet-Ducoudré, F. Pacifico, J. Pongratz, and M. Weiss (2013), Effect of anthropogenic land-use and land cover changes on climate and land carbon storage in CMIP5 projections for the 21st century, *J. Climate*, *26*, 6859–6881.

Brutsaert, W. (2011), Are the North American deserts expanding? Some climate signals from groundwater storage conditions, *Ecohydrology*, *5*(5); *doi*: 10.1002/eco.263.

CEQ (Council on Environmental Quality) (2011), Instructions for implementing climate change adaptation planning in accordance with executive order 13514, Federal Leadership in Environmental, Energy, and Economic Performance.

Chmura, D. J., P. D. Anderson, G. T. Howe, C. A. Harrington, J. E. Halofsky, D. L. Peterson, D. C., Shaw, and J. B. St Clair (2011), Forest responses to climate change in the northwestern United States: Ecophysiological foundations for adaptive management, *For. Ecol. Man.*, *261*, 1121–1142.

Coops, N. C., and R. H. Waring (2011), Estimating the vulnerability of fifteen tree species under changing climate in Northwest North America, *Ecol. Mod.*, *222*, 2119–2129.

Daly, C., D. Bachelet, J. M. Lenihan, R. P. Neilson, W. Parton, and D. Ojima (2000), Dynamic simulation of tree-grass interactions for global change studies, *Ecol. Appl.*, *10*, 449–469.

Daly, C., M. Halbleib, J. I. Smith, W. P. Gibson, M. K. Doggett, G. H. Taylor, J. Curtis, and P. P. Pasteris (2008), Physiographically sensitive mapping of climatological temperature and precipitation across the conterminous United States, *Int. J. Climatol.;* doi: 10.1002/joc.1688.

Furniss, M. J., B. P. Staab, S. Hazelhurst, C. F. Clifton, K. B. Roby, B. L. Ilhadrt, E. B. Larry, A. H. Todd, L. M. Reid, S. J. Hines, K. A. Bennett, C. H. Luce, and Edwards, P. J. (2010), Water, climate change, and forests: watershed stewardship for a changing climate, Gen. Tech. Rep. PNW-GTR-812. Portland, OR: U.S. Department of Agriculture, Forest Service, Pacific Northwest Research Station. 75 p.

Furniss, M. J., B. P. Staab, S. Hazelhurst, C. F. Clifton, K. B. Roby, B. L. Ilhadrt, E. B. Larry, A. H. Todd, L. M. Reid, S. J. Hines, K. A. Bennett, C. H. Luce, and P. J. Edwards (2010), Water, climate change, and forests: Watershed stewardship for a changing climate, Gen. Tech. Rep. PNW-GTR-812, Portland, OR, USDA, Forest Service, Pacific Northwest Research Station.

GAO (US General Accounting Office) (2007) Climate Change: Agencies Should Develop Guidance for Addressing the Effects on Federal Land and Water Resources, U.S. General Accounting Office, GAO-07-863.

Gillies, R. R., S. Y. Wang, and M. R. Booth (2012), Observational and Synoptic Analyses of the Winter Precipitation Regime Change over Utah. *J. Climate*, *25*, 4679–4698.

Gonzalez, P., R. P. Neilson, J. M. Lenihan, and R. J. Drapek (2010), Global patterns in the vulnerability of ecosystems to vegetation shifts due to climate change, *Glob. Ecol. Biogeog.*, *19*(6); doi: 10.1029/2001GB001508.

Halofsky, J. E., M. K. Creutzburg, and M. A. Hemstrom, eds. (2014), Integrating social, economic, and ecological values across large landscapes, Gen. Tech. Rep. PNW-GTR-896, Portland, Oregon, USDA Forest Service, Pacific Northwest Research Station.

Jiménez, M. A., F. M. Jaksic, J. J. Armesto, A. Gaxiola, P. L. Meserve, D. A. Kelt, and J. R. Gutiérrez (2011), Extreme climatic events change the dynamics and invasibility of semi-arid annual plant communities, *Ecol. Lett.*, *14*, 1227–1235.

Kam, J., J. Sheffield, and E. F. Wood (2014), Changes in drought risk over the contiguous United States (1901–2012): The influence of the Pacific and Atlantic oceans, *Geophys. Res. Lett.*, *41*, 2014GL060973.

Kasischke, E. S., B. D. Amiro, N. N. Barger, N. H. F. French, S. J. Goetz, G. Grosse, M. E. Harmon, J. A. Hicke, S. Liu, and J. G. Masek (2013), Impacts of disturbance on the terrestrial carbon budget of North America, *J. Geophys. Res. Biogeosci.*, *118*(1), 303–316.

Keenan, T. F., D. Y. Hollinger, G. Bohrer, D. Dragoni, J. W. Munger, H. P. Schmid, and A. D. Richardson (2013), Increase in forest water-use efficiency as atmospheric carbon dioxide concentrations rise, *Nature*, *499*, 324–327; doi: 10.1038/nature12291 DOI:10.1038/nature12291#_blank.

Kern, J. S. (1995), Geographic patterns of soils water-holding capacity in the contiguous United States, *Soil Sci. Soc. Amer.*, *59*(4), 1126–1133.

Kern, J. S. (2000), Erratum for geographic patterns of soils water-holding capacity in the contiguous United States, *Soil Sci. Soc. Amer.*, *64*, 382.

King, D. A., D. M. Bachelet, and A. J. Symstad (2013), Climate change and fire effects on a prairie-woodland ecotone: Projecting species range shifts with a dynamic global vegetation model, *Ecol. Evol.*; doi: 10.1002/ece3.877.

Kuchler, A. (1975), *Potential Natural Vegetation of the United States*, 2nd ed., American Geographical Society, New York.

Kuchler, A. W. (1964), Potential natural vegetation of the conterminous United States, American Geographical Society Special Publication 36, New York

Latta, G., H. Temesgen, and T. M. Barrett (2009), Mapping and imputing potential productivity of Pacific Northwest forests using climate variables, *Can. J. For. Res. (Revue Canadienne de Recherche Forestiere)*, *39*, 1197–1207.

Lenihan, J. M., D. Bachelet, R. J. Drapek, and R. P. Neilson (2008a), The response of vegetation distribution, ecosystem productivity, and fire in California to future climate scenarios simulated by the MC1 dynamic vegetation model, *Clim. Change*, *87*(*Supp* 1), S215–S230.

Lenihan J. M., D. Bachelet, R. P. Neilson, and R. J. Drapek (2008b), Simulated response of conterminous United States ecosystems to climate change at different levels of fire suppression, CO_2 emission rate, and growth response to CO_2, *Global Planet. Change*, *64*, 16–25.

Lenihan, J. M., C. Daly, D. Bachelet, and R. P. Neilson (1998), Simulating broad-scale fire severity in a dynamic global vegetation model, *Northwest Sci.*, *72*(2), 91–103.

Littell, J. S., E. E. O'Neil, D. McKenzie, J. A. Hicke, J. A. Lutz, R. A. Norheim, and M. M. Elsner (2010), Forest ecosystems, disturbance, and climatic change in Washington State, USA, *Clim. Change*, *102*(1–2), 129–158.

McCabe G. J., M. A. Palecki, and J. L. Betancourt (2004), Pacific and Atlantic Ocean influences on multidecadal drought frequency in the United States, *Proc. Nat. Acad. Sci. USA*, *101*(12), 4136–4141.

McCabe, G. J., J. L. Betancourt, S. T. Gray, M. A. Palecki, and H. G. Hidalgo (2008), Associations of multi-decadal sea-surface temperature variability with US drought, *Quaternary Int.*, *188*, 31–40.

Meehl, G. A., T. F. Stocker, W. Collins, P. Friedlingstein, A. T. Gaye, J. M. Gregory, A. Kitoh, R. Knutti, J. M. Murphy, A. Noda, S. C. B. Raper, I. G. Watterson, A. J. Weaver, and Z. C. Zhao (2007), Global climate projections, in *Climate Change 2007: The Physical Science Basis*, Contribution of Working Group I to the Fourth Assessment Report of the Intergovernmental Panel on Climate Change Cambridge University Press, Cambridge, UK.

Metherell, A. K., L. A. Harding, C. V. Cole, and W. J. Parton (1993), CENTURY soil organic matter model environment:

Technical documentation Agroecosystem version 4.0. Great Plains System Research Unit Technical Report No. 4, USDA-ARS, Fort Collins, CO.

Munson, S. M., J. Belnap, and G. S. Okin (2011), Responses of wind erosion to climate-induced vegetation changes on the Colorado Plateau, *Proc. Nat. Acad. Sci.*, *108*, 3854–3859.

Nayak, A., D. Marks, D. G. Chandler, and M. Seyfried (2010), Long-term snow, climate, and streamflow trends at the Reynolds Creek Experimental Watershed, Owyhee Mountains, Idaho, United States, *Water Resour. Res.*, *46*, W06519.

Neilson, R. P. (1995), A model for predicting continental scale vegetation distribution and water balance, *Ecol. Appl.*, *5*(2), 362–385.

Peterman, W., D. Bachelet, K. Ferschweiler, and T. Sheehan (2014), Soil depth affects simulated carbon and water in the MC2 dynamic global vegetation model, *Ecol. Mod.*, *294*, 84–93.

Peterson, D. L., and K. C. Ryan (1986), Modeling post-fire conifer mortality for long-range planning, *Environ. Man.*, *10*, 797–808.

Peterson, D., C. Millar, L. Joyce, M. Furniss. J. Halofsky, R. Neilson, and T. Morelli (2011), Responding to climate change in national forests: A guidebook for developing adaptation options, Gen. Tech. Rep. PNW-GTR-855, Portland, OR: US Forest Service.

Polade, S. J., A. Gershunov, D. R. Cayan, M. D. Dettinger, and D. W. Pierce (2013), Natural climate variability and teleconnections to precipitation over the Pacific-North American region in CMIP3 and CMIP5 models, *Geophys. Res. Lett.*, *40*, 2296–2301; doi:10.1002/grl.50491.

Reed, S. C., K. K. Coe, J. P. Sparks, D. C. Housman, T. J. Zelikova, and J. Belnap (2012), Changes to dryland rainfall result in rapid moss mortality and altered soil fertility, *Nature Climate Change*, *2*, 752–755.

Rehfeldt, G. E., N. L. Crookston, C. Sáenz-Romero, E. M. Campbell (2012), North American vegetation model for land-use planning in a changing climate: a solution to large-scale classification problems, *Ecol. Appl.*, *22*, 119–141.

Robinet, C., and A. Roques (2010), Direct impacts of recent climate warming on insect populations, *Integrat. Zoo.*, *5*(2), 132–142.

Rogers, B. M., R. P. Neilson, R. Drapek, J. M. Lenihan, J. R. Wells, D. Bachelet, and B. E. Law (2011), Impacts of climate change on fire regimes and carbon stocks of the U.S. Pacific Northwest, *J. Geophys. Res. Biogeosci.*, *116*, G03037; doi:10.1029/2011JG001695.

Rothermel, R. (1972), A mathematical model for fire spread predictions in wildland fuels, USDA Forest Service Research Paper INT-RP-115.

Rupp, D. E., J. T. Abatzoglou, K. C. Hegewisch, and P. W. Mote (2013), Evaluation of CMIP5 20th century climate simulations for the Pacific Northwest USA, *J. Geophys. Res. Atmos.*, *118*, 10,884–10,906.

Sherriff, R. L., R. V. Platt, T. T. Veblen, T. L. Schoennagel, and M. H. Gartner (2014), Historical, observed, and modeled wildfire severity in montane forests of the Colorado front range, *PLoS ONE*; doi: 10.1371/journal.pone.0106971.

Van Wagner, C. E. (1993), Prediction of crown fire behavior in 2 stands of jack pine, *Can. J. For. Res.*, *23*, 442–449.

Van Wagner C. E., and T. L. Pickett (1985), Equations and FORTRAN program for the Canadian Forest Fire Weather Index System, *Can. For. Serv. For. Tech. Rep.*, *33*.

Wang, W., K. Ichii, H. Hashimoto, P. Thornton, and R. Nemani (2009), A hierarchical analysis of the terrestrial ecosystem model BIOME-BGC: Model calibration and equilibrium analysis, *Ecol. Mod.*, *220*(17), 2009–2023.

Wang, S. Y., L. Hipps, R. R. Gillies, and J. -H. Yoon (2014), Probable causes of the abnormal ridge accompanying the 2013–2014 California drought: ENSO precursor and anthropogenic warming footprint, *Geophys. Res. Lett.*, *41*, 2014GL059748.

Westerling A. L., H. G. Hidalgo, D. R. Cayan, and T. W. Swetnam (2006), Warming and earlier spring increases western US forest wildfire activity, *Science*, *313*, 940–943.

Whitlock, C., S. L. Shafer, and J. Marlon (2003), The role of climate and vegetation change in shaping past and future fire regimes in the northwestern US and the implications for ecosystem management, *For. Ecol. Man.*, *178*, 5–21.

18

Sensitivity of Vegetation Fires to Climate, Vegetation, and Anthropogenic Drivers in the HESFIRE Model: Consequences for Fire Modeling and Projection Uncertainties

Yannick Le Page

ABSTRACT

Vegetation fires are a complex aspect of terrestrial ecology to model because they depend on a wide range of climate, vegetation, and anthropogenic factors. Little is known about how interactions between these fire drivers will play out under altered environmental conditions. In this paper, we explore the sensitivity of fire activity over temperate and tropical regions using the HESFIRE model, which is parameterized with an optimization method driven by observation data. The results highlight contrasting patterns of fire sensitivity across biomes and driver conditions, in line with our current understanding of fire ecology. We also provide additional insights into fire sensitivity and the resulting influence on model outputs based on runs with alternative climate and land-cover input data. Discrepancies in data inputs lead to large changes in fire activity across 11 regions, including 6 regions where average burned area varies by a factor of 2 or more across the alternative runs. We conclude that evaluating the representation of fire sensitivity in regional to global fire models is an essential step of model development, and that the accuracy of input data is paramount to model parameterization and to the interpretation of fire projections.

18.1. INTRODUCTION

Vegetation fires are a major driver of ecosystem and carbon dynamics, and are driven by a wide range of factors. It has been estimated that forests would double their extent in a world without fires [*Bond et al.*, 2005], and fires currently emit 2–4 Pg of carbon (PgC) per year [*Bowman et al.*, 2009], 20%–40% of industrial and fossil fuel emissions. These emissions are not always compensated for by ecosystem regrowth (e.g., deforestation fires, ~0.7 PgC/yr), and can reach extreme levels during specific events: tropical forest fires in Indonesia alone emitted 1–3 PgC during the El Niño drought of 1997–1998, with costs estimated at USD $10 Billion [*Schweithelm et al.*, 1999]. The incidence and severity of fires (referred to as "fire regimes"), depend on climate, vegetation characteristics, and anthropogenic activities (referred to as "fire drivers"), all of which are expected to change substantially in the future.

A good understanding of these fire drivers is thus of primary relevance to anticipate future fire regimes. A broad range of fire drivers has to be considered, and their influence depends on the ecosystem and timescale being considered [*Loehman et al.*, 2014]. Top-down drivers operate over decades to millennia, including global climate patterns, which play a key role in the distribution of vegetation biomes and their susceptibility to fires. High moisture levels in tropical rain forests and the lack of fuel in arid areas strongly limit fire occurrence, while seasonal grasslands typically burn every few years. Bottom-up drivers operate on shorter timescales, playing a key role in fire variability. Fire seasonality is largely controlled by seasonal changes in weather conditions and vegetation productivity [*Le Page et al.*, 2010b], and synoptic weather

Pacific Northwest National Laboratory, Joint Global Change Research Institute, University of Maryland, College Park, Maryland, USA

Natural Hazard Uncertainty Assessment: Modeling and Decision Support, Geophysical Monograph 223, First Edition.
Edited by Karin Riley, Peter Webley, and Matthew Thompson.

patterns such as large-scale droughts can lead to substantial interannual fire variability [*van der Werf et al.*, 2008; *Sedano and Randerson*, 2014]. In addition to these natural drivers, anthropogenic activities also have a clear footprint on fire regimes [*Le Page et al.*, 2010b], directly through anthropogenic ignitions (e.g., fire management, agricultural practices) and fire suppression efforts, and indirectly through their impacts on ecosystems (e.g., fragmentation, fuel management) and on climate.

A range of scientific approaches has been applied to explore the sensitivity of fire regimes to altered environmental conditions. Statistical models focusing on one particular driver have projected changes in fire frequency, fire size, and fire impacts under climate change, especially in boreal regions where climate is a dominant fire driver, and where most projections agree on substantial warming [*Flannigan et al.*, 2009]. However, modeling fire activity relies on depicting the interactions between climate, vegetation, and anthropogenic drivers [*Bowman et al.*, 2009; *Pechony and Shindell*, 2009]. As an illustration *Héon et al.* [2014] find that when including the role of fuel limitation in a modeling experiment, the warming-related increase in boreal fire risk does not lead to more fires. To capture such interactions, complex fire models have been coupled to dynamic vegetation models, include assumptions on anthropogenic fire practices, and estimate fire impacts [*Arora and Boer*, 2005; *Thonicke et al.*, 2010; *Li et al.*, 2013]. Overall, achieving a realistic global representation of fire regimes remains a challenge, and the sensitivity of fire projections to model parameterization and to input data has been little explored.

In this study, we use HESFIRE (a global fire model) under alternative parameterizations to characterize the sensitivity of fire incidence and interannual variability to anthropogenic and natural fire drivers. We run the model with different climate and land-cover input datasets to quantify the propagation of input data uncertainties to fire estimates. We then discuss the implications of these findings for fire modeling and for anticipating future changes in fire activity.

18.2. METHODS

18.2.1. The HESFIRE Model

HESFIRE is a global fire model at 1° resolution designed to project fire activity under future environmental and societal scenarios [*Le Page et al.*, 2015; Fig. 18.1]. It represents the influence of human activities, weather, and fuel and includes an optimization procedure to estimate

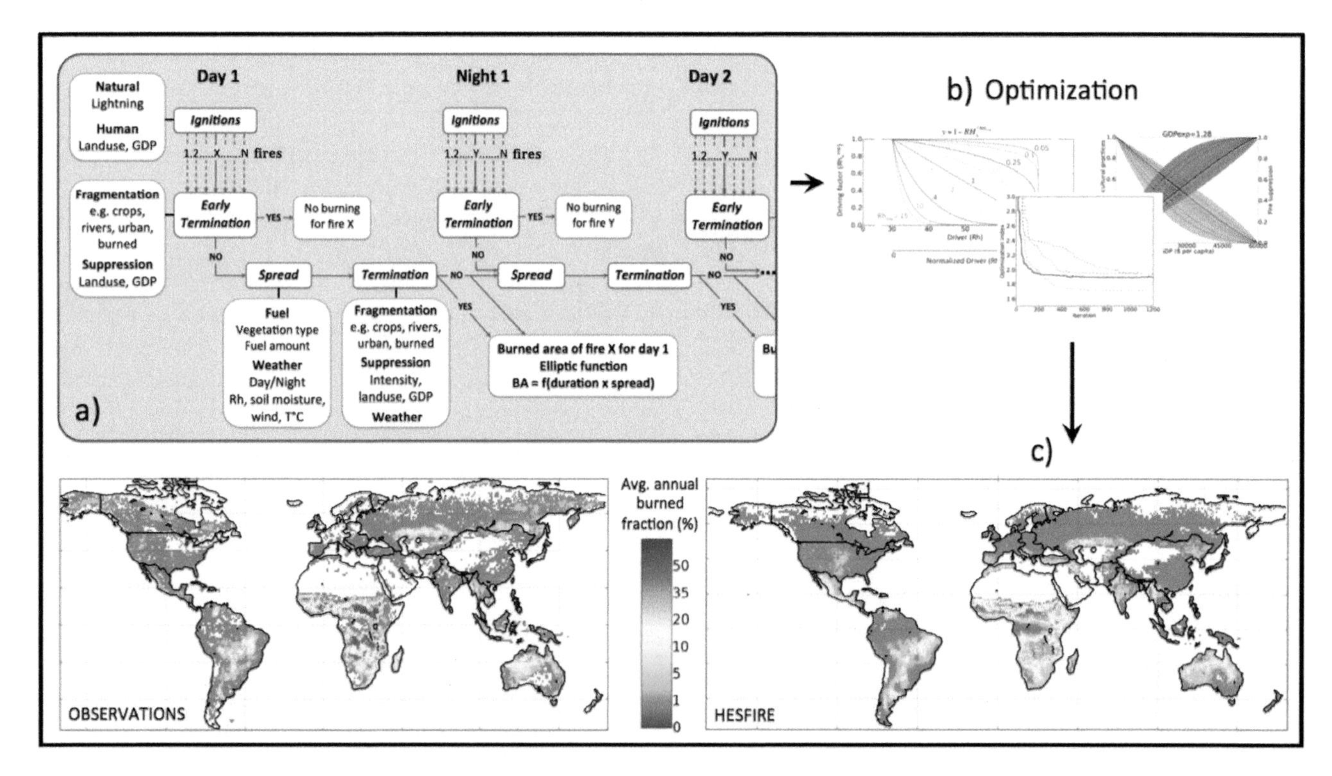

Figure 18.1 Overview of the HESFIRE model: (a) Model diagram, (b) optimization procedure, (c) comparison of average annual burned area over 1997–2011 between HESFIRE and observation-derived data (Global Fire Emission Database, GFED [*van der Werf et al.*, 2010]). Note that agricultural and deforestation fires were excluded from the GFED data, consistent with the representation of fires on natural ecosystems only in the model.

poorly constrained parameters. It also integrates a novel approach to fire spread by tracking multiday fires, a feature found in some regional models [*Finney et al.*, 2011] but not in global models of contemporary fire regimes (maximum fire duration ≤ 1 day). Multiday fires account for most of the burned area in many ecosystems (e.g., *Stocks et al.* [2002]; *Morton et al.* [2013] in boreal and tropical forests, respectively). This section includes an overview of the modeling approach, of the parameter optimization procedure, and of the main findings of the global evaluation of the model. It provides a basic understanding to support the interpretation of this study, while more detailed information is available in the original HESFIRE paper [*Le Page et al.*, 2015].

HESFIRE is organized in three modules, with specific drivers for fire ignition, spread, and termination (Fig. 18.1a):

1. Fire ignitions. Natural ignitions are a function of cloud-to-ground lightning strikes. Their occurrence is estimated from convective precipitation [*Allen and Pickering*, 2002], and a probability of ignition per strike. Human ignitions are a function of land use (cropland + urban areas) and national gross domestic product (GDP), with a single parameterization globally, independent of the type of land use (e.g., agricultural, deforestation or unintentional ignitions are not modeled explicitly).

2. Fire spread. Fire spread rate is a function of weather conditions (relative humidity, temperature, wind speed), soil moisture as a proxy for fuel moisture, and fuel structure categories (forest, shrub, and grass).

3. Fire termination. Four factors control the termination of fires: weather conditions (e.g., fires terminate when relative humidity increases above 80%), fuel availability (a function of precipitation over the last 15 months), landscape fragmentation (water bodies, bare areas, croplands, urban areas, and burned areas over the last 8 months all contribute to fragmentation), and fire suppression efforts (a function of land-use area, GDP, and climate conditions).

To account for the diurnal variability in fire spread and termination, every fire in every grid-cell is tracked individually with a 12 hr time step, from ignition to termination. HESFIRE focuses on fires in natural ecosystems: deforestation and agricultural fires are dependent on very different dynamics (controlled spread, pile burning) and thus only considered as a source of ignition for escaped fires.

Ten of the model parameters were optimized through maximizing the agreement of the model outputs with observed fire activity (Table 18.1, see details in *Le Page et al.* [2015]). These include eight shape parameters, controlling the shape of the relationship between a given driver and fires (e.g., influence of relative humidity on fire spread, influence of landscape fragmentation on fire termination), and two nonshape parameters

Table 18.1 HESFIRE Optimized Parameters, Which Were the Ones Also Selected for the Sensitivity Experiment

Parameter	Description	Original value
Ignitions		
CG_{ignp}	Cloud-to-Ground ignition probability. Average probability of ignition from a cloud-to-ground lightning strike on natural vegetation.	6.9%
LU_{ign}	Land Use ignitions. Original number of human ignitions per km² of land use per 24 hr, prior to applying density-decreasing function (see LU_{exp}).	2.3×10^{-3} km^{-1}
LU_{exp}	Land Use exponent. Shape parameter: Controls the decreasing contribution of incremental land use areas to human ignitions	14.9
GDP_{exp}	GDP exponent. Shape parameter: Impact of GDP on ignitions, through land use practices, and on the fire suppression effort.	1.28
Spread		
RH_{exp}	RH exponent. Shape parameter: Impact of relative humidity on fire spread rate.	1.18
SW_{exp}	Soil Water exponent. Shape parameter: Impact of volumetric soil moisture on fire spread rate.	1.21
T_{exp}	Temperature exponent. Shape parameter: Impact of air temperature on fire spread rate.	1.78
Termination		
$Fuel_{exp}$	Fuel exponent. Shape parameter: Impact of precipitation over –15 to –3 mo on fire termination probability, a proxy fuel buildup.	1.72
$Frag_{exp}$	Fragmentation exponent. Shape parameter: Impact of landscape fragmentation on fire termination probability.	1.81
$LUSUP_{exp}$	Land Use SUPpression exponent. Shape parameter: Impact of land use on fire termination probability, through suppression efforts, in interaction with GDP.	4.08

Note: For each of these parameters, HESFIRE is run twice, once with the value of the considered parameter increased by 50%, once with its value decreased by 50%.

(the default number of ignitions per km^2 of land use and the probability of lightning strikes to ignite a fire). The optimization was performed using a Markov Chain Monte Carlo approach on a small spatial subset (less than 2% of the world grid-cells). In order to test the model's ability to reproduce fire patterns under combinations of drivers not encountered during optimization, the spatial subset did not include any grid-cell in South America.

The evaluation of HESFIRE focused on fire incidence (Fig. 18.1c, *Le Page et al.* [2015]), seasonality, interannual variability, and on the predicted regional distribution of fire size. The main shortcomings include the underestimation of fire incidence in boreal regions partially due to biased climate inputs, uneven performances in semiarid regions due to the use of a simple cumulated-precipitation proxy for fuel load, and discrepancies with observed fire seasonality suggesting the need to better represent the timing of specific fire practices as a source of escaped fires. Overall, however, the evaluation showed strong performances regarding fire activity in most regions, including in South America, which was not part of the optimization procedure. The representation of multiday fires, the most innovative feature of HESFIRE, was also shown to lead to reasonable estimates of fire size.

18.2.2. Model Runs

18.2.2.1. Spatiotemporal Extent

The HESFIRE outputs presented here are retrieved from global, 1997–2010 runs at 1° resolution. We limit the analysis to temperate and tropical regions, given HESFIRE is not depicting realistic fire regimes in boreal regions (see Section 18.2.1).

18.2.2.2. Sensitivity to Anthropogenic and Natural Driving Parameters

We performed a parameter sensitivity experiment following the same method described in *Le Page et al.* [2015]. For each of the 10 optimized fire-model parameters (Table 18.1), HESFIRE is run twice, once with the considered parameter value increased by 50%, once with its value decreased by 50%, all other parameters being equal to the original HESFIRE configuration. These changes do not fundamentally alter the relationship depicted by shape parameters, but rather influence its nonlinearity, as illustrated in Figure 18.1b. For each grid-cell, we then extracted the parameter that generated the largest change in burned area. This approach has been applied in numerous modeling studies [*White et al.*, 2000; *Potter et al.*, 2001; *Zaehle and Friend*, 2010]. Alternative methods are discussed in [*Saltelli et al.*, 2000]. Note that wind sensitivity was not included in the analysis because unlike other drivers, its influence is modeled with several parameters, all of which were inferred from the literature rather than optimized [*Le Page et al.*, 2015]. Wind plays a major role on fire activity in some regions, however, including Mediterranean ecosystems [*Moritz et al.*, 2010; *Barros et al.*, 2012].

18.2.2.3. Sensitivity to Uncertainties in Input Data

To further characterize the sensitivity of fire projections to fire drivers, HESFIRE was run with alternative climate and land-cover input data. Originally, HESFIRE uses the Globcover version 2.3 land-cover map for 2009 [*Bontemps et al.*, 2011] and bidaily relative humidity, soil moisture, wind speed, and convective precipitation (lightning strike proxy) from the National Centers for Environmental Prediction (NCEP) reanalysis II project [*Kanamitsu et al.*, 2002]. In this analysis, it is labeled the reference run, and we perform additional runs with the Moderate Resolution Imaging Spectroradiometer (MODIS) 2009 Land Cover Type 1 product [*Friedl et al.*, 2010], and with the European Center for Medium-range Weather Forecasting (ECMWF) reanalysis ERA-Interim data [*Dee et al.*, 2011] as alternative inputs.

Uncertainties in global land-cover distribution are substantial, as illustrated by the discrepancies between the MODIS and Globcover products. In particular, grid-cell-level cropland area differs by more than 30% in many regions, including central and eastern United States, Eeastern Brazil, sub-Saharan Africa, India, China, and equatorial Asia (Fig. 18.2 a,c and *Fritz et al.* [2011]), which will impact the ignition, suppression, and fragmentation calculations of HESFIRE. Similarly, there are large discrepancies in the distribution of forests, woodlands/shrublands, and grasslands, for example in Africa [*Kaptué Tchuenté et al.*, 2011].

The NCEP and ECMWF climate data also carry relatively large uncertainties. Although global patterns and seasonal dynamics are generally consistent [*Betts et al.*, 2006], a number of studies have identified important discrepancies [e.g., *Serreze and Hurst*, 2000; *Bordi et al.*, 2006]. As an illustration, Figure 18.2 b,d shows a comparison of the average annual relative humidity in both products.

18.2.2.4. Results

Global vegetation and climate patterns have a strong influence on the regional sensitivity of fire regimes in HESFIRE (Fig. 18.3). Figure 18.3b groups the model parameters tested for sensitivity into four broad classes, indicating that fuel limitation is dominant in most arid and semiarid ecosystems, while sensitivity to landscape fragmentation extends over most grasslands and savannas of Africa and South America, as well as in the densely cultivated landscapes of Southeast Asia. Sensitivity to weather-related parameters is mostly confined to tropical forests, while the influence of anthropogenic fire practices (ignitions and suppression efforts) prevails in a few

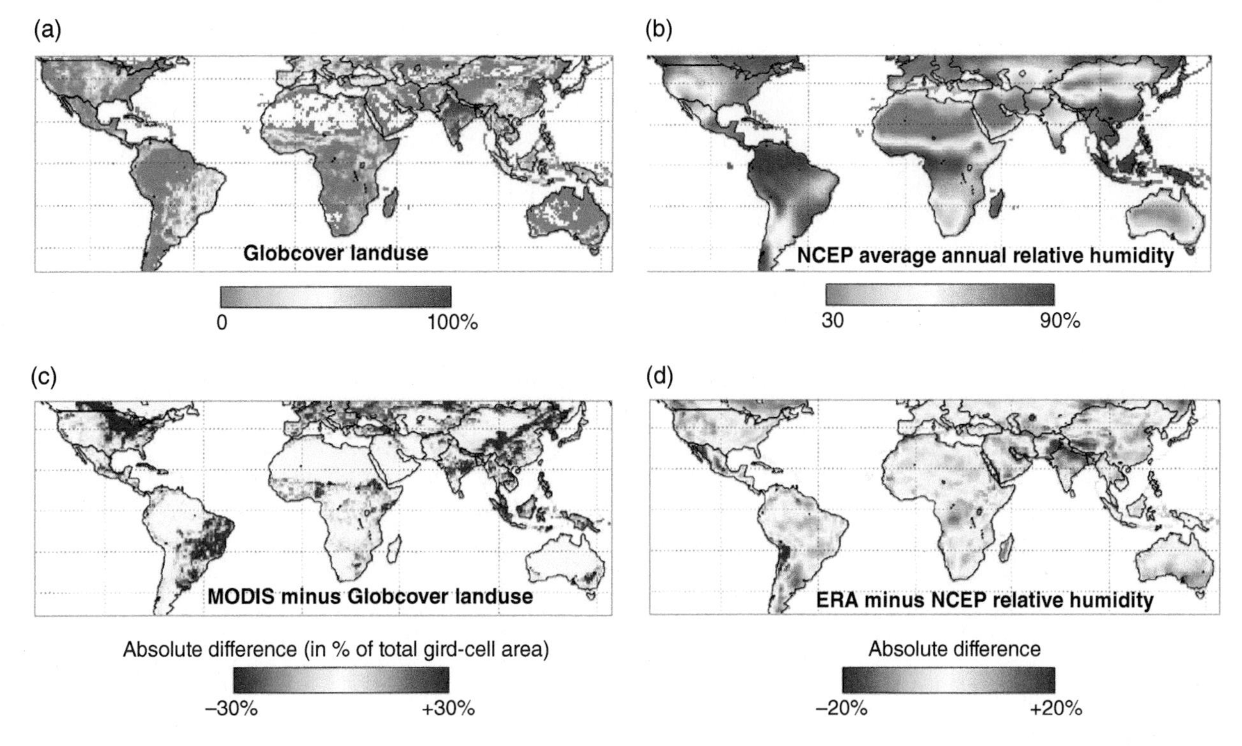

Figure 18.2 Examples of discrepancies in the input data used to run HESFIRE: (a) land use (cropland + urban areas) patterns in Globcover, (b) difference with the MODIS land cover product, (c) average annual relative humidity in NCEP, and (d) difference with the ERA data.

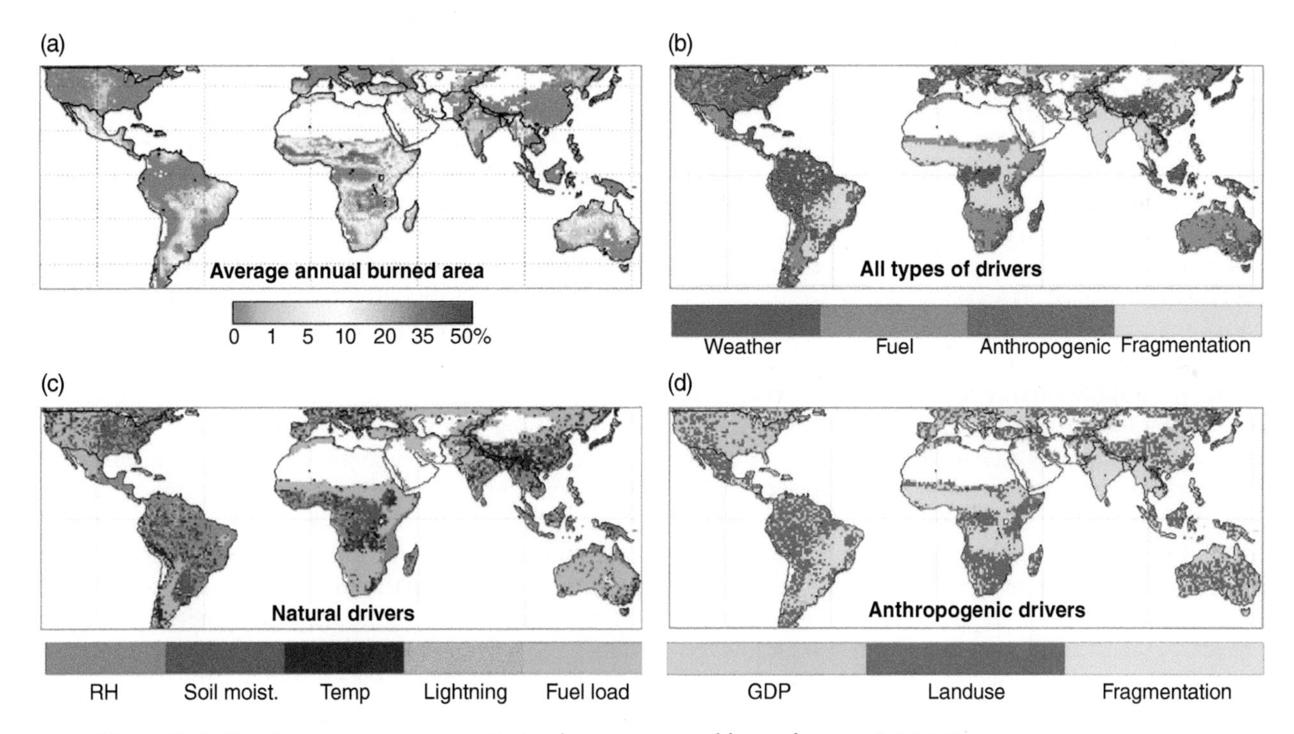

Figure 18.3 Dominant parameter-sensitivity of average annual burned area in HESFIRE.

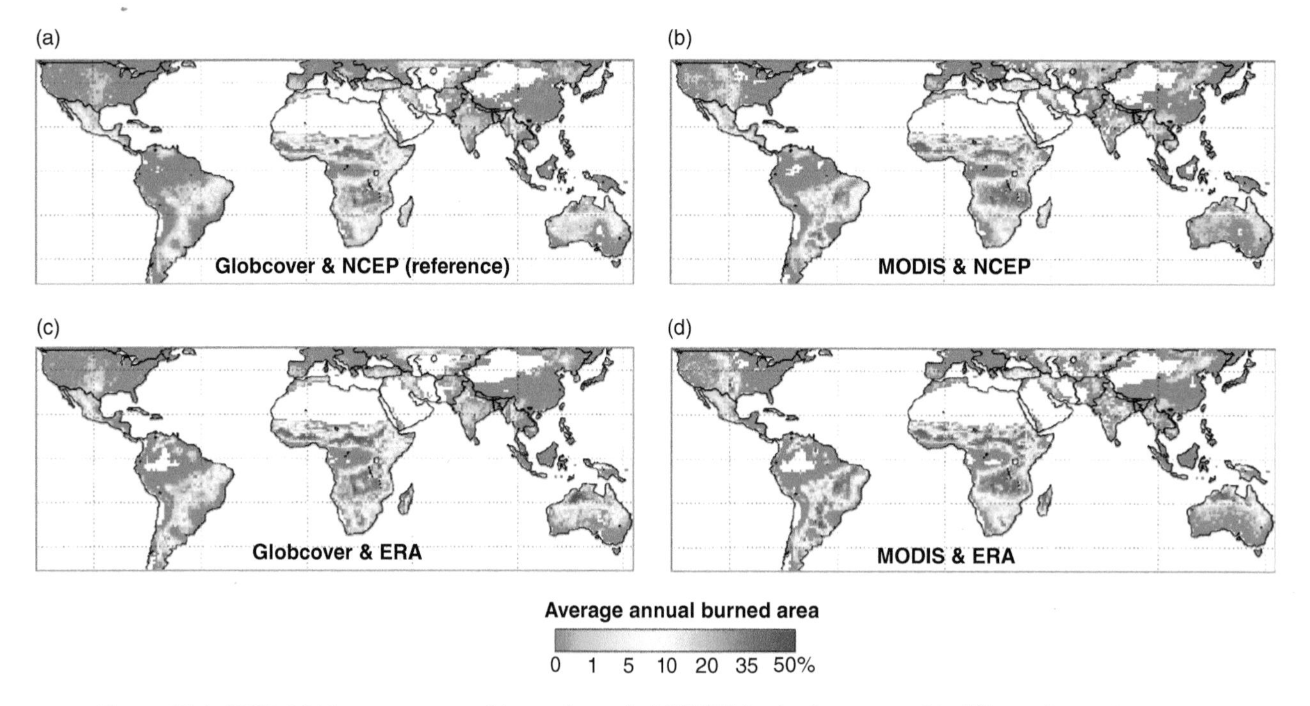

Figure 18.4 1997–2011 average annual burned area in HESFIRE in the four runs with different input data.

specific regions, namely around some populated areas of developed countries, and at the agricultural-forest frontier of the Amazon basin.

Among natural drivers, relative humidity, soil moisture, and fuel precipitation-proxy exert the largest control on fire incidence over most regions, while temperature and lightning-strike parameters are not so influential (Fig. 18.3c). For anthropogenic drivers, fire practices and fragmentation prevail over GDP (Fig. 18.3d). Note that fragmentation was included in the anthropogenic map because land use plays a major role on fragmentation in many regions, but natural factors also contribute (e.g., bare and recently burned areas).

Using alternative observation data as inputs in HESFIRE leads to very different fire projections (Fig. 18.4 and Table 18.2). When switching the land-cover data (from Globcover to MODIS), the mean annual burned area changes by a factor of 2.1 on average across 11 regions. Some regions see very large changes by a factor of 3 to 6, most notably South America (Fig. 18.4 a,b), but also the United States and equatorial Asia (Table 18.2). The change is relatively more modest when switching the climate data from NCEP to ERA (by a factor of 1.48 on average), which instead has a large influence on fire variability, with an average regional interannual correlation with the reference run of 0.55 (0.81 when switching the land-cover data). Finally, the run with both MODIS land cover and ERA climate data shows the largest discrepancies with the reference run, with a mean annual burned area changing by a factor of 3.01 and interannual correlations of 0.48, on average across the 11 regions considered.

18.3. DISCUSSION

The sensitivity of fire activity in HESFIRE changes as a function of ecosystem types, climate conditions, and anthropogenic activities. Prevalent sensitivity to weather and fuel load is widespread, but limited to biomes at both climate extremes (i.e., xeric ecosystems where fire is constrained by fuel load, and rain forests in the tropics where fire is constrained by high levels of relative humidity and soil moisture). Fire activity is found to be sensitive to landscape fragmentation over large regions, in some cases due to land-use areas (e.g., croplands in India), in others due to landscape fragmentation by early and midseason fires limiting the spread of late-season fires (e.g., sub-Saharan Africa; *Laris* [2002]). Some of this sensitivity is thus of anthropogenic origin, and adds to the area where fires are sensitive to anthropogenic fire practices, either as a dominant driver (Fig. 18.3b) or as a secondary driver (Fig. 18.3d). These patterns are consistent with the most established concepts of fire ecology, suggesting HESFIRE may provide further insights on fire drivers and on major sources of uncertainty for future fire activity.

Anticipating future fire activity and its impacts in temperate and tropical regions thus requires models including a representation of these drivers, their relative importance across regions, and their interactions.

Table 18.2 Impacts of Alternative Input Data on HESFIRE Outputs

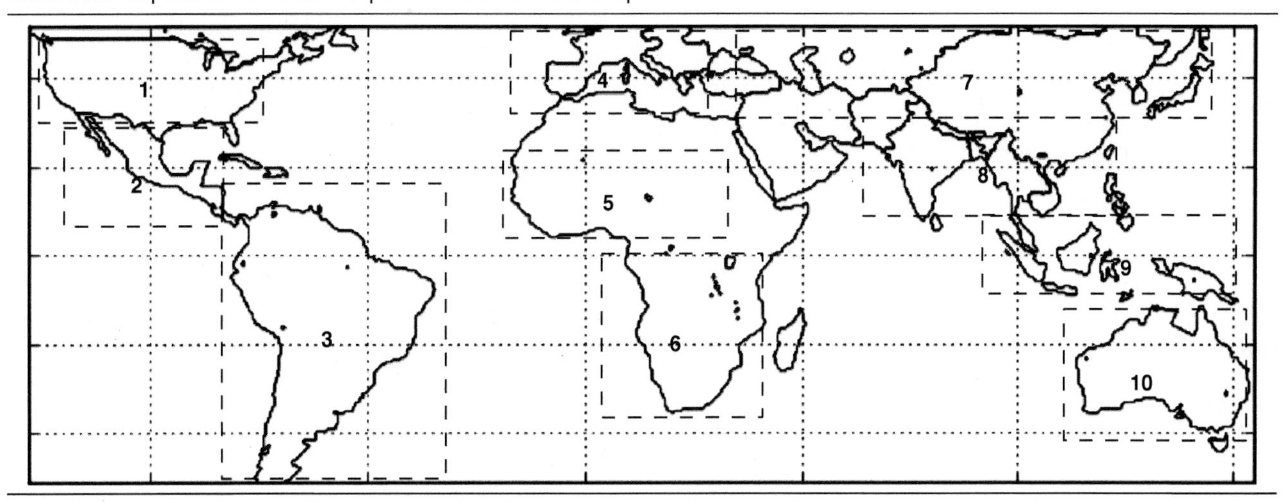

	Reference run Avg annual burned area (10^3 km^2)	Alternative data input runs Factorial change in Avg annual burned area (relative to reference run) *Correlation of regional interannual variability with reference run*		
	Globcover NCEP	Globcover ERA	MODIS NCEP	MODIS ERA
Whole area	2942	1.2 *0.49*	1.25 *0.67*	1.44 *0.20*
1	16	1.66 *−0.08*	3.14 *0.78*	5.74 *−0.04*
2	85	0.74 *0.39*	0.84 *0.93*	0.70 *0.51*
3	244	1.57 *0.45*	3.40 *0.91*	4.28 *0.41*
4	16	0.52 *0.82*	0.37 *0.93*	0.24 *0.78*
5	848	1.11 *0.81*	1.02 *0.99*	1.11 *0.82*
6	987	1.12 *0.65*	1.05 *0.86*	1.06 *0.61*
7	80	1.17 *0.46*	2.05 *0.88*	2.72 *0.41*
8	84	0.80 *0.70*	1.61 *0.88*	1.29 *0.29*
9	164	2.45 *0.96*	5.65 *1.00*	8.91 *0.97*
10	329	1.49 *0.45*	0.78 *0.96*	0.99 *0.30*
Average	--	×|÷1.48 *0.55*	×|÷2.10 *0.81*	×|÷3.01 *0.48*

Note: For the whole study area and each of 10 regions, the change in average annual burned area compared to the reference run is shown, as well as the time series interannual correlation with the reference run.

Most global models do have a representation of anthropogenic, weather, and fuel drivers, but their evaluation with observation data is generally limited to a quantitative assessment of a few statistics such as average burned area or seasonality. In-depth evaluations should include sensitivity analyses, regionally aggregated statistics, and investigating models' representation of fundamental aspects of fire ecology. Such insights are critical to ascertain that model outputs originate from realistic fire-driving mechanisms as opposed to compensative biases, and to characterize model performances and projection uncertainties.

Fire-model development and evaluation involves dealing with uncertainties in input data, however. The combination of the HESFIRE sensitivity to drivers with climate and land-cover input uncertainties significantly affects regional fire incidence and interannual variability, as shown in Table 18.2 and Figure 18.4. The magnitude of the observed changes suggests that a good knowledge of input data uncertainties is essential to the interpretation of fire outputs, and to the parameterization of the model when done through observation data assimilation methods. For fire models embedded within vegetation models, misrepresentations in vegetation cover and fuel load are likely to have a large influence on projected fire patterns. This further complicates model validation and parameterization, and could be addressed with runs forced with observation-derived land cover.

Ultimately, fire models like HESFIRE aim to project fire regimes under future scenarios of human activities and climate conditions. As reported in the literature, some regions may be particularly at risk of drastic changes, including boreal regions with warmer temperatures [e.g., *Stocks et al.*, 1998; *Flannigan et al.*, 2009] and tropical regions with further agricultural expansion in forests [*Cardoso et al.*, 2003; *Le Page et al.*, 2010a; *Silvestrini et al.*, 2011]. Sensitivity analysis provides insights into which regional environmental changes could most readily lead to altered fire regimes, independent of scenarios established in projection runs of socioeconomic, climate, and vegetation models. For example, this analysis suggests that climate change could play a major role for future fire regimes in tropical forests, with little potential for fire spread beyond the current deforestation frontiers under today's climate, even if agricultural activities were to expand in the interior Amazon. Although not conclusive given the use of a single fire model, such analysis can highlight new research opportunities to evaluate and expand on their findings, which may ultimately be relevant for land managers and policy makers.

Some of the assumptions in this first version of HESFIRE imply that the results of this study are specific to the model and further research will be needed to establish robust conclusions on the sensitivity of fire regimes. In particular, some fire drivers were modeled through proxies (e.g., soil moisture for fuel moisture), and the optimization procedure was applied at global scale, thus inferring a single value for fire-driving parameters. While the evaluation of the model suggests that this approach satisfyingly reproduces fire patterns in most ecosystems [*Le Page et al.*, 2015], fire regimes are undoubtedly influenced by mechanisms not considered in HESFIRE, specific to ecosystems (e.g., grass versus woody fuel moisture dynamics) or to anthropogenic activities (e.g., agricultural versus deforestation versus accidental fire ignitions). Consequently, the spatial variability of fire regimes sensitivity might be underestimated in HESFIRE. Observation-data analysis and regional or global models with a detailed representation of anthropogenic activities, vegetation dynamics, and fire processes could be used to evaluate and refine the results presented here.

REFERENCES

Allen, D. J., and K. E. Pickering (2002), Evaluation of lightning flash rate parameterizations for use in a global chemical transport model, *J. Geophys. Res. Atmos., 1984–2012*, *107*(D23), ACH-15.

Arora, V. K., and G. J. Boer (2005), Fire as an interactive component of dynamic vegetation models, *J. Geophys. Res. Biogeosci., 2005–2012*, *110*(G2).

Barros, A. M. G., J. M. C. Pereira, and U. J. Lund (2012), Identifying geographical patterns of wildfire orientation: A watershed-based analysis, *For. Ecol. Man.*, *264*, 98–107; doi:10.1016/j.foreco.2011.09.027.

Betts, A. K., M. Zhao, P. A. Dirmeyer, and A. C. M. Beljaars (2006), Comparison of ERA40 and NCEP/DOE near-surface data sets with other ISLSCP-II data sets, *J. Geophys. Res. Atmos.*, *111*(D22), D22S04; doi:10.1029/2006JD007174.

Bond, W. J., F. I. Woodward, and G. F. Midgley (2005), The global distribution of ecosystems in a world without fire, *New Phytol.*, *165*(2), 525–538.

Bontemps, S., P. Defourny, E. V. Bogaert, O. Arino, V. Kalogirou, and J. R. Perez (2011), GLOBCOVER 2009 Products Description and Validation Report.

Bordi, I., K. Fraedrich, M. Petitta, and A. Sutera (2006), Large-scale assessment of drought variability based on NCEP/NCAR and ERA-40 re-analyses, *Water Resour. Man.*, *20*(6), 899–915; doi:10.1007/s11269-005-9013-z.

Bowman, D. M. J. S., et al. (2009), Fire in the Earth system, *Science*, *324*(5926), 481–484; doi:10.1126/science.1163886.

Cardoso, M. F., G. C. Hurtt, B. Moore, C. A. Nobre, and E. M. Prins (2003), Projecting future fire activity in Amazonia, *Glob. Change Biol.*, *9*(5), 656–669.

Dee, D. P. et al. (2011), The ERA-interim reanalysis: Configuration and performance of the data assimilation system, *Quart. J. Roy. Meteor. Soc.*, *137*(656), 553–597; doi:10.1002/qj.828.

Finney, M. A., C. W. McHugh, I. C. Grenfell, K. L. Riley, and K. C. Short (2011), A simulation of probabilistic wildfire risk components for the continental United States, *Stochas. Environ. Res. Risk Assess.*, *25*(7), 973–1000.

Flannigan, M., B. Stocks, M. Turetsky, and M. Wotton (2009), Impacts of climate change on fire activity and fire management in the circumboreal forest, *Glob. Change Biol.*, *15*(3), 549–560; doi:10.1111/j.1365-2486.2008.01660.x.

Friedl, M. A., D. Sulla-Menashe, B. Tan, A. Schneider, N. Ramankutty, A. Sibley, and X. Huang (2010), MODIS Collection 5 global land cover: Algorithm refinements and characterization of new datasets, *Remote Sens. Environ.*, *114*(1), 168–182; doi:10.1016/j.rse.2009.08.016.

Fritz, S., L. See, I. McCallum, C. Schill, M. Obersteiner, M. van der Velde, H. Boettcher, P. Havlík, and F. Achard (2011), Highlighting continued uncertainty in global land cover maps for the user community, *Environ. Res. Lett.*, *6*(4), 044005.

Héon, J., D. Arseneault, and M.-A. Parisien (2014), Resistance of the boreal forest to high burn rates, *Proc. Nat. Acad. Sci.*, *111*(38), 13888–13893; doi:10.1073/pnas.1409316111.

Kanamitsu, M., W. Ebisuzaki, J. Woollen, S. K. Yang, J. J. Hnilo, M. Fiorino, and G. L. Potter (2002), Ncep-doe amip-ii reanalysis (r-2), *Bull. Amer. Meteor. Soc.*, *83*(11), 1631–1644.

Kaptué Tchuenté, A. T., J.-L. Roujean, and S. M. De Jong (2011), Comparison and relative quality assessment of the GLC2000, GLOBCOVER, MODIS and ECOCLIMAP land cover data sets at the African continental scale, *Int. J. Appl. Earth Obs. Geoinf.*, *13*(2), 207–219; doi:10.1016/j.jag.2010.11.005.

Laris, P. (2002), Burning the seasonal mosaic: Preventative burning strategies in the wooded savanna of southern Mali, *Hum. Ecol.*, *30*(2), 155–186.

Le Page, Y., D. Morton, B. Bond-Lamberty, J. M. C. Pereira, and G. Hurtt (2015), HESFIRE: a global fire model to explore the role of anthropogenic and weather drivers, *Biogeosciences*, *12*(3), 887–903.

Le Page, Y., D. Oom, J. Silva, P. Jönsson, and J. Pereira (2010b), Seasonality of vegetation fires as modified by human action: observing the deviation from eco-climatic fire regimes, *Glob. Ecol. Biogeogr.*, *19*(4), 575–588.

Le Page, Y., G. van der Werf, D. Morton, and J. Pereira (2010a), Modeling fire-driven deforestation potential in Amazonia under current and projected climate conditions, *J. Geophys. Res.*, *115*(G3), G03012.

Li, F., S. Levis, and D. S. Ward (2013), Quantifying the role of fire in the Earth system, Part 1: Improved global fire modeling in the Community Earth System Model (CESM1), *Biogeosciences*, *10*(4), 2293–2314; doi:10.5194/bg-10-2293-2013.

Loehman, R. A., E. Reinhardt, and K. L. Riley (2014), Wildland fire emissions, carbon, and climate: Seeing the forest and the trees–A cross-scale assessment of wildfire and carbon dynamics in fire-prone, forested ecosystems, *For. Ecol. Man.*, *317*, 9–19.

Moritz, M. A., T. J. Moody, M. A. Krawchuk, M. Hughes, and A. Hall (2010), Spatial variation in extreme winds predicts large wildfire locations in chaparral ecosystems, *Geophys. Res. Lett.*, *37*(4), L04801; doi:10.1029/2009GL041735.

Morton, D. C., Y. Le Page, R. DeFries, G. J. Collatz, and G. C. Hurtt (2013), Understorey fire frequency and the fate of burned forests in southern Amazonia, *Philos. Trans. R. Soc. B Biol. Sci.*, *368*(1619).

Pechony, O., and D. T. Shindell (2009), Fire parameterization on a global scale, *J. Geophys. Res. Atmos. 1984–2012*, *114*(D16).

Potter, C. S., et al. (2001), Comparison of boreal ecosystem model sensitivity to variability in climate and forest site parameters, *J. Geophys. Res. Atmos.*, *106*(D24), 33671–33687; doi:10.1029/2000JD000224.

Saltelli, A., S. Tarantola, and F. Campolongo (2000), Sensitivity analysis as an ingredient of modeling, *Stat. Sci.*, *15*(4), 377–395.

Schweithelm, J., T. Jessup, and D. Glover (1999), Conclusions and policy recommendations, 130–44, in *Indonesia's Fires and Haze: The Cost of Catastrophe*, edited by D. Glover and T. Jessup, Institute of Southeast Asian Studies, Singapore.

Sedano, F., and J. T. Randerson (2014), Multi-scale influence of vapor pressure deficit on fire ignition and spread in boreal forest ecosystems, *Biogeosciences*, *11*(14), 3739–3755; doi: 10.5194/bg-11-3739-2014.

Serreze, M. C., and C. M. Hurst (2000), Representation of mean Arctic precipitation from NCEP-NCAR and ERA reanalyses., *J. Climate*, *13*(1).

Silvestrini, R. A., B. S. Soares-Filho, D. Nepstad, M. Coe, H. Rodrigues, and R. Assunção (2011), Simulating fire regimes in the Amazon in response to climate change and deforestation, *Ecol. Appl.*, *21*(5), 1573–1590.

Stocks, B. J. et al. (1998), Climate change and forest fire potential in Russian and Canadian boreal forests, *Clim. Change*, *38*(1), 1–13.

Stocks, B. J., et al. (2002), Large forest fires in Canada, 1959–1997, *J. Geophys. Res. Atmos.*, *107*(D1), 8149; doi:10.1029/2001JD000484.

Thonicke, K., A. Spessa, I. C. Prentice, S. P. Harrison, L. Dong, and C. Carmona-Moreno (2010), The influence of vegetation, fire spread and fire behaviour on biomass burning and trace gas emissions: results from a process-based model, *Biogeosciences*, *7*(6), 1991–2011.

Van der Werf, G. R., J. T. Randerson, L. Giglio, G. Collatz, M. Mu, P. S. Kasibhatla, D. C. Morton, R. DeFries, Y. Jin, and T. T. van Leeuwen (2010), Global fire emissions and the contribution of deforestation, savanna, forest, agricultural, and peat fires (1997–2009), *Atmos. Chem. Phys.*, *10*(23), 11707–11735.

Van der Werf, G. R., J. T. Randerson, L. Giglio, N. Gobron, and A. J. Dolman (2008), Climate controls on the variability of fires in the tropics and subtropics, *Glob. Biogeochem. Cycles*, *22*(3); doi:10.1029/2007GB003122.

White, M. A., P. E. Thornton, S. W. Running, and R. R. Nemani (2000), Parameterization and sensitivity analysis of the BIOME-BGC terrestrial ecosystem model: Net primary production controls, *Earth Interact.*, *4*(3), 1–85; doi:10.1175/1087-3562(2000)004<0003:PASAOT>2.0.CO;2.

Zaehle, S., and A. D. Friend (2010), Carbon and nitrogen cycle dynamics in the O-CN land surface model: 1. Model description, site-scale evaluation, and sensitivity to parameter estimates, *Glob. Biogeochem. Cycles*, *24*(1), GB1005; doi:10.1029/2009GB003521.

19

Uncertainties in Predicting Debris Flow Hazards Following Wildfire

Kevin D. Hyde,[1] Karin Riley,[2] and Cathelijne Stoof[3]

ABSTRACT

Wildfire increases the probability of debris flows posing hazardous conditions where values-at-risk exist downstream of burned areas. Conditions and processes leading to postfire debris flows usually follow a general sequence defined here as the postfire debris flow hazard cascade: biophysical setting, fire processes, fire effects, rainfall, debris flow, and values-at-risk. Prediction of postfire debris flow hazards is a problem of identifying and understanding the spatial and temporal interactions within this cascade. Models exist that can predict each or several components of the cascade but no single model or prediction approach exists with capacity to link the entire sequence. Assessment of the uncertainty inherent with each of these approaches is limited and compounds if these predictive approaches are integrated or otherwise linked. We summarize present knowledge of the processes involved in this postfire debris flow hazard cascade and identify uncertainties in terms of knowledge gaps, contradictions in current process understanding, stochastic system variables, and limits to data to support hazard prediction. Understanding these uncertainties can improve delineation of areas threatened by postfire debris flows, can guide future research, and, when addressed, contribute to development of comprehensive and robust modeling and prediction systems that may ultimately reduce threats to values-at-risk.

19.1. INTRODUCTION

Debris flows are rapidly moving masses of water, fine sediments, rocks, and often woody material [*Iverson*, 1997] that constitute an extreme form of postfire erosion. Their probability is heightened following moderate- to high-severity wildfire on steep slopes in forested and shrub-dominated environments, such as sage and chaparral. These extreme events are natural processes that scour low-order channels, work as a major driver of landscape evolution [*Kirchner et al.*, 2001; *Pierce et al.*, 2004] and impact aquatic ecosystems. Depending on the location and timing of the debris flows relative to other disturbances, sediments and woody debris transported by debris flows into stream systems may periodically enhance or restore the complexity of aquatic habitats, or threaten the stability of fish populations [*Gresswell*, 1999; *Rieman et al.*, 2003].

Debris flows also pose hazardous conditions where values-at-risk exist downstream of burned areas and can threaten human life and property. While comprehensive losses are not well documented, multiple accounts report loss of life, damaged and destroyed infrastructure [*USGS*, 2015a], and impaired drinking water supplies [*Bladon et al.*, 2014; *Smith et al.*, 2011]. Losses are likely to increase given the expectation of more frequent large and severe fires [*Bachelet et al.*, Chapter 17, this volume; *Carvalho et al.*, 2010; *Holz and Veblen*, 2011; *LePage*, Chapter 18, this volume]

[1] *University of Wyoming, Laramie, Wyoming, USA*

[2] *Numerical Terradynamic Simulation Group, College of Forestry and Conservation, University of Montana, Missoula, Montana, and Rocky Mountain Research Station, US Forest Service, Missoula, Montana, USA*

[3] *Wageningen University, Wageningen, Netherlands*

Natural Hazard Uncertainty Assessment: Modeling and Decision Support, Geophysical Monograph 223, First Edition.
Edited by Karin Riley, Peter Webley, and Matthew Thompson.

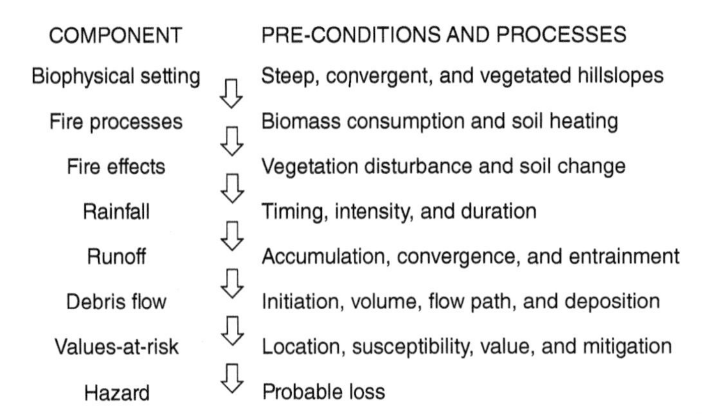

Figure 19.1 The postfire debris flow hazard cascade with the sequence of preconditions and processes leading to hazardous conditions. The down arrows reinforce the linear nature of this conceptual model where the generation of hazardous conditions is contingent upon the preceding component steps. This hazard cascade offers logic for future development of a comprehensive postfire debris flow hazard prediction framework.

and growing population and infrastructure development at the wildland-urban interface [*Calkin et al.*, 2011a; *Theobald and Romme*, 2007].

Not all burned areas subjected to heavy rainfall produce debris flows, suggesting that some combination of burn conditions, landform, and rainfall characteristics influence susceptibility to debris flows following fire [*Cannon and Gartner*, 2005]. Prediction of postfire debris flow hazards is a problem of identifying and understanding the spatial and temporal sequence of interactions of multiple preconditions and processes. We refer to this sequence as the postfire debris flow hazard cascade (Fig. 19.1). First, the biophysical setting must include flammable vegetation on a steep, concave hillslope. Second, a moderate- to high-severity fire must occur (fire processes) that removes of a significant amount of the vegetation and changes soil properties (fire effects). Next, rainfall of sufficient intensity and duration to produce overland flow (surface runoff) must follow the fire. Runoff must accumulate and converge, entraining sediments, and flowing with sufficient force to down-cut and initiate a debris flow. The flowing mass of debris must intersect values-at-risk with sufficient volume and force to cause damage and loss. Without threat to valued resources, there is no hazard.

The level of understanding of processes and process interactions varies among the hazard cascade components rendering differing levels of uncertainty in efforts to predict fire-related debris flow hazards. For example, the physics of debris flows and general debris flow behavior have been thoroughly studied [*Iverson*, 1997; *Jakob and Hungr*, 2005] while process-based understanding of fire behavior is emerging [*Finney et al.*, 2013] and process-based linkages between fire behavior and fire effects are poorly understood [*Hyde et al.*, 2013]. Further, the stochastic nature of environmental processes (e.g., fire ignitions and rainfall distribution) introduces additional uncertainty throughout the hazard cascade.

Models exist that can predict each or several components of the hazard cascade but no single model or prediction approach exists with capacity to link the entire sequence of preconditions and processes. The behavior of a fire can be modeled at one point in time given a single set of fuels and weather conditions, from which fire spread and intensity can be estimated [*Finney*, 2006]. Effective prediction of fire spread under active fire conditions can be accomplished using short-term weather predictions and a suite of weather scenarios based on historic weather patterns [*Finney et al.*, 2011b].This ensemble modeling approach accounts for the uncertainty of fire behavior under shifting weather patterns and has been incorporated into decision support systems [*Calkin et al.*, 2011b]. Burn probabilities and fire size distributions have been modeled across very large landscapes over extended time frames [*Finney et al.*, 2011a] and used to support prioritization of fuel treatments [*Ager et al.*, 2014].

Models of vegetation change from fire combined with crosswalks or rule sets link vegetation changes to alterations in ecosystem processes [*Black and Opperman*, 2005]. Empirical rainfall intensity-duration thresholds provide decision support for assessing the probability of postfire floods and debris flows throughout the western United States [e.g., *Cannon et al.*, 2008]. Probabilistic models have been developed to determine the binary occurrence of postfire debris flows based on fire severity, landform, soils, and other factors [e.g. *Gartner et al.*, 2015; *Hyde et al.*, 2015]. Several of the above-mentioned approaches have been combined, integrating prediction of fire intensity and spread and the estimation of vegetation consumption for use in probabilistic debris flow occurrence models [*Haas et al.*, Chapter 20, this volume].

General debris models estimate entrainment of materials during debris flow development, accumulated debris volume, and debris flow run-out and deposition [*Ghilardi et al.*, 2001; *Luna et al.*, 2012; *Luna et al.*, 2014]. A model developed to map the transport and impact of aqueous contaminants in streams and rivers [*Leidos Corp.*, 2014; *Samuels et al.*, 2006] is being modified to support fire management decisions including threats in the event of misapplications of fire retardant chemicals and from sediments transported from burned areas [*Hyde*, 2009]. Finally, a tool to support benefit-cost analysis for postfire emergency response treatments combines spatially explicit assessment of potential debris flow source with values-at-risk downstream with estimates of probable debris flow and flooding occurrence with probable loss [*Calkin et al.*, 2007].

Some of these approaches to predicting components of the postfire debris flow cascade have been subject to verification. However, there is limited if any assessment of the uncertainty inherent with each of these approaches, much less consideration of compounding uncertainties if these predictive approaches are integrated or otherwise linked. The many sources of uncertainty make accurate prediction of postfire debris flow occurrence complex. Further, these uncertainties have not been comprehensively identified or discussed. This information is needed to improve delineation of areas threatened by postfire debris flows and thus facilitate protection of human life and property, ecosystems, and ecosystem services.

In this chapter, we first summarize present knowledge of the processes involved in this postfire debris flow hazard cascade and then identify uncertainties in terms of knowledge gaps, contradictions in current process understanding, and stochastic system variables. We also discuss uncertainty in the data used to support current modeling approaches. The identified knowledge gaps can guide future research and, when addressed, contribute to development of comprehensive and robust modeling and prediction systems that may ultimately reduce threats to values-at-risk. We address two phases of postfire debris flow hazards: direct impacts from a debris flow mass and impaired water quality from sediments deposited into streams by debris flows. This discussion also considers two planning and analysis domains: hazard assessment following fire and prefire planning to predict both future fire and resulting postfire debris flows.

19.2. BIOPHYSICAL SETTING

The biophysical setting for postfire debris flows includes a combination of forested or shrub-covered landscapes, steep slopes (>25%), convex terrain, erodible soils, and stored debris. The probability of debris flows is much increased in watersheds that experience moderate or high-severity fire [*Cannon et al.*, 2010; *Gartner et al.*, 2008], suggesting the landscape must be prone to crown fires during which the majority of the forest canopy is consumed. Factors controlling crown fires include the spatial variability of forest structure or, more specifically, whether surface fire can ascend up "ladder fuels" into the upper canopy, a common condition in a multiage stand, and whether stands are of sufficient density. However, the physical conditions of the stand are not enough to predict crown fire; certain weather conditions, including high winds, are generally necessary for crown fire to be sustained. Where crown fires occur without surface fires, erosion risk is reduced, as undisturbed vegetation stabilizes the soil surface and severe surface erosion generally does not occur where vegetation is intact [*Jenkins et al.*, 2011; *Prosser and Williams*, 1998; *Wondzell and King*, 2003].

Under stable conditions, erodible materials (including soil, rock fragments, woody debris and other organic material) accumulate slowly over time on hillslopes and within hillslope hollows and channels [*Santi et al.*, 2008]. The amount of stored material available to be entrained and incorporated into a debris flow depends on the rock type, the local weathering conditions, and the time since the last fire [*Jenkins et al.*, 2011]. The last factor is in part determined by the fire return interval, which varies by region and vegetation type and can range from less than 10 to hundreds of years [*Oliveira et al.*, 2012; *Westerling et al.*, 2011]. Several soil properties also control the probability that debris flows will occur, such as clay content and organic matter, among others that increase or decrease occurrence probability [*Cannon et al.*, 2010].

Topography exerts two strong effects through curvature and slope. Curvature affects debris flow initiation as concave and therefore convergent contributing areas typically focus sediment-laden flows into existing swales [*Cannon*, 2001; *Hyde et al.*, 2014], although debris flows have been observed to originate on planar or flat slopes in very steep areas as well [*Neary et al.*, 2012]. Slope is particularly important in relation to the size of contributing areas, with steeper slopes generally requiring smaller upstream areas generating surface runoff to initiate debris flows than where slopes are less steep [*Stock and Dietrich*, 2006; *Hyde et al.*, 2014]. Further discussion of interactions between slope, curvature, and debris flow probability follows below in the section on initiation and mobilization.

Sources of uncertainty related to physical setting include the susceptibility of the vegetation structure to fire, time since last fire and therefore level of fuel accumulation, properties of the unburned soils, and limited understanding of the interactions of hillslope geometry (steepness and curvature) relative to fire effects. Mapping of forest stand conditions is reasonably thorough for all forested lands within the United States [*USGS*, 2009] and throughout Europe and Australia. However, given vegetation growth patterns, local conditions vary and the existence of forest structure mapping for other fire-prone regions throughout the world is uncertain. Mapping of soil properties poses similar uncertainty. Current soil maps in the United States were meant for broadscale regional comparisons [*Lathrop et al.*, 1995], not to capture the smaller-scale variability of critical properties of soils within the upper hillslope regions where debris flows initiate. European soil databases are more detailed and can more reliably provide soil property information to parameterize hazard prediction models.

19.3. FIRE PROCESSES

Fire processes and how fire behaves on the landscapes control the spatial variability and magnitude of fire effects. Fire propagates or spreads by a physical process that spawns a series of ignitions across the landscape. This physical process of surface fire spread has been widely modeled as propagating by thermal radiation where heat energy is transferred by electromagnetic waves [e.g., *Rothermel*, 1972]. However, recent work has shown that radiation is inadequate to ignite fine fuels, and that direct flame contact and more specifically ignition by buoyant gases are the more likely mechanism [*Finney et al.*, 2010; *Finney et al.*; 2015]. In addition, the mechanisms of crown fire spread (the spread of fire in the vegetation canopy usually leading to full canopy loss) are not well understood. Further, in current models, there is a degree of uncertainty in the rate of spread, fire intensity, and flame length calculations, factors that influence fire extent and degree of surface vegetation loss and soil heating, resulting in part from a number of questionable modeling assumptions. For example, fire spread models assume homogeneity of fuels when fuels are commonly not homogenous, use a discrete set of fuel models to map the landscape when a nearly infinite array of combinations of fuels is possible, and use a constant wind speed and direction when wind speed and directions typically vary across small distances and time steps.

Regardless of the mechanism for fire spread and these uncertainties, current models have been empirically tuned so that they predict physical properties of the fire (such as intensity, heat release, and rate of spread) well enough for the models to be useful during wildfire incidents and for landscape assessment [*Alexander and Cruz*, 2011; *Finney et al.*, 2011b]. While fire physics are not well understood [*Finney et al.*, 2010], a suite of models allows for reasonable prediction of the number of fires, their size, and intensity [*Finney et al.*, 2011a]. Because these models can be run over long time frames, simulating tens of thousands of possible fire seasons, a probability distribution for flame lengths (which depend on wind, fuel moisture, and direction of the fire) at each cell of a rasterized landscape can be generated. Translating these flame lengths into fire effects and impacts on highly valued resources is more difficult, and currently depends on expert opinion [*Thompson et al.*, 2010] introducing uncertainty related to decisions regarding ecological effects as well as valuation. In the postfire landscape, the extent of the fire and its behavior are already known, reducing overall uncertainty of fire behavior and fire effects. However, in the planning context, the extent and behavior of future fires are highly uncertain since fires will occur in response to stochastic and therefore very unpredictable patterns of ignition and weather conditions [*Riley and Thompson*, Chapter 13, this volume].

19.4. FIRE EFFECTS

The processes of fire behavior are distinct from fire effects, the resulting changes to biomass and soils [*Hyde et al.*, 2013; *Reinhardt and Dickinson*, 2010]. Fire consumes live and decaying vegetation or biomass above ground, and in some cases below ground, such as soil organic matter and roots (Fig. 19.2). The canopy, subcanopy, shrub, and herb layer, litter, and duff may be partially or completely killed depending on fire behavior and fuel moisture. Under severe fire conditions, forest canopy and ground cover are fully consumed, with only the trunks and larger branches of trees and shrubs remaining. Under less severe conditions, residual amounts of partially consumed vegetation, litter and duff form irregular patchworks or burn mosaics. The degree of vegetation disturbance and soil impacts is expressed as fire severity and soil-burn severity, respectively [*Keeley*, 2009].

Fire impact on soils and thus belowground fire effects are to a large degree determined by prefire soil characteristics and conditions, the soil temperature reached, and the duration of soil heating [*Keeley*, 2009; *Mataix-Solera et al.*, 2011; *Stoof et al.*, 2010]. Effects of fire on soil physical properties can range from increases in bulk density to decreases in soil organic matter, aggregate stability, infiltration capacity, and soil-water retention [*Certini*, 2005; *Ebel*, 2012; *Mataix-Solera et al.*, 2011; *Stoof et al.*, 2010]. Fire can also induce or enhance soil-water repellency, both due to soil heating [*DeBano*, 2000] and due to drier topsoils because of vegetation removal [*Stoof et al.*, 2011]. However, fire can also reduce or not change soil-water repellent properties [*Shakesby and Doerr*, 2006]. While soil physical changes only occur in fires where soil heating is pronounced, surface cover changes due to deposition of ash and char also occur in "cooler" fires. Like soil-burn severity, postfire ash and char cover form a mosaic reflecting prefire fuel

Figure 19.2 Conceptualization of forest hillslope structure before and after high-severity wildfire where all biomass is consumed, emphasizing the role of vegetation disturbance in postfire erosion processes and debris flow generation. Loss of vegetative cover including canopy, litter, and duff increases effective rainfall. Though coniferous trees are illustrated, the processes are similar in deciduous forests and shrublands. The presence of a soil hydrophobic layer is uncertain as fire may increase, decrease, or destroy water-repellent conditions [*Shakesby and Doerr*, 2006].While vegetation disturbance is recognized as exerting significant influence over postfire erosion and debris flow processes, research has substantially focused on the role of soil changes [*Moody et al.*, 2013; *Shakesby*, 2011] with limited study of effects of vegetation disturbance, leaving significant knowledge gaps.

loads and vegetation patterns [*Bodí et al.*, 2014]. After fire, ash and char are quickly redistributed across the landscape by wind and water.

The degree of vegetation disturbance by fire influences the hydrogeomorphic response and the occurrence of gully rejuvenation and debris flows following fire [*Hyde et al.*, 2007; *Hyde*, 2013]. Changes to vegetation by fire exert similar or greater control over postfire erosion response than fire effects on soils [*Doerr et al.*, 2009; *Hyde*, 2013; *Larsen et al.*, 2009]. As vegetation cover returns, the soil surface restabilizes. Vegetation recovery can begin within days after a fire, and can take from a few years to many decades for complete recovery, depending on weather and climate conditions and the regeneration potential of remaining and/or colonizing vegetation and seed stocks.

Fire effects–related uncertainties relevant in the debris flow hazard cascade can be grouped into four broad topics, related to the characterization and prediction of (1) spatial patterns and variability of landscape-scale soil heating and vegetation disturbance, (2) fire-induced changes to hydrological response, (3) spatial and temporal patterns of vegetation recovery, and (4) challenges to quantifying and mapping fire effects. Prediction of debris flow occurrence requires knowledge of the extent and duration of soil heating and related changes in soil structure in order to understand changes in erodibility and soil-water repellency. Yet, most fire behavior research is focused on the upward heat component, with little focus on the downward heat fluxes. Understanding of the relationship between upward and downward heat components is currently incomplete [*Stoof et al.*, 2013a], but essential for prefire assessment of landscape vulnerability to fire-induced soil changes. As not all fires cause soil changes, and fire effects with hydrological impact are not limited only to soil physical changes, it is important to assess the hydrological and erosion effects of soil changes in the context of the effects of vegetation removal and soil-water repellency. Also, ash is a factor to take into account, certainly above ground but also below ground. Ash can mitigate surface runoff by absorbing rainfall [*Bodí et al.*, 2014; *Cerdà and Doerr*, 2008; *Woods and Balfour*, 2008], but little is known about factors controlling ash production, composition, and downward movement into soils. Below ground, pore-clogging by ash has often been cited as cause of the reduction in infiltration observed after fire [*Bodí et al.*, 2014; *Etiegni and Campbell*, 1991; *Woods and Balfour*, 2010], though evidence for this is lacking. On the contrary, recent work by *Stoof et al.* [2016] indicates that the mere presence of ash in pores does not automatically lead to pore-clogging to the point that infiltration is hampered and ponding occurs. In short, fire-induced changes to soil produce a major source of uncertainty in prediction of debris flows in both the immediate postfire and landscape assessment contexts.

With the high spatial variability of fire effects in the postfire landscape, postfire hydrological response is also highly heterogeneous. The highly heterogeneous landscape after fire is a mosaic of fire severity, soil-burn severity, partially to fully consumed litter and duff layers, and ash and char layers of varying thickness and characteristics. This implies that there is strong variability in a range of factors including canopy interception and raindrop impact, surface water storage in litter/duff/ash layers, soil physical properties, water repellency, erodibility, and so on. The interactions and relative hydrological importance of effects of ash, vegetation removal, soil-water repellency, and the various soil physical changes that can occur are poorly understood [*Doerr et al.*, 2009; *Larsen et al.*, 2009; *Moody et al.*, 2013]. The hydrological impact of fire in such a variable landscape is influenced by the location of the most severe hydrological effects and probably by the hydrological connectivity between the various patches [*Hyde et al.*, 2014; Reaney *et al.*, 2013] and other features such as the location of macropores [*Nyman et al.*, 2010]. A major uncertainty here is the specific threshold for plot-, hillslope-, and watershed-scale connectivity related to erosion response [*Reaney et al.*, 2007; *Reaney et al.*, 2013].

It is important to note that the above-mentioned heterogeneity is present not only spatially, but also temporally. Burned areas are also highly dynamic in time, for instance in terms of the rapid redistribution of ash and char as well as the evolution of soil-water repellency, runoff and erosion, and vegetation cover [*Pereira et al.*, 2013; *Pierson et al.*, 2008; *Shakesby et al.*, 2013; *Stoof et al.*, 2013b; *Woods et al.*, 2007]. The rate of ecosystem recovery is highly relevant in the debris flow hazard cascade, in particular when severe rain events occur following fire before vegetation regenerates or newly colonizes and restabilizes the soil surface. Many factors control natural revegetation of burned areas, including remaining seed stocks, seed environment, meristem tissue of remaining flora and sprouting potential, and weather conditions favoring germination [*Brown and Smith*, 2000]. Existing research focuses on the effects of vegetation recovery on runoff and erosion [e.g., *Gimeno-García et al.*, 2007; *Shin et al.*, 2013; *Wittenberg et al.*, 2014]. Work remains to directly address uncertainties about regeneration or advancing understanding of revegetation processes.

Several factors present major limitations to quantitatively assessing fire effects. Existing field methods to quantify and map fire effects are arguably subjective relying heavily on visual assessment [see *Key and Benson*, 2006; *Parsons et al.*, 2010] and raising questions of interobserver reliability [*Gwet*, 2001]. Commonly used methods to map fire severity using remotely sensed imagery are designed to capture effects of fire on vegetation [*Chafer*, 2008; *Epting et al.*, 2005; *Hudak et al.*, 2007] but are routinely used as the basis for field mapping of fire effects on soils [see *Parsons et al.*, 2010]. Uncertainties in remotely sensed methods to map fire severity include highly variable accuracy of signal discrimination between types and levels of fire severity and concerns that the wavelengths chosen to characterize fire severity may not be sufficient to best measure fire effects [*Roy et al.*, 2006] and soil-burn severity. Furthermore, understanding of the linkages between the information about fire effects in the remotely sensed signal and hydrogeomorphic processes related to debris flows is limited [*Hyde et al.*, 2013]. Probable interaction of multiple disturbance processes (e.g., land use, fire history, and mortality by insects and disease) further confounds understanding of the degree to which fire effects on soil determine postfire erosion processes and the generation of debris flows [*Ebel and Mirus*, 2014].

19.5. RAINFALL TRIGGERS

During the recovery period, rainfall occurring over the disturbed landscape meets little or no resistance compared to vegetated hillslopes. The effect of rainfall on the burned landscape depends on storm timing relative to vegetation recovery, intensity, and duration [*Moody et al.*, 2013; *Shakesby et al.*, 2013]. Canopy loss decreases rainfall interception, both regarding its quantity [*Stoof et al.*, 2012] and its erosive power [*Gabet and Dunne*, 2003]. In burned landscapes, the erosive power of rainfall is typically greater than in unburned landscapes as more rain impacts exposed surfaces than under vegetated conditions. Combined with possible soil-water repellency, the net result may be the generation of overland flow. As illustrated in Figure 19.2, loss of surface vegetation, litter. and duff can permit rapid accumulation of surface flow and entrainment of fine material including vegetative ash, bulking the flow and potentially increasing its force [*Gabet and Sternberg,* 2008].

The uncertainties associated with rainfall triggers include the stochastic nature of rainfall events, the timing of rainfall relative to vegetation recovery, limited knowledge of changes in rainfall energy relative to vegetation loss, and how these factors interact with landscape steepness, curvature, and the mosaic of fire effects. Further uncertainties exist about thresholds of rainfall intensity and duration needed to initiate a debris flow and how these vary by terrain, soil properties, and fire severity patterns [e.g., *Cannon et al.*, 2008]. Rainfall presents a source of uncertainty from stochastic variability, since the timing and intensity of rainfall cannot be predicted. RADAR technology may provide some real-time warning of locations of potentially threatening rainfall that could trigger debris flows [*Nikolopoulos et al.*, Chapter 21, this volume]. However, in mountainous environments where wildfires typically occur, the capacity of RADAR to accurately identify storm intensity is substantially compromised [*Young et al.*, 1999].

Rainfall can be estimated statistically, by using historic distributions of intensity, duration, and timing. However, changing climate adds other layers of uncertainty to the predictability of the timing, magnitude, and spatial distribution of rainfall events [*Beniston*, 2006]. Assuming reliable rainfall estimates can be derived, these can be combined with maps of topography and fire effects to produce a probability of debris flow occurrence [*Cannon et al.*, 2010]. The reliability of these combined prediction tools is limited by the previously identified limits to consistently and objectively mapping fire effects.

19.6. DEBRIS FLOW INITIATION, MOBILIZATION, AND DEPOSITION

Debris flow initiation commences through two primary processes, progressive sediment bulking and landslides caused by saturation-induced failure. Progressive sediment bulking [*Cannon and Gartner*, 2005] is the most commonly recognized process following fire, and typically occurs within the first 5 yr postfire [*Riley et al.*, 2013]. Runoff generated by rainfall spawns sediment-laden flows that accumulate in headwater hollows and continue down-channel in first- or second-order catchments. Overland flow increases in volume and sediment content until some threshold at which down-cutting commences, the flow becomes highly viscous, and the downslope velocity rapidly increases. Initiation commonly occurs in low-order catchments near watershed divides and topography exerts strong effects through curvature and slope steepness [*Hyde et al.*, 2014]. Curvature affects debris flow initiation as concave contributing areas typically focus sediment-laden flows into existing swales, although debris flows have been observed to originate on planar slopes in very steep areas as well [*Neary et al.*, 2012]. Slope steepness is particularly important in relation to the size of contributing areas, with steeper slopes generally requiring smaller contributing areas for debris flow initiation than where slopes are shallower [*Stock and Dietrich*, 2006]. This inverse slope-area relationship is sensitive to fire severity [*Hyde et al.*, 2014; *Pelletier and Orem*, 2014], as debris flows may initiate from smaller and less-steep conditions where fire effects are severe. Saturation-induced failures related to fire have been identified in the US Pacific Northwest and have been observed as a delayed response that occurs 10 to 15 yr following fire [*Benda and Dunne*, 1997; *Roering and Gerber*, 2005; *Roering et al.*, 2003] where root strength fails as fire-killed trees decay.

Debris flow volume increases as the mass gains speed and moves downslope, typically within constrained channels in the case of postfire debris flows. The force of the flowing mass destabilizes the channel bed and scours material by abrading, dislocating, and plucking rocks and entraining soil and organic matter into the flow [*Hungr et al.*, 2005; *Stock and Dietrich*, 2006]. Entrainment exerts a positive feedback as the bulking mass gains velocity and becomes more erosive [*Iverson et al.*, 2011]. The cohesion of a debris flow mass influences the speed and travel distance down channel and depends on the availability of clay and probably wildfire ash [*Burns and Gabet*, 2014; *Gabet and Sternberg*, 2008]. Debris flows typically occur in surges interspaced with sediment-laden flood waters and hyperconcentrated flows [*Iverson*, 1997]. The flowing mass will slow and come to rest as a debris deposit as the steepness of the flow path decreases. This can occur within confined channels or on the debris fan, where flow typically transitions from a constrained channel to the broader depositional plane of the fan [*Rickenmann*, 2005]. Debris fans are formed by previous debris flow events over geologic time [*Kirchner et al.*, 2001; *Pierce et al.*, 2004]. Debris flow deposition occurs within low-gradient channels or on the debris fan as levees form on the edges of less viscous flow. Levees channelize debris flows thereby extending debris flow runout at various distances down the fan face. The debris fan is typically the primary location where a debris flow directly impacts values-at-risk. Obstacles within depositional areas may be inundated by a debris flow or change the flow direction and final area of deposition, adding another element of uncertainty.

Confidence in the prediction of debris flow initiation is confounded by the uncertainties of the interactions between the spatial patterns and connectivity of fire effects and landform leading to the accumulation and convergence of rainfall runoff [*Hyde et al.*, 2014; *Moody et al.*, 2007]. Further, while wildfire ash influences sediment bulking processes and debris flow formation by increasing viscosity [*Burns and Gabet*, 2014; *Gabet and Sternberg*, 2008], the influence of ash relative to other initiation factors remains to be established. Methods using light detection and ranging (LiDAR) technology to model erosion from burned landscapes provide insights into sediment source and deposition areas relative to hillslope structure [*Harman et al.*, 2014; *Pelletier and Orem*, 2014] but work remains to incorporate this knowledge into process-based models. Patterns of local controls, such as exposed bedrock and the existence of midslope seeps also influence the probability of debris flow occurrence [*Hyde*, 2013], yet the relative significance remains uncertain. Multiple methods may be employed to mitigate hillslope runoff and erosion leading debris flow initiation [*deWolfe et al.*, 2008; *Santi et al.*, 2007]. Treatment effectiveness varies [*deWolfe et al.*, 2008; *Robichaud et al.*, 2013] and treatments may not be feasible in remote locations or warranted where treatment costs exceed potential benefits [*Calkin*, 2007].

Entrainment processes introduce additional uncertainty related to flow kinematics or the physics of the motion of the entire mass and the debris within the mass, the volume of material available to be mobilized, and the rate and duration of bulking as the mass moves down channel before deposition occurs. The location of obstructions relative to decreasing flow-path steepness introduces uncertainty about where deposition occurs as well as destructive potential. The final depositional zone is also influenced by the intersection with prior debris flow pathways and may shift in unpredictable ways as surges of muddy flood waters, hyperconcentrated flows, and additional debris flows follow during the same event.

The area inundated by debris flows can be predicted by several approaches in terms of the total travel distance and runout length based on the volume of transported sediment and topography of the debris flow pathway [*Rickenmann*, 2005]. Most of these methods are statistically derived empirical models, limiting their application to locations with similar conditions. Modeling predictions such as these can also be significantly affected by uncertainty in inputs, for example, the choice of digital elevation model [*Anderson et al.*, Chapter 11, this volume].

19.7. VALUES-AT-RISK

Debris flows present hazards downstream where the flowing mass may impact values-at-risk (Fig. 19.3). The potential hazard posed by debris flow to highly valued resources can be derived from the mapped location and value of resources in the inundated area and susceptibility of the resources to a debris flow event [*Calkin et al.*, 2007]. Direct debris flow impacts may threaten life and damage or destroy infrastructure in populated areas. Fine sediments carried in the sediment-laden flows that often continue downstream from debris flows can compromise water quality [*Smith et al.*, 2011], rendering it unsuitable for intended uses (Fig. 19.3). Depending on the condition of the ecosystem, the volume of the debris flow, and the grain-size distribution of the debris flow, an ecosystem can be either positively or negatively affected by debris flow impacts. Debris flows may in some cases enhance stream habitat and riparian ecosystems by restoring system complexity [*Dunham et al.*, 2007; *Rieman et al.*, 2003]. Ecosystem structure and functions may be impaired, depending on the condition of a population or habitat

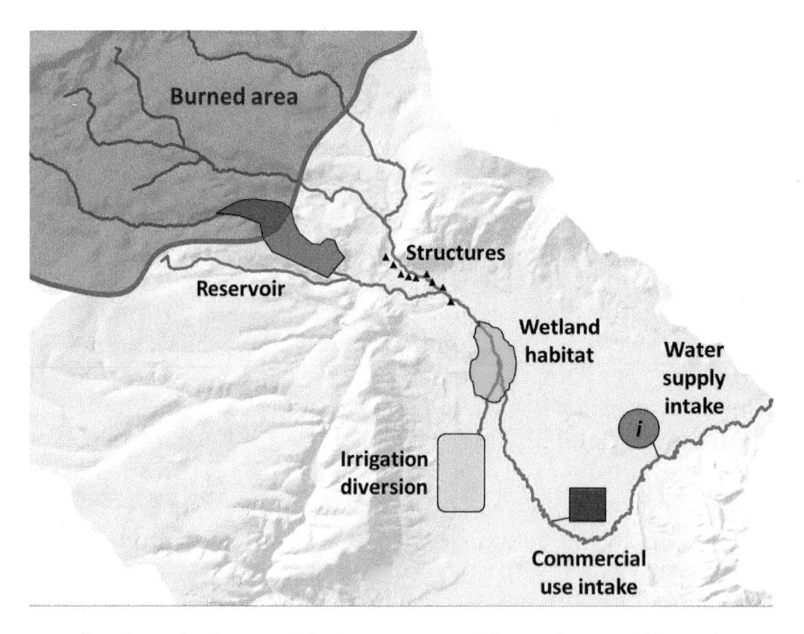

Figure 19.3 Conceptualization of values-at-risk downstream of burned areas, illustrating structures and infrastructure that can be damaged or destroyed, reflecting two phases of postfire debris flow hazards: direct impacts from a debris flow mass, and impaired water quality from sediments deposited into streams by debris flows. Sediments may compromise intended water use in reservoirs and at irrigation, commercial, and water supply intakes. While wetland habitat may be enhanced, it can also be disturbed beyond ecosystem tolerances depending on the resilience of the system, debris flow magnitude, and history of prior disturbances.

prior to the debris flow impact [*Gresswell*, 1999; *Rieman et al.*, 2003]. Where systems are already impaired, the additional disturbance may critically compromise ecosystem components and functions.

The uncertainties of predicting threats to values-at-risk begin with identifying if, in fact, valued resources are located along the relatively narrow paths of potential debris flows or within the areas on debris fans where they may potentially deposit. In the United States, multiple datasets exist to identify structures and infrastructure within or adjacent to stream channels. These data have been used to support wildland fuel reduction programs [*Stockmann et al.*, 2010], support active fire management [*Calkin et al.*, 2011b], and plan postfire emergency assessment and response [*Calkin et al.*, 2007]. However, the scale of existing geospatial data is often too coarse and locations are too imprecise for accurate discrimination of most threatened resources [*Calkin et al.*, 2011a; *Calkin et al.*, 2007; *Zerger and Smith*, 2003]. Data may be sufficient to conduct initial assessments and guide field observations to confirm locations. However, this time-consuming process can be impractical during especially large fires or busy fire seasons where personnel resources are limited.

Nonmarket resources, ecosystem components, and services without defined economic value are difficult to identify and assess [*Calkin et al.*, 2008; *Thompson and Calkin*, 2011], and are subject to uncertain valuation as decision makers decide which resources matter, and how they should make this decision. Mapping of critical habitat is generally inconsistent, often limited to public lands, and varies by land management agency. Different techniques, such as nonmarket valuation, exist to quantify debris flow impacts [*Calkin et al.*, 2008], but most rely on translating natural values into monetary values [*Sagoff*, 2011]. The commodification of ecosystem services is controversial and potentially counterproductive for environmental sustainability [*Gómez-Baggethun and Ruiz-Pérez*, 2011].

Assuming valued resources can be identified with reasonable certainty, the next challenge is to determine the probability that a debris flow of a certain force and volume will reach valued resources. Debris flow volumes vary widely [*Riley et al.*, 2013] and if a debris flow reaches a valued resource, losses can vary. A house could, for instance, be completely destroyed or the effect limited to a layer of mud and debris covering a portion of a yard or driveway. This uncertainty confounds the challenge of predicting potential losses. If the potential for loss has been identified, then the uncertainty arises of whether or not losses can be avoided or minimized through preventative mitigation. Possible strategies to mitigate debris flow impacts include engineering solutions such as check dams, deflection berms, debris racks, and debris basins [*deWolfe et al.*, 2008; *Santi et al.*, 2007]. These mitigation structures are typically engineered to meet expected forces of a design storm: a rainfall event of a certain magnitude, duration and return interval expected to produce runoff conditions sufficient to initiate debris flow [*Robichaud et al.*, 2000]. The effectiveness of mitigation structures depends on proper placement and installation [*deWolfe et al.*, 2008; *Robichaud et al.*, 2000]. However engineered solutions are costly [*deWolfe et al.*, 2008; *Santi et al.*, 2007] and might not be warranted where treatment costs exceed potential benefits [*Calkin et al.*, 2007].

Table 19.1 Summary of Uncertainties Associated with the Components of the Postfire Debris-Flow Hazard Cascade

Component	Uncertainties
Biophysical setting	Forest stand structure and susceptibility of vegetation to fire; interactions of hillslope geometry: curvature, steepness; soil properties
Fire processes	Knowledge gaps in fire physics; fire spread processes; fire intensity relative to nonuniform fuel beds; translating fire intensity to fire effects; ignition when planning for future fires
Fire effects	Process linkages between effects and hydrogeomorphic response; temperature reached by soils; changes in soil properties; macropore location, extent, and contribution; degree and spatial pattern; interaction with landscape geometry; mapping and quantifying fire effects; relative role of soil versus vegetation effects; timing and rate of vegetation recovery
Rainfall triggers	Distribution, intensity, and duration and rainfall thresholds; time since fire relative to vegetation recovery; interaction with fire effects: soil and vegetation disturbances; measurement of rainfall over burned areas; antecedent rainfall accumulations during prior season
Debris flow	Spatial pattern of fire effects and hydrologic connectivity relative to flow accumulation and convergence; role of ash and char in mobilization; rate and duration of bulking; flow path and depositional zone; effectiveness of landscape treatments and mitigation structures; destructive capacity
Values-at-risk	Location of valued resources relative to debris flow; availability, accuracy, and sufficiency of spatial inventories; susceptibility to debris impacts; magnitude of potential loss; valuation of nonmarket resources; effectiveness of mitigation

19.8. CONCLUSIONS

Prediction of debris flows can help protect and preserve values-at-risk (Fig. 19.3) but is currently hampered by a range of uncertainties. These uncertainties are related to each step of the debris flow hazard cascade (biophysical setting, fire processes, fire effects, rainfall, debris flow, and values-at-risk) and are related to knowledge gaps, variability driven by the stochasticity of natural systems, the uncertainties inherent in human decision-making processes, and inconsistent availability of data required for hazard prediction (Table 19.1). Awareness of these uncertainties and knowledge gaps can not only shed light on the potential error in current debris flow prediction models, but also highlight needs for fundamental and applied research to improve future models and build toward a comprehensive and integrated prediction approach. Arguably the greatest gaps may be closed through developing broadscale, integrated, and physically based analysis of fire effects and process-based understanding of the interactions between fire effects, especially vegetation disturbance, and terrain geometry. Deliberate team efforts will be needed to aggregate individual prediction components, those currently available and to be developed, and to build an integrated framework to predict postfire debris flow hazards. Clearly, uncertainties, both within each component and as compounded with integration, will require explicit articulation to identify and prioritize those that can be resolved and to define confidence boundaries on those uncertainties that are indeterminate.

REFERENCES

Ager, A. A., M. A. Day, C. W. McHugh, K. Short, J. Gilbertson-Day, M. A. Finney, and D. E. Calkin (2014), Wildfire exposure and fuel management on western US national forests, *J. Environ. Man.*, in review.

Alexander, M. E., and M. G. Cruz (2011), Interdependencies between flame length and fireline intensity in predicting crown fire initiation and crown scorch height, *Int. J. Wildland Fire*, *21*, 95–113.

Benda, L., and T. Dunne (1997), Stochastic forcing of sediment supply to channel networks from landsliding and debris flow, *Water Resour. Res.*, *33*(12), 2849–2863.

Beniston, M. (2006), Mountain weather and climate: a general overview and a focus on climatic change in the Alps, *Hydrobiologia*, *562*(1), 3–16.

Black, A., and T. Opperman (2005), Fire Effects Planning Framework: A user's guide, Rep. GTR-RMRS-163WWW, USDA Forest Service, Rocky Mountain Research Station, Fort Collins, CO.

Bladon, K. D., M. B. Emelko, U. Silins, and M. Stone (2014), Wildfire and the future of water supply, *Environ. Sci. Tech.*, *48*(16), 8936–8943.

Bodí, M. B., D. A. Martin, V. N. Balfour, C. Santín, S. H. Doerr, P. Pereira, A. Cerdà, and J. Mataix-Solera (2014), Wildland fire ash: Production, composition and eco-hydro-geomorphic effects, *Earth-Science Rev.*, *130*, 103–127.

Brown, J. K., and J. K. Smith, eds. (2000), Wildland Fire in Ecosystems: Effects of Fire on Flora, Rep. RMRS-GTR-42 v.2, USDA Forest Service, Rocky Mountain Research Station, Ogden, UT.

Burns, K., and E. J. Gabet (2014), The effective viscosity of slurries laden with vegetative ash, *CATENA*, *135*, 350–357; doi:10.1016/j.catena.2014.06.008

Calkin, D., G. Jones, and K. Hyde (2008), Nonmarket resource valuation in the postfire environment, *J. Forestry*, *106*(6), 305–310.

Calkin, D. E., J. D. Rieck, K. D. Hyde, and J. D. Kaiden (2011a), Built structure identification in wildland fire decision support, *Int. J. Wildland Fire*, *20*, 78–90.

Calkin, D. E., K. D. Hyde, P. R. Robichaud, J. G. Jones, L. E. Ashmun, and L. Dan (2007), Assessing post-fire values-at-risk with a new calculation tool, Rep. RMRS-GTR-205, USDA Forest Service, Rocky Mountain Research Station, Fort Collins, CO.

Calkin, D. E., M. P. Thompson, M. A. Finney, and K. D. Hyde (2011b), A real-time risk assessment tool supporting wildland fire decision making, *J. Forestry*, *109*(5), 274–280.

Cannon, S. H. (2001), Debris-flow generation from recently burned watersheds, *Environ. Eng. Geosci.*, *7*(4), 321–341.

Cannon, S. H., and J. E. Gartner (2005), Wildfire-related debris flow from a hazards perspective, 363–385, in *Debris-flow Hazards and Related Phenomena*, edited by M. Jakob and O. Hungr, Springer, Berlin.

Cannon, S. H., J. E. Gartner, M. G. Rupert, J. A. Michael, A. H. Rea, and C. Parrett (2010), Predicting the probability and volume of postwildfire debris flows in the intermountain western United States, *GSA Bulletin*, *122*, 127–144; doi: 110.1130/B26459.26451.

Cannon, S. H., J. E. Gartner, R. C. Wilson, J. C. Bowers, and J. L. Laber (2008), Storm rainfall conditions for floods and debris flows from recently burned areas in southwestern Colorado and southern California, *Geomorphology*, *96*, 250–269.

Carvalho, A., M. D. Flannigan, K. A. Logan, L. M. Gowman, A. I. Miranda, and C. Borrego (2010), The impact of spatial resolution on area burned and fire occurrence projections in Portugal under climate change, *Clim. Change*, *98*(1–2), 177–197.

Cerdà, A., and S. H. Doerr (2008), The effect of ash and needle cover on surface runoff and erosion in the immediate post-fire period, *CATENA*, *74*(3), 256–263.

Certini, G. (2005), Effects of fire on properties of forest soils: A review, *Oecologia*, *143*(1), 1–10.

Chafer, C. J. (2008), A comparison of fire severity measures: An Australian example and implications for predicting major areas of soil erosion, *CATENA*, *74*, 235–245.

DeBano, L. F. (2000), The role of fire and soil heating on water repellency in wildland environments: A review, *J. Hydrol.*, *231-232*, 195–206.

deWolfe, V. G., P. M. Santi, J. Ey, and J. E. Gartner (2008), Effective mitigation of debris flows at Lemon Dam, La Plata County, Colorado, *Geomorphology*, *96*(3), 366–377.

Doerr, S. H., R. A. Shakesby, and L. H. MacDonald (2009), Soil water repellency: A key factor in post-fire erosion, 197–223, in *Fire Effects on Soils and Restoration Strategies*, edited by A. Cerda and P. R. Robichaud, Science Publishers, Enfield, HN.

Dunham, J. B., A. E. Rosenberger, C. H. Luce, and B. E. Rieman (2007), Influences of wildfire and channel reorganization on spatial and temporal variation in stream temperature and the distribution of fish and amphibians, *Ecosystems*, *10*(2), 335–346.

Ebel, B. A. (2012), Wildfire impacts on soil-water retention in the Colorado Front Range, United States, *Water Resour. Res.*, *48*(12), W12515.

Ebel, B. A., and B. B. Mirus (2014), Disturbance hydrology: challenges and opportunities, *Hydrolog. Processes*, *28*(19), 5140–5148.

Epting, J., D. Verbyla, and B. Sorbel (2005), Evaluation of remotely sensed indices for assessing burn severity in interior Alaska using Landsat TM and ETM+, *Rem. Sens. Environ.*, *96*, 328–339.

Etiegni, L., and A. G. Campbell (1991), Physical and chemical characteristics of wood ash, *Bioresource Tech.*, *37*, 173–178.

Finney, M. (2006), An overview of FlamMap fire modeling capabilities, Paper presented at Fuels Management–How to Measure Success, USDA Forest Service, Rocky Mountain Research Station, Portland, OR.

Finney, M., C. McHugh, I. Grenfell, K. Riley, and K. Short (2011a), A simulation of probabilistic wildfire risk components for the continental United States, *Stochas. Environ. Res. Risk Assess.*, *25*(7), 973–1000.

Finney, M. A., I. C. Grenfell, C. W. McHugh, R. C. Seli, D. Trethewey, R. D. Stratton, and S. Brittain (2011b), A method for ensemble wildland fire simulation, *Environ. Mod. Assess.*, *16*(2), 153–167.

Finney, M. A., J. D. Cohen, I. C. Grenfell, and K. M. Yedinak (2010), An examination of fire spread thresholds in discontinuous fuel beds, *Int. J. Wildland Fire*, *19*(2), 163–170.

Finney, M. A., J. D. Cohen, J. M. Forthofer, S. S. McAllister, M. J. Gollner, D. J. Gorham, K. Saito, N. K. Akafuah, B. A. Adam, and J. D. English (2015), Role of buoyant flame dynamics in wildfire spread, *Proc. Nat. Acad. Sci.*, 201504498.

Finney, M. A., J. D. Cohen, S. S. McAllister, and W. M. Jolly (2013), On the need for a theory of wildland fire spread, *Int. J. Wildland Fire*, *22*(1), 25–36.

Gabet, E. J., and P. Sternberg (2008), The effects of vegetative ash on infiltration capacity, sediment transport, and the generation of progressively bulked debris flows, *Geomorphology*, *101*(4), 666–673.

Gabet, E. J., and T. Dunne (2003), Sediment detachment by rain power, *Water Resour. Res.*, *39*(1), 1002.

Gartner, J., P. Santi, and S. Cannon (2015), Predicting locations of post-fire debris-flow erosion in the San Gabriel Mountains of southern California, *Nat. Hazards*, *77*(2), 1305–1321.

Gartner, J. E., S. H. Cannon, P. M. Santi, and V. G. Dewolfe (2008), Empirical models to predict the volumes of debris flows generated by recently burned basins in the western U.S., *Geomorphology*, *96*, 339–354.

Ghilardi, P., L. Natale, and F. Savi (2001), Modeling debris flow propagation and deposition, Physics and Chemistry of the Earth, Part C: Solar, *Terr. Planet. Sci.*, *26*(9), 651–656.

Gimeno-García, E., V. Andreu, and J. L. Rubio (2007), Influence of vegetation recovery on water erosion at short and medium-term after experimental fires in a Mediterranean shrubland, *CATENA*, *69*(2), 150–160.

Gómez-Baggethun, E., and M. Ruiz-Pérez (2011), Economic valuation and the commodification of ecosystem services, *Prog. Phys. Geog.*, *35*(5), 613–628.

Gresswell, R. E. (1999), Fire and aquatic ecosystems in forested biomes of North America, *Trans. Amer. Fish. Soc.*, *128*, 193–221.

Gwet, K. (2001), *Handbook of Interrater Reliability*, Gaithersburg, MD, STATAXIS Publishing Company, 223–246.

Harman, C. J., K. A. Lohse, P. A. Troch, and M. Sivapalan (2014), Spatial patterns of vegetation, soils and microtopography from terrestrial laser scanning on two semi-arid hillslopes of contrasting lithology, *J. Geophys. Res. Biogeosci.*, *199*(2), 163–180.

Holz, A., and T. T. Veblen (2011), Variability in the Southern Annular Mode determines wildfire activity in Patagonia, *Geophys. Res. Lett.*, *38*(14), doi: 10.1029/2011GL047674.

Hudak, A. T., P. Morgan, M. J. Bobbitt, A. M. S. Smith, S. A. Lewis, L. B. Lentile, P. R. Robichaud, J. T. Clark, and R. A. McKinley (2007), The Relationship of Multispectral Satellite Imagery to Immediate Fire Effects, *Fire Ecol.*, *3*(1), 64–90.

Hungr, O., S. McDougall, and M. Bovis (2005), Entrainment of material by debris flows, 135–158, in *Debris-Flow Hazards and Related Phenomena*, edited by M. Jakob and O. Hungr, Springer, Berlin.

Hyde, K., D. Ryan, M. Finney, and W. Samuels (2009), Emerging technologies: Hydro-geomorphic tools for wildland fire decision support, USA, Paper presented at Seventh International Conference on Geomorphology, Melbourne, Australia.

Hyde, K., M. B. Dickinson, G. Bohrer, D. Calkin, L. Evers, J. Gilbertson-Day, T. Nicolet, K. Ryan, and C. Tague (2013), Research and development supporting risk-based wildfire effects prediction for fuels and fire management: Status and needs, *Int. J. Wildland Fire*, *22*(1), 37–50.

Hyde, K., S. W. Woods, and J. Donahue (2007), Predicting gully rejuvenation after wildfire using remotely sensed burn severity data, *Geomorphology*, *86*(3–4), 496–511.

Hyde, K. D. (2013), Control by Vegetation Disturbance on Gully Rejuvenation Following Wildfire, Dissertation thesis, The University of Montana, Missoula, MT.

Hyde, K. D., A. C. Wilcox, K. Jencso, and S. Woods (2014), Effects of vegetation disturbance by fire on channel initiation thresholds, *Geomorphology*, *214*, 84–96.

Hyde, K. D., A. C. Wilcox, K. Jencso, and S. Woods (2015), Influences of vegetation disturbance on hydrogeomorphic response following wildfire, *Hydrological Processes*, doi: 10.1002/hyp.10691.

Iverson, R. M. (1997), The physics of debris flows, *Rev. Geophys.*, *35*(3), 245–296.

Iverson, R. M., M. E. Reid, M. Logan, R. G. LaHusen, J. W. Godt, and J. P. Griswold (2011), Positive feedback and momentum growth during debris-flow entrainment of wet bed sediment, *Nat. Geosci.*, *4*, 116–121.

Jakob, M., and O. Hungr (2005), *Debris-flow Hazards and Related Phenomena*, Springer, Berlin, *795*.

Jenkins, S. E., C. H. Sieg, D. E. Anderson, D. S. Kaufman, and P. A. Pearthree (2011), Late Holocene geomorphic record of fire in ponderosa pine and mixed-conifer forests, Kendrick Mountain, northern Arizona, USA, *Int. J. Wildland Fire*, *20*(1), 125–141.

Keeley, J. E. (2009), Fire intensity, fire severity and burn severity: A brief review and suggested usage, *Int. J. Wildland Fire*, *18*, 116–126.

Key, C. H., and N. C. Benson (2006), Landscape assessment: Ground measure of severity, the Composite Burn Index; and remote sensing of severity, the normalized burn ratio, Rep. General Technical Report RMRS-GTR-164-CD, USDA Forest Service, Rocky Mountain Research Station, Fort Collins, CO.

Kirchner, J. W., R. C. Finkel, C. Riebe, S. Granger, Darryl E., J. L. Clayton, J. King, G., and W. F. Megahan (2001), Mountain erosion over 10 yr., 10 k.y., and 10 m.y. time scales, *Geology*, *29*, 591–594.

Larsen, I. J., L. H. MacDonald, E. Brown, D. Rough, and M. J. Welsh (2009), Causes of post-fire runoff and erosion: Water repellency, cover, or soil sealing?, *Soil Sci. Soc. Amer. J.*, *73*, 1393–1407.

Leidos Corp. (2014), Incident Command Tool for Drinking Water Protection (ICWater), General Overview of ICWater, Leidos Corp., Reston, VA.

Luna, B. Q., A. Remaître, T. W. van Asch, J. -P. Malet, and C. Van Westen (2012), Analysis of debris flow behavior with a one dimensional run-out model incorporating entrainment, *Eng. Geol.*, *128*, 63–75.

Luna, B. Q., J. Blahut, C. Camera, C. van Westen, T. Apuani, V. Jetten, and S. Sterlacchini (2014), Physically based dynamic run-out modelling for quantitative debris flow risk assessment: A case study in Tresenda, northern Italy, *Environ. Earth Sci.*, *72*(3), 645–661.

Mataix-Solera, J., A. Cerdà, V. Arcenegui, A. Jordán, and L. M. Zavala (2011), Fire effects on soil aggregation: A review, *Earth-Science Rev.*, *109*(1–2), 44–60.

Moody, J. A., D. A. Martin, S. L. Haire, and D. A. Kinner (2007), Linking runoff response to burn severity after a wildfire, *Hydrolog. Processes*, *22*, 2063–2074.

Moody, J. A., R. A. Shakesby, P. R. Robichaud, S. H. Cannon, and D. A. Martin (2013), Current research issues related to post-wildfire runoff and erosion processes, *Earth-Science Rev.*, *122*, 10–37.

Neary, D. G., K. A. Koestner, A. Youberg, and P. E. Koestner (2012), Post-fire rill and gully formation, Schultz Fire 2010, Arizona, USA, *Geoderma*, *191*, 97–104.

Nyman, P., G. Sheridan, and P. N. Lane (2010), Synergistic effects of water repellency and macropore flow on the hydraulic conductivity of a burned forest soil, south-east Australia, *Hydrolog. Processes*, *24*(20), 2871–2887.

Oliveira, S. L., J. M. Pereira, and J. M. Carreiras (2012), Fire frequency analysis in Portugal (1975–2005), using Landsat-based burnt area maps, *Int. J. Wildland Fire*, *21*(1), 48–60.

Parsons, A., P. R. Robichaud, S. A. Lewis, C. Napper, and J. T. Clark (2010), Field guide for mapping post-fire soil burn severity, Rep. RMRS-GTR-243, USDA Forest Service, Rocky Mountain Research Station, Fort Collins, CO.

Pelletier, J. D., and C. A. Orem (2014), How do sediment yields from post-wildfire debris-laden flows depend on terrain slope, soil burn severity class, and drainage basin area? Insights from airborne-LiDAR change detection, *Earth Surf. Process. Landforms*, *39*(13), 1822–1832.

Pereira, P., A. Cerdà, X. Úbeda, J. Mataix-Solera, V. Arcenegui, and L. M. Zavala (2013), Modelling the impacts of wildfire on ash thickness in a short-term period, *Land Degrad. Dev.*, *26*(2), 180–192.

Pierce, J. L., G. A. Meyer, and A. J. T. Jull (2004), Fire-induced erosion and millennial scale climate change in northern ponderosa pine forests, *Nature*, *432*, 87–90.

Pierson, F. B., P. R. Robichaud, C. A. Moffet, K. E. Spaeth, C. J. Williams, S. P. Hardegree, and P. E. Clark (2008), Soil water repellency and infiltration in coarse-textured soils of burned and unburned sagebrush ecosystems, *CATENA*, *74*(2), 98–108.

Prosser, I. P., and L. Williams (1998), The effect of wildfire on runff and erosion in native eucalyptus forest, *Hydrolog. Processes*, *12*, 251–265.

Reaney, S., L. Bracken, and M. Kirkby (2007), Use of the connectivity of runoff model (CRUM) to investigate the influence of storm characteristics on runoff generation and connectivity in semi-arid areas, *Hydrolog. Processes*, *21*(7), 894–906.

Reaney, S. M., L. J. Bracken, and M. J. Kirkby (2013), The importance of surface controls on overland flow connectivity in semi-arid environments: Results from a numerical experimental approach, *Hydrolog. Processes*; doi: 10.1002/hyp.9769.

Reinhardt, E. D., and M. B. Dickinson (2010), First-order fire effects models for land management: overview and issues, *Fire Ecol.*, *6*(1), 131–142.

Rickenmann, D. (2005), Runout prediction methods, 305–324, in *Debris-Flow Hazards and Related Phenomena*, edited by M. Jakob and O. Hungr, Springer, Berlin.

Rieman, B., D. Lee, D. Burns, R. Gresswell, M. Young, R. Stowell, J. Rinne, and P. Howell (2003), Status of native fishes in the western United States and issues for fire and fuels management, *For. Ecol. Man.*, *178*, 197–211.

Riley, K. L., R. Bendick, K. D. Hyde, and E. J. Gabet (2013), Frequency-magnitude distribution of debris flows compiled from global data, and comparison with post-fire debris flows in the western U.S, *Geomorphology*, *191*, 118–128.

Robichaud, P. R., J. L. Beyers, and D. G. Neary (2000), Evaluating the effectiveness of postfire rehabilitation treatments, Rep. RMRS-GTR-63, USDA Forest Service, Rocky Mountain Research Station, Fort Collins.

Robichaud, P. R., J. W. Wagenbrenner, S. A. Lewis, L. E. Ashmun, R. E. Brown, and P. M. Wohlgemuth (2013), Post-fire mulching for runoff and erosion mitigation Part II: Effectiveness in reducing runoff and sediment yields from small catchments, *CATENA*, *105*, 93–111.

Roering, J. J., and M. Gerber (2005), Fire and the evolution of steep, soil-mantled landscapes, *Geology*, *33*(5), 349–352.

Roering, J. J., K. M. Schmidt, J. D. Stock, W. E. Dietrich, and D. R. Montgomery (2003), Shallow landsliding, root reinforcement, and the spatial distribution of trees in the Oregon Coast Range, *Can. Geotech. J.*, *40*(2), 237–253.

Rothermel, R. C. (1972), A mathematical model for predicting fire spread in wildland fuels, Rep. RP INT-115, USDA Forest Service, Intermountain Forest and Range Experimental Station, Ogden, UT.

Roy, D. P., L. Boschetti, and S. N. Trigg (2006), Remote sensing of fire severity: Assessing the performance of the normalized burn ratio, *Geosci. Rem. Sens. Lett., IEEE*, *3*(1), 112–116.

Sagoff, M. (2011), The quantification and valuation of ecosystem services, *Ecol. Econ.*, *70*(3), 497–502.

Samuels, W. B., D. E. Amstutz, R. Bahadur, and J. M. Pickus (2006), RiverSpill: A national application for drinking water protection, *J. Hydraul. Eng.*, *132*, 393–403.

Santi, P. M., V. G. deWolfe, J. D. Higgins, S. H. Cannon, and J. E. Gartner (2007), Effectiveness of debris flow mitigation methods in burned areas, Paper presented at First North American Landslide Conference, Vail, CO.

Santi, P. M., V. G. deWolfe, J. D. Higgins, S. H. Cannon, and J. E. Gartner (2008), Sources of debris flow material in burned areas, *Geomorphology*, *96*, 310–321.

Shakesby, R. (2011), Post-wildfire soil erosion in the Mediterranean: Review and future research directions, *Earth-Science Rev.*, *105*(3), 71–100.

Shakesby, R. A., and S. H. Doerr (2006), Wildfire as a hydrological and geomorphological agent, *Earth-Science Rev.*, *74*(3–4), 269–307.

Shakesby, R. A., C. P. M. Bento, C. S. S. Ferreira, A. J. D. Ferreira, C. R. Stoof, E. Urbanek, and R. P. D. Walsh (2013), Impacts of prescribed fire on soil loss and soil quality: An assessment based on an experimentally-burned catchment in central Portugal, *CATENA*, *128*, 278–293.

Shin, S. S., S. D. Park, and K. S. Lee (2013), Sediment and hydrological response to vegetation recovery following wildfire on hillslopes and the hollow of a small watershed, *J. Hydrol.*, *499*, 154–166.

Smith, H. G., G. J. Sheridan, P. N. J. Lane, P. Nyman, and S. Haydon (2011), Wildfire effects on water quality in forest catchments: A review with implications for water supply, *J. Hydrol.*, *396*(1–2), 170–192.

Stock, J. D., and W. E. Dietrich (2006), Erosion of steepland valleys by debris flows, *GSA Bull.*, *118*(9/10), 1125–1148.

Stockmann, K., K. Hyde, G. Jones, D. Loeffler, and R. Silverstein (2010), Integrating fuel treatment into ecosystem management: A proposed project planning process, *Int. J. Wildland Fire*, *19*, 725–736.

Stoof, C., R. Vervoort, J. Iwema, E. Van Den Elsen, A. Ferreira, and C. Ritsema (2012), Hydrological response of a small catchment burned by experimental fire, *Hydrol. Earth Syst. Sci.*, *16*(2), 267.

Stoof, C. R., A. Gevaert, C. Baver, D. Martin, B. Hassanpour, V. Morales, and T. Steenhuis (2013b), Can pore-clogging by ash explain post-fire runoff and erosion?, 153–154, in *AGU Chapman Conference on Postwildfire Runoff and Erosion Responses*, 25–31 August 2013, edited by J. A. Moody and D. A. Martin, available at http://chapman.agu.org/post-wildfire/files/2013/2012/Collected-Abstracts.pdf, Estes Park, CO, USA.

Stoof, C. R., D. Moore, C. J. Ritsema, and L. W. Dekker (2011), Natural and fire-induced soil water repellency in a Portuguese shrubland, *Soil Sci. Soc. Amer. J.*, *75*(6), 2283–2295.

Stoof, C. R., D. Moore, P. M. Fernandes, J. J. Stoorvogel, R. E. Fernandes, A. J. Ferreira, and C. J. Ritsema (2013a), Hot fire, cool soil, *Geophys. Res. Lett.*, *40*(8), 1534–1539.

Stoof, C. R., J. G. Wesseling, and C. J. Ritsema (2010), Effects of fire and ash on soil water retention, *Geoderma*, *159*(3), 276–285.

Stoof, C. R., A. I. Gevaert, C. Baver, B Hassanpour, V. L. Morales, W. Zhang, D. Martin, S. K. Giri, and T. S. Steenhuis (2016), Can pore-clogging by ash explain post-fire runoff? *International Journal of Wildland Fire*, *25*(3), 294–305 doi: http://dx.doi.org/10.1071/WF15037.

Theobald, D. M., and W. H. Romme (2007), Expansion of the US wildland-urban interface, *Landscape Urb. Plan.*, *83*(4), 340–354.

Thompson, M. P., and D. E. Calkin (2011), Uncertainty and risk in wildland fire management: A review, *J. Environ. Man.*, *92*(8), 1895–1909.

Thompson, M. P., D. E. Calkin, J. W. Gilbertson-Day, and A. A. Ager (2010), Advancing effects analysis for integrated, large-scale wildfire risk assessment, *Environ. Mon. Assess.*, 1–23.

USGS (2009), The National Map LANDFIRE, available at: http://www.landfire.gov/, accessed 15 November 2011.

USGS (2015a), Postwildfire Landslide Hazards, http://landslides.usgs.gov/research/wildfire/, accessed 23 March 2015.

USGS (2015b), Emergency Assessment of Post-Fire Debris-Flow Hazards, http://landslides.usgs.gov/hazards/postfire_debrisflow/, accessed 23 March 2015.

Westerling, A. L., M. G. Turner, E. A. Smithwick, W. H. Romme, and M. G. Ryan (2011), Continued warming could transform Greater Yellowstone fire regimes by mid-21st century, *Proc. Nat. Acad. Sci.*, *108*(32), 13165–13170.

Wittenberg, L., D. Malkinson, and R. Barzilai (2014), The differential response of surface runoff and sediment loss to wildfire events, *CATENA*, *121*, 241–247.

Wondzell, S. M., and J. King, G. (2003), Postfire erosional processes in the Pacific Northwest and Rocky Mountain regions, *For. Ecol. Man.*, *178*, 75–87.

Woods, S. W., and V. N. Balfour (2008), The effect of ash on runoff and erosion after a severe forest wildfire, Montana, USA, *Int. J. Wildland Fire*, *17*, 535–548.

Woods, S. W., and V. N. Balfour (2010), The effects of soil texture and ash thickness on the post-fire hydrological response from ash-covered soils, *J. Hydrol.*, *393*, 274–286.

Woods, S. W., A. Birkas, and R. Ahl (2007), Spatial variability of soil hydrophobicity after wildfires in Montana and Colorado, *Geomorphology*, *86*(3–4), 465–479.

Young, C. B., B. R. Nelson, A. A. Bradley, J. A. Smith, C. D. Peters-Lidard, A. Kruger, and M. L. Baeck (1999), An evaluation of NEXRAD precipitation estimates in complex terrain, *J. Geophys. Res.*, *104*(D16), 19,691–19,703.

Zerger, A., and D. I. Smith (2003), Impediments to using GIS for real-time disaster decision support, *Comput. Environ. Urb. Syst.*, *27*, 123–141.

20

Capturing Spatiotemporal Variation in Wildfires for Improving Postwildfire Debris-Flow Hazard Assessments

Jessica R. Haas,[1] Matthew Thompson,[1] Anne Tillery,[2] and Joe H. Scott[3]

ABSTRACT

Wildfires can increase the frequency and magnitude of catastrophic debris flows. Integrated, proactive natural-hazard assessment would therefore characterize landscapes based on the potential for the occurrence and interactions of wildfires and postwildfire debris flows. This chapter presents a new modeling effort that can quantify the variability surrounding a key input to postwildfire debris-flow modeling, the amount of watershed burned at moderate to high severity, in a prewildfire context. The use of stochastic wildfire simulation captures variability surrounding the timing and location of ignitions, fire weather patterns, and ultimately the spatial patterns of watershed area burned. Model results provide for enhanced estimates of postwildfire debris-flow hazard in a prewildfire context, and multiple hazard metrics are generated to characterize and contrast hazards across watersheds. An area in northern New Mexico, USA, is presented as a case-study location, where postwildfire debris flows are a salient hazard and where land managers are actively pursuing mitigation efforts. Modeling results are described in terms of informing mitigation efforts. Limitations and future directions are presented.

20.1. INTRODUCTION

Wildfires and debris flows are two natural hazards that can occur in overlapping geographic areas. Debris flows are a high-density mix of water, soil, rocks, and debris that may have enormous destructive power with potentially catastrophic outcomes. Large, severe wildfires can pose serious threats to human communities, from the destruction of natural and built infrastructure within the fire perimeter to the degradation of watershed health and water supplies, as well as damage to other water-related infrastructure of downstream cities. The ecological and economic impacts of wildfires and debris flows on values, such as clean surface water, may continue for many years after the event occurs, as is evident from the Hayman fire in Colorado [*Rhoades et al.*, 2011] and the Las Conchas and Cerro Grande fires in New Mexico [*Harpold et al.*, 2014; *China et al.*, 2013].

Wildfires can increase the probability and magnitude of a subsequent debris flow for several years after the fire event, with most postwildfire debris flows occurring within the first 2 yr following a wildfire [*Cannon et al.*, 2011; *Smith et al.*, 2011]. The joint probability of a wildland fire event followed by a debris flow is driven by many factors. The wildfire potential is primarily a factor of weather, fuel characteristics of the landscape, and topography [*Parks et al.*, 2011]. The postwildfire debris-flow potential is a function of the percent area of a watershed that burned at moderate to high severity (MHS), the topography and soil characteristics of the watershed, and the intensity and duration of a rainfall event [*Cannon et al.*, 2010]. Burn severity is a measure of the relative changes in prewildfire and postwildfire vegetation cover [*Parsons et al.*, 2002; *Keeley*, 2009]. In recently burned areas, soils, especially on steep slopes, can be highly erodible

[1] *Rocky Mountain Research Station, US Forest Service, Missoula, Montana, USA*
[2] *US Geological Survey, Albuquerque, New Mexico, USA*
[3] *Pyrologix LLC, Missoula, Montana, USA*

Natural Hazard Uncertainty Assessment: Modeling and Decision Support, Geophysical Monograph 223, First Edition.
Edited by Karin Riley, Peter Webley, and Matthew Thompson.

because of decreased vegetation coverage, which increases sediment-laden runoff [*Cannon and Gartner*, 2005]. It is the progressive accumulation of sediment-laden runoff that triggers postwildfire debris flows, rather than discrete slope failure [*Meyer and Wells*, 1997; *Cannon et al.*, 2001].The potential for postwildfire debris-flow initiation changes through time as a function of the rate of vegetation recovery, as well as the timing and intensity of rainfall storms in the area.

Postwildfire debris flows are particularly dangerous because they can be generated in areas that were previously stable and/or in response to rainstorms that would otherwise be considered typical for an area [*Cannon et al.*, 2008]. For example, following the 2003 Grand Prix and Old fires in San Bernardino County, California, 16 people died and mountain roads were cut off by postwildfire debris flows initiated from a typical rainstorm [*Chong et al.*, 2004]. Given the threat to human life, property, and highly valued resources from wildland fires and postwildfire debris flows, there is expanding interest in identifying efficient opportunities to prioritize hazard and risk mitigation efforts [*Warziniack and Thompson*, 2013; *Nature Conservancy,* 2014]. In a postwildfire environment, hazard and risk mitigation efforts typically focus on watershed rehabilitation and soil stabilization [*DeGraff et al.*, 2007]. In these contexts, the extent, location, and severity of the wildfire event is observable, and managers can use predictive models of debris-flow probabilities and volumes to prioritize rehabilitation efforts within and across affected watersheds. These efforts, however, can be challenging because of uncertainty regarding storm patterns, among other factors [*Nyman et al.*, 2013].

Planning and prioritizing mitigation measures becomes even more challenging in a prewildfire environment, because of added uncertainty regarding the location, extent, and severity of wildfire within and across watersheds susceptible to postwildfire debris flows. Nevertheless, prewildfire hazard assessments are a critical component in risk management, especially for prioritizing preventative investments such as hazardous fuel reduction [*Thompson and Calkin*, 2011]. Fuel management, including thinning overly dense forests to reduce wildfire hazard and risk, is often a priority for land managers. While predicting the precise location of a wildfire is impossible, it is possible to identify areas where treatments are likely to be more effective at reducing wildfire hazard and risk.

An integrated hazard assessment would ideally take into account possible interactions of wildfires with postwildfire debris flows, and present a more accurate picture of threats across a landscape. Most research, however, has focused on assessing wildfire and debris-flow hazards individually (although see *Jones et al.* [2014]; *Tillery et al.* [2014], for counterexamples). Wildfire hazard and risk assessments are increasingly used in a wide range of geographic settings and planning contexts [*Haas et al.*, 2013; *Thompson et al.*, 2013a; *Thompson et al.*, 2015a], including watershed-specific wildfire risk [*Thompson et al.*, 2013b; *State of California Sierra Nevada Conservancy*, 2014], to predict the likelihood, intensity, and consequences of wildfire across a landscape. Likewise, researchers often use postwildfire debris-flow hazard assessments to map the likelihood and volume of a debris flow from a given watershed, after a wildland fire has already occurred [*Tillery et al.*, 2011; *Staley*, 2013; see also http://landslides.usgs.gov/hazards/postfire_debris flow/]. In existing models, the primary variable connecting a wildfire and a subsequent debris flow is the amount of a watershed that burns with MHS [*Cannon et al.*, 2010]. This relationship therefore forms the centerpiece of the current modeling effort, building directly from recent work using simulated fire perimeters [*Thompson et al.*, 2013c; *Ager et al.*, 2014; *Haas et al.*, 2014; *Scott et al.*, 2015; *Thompson et al.*, 2015b]. Here we use fire perimeters to generate distributions of watershed area burned, and link these outputs with spatially resolved estimates of burn severity to generate distributions of watershed area burned at MHS.

The objective of this chapter is to develop and apply an integrated modeling framework that can be used to support postwildfire debris-flow hazard assessment and mitigation prioritization, before a wildfire occurs. The major methodological innovation is modeling the sizes, shapes, locations, and burn severity patterns of thousands of wildfires, and subsequently treating each as an observed wildfire event with which to estimate postwildfire debris flow probability and volume. As a case study, we apply the framework to seven watersheds on a real-world landscape in northern New Mexico, USA. We demonstrate how this framework can provide information on which watersheds might pose the most serious postwildfire debris-flow hazard, and ultimately to support decisions related to hazard and ultimately risk mitigation. To conclude, we discuss applications, strengths, limitations, and future research opportunities.

20.2. METHODS

20.2.1. Study Area

Our case study focuses on seven watersheds situated along the western slopes to the northeast of Albuquerque in the Sandia Mountains, hereafter referred to as the "Sandias." The watersheds used in this study were delineated from major drainages into watersheds no larger than approximately 11 km^2, the maximum size represented with reasonable confidence in the original database [*Gartner et al.*, 2005] used to generate the *Cannon et al.* [2010] model used in this study (Table 20.1; see Section 20.2.5).

Table 20.1 Watershed Inputs for Debris-Flow Model

Watershed ID	Area (km^2)	Percentage of area with slope ≥ 30%	Ruggedness	Percent clay content	Liquid limit
1	7.17	70	0.54	21.1	26.2
2	0.7	29	0.57	9.4	23.0
3	1.44	22	0.43	9.4	23.0
4	1.52	66	0.79	21.1	26.2
5	2.42	34	0.41	9.4	23.0
6	8	68	0.49	21.1	26.2
7	11.09	64	0.38	21.1	26.2

These watersheds drain directly into communities surrounding the City of Albuquerque, and therefore the threat of wildfire and postwildfire debris flows within them are of particular concern. Figure 20.1 depicts the location of the seven study watersheds, as well as the location of critical water infrastructure. These watersheds vary in terms of size, topography, and soil characteristics used in the debris-flow models.

Vegetation type follows an elevation gradient, beginning with grassland at the lowest elevations and transitioning into pinyon-juniper woodland, ponderosa pine dominated forests, and finally spruce-fir forests at the highest elevations [*Julyan and Stuever*, 2005]. The terrain tends to be steep in this area, with slopes greater than 30% accounting for an average of half of the areal extent of the seven basins. The Sandias have not experienced any large wildfires in recent years. The latest fire of substantial size was in 2011, and burned only 42 acres [*Short*, 2014]. The Monitoring Trends in Burn Severity (MTBS) dataset [*Eidenshink et al.*, 2007; *US Geological Survey and USDA Forest Service*, 2013] indicates no fires within the Sandias that were over 1000 acres during the period from 1984 to 2011. Because of the lack of recent large fires or disturbances, the vegetation structure of the area is tending toward late succession, which is characterized by an increase in surface and crown fuels [*Keane et al.*, 2000]. Under the right weather conditions, the fuel buildup could lead to a wildfire that would be difficult to suppress given the steep terrain.

20.2.2. Modeling Framework

Figure 20.2 presents the conceptual overview of postwildfire debris-flow modeling framework. The primary output of interest is the estimation of postwildfire debris-flow probability and volume, but our primary modeling interest is to better capture wildfire–debris-flow hazard interactions. Starting from the upper right of Figure 20.2, landscape conditions relating to watershed size, topography, and soil characteristics are held constant as inputs into the debris-flow model. For reasons described below, the spatial pattern of potential burn severity is also held constant. The upper left side of Figure 20.2 describes the inputs that vary in modeling the postwildfire debris flows. Stochastic simulation is used to output fire perimeter polygons (i.e., the wildfire event set), which are then overlain with watershed boundaries to generate distributions of watershed area burned. For each wildfire event, the storm characteristics for six recurrence interval storms were used in the model, resulting in six probabilities and six volumes for each wildfire event. The two grey boxes represent inputs that we modeled using fire behavior and fire perimeter growth models, described below. These four boxes collectively represent inputs to statistical models of debris-flow probability and volume. Last, debris-flow model outputs are analyzed to characterize watershed-level hazards.

20.2.3. Wildfire Modeling

The primary purpose of modeling wildfire in this study is to provide the area of a watershed that could burn at MHS for a set of possible wildfire events. We used two separate fire simulation models to achieve this aim. To generate the sizes, shapes, and locations of simulated perimeters, we used the FSim fire modeling system [*Finney et al.*, 2011]. Unfortunately, because of storage and processing limitations, FSim does not retain information on fire behavior for every simulated perimeter, precluding direct estimation of burn severity. To estimate severity, we instead used the FlamMap [*Finney*, 2006] fire-behavior model. These fire models use the same basic set of landscape inputs, but vary in function and capacity, as described in the following subsections. The primary difference between FlamMap and FSim is that rather than using a single set of fire-behavior outputs for each fire event, FSim invokes FlamMap to generate many possible fire-behavior outputs, based on the historical weather combinations of wind speed, direction, and three levels of fuel dryness as measured by the 80th, 90th, and 97th percentile Energy Release Component (ERC) [*Bradshaw et al.*, 1983; *Finney et al.*, 2011]. Additionally, in FlamMap, fire growth is set for a specific burn period, where with FSim fires may continue to grow for multiple burn periods.

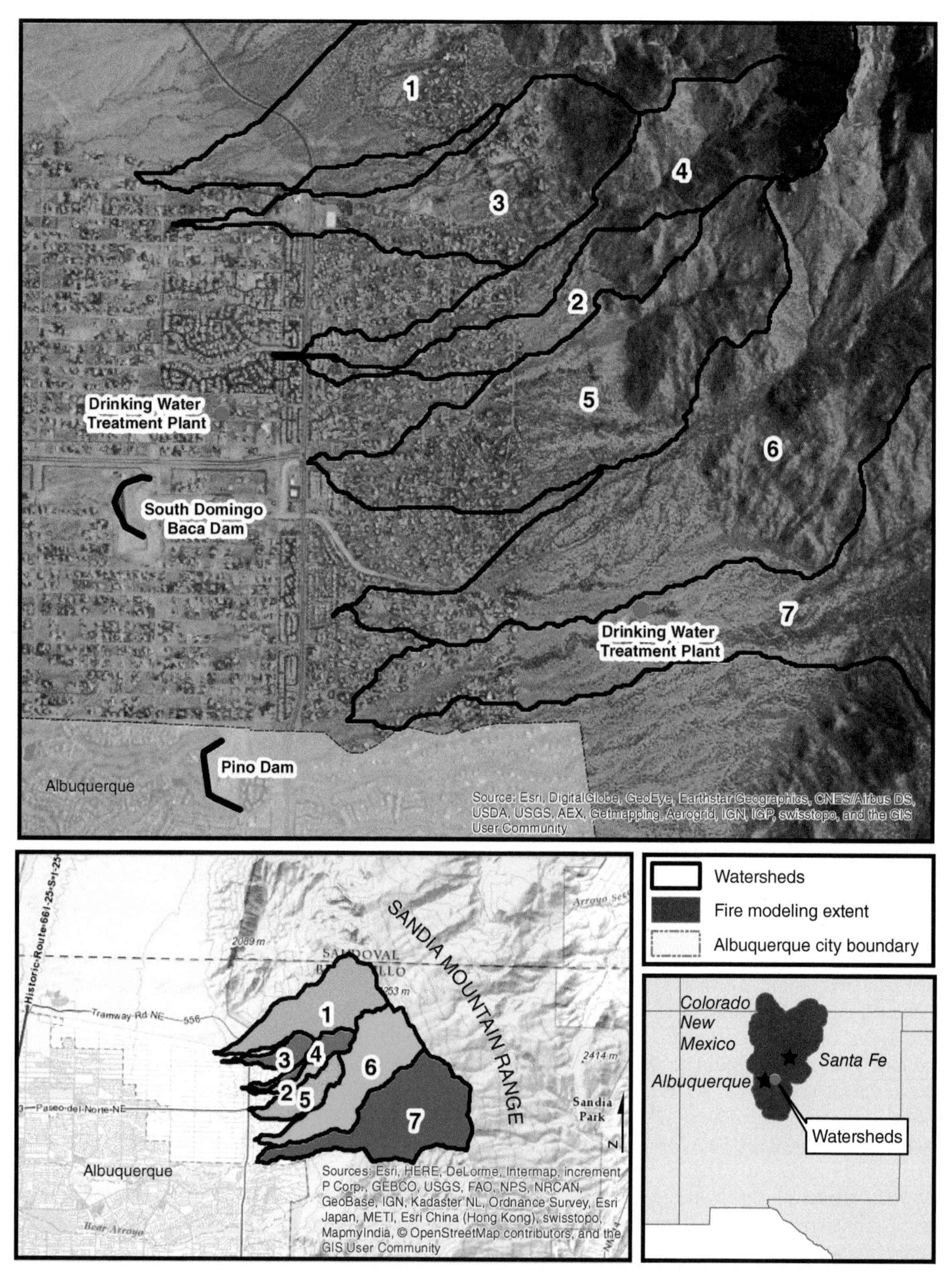

Figure 20.1 Study area with the selected watersheds identified. The top panel shows a zoomed in detailed view of the watershed outlets into the adjacent urban interface, and identifies key water-infrastructure located in the vicinity. In some cases, there is urban development within the watershed itself (i.e., watershed 1). The bottom left panel shows complete delineation of the watersheds and their more general location within the Sandia Mountain range. The color of each watershed corresponds to the colors used to distinguish between watersheds in the results figures. The bottom right panel shows the location of the watersheds with respect to the greater landscape used in modeling the fire inputs.

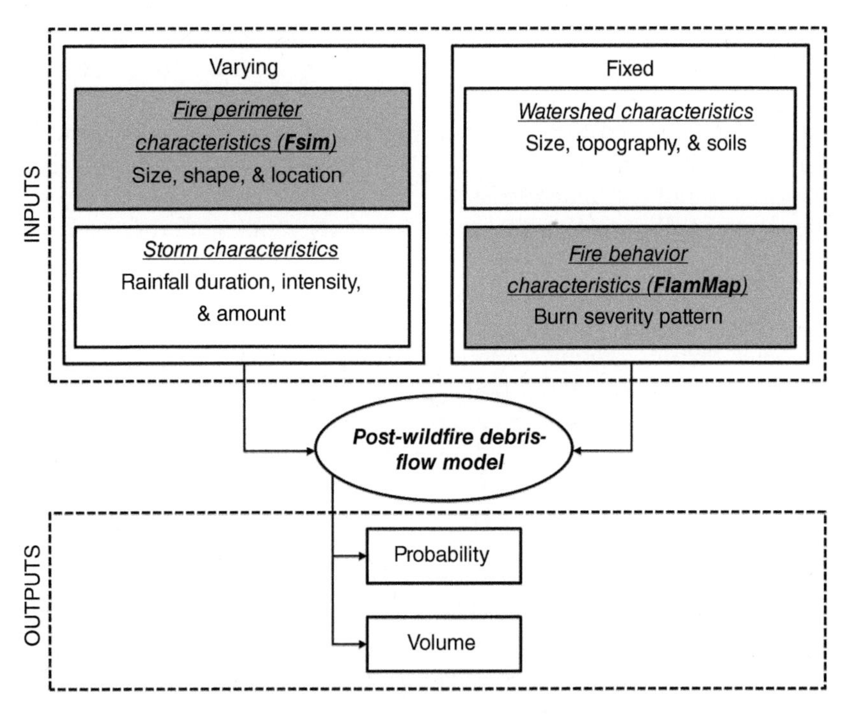

Figure 20.2 Overview of modeling framework, showing linkages between the various models (named in bold), which provides the ultimate outputs characterizing watershed-level risks (debris-flow probability and volume). The two grey boxes represent inputs modeled using fire-behavior and growth models.

20.2.3.1. FSim Modeling: Watershed Area Burned

The FSim model is focused on simulating relatively rare "large" (greater than 100 ha) wildfires that account for a small fraction of total number of fires, but that account for the vast majority of area burned [*Short*, 2013]. The FSim model uses historical information on fire occurrence, weather, and fire suppression efforts to simulate thousands of hypothetical fire seasons. In the simulation, fires are probabilistically ignited according to the relationship between the fire weather and fire occurrence derived from the historical data. Each fire then "grows" in accordance with fire-behavior outputs and Minimum Travel Time fire spread theory [*Finney*, 2002]. Each simulated fire may continue to grow for many burn periods (i.e., days). The fire behavior and growth for each burn period is determined by the fire weather generated for that day and can therefore vary throughout the duration of the fire. The fire continues to burn until it is extinguished by a statistical suppression algorithm [*Finney et al.*, 2009]. The suppression algorithm probabilistically contains fires based on the relationship between large fire containment and duration, fire growth, and fuel type. The landscape size used in the FSim and FlamMap model is much larger than the seven watersheds included in this study. We ran the FSim and FlamMap models on a landscape of 35,000 km^2 (Fig. 20.1). In addition to capturing the broader fire occurrence, this larger landscape allows fires to ignite remotely and burn into the watersheds, as well as to ignite within the watersheds and burn out of them.

The outputs are calibrated to historical fire occurrence, in terms of the mean annual area burned and mean number of large fires per year. For this effort, 30,000 unique fire seasons were simulated, with the principal output of interest being the set of final fire perimeters from each simulated season.

20.2.3.2. FlamMap Modeling: Spatial Burn Severity Patterns

Currently, fire-behavior models do not calculate burn severity. However, fire-behavior models do output fire intensity. Fire intensity is the amount of energy released when a fuel burns, and is measured in either kilowatts per meter or as flame height in meters [*Alexander and Cruz*, 2012]. This measure depends primarily on the weather, topography, and fuel characteristics at the time of burning. Through empirical modeling, intensity is used in generating fire-behavior outputs such as rate of spread and crown-fire initiation [*Rothermel*, 1972; *Scott and Burgan*, 2005; *Alexander and Cruz*, 2012]. Intensity is highly correlated with severity, especially in forested landscapes, where high fire intensity will result in crown fire [*Keeley*, 2009]. Crown fire is the movement of fire into and throughout the forest canopy, which results in relatively high levels of vegetation consumption and mortality and is therefore a useful proxy for MHS in forested landscapes.

The fire-behavior model FlamMap [*Finney*, 2006] incorporates spatial information on topography and fuel models, the ERC, and wind data to estimate a single set of fire-behavior outputs, including crown-fire potential. The crown-fire potential is generated spatially on a landscape from a set of topographic and fuel characteristics, collectively known as the landscape file, generated for this study from input data obtained from the LANDFIRE LF 1.3 dataset [*LANDFIRE*, 2012]. These characteristics include slope, elevation, aspect, crown base height, canopy base height, canopy bulk density, canopy coverage, and surface fuel model. We calculated crown-fire potential using the Scott and Reinhardt crown-fire method [*Scott and Reinhardt*, 2001]. Wind inputs were based on the most common combination of historically observed wind speed (16 km/h) and direction (270 degrees) from the same weather inputs used in the FSim model discussed above. We generated the crown-fire potential for the 90th percentile ERC, a measure of fuel moisture. The 90th percentile ERC represents fuel dryness resulting in common fire behavior during large fires, rather than the 97th percentile ERC, which would result in uncommon, extreme fire behavior, or 80th percentile ERC, which represents minimal fire growth and behavior. In forested landscapes, we use the resulting crown-fire potential as a surrogate for MHS.

In nonforested landscapes, such as shrub or grasslands, the surface-fuel model is a better indicator of MHS than crown-fire potential, because these landscapes do not possess a canopy. A surface-fuel model is the complete set of fuel inputs needed to calculate a mathematical representation of fire intensity and wildfire spread [*Rothermel*, 1972]. In grasslands, even a high-intensity fire typically results in low burn severity because of the rapid regeneration of grasses after a fire. Additional studies have shown that grassland fires typically burn under low severity and result in little to no erosion response [*Johansen et al.*, 2001]. Therefore, areas mapped as a grassland surface-fuel model were not considered prone to MHS fire. In shrublands, high-intensity fires are highly correlated with MHS fire [*Pyne et al.*, 1996; *Keeley et al.*, 2008]. Because fuel models define the intensity at which the fuel burns, high-intensity shrub-fuel models can be used as a surrogate for MHS fire potential in these communities. The shrub surface fuel models SH5 and SH7, from the Scott and Burgan 40 fire behavior fuel models [*Scott and Burgan*, 2005], have potential for high-intensity fire behavior. We used the presence of these fuel models to identify the areas with potential for MHS in shrublands.

Total MHS fire potential for the study area was calculated as the sum of the predicted crown-fire activity areas and the high-intensity shrub fuel areas. The resulting output was calibrated to observed burn severity using the Monitoring Trends in Burn Severity data [*US Geological Survey and USDA Forest Service*, 2013] for geographically relevant historical fires, based on the methods described in *Tillery et al.* [2014, Appendix]. Crown-fire activity was predicted to occur on approximately 20% of the land area of the seven watersheds; an artifact of relatively high potential for crown fire in timbered areas (66%) across a relatively small share of the landscape (30%). The two shrub-fuel models account for 7% of the watersheds, for a total of 27% of the watersheds having the potential for MHS fire. The remaining shrubs and grasslands, which are the primary vegetation types, tend to carry fast-moving surface fires resulting in a low-severity fire, as discussed above.

20.2.4. Watershed Area Burned at MHS

Figure 20.3 illustrates how fire perimeters and burn-severity patterns are combined and overlaid with watershed perimeters. While the spatial MHS patterns are static within a watershed, the location, shape, and size of each fire perimeter and its relation to watershed boundaries result in an individual calculation of watershed area burned at MHS for every simulated fire. Note that, for this particular landscape, results can exhibit substantial variation in fire size, severity, and watershed area burned. Even when the watershed experiences fire, it is not necessarily the case that the entire watershed burns (although multiple watersheds can be burned by the same fire). Further, even if the entire watershed does burn, MHS does not necessarily occur everywhere. The maximum area of a watershed that can burn with MHS occurs only when a fire perimeter burns the entire watershed, and is then limited by the total area within the watershed mapped as having the potential for MHS. This approach captures the variation surrounding potential future wildfire events with respect to the inputs required for debris-flow modeling.

20.2.5. Debris-Flow Hazard Assessment

Debris-flow modeling efforts use the statistical models for postwildfire debris-flow probability and volume developed by *Cannon et al.* [2010]. The probability of debris-flow occurrence is a function of terrain, soil characteristics, the percentage of the watershed that burned at MHS, and storm intensity. Equation 20.1 is used to calculate debris-flow probability:

$$Probability\left(DF \mid f, s\right) = \frac{e^x}{1+e^x} \tag{20.1}$$

where: $Probability\ (DF \mid f, S)$ is the probability of a post-fire debris flow, given a fire event f and a storm event S.

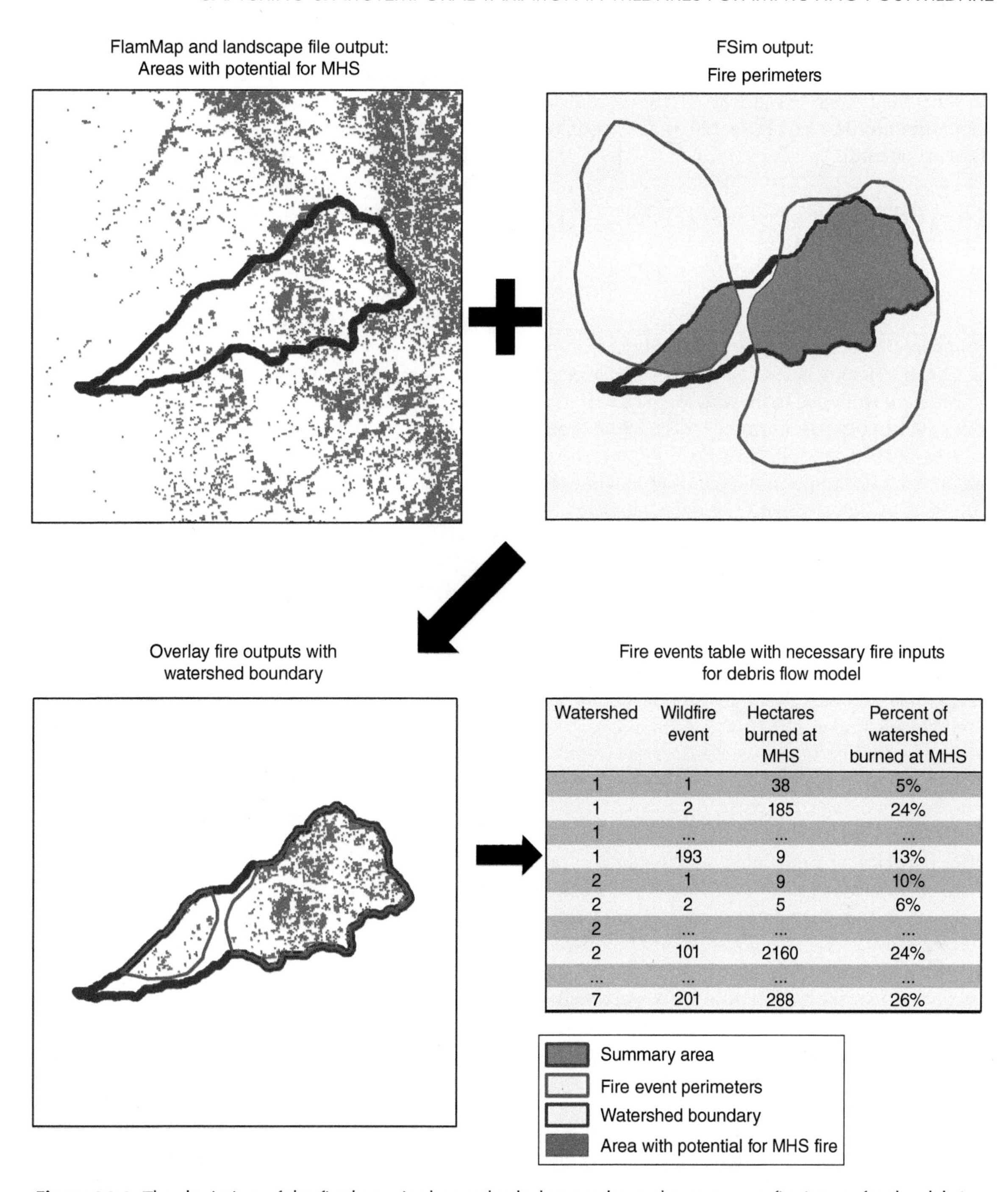

Watershed	Wildfire event	Hectares burned at MHS	Percent of watershed burned at MHS
1	1	38	5%
1	2	185	24%
1	...	...	...
1	193	9	13%
2	1	9	10%
2	2	5	6%
2	...	...	...
2	101	2160	24%
...	...	...	...
7	201	288	26%

Figure 20.3 The depiction of the final step in the methods that produces the necessary fire inputs for the debris-flow model. The spatial intersection of the FlamMap severity output and the FSim fire-perimeter output with watershed boundaries jointly determine the distribution of watershed area and percent burned at MHS.

Equation 20.2 is used to calculate x:

$$x = -0.7 + 0.03(\%SG30) - 1.6(R) + 0.06(\%AB) + 0.07(I) + 0.2(\%C) - 0.4(LL) \quad (20.2)$$

where %SG30 is the percentage of the basin area with a slope equal to or greater than 30%.

R is the basin ruggedness, the change in basin elevation (m) divided by the square root of the basin area (m^2), $\%AB$ is the percentage of basin area burned at MHS, I is the average storm intensity of storm event S, $\%C$ is the percentage clay content of soil, LL is the liquid limit of soil (the percentage of soil moisture by weight at which a solid begins to behave as a liquid).

The potential volume of a debris-flow occurrence is a function of a similar set of variables, but instead considers total watershed area burned at MHS, as well as total storm rainfall. The equation (20.3) for modeling debris-flow volume is:

$$Ln\left(Volume\left(DF \mid f, S\right)\right) = 7.2 + 0.6\left(Ln\left(SG30\right)\right) + 0.7\left(AB\right)^{0.5} + 0.2\left(T\right)^{0.5} + 0.3 \quad (20.3)$$

where: $Volume(DF \mid f, S)$ is the debris-flow volume (in cubic meters) given fire event f and storm event S; $SG30$ is the area of drainage basin with slopes equal to or greater than 30% (in square kilometers); AB is the drainage basin area burned at MHS (in square kilometers); T is the total storm rainfall (in millimeters) of storm event S; and, 0.3 is a bias correction factor that changes the predicted estimate from a median to a mean value [*Cannon et al.*, 2010; *Helsel and Hirsch*, 2002].

Debris-flow probabilities and volumes were calculated at the watershed outlets or pour points, for each of the 30,000 fire seasons. When a fire burned the watersheds during a simulated fire season, the percent area burned under MHS that was calculated in Section 20.2.4 was used for the $\%AB$ and the AB values needed above.

20.2.6. Characterizing Watershed Integrated Hazard

To characterize watershed-level hazard, we calculated postwildfire debris-flow probability and volume over a range of storm scenarios (Table 20.2). Results are conditional in the sense that they depend on a specific recurrence interval storm. A more complex analysis would simultaneously consider a range of storms and their respective occurrence probabilities, a topic we revisit in the discussion. Because of orographic effects of the mountainous terrain, rainfall totals and rainfall intensities will slightly vary over the extent of the study area. Therefore, for this study, we used the NOAA Atlas 14 gridded precipitation frequency estimates [*Bonnin et al.*, 2004] to derive unique storm intensities for each watershed, for six recurrence interval 30 min duration storms. When more than one gridded value was located in a watershed, we use the maximum precipitation value within the watershed for each recurrence interval, providing the most conservative, or highest, estimate of storm intensity and amount. Table 20.2 shows the range of storm intensities across the watersheds for the recurrence intervals.

20.3. RESULTS

20.3.1. Wildfire Simulation Results

We calibrated the FSim run to match historical rates of annual hectares burned and mean number of large fires per year. Across the study area landscape, the FSim wildfire simulation burned an average of 10,090 ha annually, relative to the historical average of 9,636 ha [*Short,* 2014]. We simulated an average of 4.3 large fires per year, compared with the historical average of 4.9 large fires per year [*Short,* 2014]. The majority of the simulated fire seasons did not produce a wildfire that burned any portion of any of the watersheds, meaning the precursor event to a postwildfire debris flow is rare (Table 20.3). Only 280 fires burned at least a portion of one of the watersheds, and there were no seasons with multiple fires burning in the watersheds. Out of 30,000 simulated fire seasons, the watershed that experienced the most wildfires (WS7) only experienced 201 wildfire events, which equates to a 0.07% chance of at least a portion of the watershed burning annually. The maximum percentage of a watershed that burned at MHS ranged from 6.5% (WS2) to 36.0% (WS7), and the mean ranged from 3.2% (WS1) to 16.5% (WS7).

20.3.2. Postwildfire Debris-Flow Results

Figure 20.4 presents scatterplots of the individual simulated fire events and their corresponding conditional debris-flow probabilities and volumes, for each of the six storm recurrence intervals. The probabilities and volumes are conditional upon a storm occurrence. Each point on each scatterplot represents an individual wildfire event,

Table 20.2 Attributes of Analysis Recurrence Interval Storms

Recurrence level	Duration (minutes)	Average storm intensity across watersheds (mm/hr)	Range of storm intensities across watersheds (mm/hr)
2 yr	30	34	34–35
5 yr	30	46	45–47
10 yr	30	55	53–55
50 yr	30	76	74–77
100 yr	30	85	84–86
500 yr	30	108	107–109

Note: Storm values are assumed to be constant across each watershed.

Table 20.3 Wildfire Events Summary Data for Each Watershed

Watershed ID	Number of fire events (F)	Percent annual probability of wildfire	Max and mode percent WS burned	Max. and mode percent MHS	Mean percent WS burned	Mean percent MHS
1	193	0.06	100	24.6	51.5	14.4
2	101	0.03	100	6.5	48.1	3.2
3	193	0.06	100	17.2	53.0	9.2
4	129	0.04	100	19.4	58.8	12.7
5	193	0.06	100	13.9	48.5	6.7
6	182	0.06	100	26.4	50.9	15.0
7	201	0.07	100	36.0	44.0	16.5

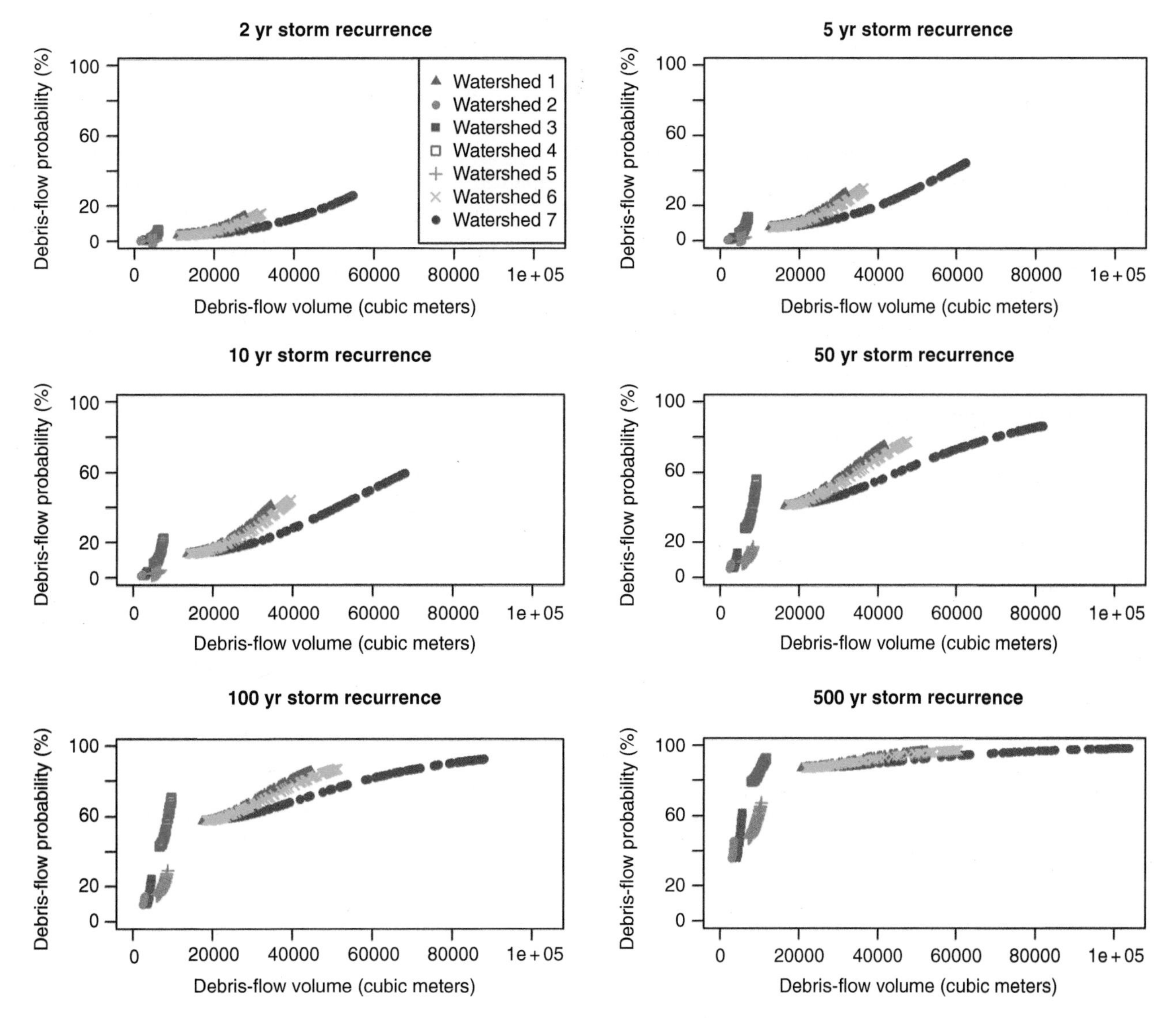

Figure 20.4 Debris-flow probability versus debris-flow volume for the six storm recurrence intervals for each watershed. The individual dots represent the individual fire events associated with each watershed and the range of variability surrounding the potential fire events. Each dot may represent more than one fire event because of the limited spatial combinations of percent area burned at MHS.

and points are colored and grouped by watershed. The colors in Figure 20.4 match those found in the lower left panel of Figure 20.1, for ease of comparison. The variation among events is caused by the simulated differences in the area of a watershed that burned under MHS. Because of the limited combinations of percent area burned at MHS, some perimeters that burned in different locations or with different final fire sizes burned the same proportion of the watershed at MHS, leading to identical debris-flow results (i.e., some points on the scatter plot may actually represent multiple fire events). As storm recurrence intervals increase, the probability and volume of postwildfire debris flows also increase.

For a given debris-flow probability, watershed 7 consistently yielded the largest overall debris-flow volume regardless of storm recurrence level. Watershed 7 also yielded the largest range of variability in debris-flow volume across wildfire events. For a given debris-flow volume, watershed 1 consistently yielded the highest overall debris-flow probability, mirrored closely by watershed 6 and to a lesser extent by watershed 7. Watersheds 2, 3, 4, and 5 tended to yield consistently lower debris-flow probabilities and volumes than watersheds 1, 6, and 7, with watershed 4 presenting the steepest increase in debris-flow probability as storm recurrence interval increased.

How watersheds might be ranked in terms of hazard depends on whether the ranking focuses on debris-flow probability or volume. For example, conditioned upon the 50 yr recurrence interval storm, watersheds 1, 4, 6, and 7 all have a simulated fire event (or set of events) that resulted in a 55% probability of a debris flow. However, the debris-flow volume for each of these watersheds at that probability varies, with watershed 4 having the lowest volume (9,066 m^3) and watershed 7 having almost four times that volume (39,000 m^3). Similarly, for the same recurrence interval, watersheds 1, 6, and 7 all had an event that produced a relatively large-volume debris flow (40,000 m^3). However, the debris-flow probabilities varied between watersheds, with watershed 7 having the lowest likelihood of a 40,000 m^3 debris flow (56%) and watershed 1 having the highest probability of this relatively large-volume debris flow (73%). This example reiterates the need for assessing the range of possible events when analyzing the potential for joint hazards.

Figure 20.5 displays histograms of simulated postwildfire debris-flow volumes for each watershed, for the 50 yr recurrence interval storm. The histogram can better capture variability across the low frequency, large-volume events and the high frequency, small-volume events. This is because, for reasons described above, each point in Figure 20.4 may represent more than one wildfire event thereby masking the total number of simulated postwildfire debris-flow events (see Table 20.3). Examining the watersheds with the greatest probability of debris flows, both watershed 1 and 6 have a similar range of predicted debris-flow volumes, however the frequency distribution of events is slightly different, with watershed 6 primarily having more large volumes events, and watershed 1 having a greater mix of volume events. Watershed 7 again shows the greatest variability where most of the events are either low-volume events or high-volume events, with very few moderate events. Notably, all three watersheds have the highest frequency in the largest volume bin because when a simulated fire burned a portion of a watershed, it most often burned the entire watershed (see Table 20.3). In other words, the maximum event (i.e., worst-case scenario) was the same as the most common event (i.e., the mode event) for all the watersheds in the study area.

While exploring the variation in probability and volume is necessary, providing a single metric of hazard across all simulated events is useful when trying to rapidly identify the watershed with the greatest hazard. This becomes particularly important when evaluating larger landscapes and comparing across a much larger set of watersheds. There are different ways to rank hazards when varying events lead to varying outcomes. One method is to take the ensemble event (the mean event), another is to use the event that occurred most often (the mode event), and yet another is to investigate the worst-case scenario (the maximum event). The reason for choosing one option or another depends on the goals of the researcher or land manager. Using the mode event and the 100 yr recurrence interval storm, watersheds can be ranked from greatest to least hazard as 7, 6, 1, 4, 5, 3, and 2.

Figure 20.6 displays results on wildfire probability, burn severity, and debris-flow probability and volume for each watershed for the 100-year recurrent interval storm. This information helps decompose and graphically display the underlying factors driving hazard assessment in each watershed. Watershed 7 ranked highest for all of the individual components, and clearly presents the greatest hazard consistent with the mode event rankings. Watersheds 1 and 6 have identical estimates of experiencing a large wildfire, and while watershed 6 has a lower area that can burn with MHS, it has slightly greater mode and mean debris-flow probability and volume estimates. This suggests that, relative to wildfire probability and burn severity, differences in watershed size along with soil and topography characteristics may be greater drivers of postwildfire response.

For fuels-mitigation planning, we compare the worst case scenario wildfire event (the maximum area with potential for MHS is burned) to the best case scenario event (none of the watershed area with potential for MHS is burned). The difference between these two events would represent the decrease in debris-flow hazard if fuel treatments could be 100 percent successful in eliminating MHS fire. In reality, this assumption rarely holds true; however, the comparison can give insight as to which

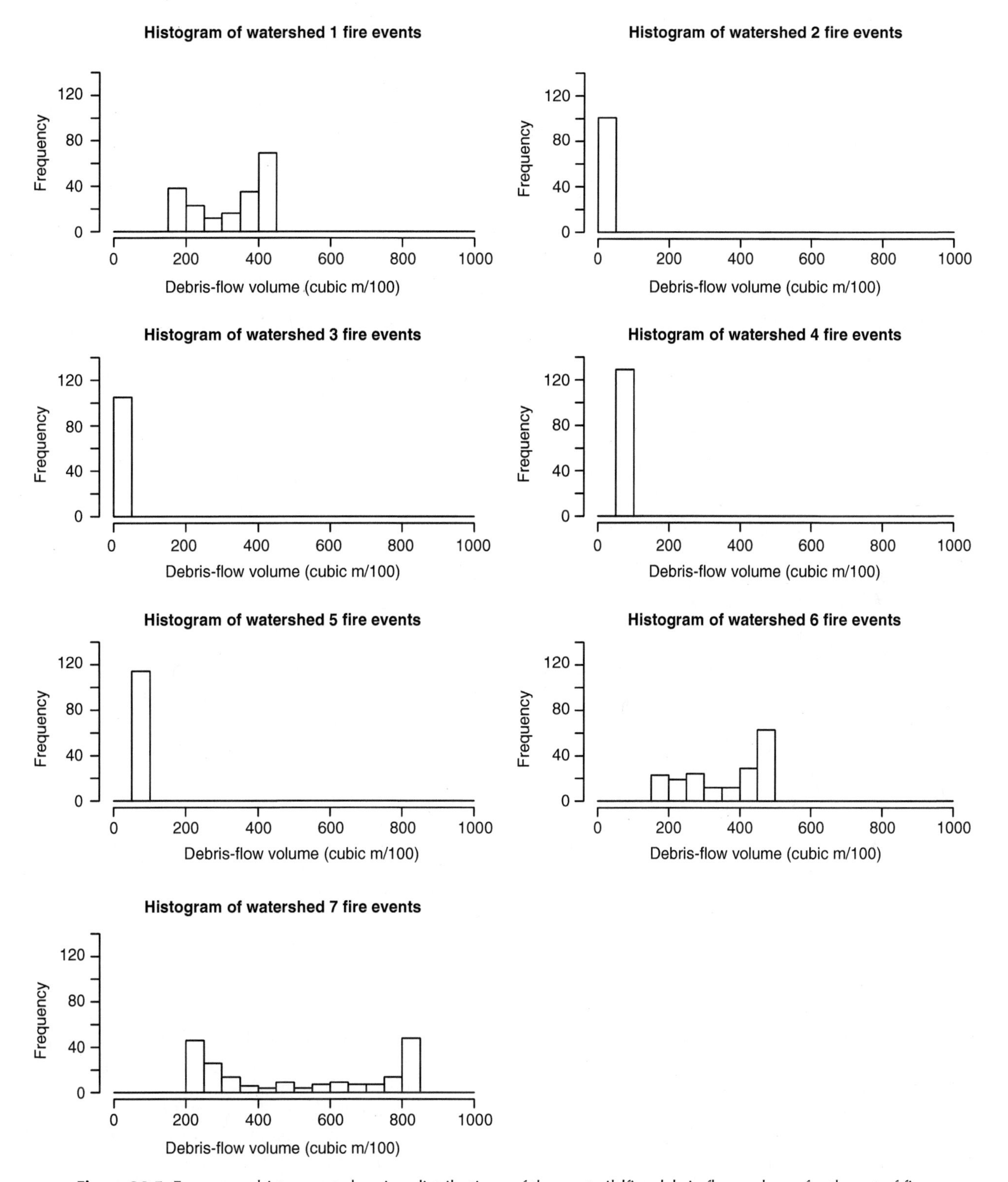

Figure 20.5 Frequency histograms showing distributions of the postwildfire debris-flow volume for the set of fire events. The histograms allow comparison of watershed hazard by frequency of event, which can distinguish watersheds with low-frequency large-volume events from those with high-frequency small-volume events, which can result in similar combined hazard metrics.

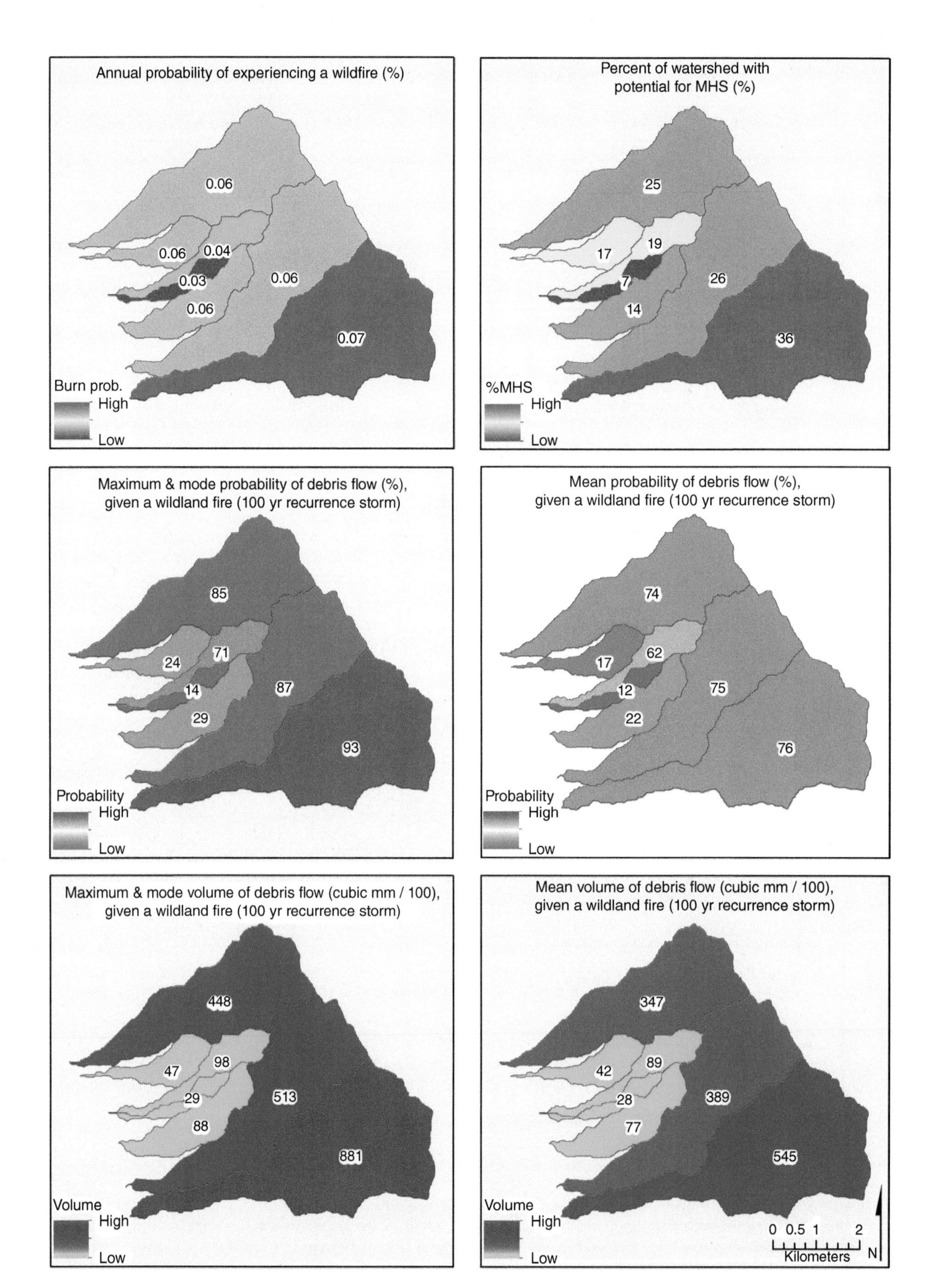

Figure 20.6 The summary statistics for the wildfire events by watershed. The top left panel shows the annual burn probability as a percentage for each watershed. The top right panel shows the percentage of each watershed that would burn at MHS as modeled in this study, if the entire watershed burned. The probability of debris flow (center panels) and volume of debris flow (bottom panels) show the maximum and mode (left, same value for each watershed) and mean (right) values, given a wildfire event. The colors represent the same value between the mean and the mode event for the two variables, allowing for a direct visual comparison between the modal statistics.

Table 20.4 A Comparison of Postwildfire Debris Flow Hazards for the Worst-Case Scenario Event Versus a Best-Case Scenario Event

	Debris-flow probability (percent)				Debris-flow volume (m3)			
	Percentage of area with potential for MHS fire burned				Percentage of area with potential for MHS fire burned			
Watershed ID	100%	0%	Absolute reduction	Percent reduction	100%	0%	Absolute reduction	Percent reduction
1	85%	57%	28	−33%	44,800	27,100	17,700	−40%
2	14%	10%	4	−29%	2,900	400	2,500	−86%
3	24%	10%	14	−58%	4,700	1,300	3,400	−72%
4	71%	44%	27	−38%	9,800	3,200	6,600	−67%
5	29%	15%	14	−48%	8,800	2,800	6,000	−68%
6	87%	58%	29	−33%	51,300	32,700	18,600	−36%
7	93%	59%	34	−37%	88,100	66,400	21,700	−25%

Note: Worst-case scenario event = 100% of area with potential for MHS fire burned; Best-case scenario event = 0% of area with potential for MHS fire burned.

watersheds would see the most reduction in postwildfire debris-flow hazard from fuel treatments designed to reduce severity. By examining the 100-year recurrence interval storm, Table 20.4 shows that if MHS fire could be eliminated from watershed 7 (the watershed with the greatest hazard ranking), debris-flow probability would be reduced from 93 percent to 59 percent, a 37 percent reduction. Additionally, debris-flow volume would be reduced from an estimated 88,100 m^3 to 66,400 m^3, a 25 percent volume reduction. Watershed 2 has the greatest percent reduction for volume (86 percent) and watershed 3 has the greatest percent reduction for probability (58 percent). Using these methods, the maximum event can be compared to other events that match a more realistic expected reduction in MHS fire if fuel treatments were implemented. This reduction would depend on the type and extent of treatment designed. Fuel treatments not only influence the severity of a fire, they influence the rate of spread, and therefore the final fire perimeters. A true analysis of fuel treatment effectiveness would be much more complex and involve incorporating the fuel-treatment effects on not only potential burn severity, but fire spread, fire size, and total watershed area burned. Regardless, the above exercise provides a quick way to assess the reduction of debris-flow hazard, given fuel-treatments designed to eliminate or reduce MHS fire.

20.4. DISCUSSION

This research effort sought to expand the horizons of postwildfire debris-flow modeling in a prewildfire context, with the ultimate aim of providing actionable information that supports hazard-mitigation. The analysis used stochastic simulation to proactively estimate a range of possible wildfire and debris-flow outcomes. In particular, we keyed in on the primary fire-related variable in the debris-flow model (watershed area burned at MHS) and did so by overlaying simulated spatial fire perimeters with modeled burn severity patterns and watershed boundaries. Given the inherent variability in the location and extent of wildfires and their interactions with watersheds, this approach provides critical spatial information on areas of high fire likelihood and severity that can lead to increased debris-flow hazards. We provided a range of hazard metrics that offer a relatively simple and graphically intuitive way to compare multiple hazards across watersheds and under a range of design storms. Our analysis incorporated information on the annual likelihood of a wildfire event, the amount of area in a watershed prone to MHS fire, watershed soil and topography characteristics, and design storm characteristics. Results can guide mitigation efforts by allowing planners to identify which factors may be contributing the most to the hazard rankings of watersheds.

20.4.1. Uncertainties and Limitations

Prediction of the exact timing, location, magnitude, and extent of postwildfire debris flows is inherently difficult if not impossible, and subject to a cascading chain of uncertainties [*Hyde et al.*, Chapter 19, this volume]. Analysis of factors influencing debris-flow initiation, however, can reveal substantial differences in spatial patterns of postwildfire debris-flow probability and volume. Leveraging state-of-the-art spatial fire models with widely used postwildfire debris-flow models can generate useful information for managers who are proactively seeking to understand, prioritize, and mitigate hazards. Hazard modeling results can directly inform mitigation efforts across a variety of planning contexts. Where watershed-level results are highly sensitive to area burned, as was illustrated in some cases, then fuel-treatment planning may focus on strategically locating treatments to interrupt fire spread pathways and/or to increase suppression efficacy. To reduce burn severity, fuel-treatment planning may target areas likely to experience crown fire. If fuel-treatment opportunities are limited but hazards are sufficiently high, downstream investments in water treatment infrastructure may be the more cost-effective option. In either case, a broader suite of the costs and benefits to values need to be weighed and balanced, such as enhanced ecosystem services from watershed restoration (in the case of forest management) or broadened defense against water-quality effects from other pollutants/stressors (in the case of water-treatment capacity investment). In some if not many cases, the relative rarity of MHS wildfire in any given location coupled with the probabilities of storm occurrence and debris-flow initiation would suggest a low return on investment for hazardous fuels treatments targeted specifically at mitigating debris-flow hazard. In other cases, where factors align to indicate high hazard, debris-flow-mitigation investments may be worthwhile, especially if implemented as part of a broader restoration strategy with other economic and environmental objectives. Capturing information on the low-probability high-consequence events can be particularly informative for evaluating mitigation options.

The modeling approach does have notable limitations, and a number of research directions are evident. The combination of models always carries potential for compounding uncertainty, which in this case would center around burn severity patterns and simulated perimeters fed into the debris-flow model [*Cannon et al.*, 2010]. As described in the results section, fire-modeling results were carefully calibrated, and modeled variation in debris-flow hazard across watersheds appears to capture underlying differences in hazard rather than artifacts of uncertainty in the models. At a basic level, future research directions include expanding the empirical base for modeling debris flows [*Riley et al.*, 2013]. Debris-flow models may misrepresent the hazard because of an imperfect knowledge of the system [*Smith et al.*, 2011] as well as a limited set of empirical observations. This latter concern is particularly relevant as the dataset upon which the *Cannon et al.* [2010] model was built included no observations from the current study area. Although the debris-flow model is being updated with more observations including debris flows in New Mexico, this limitation would nevertheless need to be addressed for any operational use of the modeling framework demonstrated here. Wildfire modeling limitations include knowledge gaps related to fire severity prediction [*Hyde et al.*, 2012] and fire-spread modeling [*Finney et al.*, 2012; *Riley and Thompson*, Chapter 13, this volume].

20.4.2. Future Research Directions

There are at least three logical extensions to the modeling framework developed here: (1) storm-occurrence probabilities, (2) burn-severity probabilities, and (3) risk assessment considering consequences and downstream values. First, while the results here focused on the conditional occurrence of a given storm, future efforts could directly incorporate the distribution of different storm intensities to generate unconditional results, for instance through an expanded Monte Carlo simulation approach. A complete analysis would consist of not only simulating the yearly fire season as done here, but the subsequent storm events for the next 3 yr when postwildfire debris flows are most likely to occur. This would allow for information related to temporal synchronicity of extreme fire weather, followed by extreme rain events to be more fully incorporated into the analysis.

Second, future efforts could adopt a more probabilistic approach to burn severity rather than the static burn-severity potential map used here. A single indicator of burn severity for a given pixel does not capture the true complexity of burn-severity patterns across a landscape. If technical details of fire-modeling systems evolve, it may also be possible to simulate patterns of burn severity for each simulated fire, for instance capturing the possibility for widespread MHS fire under extreme weather conditions. Additionally, modeling the probability and volume of a debris flow in the absence of fire would provide useful information; however, currently there is no model that simulates both prewildfire and postwildfire debris-flow hazard in a consistent manner. If such information were made available, one could decipher how much of the total debris-flow hazard was caused by wildfire compared to the inherent characteristics of the watershed and storm intensity. This information would further guide mitigation efforts.

Last, it would be useful to expand the hazard analysis into a full risk assessment. The risk assessment incorporates information on the impacts from the hazards on values of concern to society. For example, a risk analysis could incorporate the consequences of the wildland fire and the debris flow on measures of population served, the location of water-conveyance infrastructure, and sediment delivery pathways. This type of analysis would need to model sediment transport, as well as the probability and volume of a debris flow. Future efforts in the near term can focus on how to leverage these results into real-world mitigation planning on landscapes in New Mexico and elsewhere. Efforts can also focus on more strongly incorporating temporal elements and underlying forest dynamics, ideally better capturing the spatial and temporal coincidence of burned areas and storms.

20.5. CONCLUSION

This work greatly expands the information available to land managers and planners with hazard-mitigation duties by providing a richer way of investigating the variability surrounding the integration of multiple hazards, and a method for comparing hazards across watersheds. By mitigating the negative effects of postwildfire debris flows before a fire occurs, communities can better prepare for the natural hazards that occur in their area, and ultimately reduce the risks associated with these hazards. Under limited time and budgets, our work can help identify watersheds that have the highest combined hazard, and therefore aid in prioritizing where mitigation efforts would be the most beneficial. This work has already informed stakeholders involved in the Rio Grande Water Fund of the postwildfire debris-flow hazards in the Sandia Mountain area. Additional work will expand the analysis area north to include the Jemez and Sangre de Cristo mountains of New Mexico, to better capture the risk of postwildfire debris flows on the Rio Grande watershed.

REFERENCES

Ager, A. A., M. A. Day, M. A. Finney, K. Vance-Borland, and N. M. Vaillant (2014), Analyzing the transmission of wildfire exposure on a fire-prone landscape in Oregon, USA, *For. Ecol. Man.*, *334*, 377–390.

Alexander, M. E., and M. G. Cruz (2012), Interdependencies between flame length and fireline intensity in predicting crown fire initiation and crown scorch height, *Int. J. Wildland Fire*, *21*, 95–113.

Bonnin, G. M., D. Martin, B. Lin, T. Parzybok, M. Yekta, and D. Riley (2004), Precipitation-frequency atlas of the United States, Volume 1 Version 5.0, Semiarid Southwest (Arizona, Southeast California, Nevada, New Mexico, Utah), National Oceanic and Atmospheric Administration Atlas 14, National Oceanic and Atmospheric Administration, Silver Spring, MD, 271.

Bradshaw, L. S., R. E. Burgan, J. D. Cohen, and J. E. Deeming (1983), The 1978 National Fire Danger Rating System: Technical documentation, USDA Forest Service; Intermountain Forest and Range Experiment Station, General Technical Report INT-169, Ogden, UT.

Cannon, S. H., and J. E. Gartner (2005), Wildfire-related debris flow from a hazards perspective, 363–385, in *Debris-Flow Hazards and Related Phenomena*, edited by M. Jacob and O. Hungr, Praxis, Springer-Verlag, Berlin.

Cannon, S. H., E. M. Boldt, J. L. Laber, J. W. Kean, and D. M. Staley (2011), Rainfall intensity–duration thresholds for postfire debris-flow emergency-response planning, *Nat. Hazards*, *59*(1), 209–236.

Cannon, S. H., E. R. Bigio, and E. Mine (2001), A process for fire-related debris flow initiation, Cerro Grande fire, New Mexico, *Hydrolog. Processes*, *15*(15), 3011–3023.

Cannon, S. H., J. E. Gartner, M. G. Rupert, J. A. Michael, A. H. Rea, and C. Parrett (2010), Predicting the probability and volume of postwildfire debris flows in the intermountain western United States, *GSA Bull.*, *122*(1–2), 127–144.

Cannon, S. H., J. E. Gartner, R. C. Wilson, J. C. Bowers, and J. L. Laber (2008), Storm rainfall conditions for floods and debris flows from recently burned areas in southwestern Colorado and southern California, *Geomorphology*, *96*(3–4), 250–269.

China, S., C. Mazzoleni, K. Gorkowski, A. C. Aiken, and M. K. Dubey (2013), Morphology and mixing state of individual freshly emitted wildfire carbonaceous particles, *Nat. Comm.*, *4*(2122); doi:10.1038/ncomms3122.

Chong, J., J. Renaud, and E. Ailsworth (2004), Flash floods wash away lives, dreams, *Los Angeles Times* (3 January 2004), B1.

DeGraff, J. V., S. H. Cannon, and A. J. Gallegos (2007), Reducing post-wildfire debris flow risk through the Burned Area Emergency Response (BAER) process, Conference Presentations from 1st North American Landslide Conference, Vail Colorado, AEG Special Publication, 23.

Eidenshink, J., B. Schwind, K. Brewer, Z. Zhu, B. Quayle, and S. Howard (2007), A project for monitoring trends in burn severity, *Fire Ecol.*, *3*(1), 3–21.

Finney, M. A. (2002), Fire growth using minimum travel time methods, *Can. J. For. Res.*, *32* (8), 1420–1424.

Finney, M. A. (2006), FlamMap 3.0. USDA Forest Service, Rocky Mountain Research Station, Fire Sciences Laboratory, Missoula, MT, Rep., USDA Forest Service, Rocky Mountain Research Station, Portland, OR, 213–220.

Finney, M. A., C. W. McHugh, I. C. Grenfell, K. L. Riley, and K. C. Short (2011), A simulation of probabilistic wildfire risk components for the continental United States, *Stochas. Environ. Res. Risk Assess.*, *25*(7), 973–1000.

Finney, M. A., I. C. Grenfell, C. W. McHugh (2009), Modeling containment of large wildfires using generalized linear mixed-model analysis, *For. Sci.*, *55*(3), 249–255.

Finney, M. A., J. D. Cohen, S. S. McAllister, and W. M. Jolly (2012), On the need for a theory of wildland fire spread, *Int. J. Wildland Fire*, *22*(1), 25–36.

Gartner, J. E., S. H. Cannon, E. R. Bigio, N. K. Davis, C. Parrett, K. L. Pierce, M. G. Rupert, B. L. Thurston, M. J. Trebish, S. P. Garcia, and A. H. Rea (2005), Compilation of data relating to the erosive response of 608 recently burned basins in the Western United States, US Geological Survey Open-File Report, 2005–1218.

Haas, J. R., D. E. Calkin, and M. P. Thompson (2013), A national approach for integrating wildfire simulation modeling into wildland urban interface risk assessments within the United States, *Landscape Urb. Plan.*, *119*, 44–53.

Haas, J. R., D. E. Calkin, and M.P. Thompson (2014), Wildfire risk transmission in the Colorado front range, USA, *Risk Anal.*; doi: 10.1111/risa.12270.

Harpold, A. A., J. A. Biederman, K. Condon, M. Merino, Y. Korgaonkar, T. Nan, L. L. Sloat, M. Ross, P. D, Brooks (2014), Changes in snow accumulation and ablation following the Las Conchas forest fire, New Mexico, USA, *Ecohydrology*, *7*(2), 440–452.

Helsel, D. R., and R. M. Hirsch (2002), Statistical methods in water resources, 510, in *Techniques of Water-Resources Investigations of the United States Geological Survey, Book 4, Hydrological Analysis and Interpretation*, Elsevier, Reston, Virginia.

Hyde, K., M. B. Dickinson, G. Bohrer, D. Calkin, L. Evers, J. Gilbertson-Day, T. Nicolet, K. Ryan, and C. Tague (2012), Research and development supporting risk-based wildfire effects prediction for fuels and fire management: status and needs, *Int. J. Wildland Fire*, *22*(1), 37–50.

Johansen, M. P., T. E. Hakonson, and D. D. Breshears (2001), Postfire runoff and erosion from rainfall simulation: Contrasting forest with shrublands and grasslands, *Hydrolog. Processes*, *15*, 2953–2965.

Jones, O. D., P. Nyman, and G. J. Sheridan (2014), Modelling the effects of fire and rainfall regimes on extreme erosion events in forested landscapes, *Stochas. Environ. Res. Risk Assess.*, *28*, 2015–2025.

Julyan, R. H., and M. Stuever (2005), *Field Guide to the Sandia Mountains*, University of New Mexico Press, Albuquerque, NM.

Keane, R. E., S. A. Mincemoyer, K. M. Schmidt, D. G. Long, and J. L. Garner (2000), Mapping vegetation and fuels for fire management on the Gila National Forest Complex, New Mexico [CD-ROM], Gen. Tech. Rep. RMRS-GTR-46-CD, Ogden, UT, USDA Forest Service, Rocky Mountain Research Station.

Keeley, J. E. (2009), Fire intensity, fire severity and burn severity–A brief review and suggested usage, *Int. J. Wildland Fire*, *18*(1), 116–126.

Keeley, J. E., T. Brenna, and A. H. Pfaff (2008), Fire severity and ecosystem responses following crown fires in California shrublands, *Ecol. Appl.*, *18*(6), 1530–1546.

LANDFIRE (2012), LANDFIRE 1.3 LCP layer, US Department of Interior Geological Survey [Online], http://landfire.cr.usgs.gov/viewer/ (2015, January).

Meyer, G. A., and S. G. Wells (1997), Fire-related sedimentation events on alluvial fans, Yellowstone National Park, USA, *J. Sediment. Res.*, *67*(5), 776–791.

Nature Conservancy (2014), Rio Grande water fund: Comprehensive plan for wildfire and water source protection, accessed 4 September 2013, at http://www.nmconservation.org/RGWF/RGWF_CompPlan.pdf.

Nyman, P., G. J. Sheridan, and P. N. J. Lane (2013), Hydro-geomorphic response models for burned areas and their applications in land management, *Prog. Phys. Geog.*, *37*(6), 787–812.

Parks, S. A., M. Parisien, and C. Miller (2011), Multi-scale evaluation of the environmental controls on burn probability in a southern Sierra Nevada landscape, *Int. J. Wildland Fire*, *20*, 815–828.

Parsons, A., B. Jarvis, and A. Orleman (2002), Mapping of post-wildfire burned severity using remote sensing and GIS, in 22nd Annual Esri Conference, September, 2002: Redlands, CA, Proceedings, Environmental Systems Research Institute, Inc., also available at http://proceedings.esri.com/library/userconf/proc02/pap0431/p0431.htm.

Pyne, S. J., P. L. Andrews, and R. D. Laven (1996), *Introduction to Wildland Fire*, 2 ed., John Wiley and Sons, New York,

Rhoades, C. C., D. Entwistle, D. Butler (2011), The influence of wildfire extent and severity on streamwater chemistry, sediment and temperature following the Hayman Fire, Colorado, *Int. J. Wildland Fire*, *20*(3), 430–442.

Riley, K. L., R. Bendick, K. D. Hyde, and E. J. Gabet (2013), Frequency-magnitude distribution of debris flows compiled from global data, and comparison with post-fire debris flows in the western US, *Geomorphology*, *191*, 118–128.

Rothermel, R. C. (1972), A mathematical model for predicting fire spread in wildland fuels, Res. Pap. INT-115, USDA Forest Service, Intermountain Forest and Range Experiment Station, Ogden, UT.

Scott, J. H., and E. D. Reinhardt (2001), Assessing crown fire potential by linking models of surface and crown fire behavior, Res. Pap. RMRS-RP-29., USDA Forest Service, Rocky Mountain Research Station, Fort Collins, CO.

Scott, J. H., and R. E. Burgan (2005), Standard fire behavior fuel models: A comprehensive set for use with Rothermel's surface fire spread model, Gen. Tech. Rep. RMRS-GTR-153, US Department of Agriculture, Forest Service, Rocky Mountain Research Station, Fort Collins, CO.

Scott, J. H., M. P. Thompson, and J. W. Gilbertson-Day (2015), Exploring how alternative mapping approaches influence fireshed assessment and human community exposure to wildfire, *GeoJournal, 1–15*; doi: 10.1007/s10708-015-9679-6.

Short, K. C. (2013), A spatial database of wildfires in the United States, 1992–2011, *Earth Syst. Sci. Data Discuss.*, *6*(2), 297–366.

Short, K. C. (2014), Spatial wildfire occurrence data for the United States, 1992–2012 (FPA_FOD_20140428) 2nd, USDA Forest Service, Rocky Mountain Research Station, Fort Collins, CO, http://dx.doi.org/10.2737/RDS-2013-0009.

Smith, H. G., G. J. Sheridan, P. N. J. Lane, P. Nyman, and S. Haydon (2011), Wildfire effects on water quality in forest catchments: A review with implications for water supply, *J. Hydrol.*, *396*, 170–192.

Staley, D. M. (2013), Emergency assessment of post-fire debris-flow hazards for the 2013 Rim Fire, Stanislaus National Forest and Yosemite National Park, California: US Geological Survey Open-File Report 2013–1260, http://pubs.usgs.gov/of/2013/1260/.

State of California Sierra Nevada Conservancy (2014), Mokelumne Watershed Avoided Cost Analysis., http://www.sierranevada.ca.gov/our-work/mokelumne-watershed-analysis, last accessed 13 November 2014.

Thompson, M. P., and D. E. Calkin (2011), Uncertainty and risk in wildland fire management: a review, *J. Environ. Man.*, *92*(8), 1895–1909.

Thompson, M. P., J. R. Haas, J. W. Gilbertson-Day, J. H. Scott, P. Langowski, E. Bowne, and D. E. Calkin (2015a), Development and application of a geospatial wildfire exposure and risk calculation tool, *Environ. Mod. Soft.*, *63*, 61–72.

Thompson, M. P., J. Scott, D. Helmbrecht, and D. E. Calkin (2013a), Integrated wildfire risk assessment: Framework development and application on the Lewis and Clark National Forest in Montana, USA, *Integrat. Environ. Assess. Man.*, *9*(2), 329–342.

Thompson, M. P., J. Scott, J. D. Kaiden, and J. W. Gilbertson-Day (2013c), A polygon-based modeling approach to assess exposure of resources and assets to wildfire, *Nat. Hazards*, *67*(2), 627–644.

Thompson, M. P., J. Scott, P. G. Langowski, J. W. Gilbertson-Day, J. R. Haas, and E. M. Bowne (2013b), Assessing watershed-wildfire risks on National Forest System lands in the Rocky Mountain Region of the United States, *Water*, *5*(3), 945–971.

Thompson, M. P., J. W. Gilbertson-Day, and J. H. Scott (2015b), Integrating pixel-and polygon-based approaches to wildfire risk assessment: Application to a high-value watershed on the Pike and San Isabel national forests, Colorado, USA, *Environ. Mod. Assess.*, 1–15.

Tillery, A. C., J. R. Haas, L. W. Miller, J. H. Scott, and M. P. Thompson (2014), Potential postwildfire debris-flow hazards: A prewildfire evaluation for the Sandia and Manzano mountains and surrounding areas, Central New Mexico, US Geological Survey Scientific Investigations Report 2014–5161, http://dx.doi.org/10.3133/sir20145161.

Tillery, A. C., M. J. Darr, S. H. Cannon, and J. A. Michael (2011), Postwildfire preliminary debris flow hazard assessment for the area burned by the 2011 Las Conchas Fire in north-central New Mexico, US Geological Survey Open-File Report 2011–1308.

US Geological Survey and USDA Forest Service (2013), Monitoring Trends in Burn Severity, Fire Level Geospatial Data, MTBS Project accessed June 2013, at http://mtbs.gov/data/individualfiredata.html.

Warziniack, T., and M. Thompson (2013), Wildfire risk and optimal investments in watershed protection, *West. Econ. For.*, *12*(2), 19–28.

21

Uncertainty in Estimation of Debris-Flow Triggering Rainfall: Evaluation and Impact on Identification of Threshold Relationships

E. I. Nikolopoulos,[1] F. Marra,[2] and M. Borga[1]

ABSTRACT

Operational debris-flow warning systems are often based on the use of empirical rainfall thresholds derived from rain gauge observations. However, rain gauges are usually located away from the actual debris-flow locations thus estimation of triggering rainfall properties from rain gauges can be associated with considerable uncertainty. This work examines the uncertainty in gauge-based estimation of debris-flow triggering rainfall and evaluates its impact on the identification of rainfall thresholds used for debris-flows prediction. These issues are assessed by using high-resolution radar data to represent "actual" space-time patterns of precipitation at and around the debris-flow initiation points. Rain-gauge network sampling is simulated by randomly sampling radar-rainfall fields. Rainfall is estimated by using three rainfall interpolation methods: nearest neighbor (NN), inverse distance weighting (IDW), and ordinary kriging (OK). Comparison of results from these three methods shows that no particular benefit in intensity-duration threshold estimation is obtained by using approaches that are more complex than the NN method. NN provides estimates with smaller bias than IDW and OK but larger estimation variance. On average, decrease in gauge density leads to increased underestimation of debris-flow rainfall and subsequently this results in large underestimation of the intensity-duration thresholds.

21.1. INTRODUCTION

Debris flows are water-laden masses of soil and fragmented rock that rush down mountainsides, funnel into stream channels, and entrain objects in their paths [*Iverson*, 2005]. Because they have volumetric sediment concentrations that exceed 40%, maximum speeds that surpass 10 m/s, and sizes that can range up to ~10^9 m^3, debris flows can denude slopes, bury floodplains, and devastate people and property [*Petley*, 2012]. Most debris flows originate from discrete or distributed source areas where slopes steeper than about 25° to 30° are mantled with low-cohesion soil and/or fragmented rock. This marginally stable debris becomes at least partly saturated through a rapid introduction of surface water or groundwater, commonly as a result of intense rainfall or snowmelt. Water-laden debris starts to shear and move downslope when at some depth frictional forces cannot resist driving forces, irrespective of whether the debris is positioned on slopes or in water-filled channels [*Iverson*, 2005; *Borga et al.*, 2014].

Debris-flow hazard assessments must address two kinds of questions. First, where and when will debris flows occur, and how large will they be? And second, how fast will debris flows travel, and what areas will be impacted downstream? Answers to the first type of question require knowledge of hydrological conditions favoring the development of debris flows, whereas answers to the second type require knowledge on behavior during flow runout [*Borga et al.*, 2014]. This chapter is focused on the first key question.

Forecasting of debris-occurrence timing is key for issuing hazard warnings, and focuses largely on rainfall as a

[1] *Department of Land, Environment, Agriculture and Forestry, University of Padova, Legnaro, Italy*

[2] *Department of Geography, Hebrew University of Jerusalem, Israel*

Natural Hazard Uncertainty Assessment: Modeling and Decision Support, Geophysical Monograph 223, First Edition.
Edited by Karin Riley, Peter Webley, and Matthew Thompson.

triggering agent. Debris-flow forecasting relies very often on approaches based on the identification of combinations of rainfall intensity and duration that provoke widespread debris flows [e.g., *Caine*, 1980; *Wieczorek*, 1996; *Deganutti et al.*, 2000; *Guzzetti et al.*, 2008; *David-Novak et al.*, 2004; *Frattini et al.*, 2009; *Borga et al.*, 2014]. Usually these combinations consist of a relationship that links rainfall duration D (hr) to the average rainfall intensity I (mm hr^{-1}) [*Aleotti*, 2004; *Guzzetti et al.*, 2007] such as:

$$I = \alpha D^{-\beta} \tag{21.1}$$

where α and β are the scale and the shape parameters, respectively, that adapt the power-law model to the empirical data. The identification of the rainfall intensity-duration (*ID*, hereinafter) threshold is based most often on the availability of rain gauge rainfall data and debris-flows data occurrence [*Guzzetti et al.*, 2008] and as such it is affected by a number of uncertainties [*Nikolopoulos et al.*, 2014]. With the exception of the very rare cases when rain gauge data are available at the debris-flow initiation site, rainfall amounts and intensities responsible for debris-flow events are estimated based on data from more or less remote neighboring stations. This rainfall estimation problem affects the identification of the intensity-duration thresholds in two main ways. First, recent work [*Nikolopoulos et al.*, 2014] has shown that the variance in rainfall estimation (hereinafter refer to as estimation variance) due to random errors (i.e., errors with zero mean), which is affecting the rainfall interpolation over the debris-flow sites, translates into a systematic error in the estimated intensity-duration threshold. Given that the intensity-duration threshold is estimated as a lower boundary of the estimated debris-flow triggering rainfalls, it is expected that any increase in the rainfall estimation variance will increase the negative bias in the identification of the *ID* threshold. Second, the unknown precipitation must be estimated at points where by design it exceeds a threshold [*Marra et al.*, 2014]. Then, the estimated rainfall at the debris-flow initiation points not only is affected by a random error, but also is inherently negatively biased (i.e., affected by a systematic error with nonzero mean). This bias will add to any other bias in the identification of the intensity-duration threshold.

Inspection of the literature shows that the quantitative analysis of these two problems has been so far rarely, if ever, approached [*Nikolopoulos et al.*, 2014]. This seems to be due mainly to the lack of reliable rainfall data, which are required both at the debris-flow initiation points (these data are available for debris-flow experimental catchments [*Comiti et al.*, 2014]) and in the surrounding region. The recent availability of accurate radar rainfall estimates for debris-flow triggering rain events [*Tiranti et al.*, 2008; *Tiranti et al.*, 2014; *Marra et al.*, 2014] open new perspectives in the examination of the sources of uncertainties affecting the identification of intensity-duration thresholds.

Given this background, this work has two objectives. First, we aim to quantify the dependence of debris-flow-triggering rainfall estimation accuracy on the rain gauge density and on the use of various rainfall estimation procedures. Second, we evaluate the impact of the rainfall estimation errors on the identification of the intensity-duration thresholds used for the prediction of debris-flow occurrence. These questions are examined by using high-resolution, carefully corrected radar rainfall data to represent space-time patterns of true precipitation at the debris-flow initiation points and in the surrounding area. These rainfall fields are sampled by simulated rain gauge networks, stochastically generated with varying gauge densities. This approach is ideally suited to assess the properties of rainfall estimation errors in areas with strong gradients, where alternative methods are lacking.

The present work is based on the availability of accurate radar-based estimates of rainfall fields for 10 storms, which triggered 82 debris flows in a mountainous region in the eastern Italian Alps (the Upper Adige River basin).

21.2. METHODS

21.2.1. Study Area and Data

The area of study is part of the Upper Adige River basin in northern Italy (Fig. 21.1), a mountainous region with more than 64% of the area located above 1500 m a.s.l., whereas only 4% of the area is located below 500 m a.s.l. The southeastern sector of the region belongs to the Dolomites, the northeastern part to the Noric Alps and the western sector to the eastern Rhaetian Alps, including the highest peak of Mount Ortles (3902 m) [*Norbiato et al.*, 2009; *Nikolopoulos et al.*, 2014]. The region is monitored by a network of 120 rain gauges with an average spatial density of about 1/80 km^{-2}, and by a C-band, Doppler weather radar located at 1860 m a.s.l. on the top of Mount Macaion, a central position in the study area (Fig. 21.1). Quantitative precipitation estimates, derived from the radar reflectivity observations, are available at high spatial (1 km) and temporal (5 min) resolution. A database reporting the location and the date of debris-flow occurrence in the study area is available for more than 400 debris flows during period 2000–2012, with high spatial and temporal accuracy.

Ten debris-flow triggering rainfall events that occurred in the study area between 2005 and 2010 are examined. This period was selected based on the availability of high-resolution, quality-controlled radar rainfall estimates. The 10 events selected are among the most severe in the region during this period and triggered a total of 82

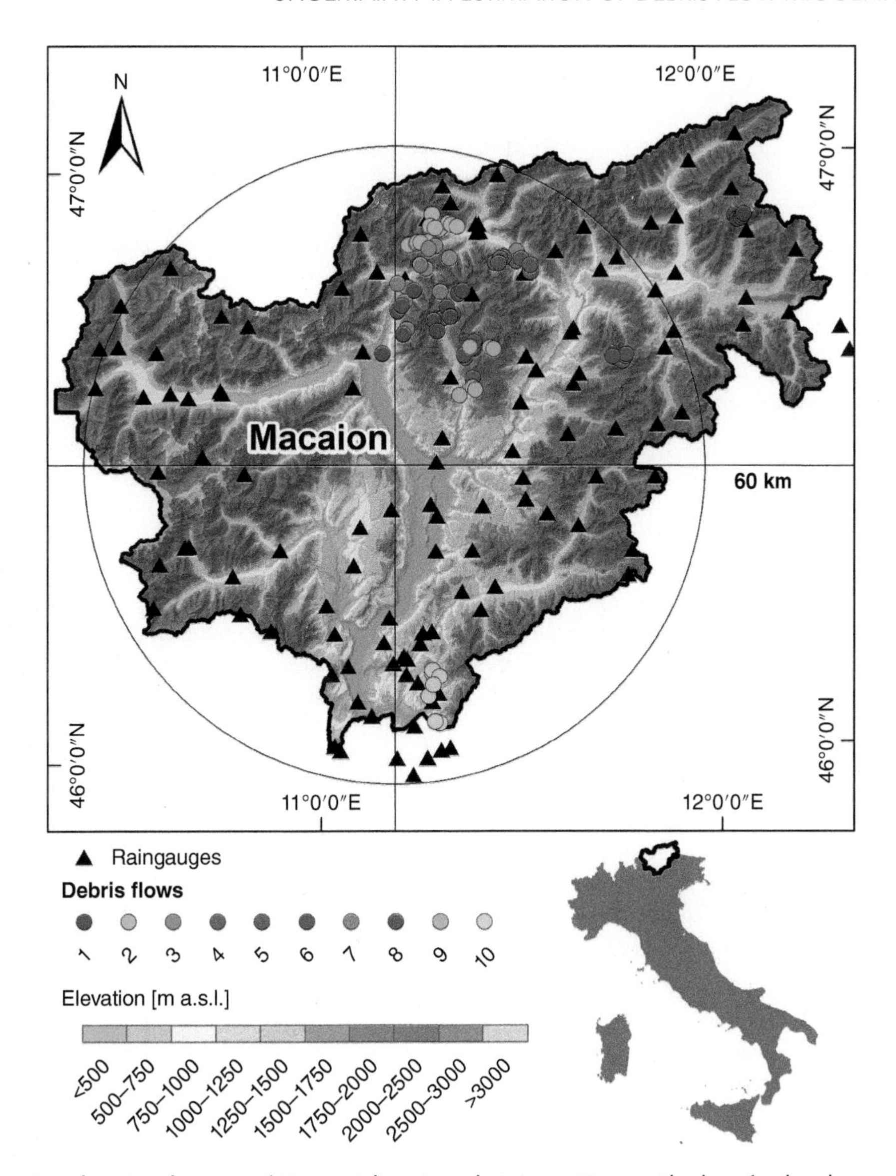

Figure 21.1 Map showing the area of Upper Adige river draining at Trento. Shades of color show terrain elevation. Triangles show locations of available rain gauges in the region. Dot symbols correspond to the locations of the 82 debris flows analyzed. Location of the Macaion weather radar with the range circle at 60 km is also shown.

debris flows (termed DF hereinafter) (Fig. 21.1) that caused significant damage to people and infrastructures. All events occurred during the summer months and thus are considered to be representative of the DF regime and its seasonality characteristics [*Nikolopoulos et al.*, 2014]. Most of the considered events are due to channel bed mobilization [*Berti and Simoni*, 2005; *Gregoretti and Dalla Fontana*, 2008], and as such their dependence on initial soil moisture conditions is less marked than for landslide-initiated debris flows [*Berti and Simoni*, 2005]. This suggests that a simple rainfall threshold model may be adequate to identify the triggering events. For all events, eyewitnesses reported that the debris flows occurred in response to periods of high-intensity rainfall during the storm. Location of the initiation point of the individual debris flows is georeferenced with accuracy better than ±50 m.

Radar data used in this study have been carefully processed and corrected for error sources associated with (1) signal attenuation in heavy rain, (2) wet radome attenuation, (3) beam blocking, and (4) vertical profile of reflectivity, according to the procedures described by *Marra et al.* [2014]. In addition, the corrected radar-rainfall fields are further adjusted according to the mean field bias derived on an event-basis by comparing rain gauge data and radar observations. For each event, an analysis domain was established as a rectangular subset of the whole radar umbrella (Fig. 21.2) that is centered

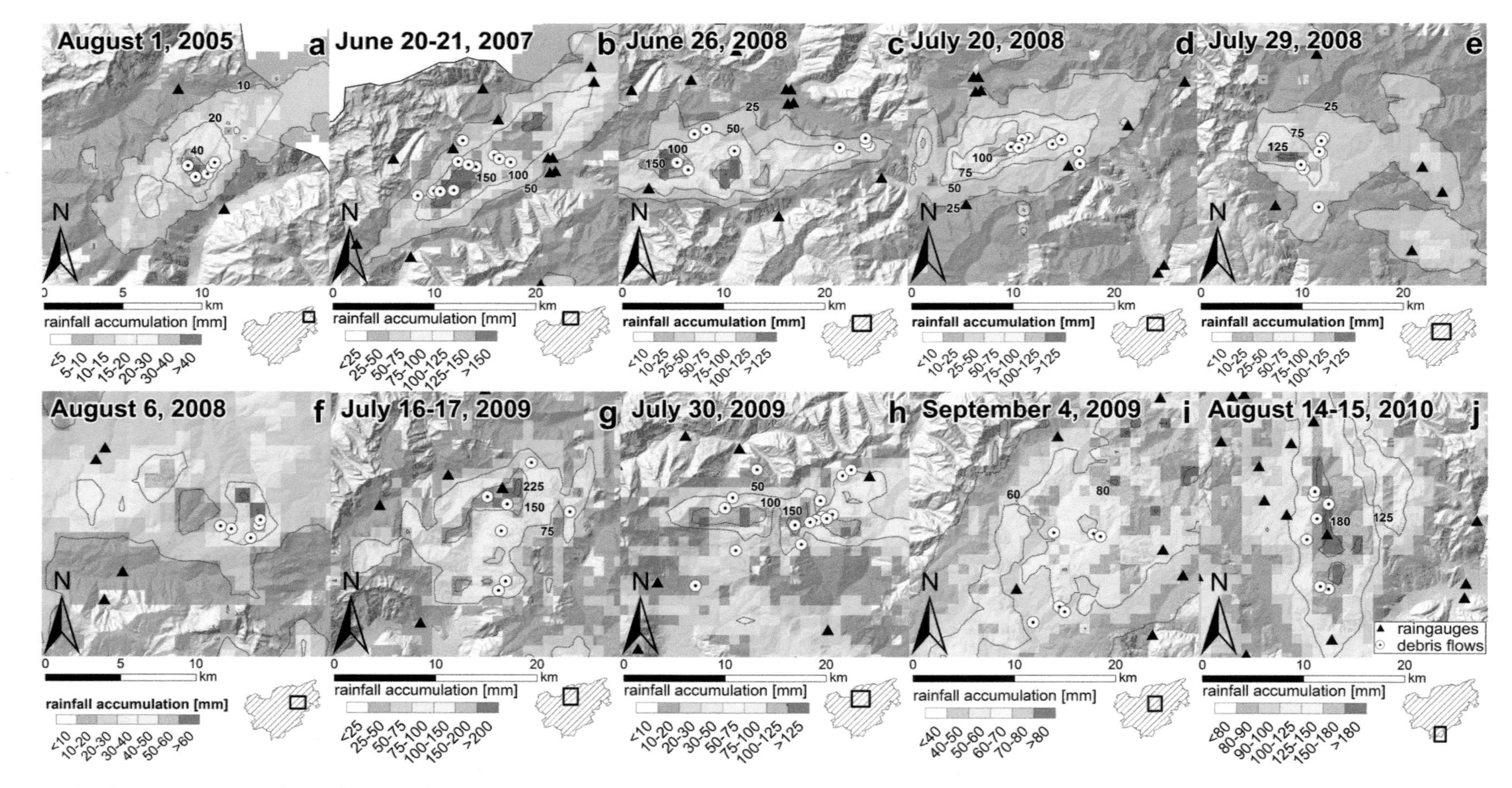

Figure 21.2 Maps of rainfall accumulation for the 10 rainfall events (a–j). Dotted circles represent the location of triggered debris flows, black triangles show the location of rain gauges. Events are presented in chronological order: (a) 1 August 2005; (b) 20–21 June 2007; (c) 26 June 2008; (d) 20 July 2008; (e) 29 July 2008; (f) 6 August 2008; (g) 16–17 July 2009; (h) 30 July 2009; (i) 4 September 2009; (j) 14–15th August 2010. Location of each rainfall map relative to the basin is shown using a black rectangular box superimposed on a minimap of the basin.

over the location of DF and large enough to include the triggering storm event. The size of the domain varied from ~500–1200 km^2 and increased according to the spreading of DF locations for each case. Figure 21.2 shows that several DF locations correspond to the main rainfall accumulation peaks. The figure also provides an appreciation of the large rainfall spatial variability characterizing these events and of the corresponding rain gauge sampling problem. Duration of rainfall events varied from 2 hr (1 August 2005) to 27 hr (16–17 July 2009) and the cumulated rainfall at debris-flow locations ranged from 14 mm (30 July 2009) to 180 mm (14–15 August 2010).

21.2.2. Methodology

A main assumption of the methodology followed in this study was that the radar data set provided the "true" rainfall and was therefore used as the reference for the comparative analysis carried out throughout this work. A specific work is ongoing to explore the implication of this assumption. In essence, this allowed us to represent empirically the issues related to the spatial sampling error of rainfall fields by comparing variations in precipitation (as observed by radar and gauges). Hence, use of observed data is beneficial and needed to elucidate the impact of the sampling errors.

The methodological framework we followed is the one provided by *Nikolopoulos et al.* [2015] and is briefly described in the sections below. For more details, the interested reader is referred to *Nikolopoulos et al.* [2015].

21.2.2.1. Rain Gauge Network Configurations

A numerical exercise was carried out to simulate the sampling of the rainfall field from a rain gauge network of a given density. This involved two main steps. First, the location of virtual gauges was selected at random within the study region. Second, the radar rainfall values corresponding to the simulated gauge locations were extracted. The number of rain gauges (i.e., radar pixels) selected for the construction of hypothetical rain gauge networks corresponds to predefined sampling density equal to $1/Ar$ (gauge/km^2) with area Ar of 10, 20, 50, and 100 km^2. Clustering of two or more gauges in close proximity was avoided by conditioning the random sampling according to a threshold on minimum intergauge distance, which after visual inspection of generated networks is set equal to $0.5\ A_r^{0.5}$. A total of 100 realizations of simulated rain gauge networks was generated for each case to account for the effect of variability in network geometry.

21.2.2.2. Rainfall Estimation Methods

We applied three widely used interpolation methods for estimating rainfall at DF locations according to estimates obtained at simulated rain gauge locations. The interpolation methods include the (1) nearest neighbor (NN), (2) inverse distance weighting (IDW), and (3) ordinary kriging (OK). A short description of the methods and some details on their application is provided in *Nikolopoulos et al.* [2015]. These authors have shown that there is no systematic relationship between rainfall depth and elevation for the examined cases. Owing to this reason, we haven't used elevation as a secondary variable in a multivariate interpolation scheme to improve performance in rainfall estimation. The NN method is frequently used for estimation of debris-flow triggering rainfall (e.g., *Jakob and Weatherly,* 2003; *Aleotti,* 2004; *Godt et al.*, 2006; *Brunetti et al.*, 2010; *Berti et al.*, 2012).

21.2.2.3. Identification of Critical Rainfall Threshold

In this study, we used the intensity-duration (*ID*) model presented in equation (21.1). Estimation of *ID* threshold parameters α and β was based on the *frequentist* approach proposed by *Brunetti et al.* [2010]. The method can be used to identify *ID* thresholds at different levels of exceedance probabilities. In this work, we adopted a 10% exceedance level, which means that the probability of a DF- triggering rainfall event (*I, D* pair) to be under the estimated *ID* threshold is less than 10%. Figure 21.3 summarizes the *I, D* pairs for the 82 DFs analyzed in this study, and the corresponding estimated *ID* threshold at 10% exceedance, $I = 13.92\ D^{-0.66}$. This relationship is termed "reference ID threshold" in the following sections.

21.3. RESULTS

For each realization of simulated rain gauge networks, rainfall estimates at DF locations were obtained from the three interpolation methods (NN, IDW, OK) examined. Interpolated values were then compared against the radar rainfall estimates at the actual DF location, which serves as reference throughout the analysis. Error in rainfall estimation is expressed in terms of relative error, ε_r, defined as

$$\varepsilon_r = \frac{Rtot_{iDF} - Rtot_{DF}}{Rtot_{DF}} \tag{21.2}$$

where $Rtot_{iDF}$ and $Rtot_{DF}$ are interpolated and reference total event rainfall at DF location, respectively. Recall that the data sample used in the analysis comprises 82 DF events with 100 realizations of simulated rain gauge networks generated for each event (i.e., a total of 8200 values). The mean and standard deviation of relative error is shown in Figure 21.4 for the different rain gauge densities and interpolation algorithms examined. Note that positive (negative) error means that interpolated rainfall estimates overestimate (underestimate) rainfall at DF locations.

There are two main conclusions that can be drawn from the results presented in Figure 21.4. First, the mean error is negative (i.e., rainfall at DF locations is consistently

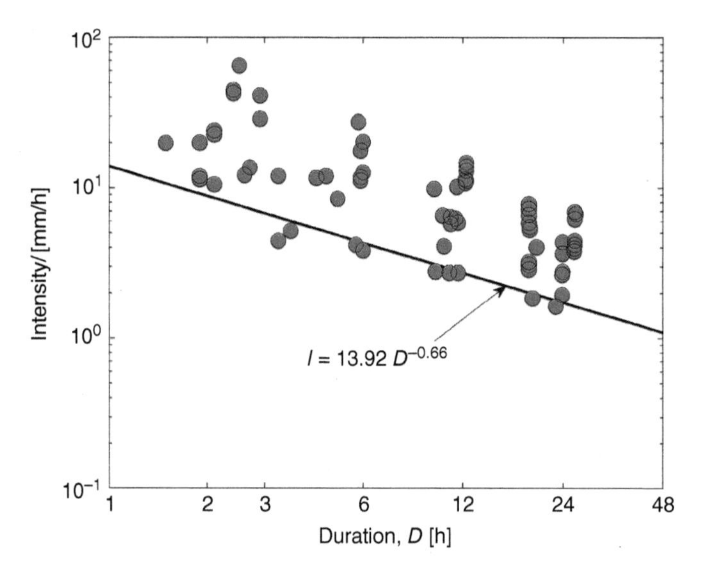

Figure 21.3 Rainfall duration (D, in hours) and mean rainfall intensity (I, in mm h^{-1}) of debris flow triggering rainfall events, according to radar observations. Black line is the estimated ID rainfall threshold at 10% exceedance level.

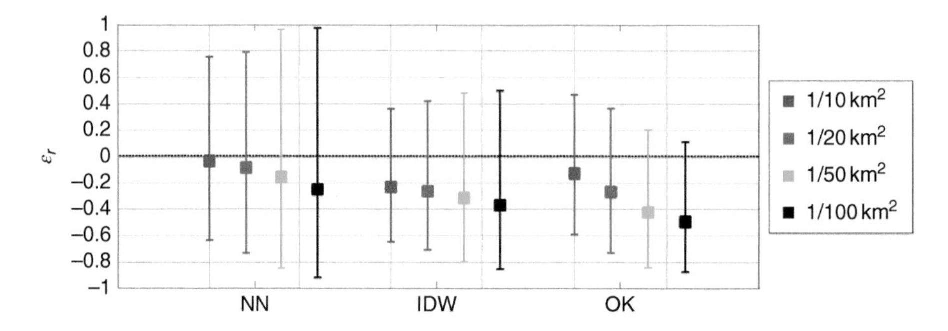

Figure 21.4 Relative error in rainfall estimation for different interpolation methods and gauge densities examined. Note that square symbols correspond to mean values and vertical bars extend from 5th to 95th percentile.

underestimated) for all interpolation methods and rain gauge density scenarios examined. Second, underestimation increases with decreasing rain gauge density. This pattern is consistent for all interpolation methods but the rate of increase of error with rain gauge density varies among methods, with OK being associated with the higher rate (manifested as steeper slope of mean error in Fig. 21.4). Overall, the range of mean error for the different interpolation methods is 3%–25% (NN), 23%–37% (IDW), and 13%–50% (OK). Therefore, in terms of mean error, NN method provides consistently the best results, while OK performs better than IDW only for the highest density examined. However, the variance of error (using as proxy the 5th–95th quantile range) is significantly higher for NN than IDW and OK.

Overall, these results contradict previous evaluations of methods of rainfall estimation at a point, which showed clearly the advantage of geostatistical techniques over IDW and NN [*Tabios and Salas*, 1985; *Creutin and Obled*, 1982; *Seo*, 1998]. Moreover, earlier results showed that the natural variability of rainfall makes the use of kriging attractive at low-gauge density, while at high-gauge density, the increase in accuracy of estimated field is lower [*Borga and Vizzaccaro*, 1997]. It is important to note here that while earlier results were obtained by evaluating the methods over the whole distribution of rainfall values, in the present work we analyzed rainfall estimates over DF locations, where rainfall is likely to exceed a threshold. *Seo* [2013] has shown that kriging, as well as most of the deterministic interpolation procedures, tends to significantly underestimate precipitation over a given threshold, and that the underestimation increases with increasing the threshold. For this reason, the predictor (among those examined here) that minimizes the bias in rainfall estimation is the one based on the closest rain gauge.

We also analyzed the propagation of rainfall estimation error in the estimation of ID thresholds. Figure 21.5 shows how the variability in the estimation of ID parameters,

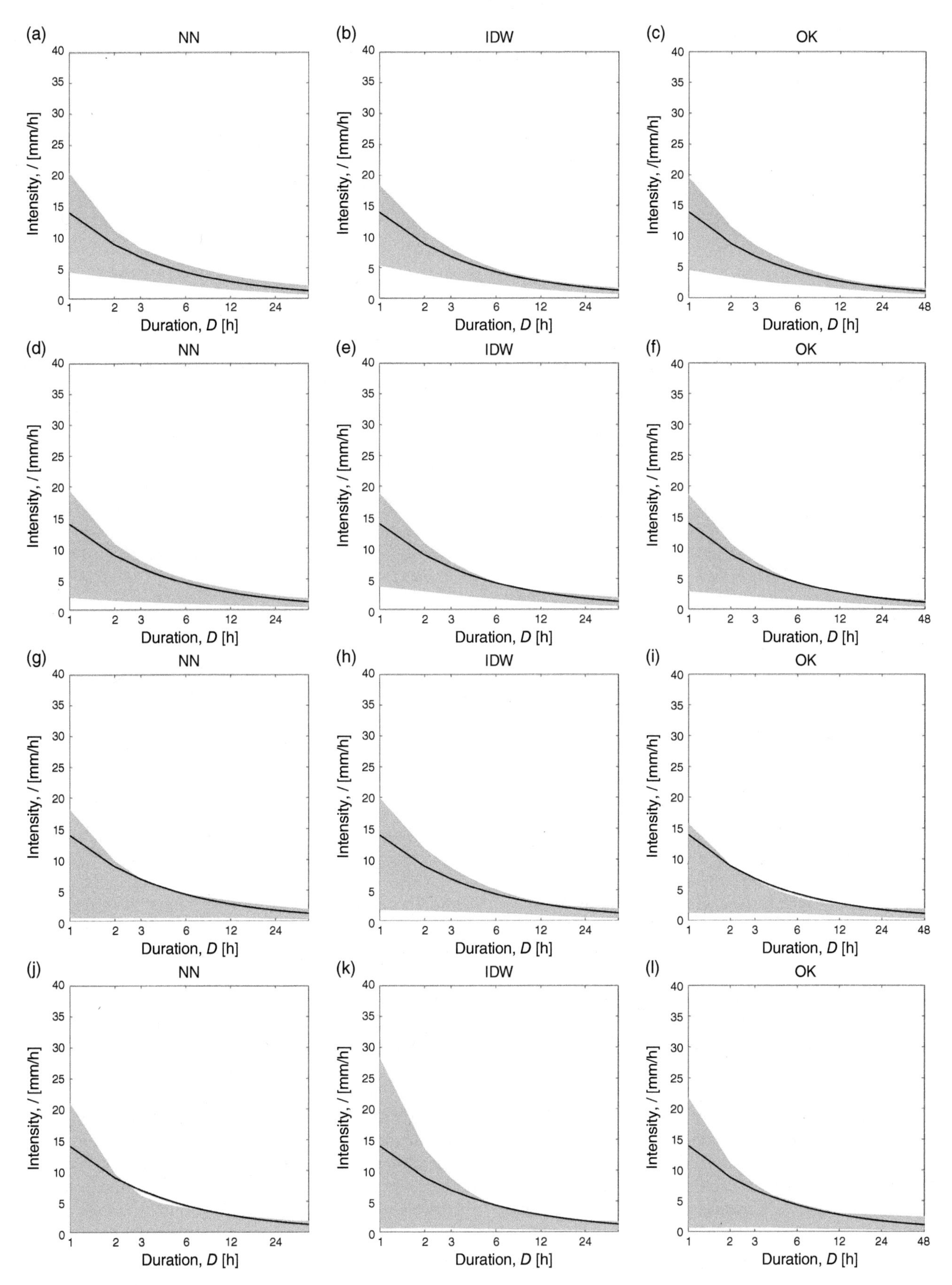

Figure 21.5 Distribution of *ID* thresholds derived according to different estimation methods and for different gauge densities. Color area corresponds to range of distribution of resulted *ID* thresholds. Reference *ID* threshold (black line), based on radar, is superimposed. Panels from top to bottom correspond to gauge densities of one station over 10, 20, 50, and 100 km^2, respectively.

obtained for each set of virtual gauges, translates into a distribution of *ID* thresholds. The *ID* spread, which is essentially a representation of the uncertainty associated with the estimation of *ID* threshold, increases considerably with decreasing rain gauge density, for all estimation methods. These results highlight the particular importance of rainfall field sampling on *ID* threshold estimation. Comparison of results with respect to different interpolation methods used, shows that no particular benefit in *ID* estimation was obtained by using approaches that are more complex than the NN method.

Overall, these results confirm the findings reported by *Nikolopoulos et al.* [2014], which showed that the statistical properties of the *ID* relationships are controlled by the left tail of the rainfall estimates distribution for DF locations. Both negative bias (i.e., mean error) and estimation variance determine the left tail distribution, and, as reported above, they compensate each other among the various interpolation methods. NN, which is the method characterized by the lowest negative bias (NN), has the largest estimation variance, whereas the other two methods (IDW and OK), which have lesser variance, are characterized by the largest negative bias. These combinations lead to an overall similarity in the left tail distribution among the three different estimators. This similarity implies that the differences in performance among the three interpolation methods are negligible.

21.4. DISCUSSION AND CONCLUSIONS

The overall objective of this work is to examine the uncertainty in estimation of debris-flow triggering rainfall and evaluate its impact on the identification of rainfall thresholds used for debris-flows prediction. We focused on the uncertainty related to rain gauge–based estimation of debris-flow triggering rainfall, which arises from limitations on the sampling of the rainfall field. We analyzed first the effect of rain-gauge network density and spatial interpolation methods on the estimation of rainfall at the DF locations and then investigated the propagation of the rainfall estimation error in the identification of *ID* thresholds. We have examined these issues by simulating rain-gauge networks based on a random sampling exercise of high-resolution radar rainfall fields corresponding to 10 major DF-triggering storm events in the area of Upper Adige River basin, North Italy.

We found that the mean of the relative error in rainfall estimation at DF location, obtained by using the NN method, is negative (underestimation) and steadily increases in magnitude with the distance from the DF to the nearest rain gauge. The underestimation is remarkable and is up to 40% at 6–7 km distance. This illustrates well the characteristics of the specific rainfall estimation problem, which requires the rainfall interpolation at sites where it exceeds some (relatively large) thresholds. These results imply that the method that minimizes the bias is the NN technique, simply because it uses the closest rain gauge to the DF. Being the area of heavy rainfall localized compared to the station density, using additional stations, as it is done by IDW and OK, tends to sample areas characterized by lower rainfall amounts, introducing larger biases. Similar results are very likely to apply to all other methods, which are based on use of multiple rain-gauge data. On the other hand, since the estimation variance decreases with the number of rain gauges used in the interpolation, NN is characterized also by the maximum value of estimation variance. The other two methods, which use multiple rain gauges, are characterized by markedly smaller estimation variances. The effect of decreasing the rain-gauge density is to increase both bias and estimation variance, with all three interpolation methods.

Overall, this shows the advantages and the limitations of the reference rain-gauge method, so often used in DF investigations. The method, in the long term, provides estimates that are characterized by comparatively reduced estimation bias; however, the estimation variance is also very high.

Examination of the propagation of the rainfall estimation errors into the identification of *ID* relationship showed that the largest part of resulted *ID* distribution is located below the reference *ID*, which suggests that *ID* threshold relationships are generally underestimated. This is shown to be consistent for all methods and gauge densities examined. The rain gauge density appears to have a strong influence on the estimation of *ID* thresholds and it is worth noting that even for the highest gauge density examined (1/10 gauge/km^2), underestimation of ID thresholds is still considerable. Comparison of results with respect to different interpolation methods used shows that no particular benefit in *ID* estimation is obtained by using approaches that are more complex than the NN method.

We acknowledge that these results may reflect specific characteristics of the studied rainfall and debris-flow events. For example, the convective nature of most of the storm events examined, together with their small space-time scales, may have weighted in the large bias affecting both IDW and OK. Further investigations are required to elucidate how bias and estimation variance are controlled by the time-space variability of the rainfall fields and colocated debris flows. Also, we should acknowledge that our decision to use rainfall totals and storm durations from debris-flow producing storms may have an impact on our results. Some researchers have evaluated periods of rainfall within storms, or more optimally, the storm rainfall leading up to a known time of failure [*Cannon et al.*, 2008, 2011]. An investigation is ongoing to evaluate the impact of rainfall estimation uncertainty considering storm rainfall up to the known time of failure. Nevertheless, the results presented deliver a clear message regarding the impact of rainfall estimation uncertainty on *ID* thresholds. Therefore, advancing the use of rainfall thresholds for debris-flow prediction requires accounting for this important source of uncertainty. Research should address the propagation of the rainfall estimation errors through the use of identified rainfall thresholds for debris-flows forecasting. Statistical characterization of the uncertainty and development of probabilistic thresholds can potentially improve the effectiveness of threshold-based prediction systems. Finally, we stress that systematic use of high-resolution radar rainfall observations should be considered for improving the effectiveness of rainfall thresholds for debris-flow warning purposes.

ACKNOWLEDGMENTS

This work is supported from EU FP7 Marie Curie Actions IEF project PIEF-GA-2011-302720 (HYLAND, http://intra.tesaf.unipd.it/cms/hyland/default.asp). We sincerely acknowledge Ripartizione Opere Idrauliche, Autonomous Province of Bolzano (Italy), for providing access to the archive of debris flows; Ufficio Idrografico, Autonomous Province of Bolzano (Italy), for providing hydrological and meteorological data and for providing access to the weather radar information.

REFERENCES

Aleotti, P. (2004), A warning system for rainfall-induced shallow failures, *Eng. Geol.*, *73*(3–4), 247–265; doi:10.1016/j.enggeo.2004.01.007.

Berti, M., and A. Simoni (2005), Experimental evidences and numerical modelling of debris flow initiated by channel runoff, *Landslides*, *2*, 171–182.

Borga, M., and A. Vizzaccaro (1997), On the interpolation of hydrologic variables: Formal equivalence of multiquadratic surface fitting and kriging, *J. Hydrol*, *195*(1), 160–171.

Borga, M., M. Stoffel, L. Marchi, F. Marra, and M. Jakob (2014), Hydrogeomorphic response to extreme rainfall in headwater systems: Flash floods and debris flows, *J. Hydrol*, *518*, 194–205; doi:10.1016/j.jhydrol.2014.05.022.

Brunetti, M. T., S. Peruccacci, M. Rossi, S. Luciani, D. Valigi, and F. Guzzetti (2010), Rainfall thresholds for the possible occurrence of landslides in Italy, *Nat. Hazards Earth Syst. Sci.*, *10*(3), 447–458; doi:10.5194/nhess-10-447-2010.

Caine, N. (1980), The rainfall intensity: duration control of shallow landslides and debris flows, *Geografiska Annaler, Series A. Physical Geography*, 23–27.

Cannon, S. H., E. M. Boldt, J. J. Laber, J. W. Kean, and D. M. Staley (2011), Rainfall intensity-durations for post-fire debris-flow emergency-response planning, *Nat. Hazards*, *59*, 209–236.

Cannon, S. H., J. E. Gartner, R. C. Wilson, and J. L. Laber (2008), Storm rainfall conditions for floods and debris flows from recently burned areas in southwestern Colorado and southern California, *Geomorphology*, *96*, 250–269.

Comiti, F., et al. (2014), A new monitoring station for debris flows in the European Alps: first observations in the Gadria basin, *Nat Hazards*, *73*(3), 1175–1198.

Creutin, J., and C. Obled (1982), Objective analyses and mapping techniques for rainfall fields: An objective comparison, *Water Resour. Res.*, *18*(2), 413–431.

David-Novak, H. B., E. Morin, and Y. Enzel (2004) Modern extreme storms and the rainfall thresholds for initiating debris flows on the hyperarid western escarpment of the Dead Sea, Israel, *Bull. Geolog. Soc. Amer.*, *116*(5–6), 718–728.

Deganutti, A. M., L. Marchi, and M. Arattano (2000), Rainfall and debris-flow occurrence in the Moscardo basin (Italian Alps), *Debris-flow hazards mitigation: Mechanics, prediction, and assessment*, 67–72.

Frattini, P., G. Crosta, and R. Sosio (2009), Approaches for defining thresholds and return periods for rainfall-triggered shallow landslides, *Hydrolog. Processes*, *23*(10), 1444–1460; doi:10.1002/hyp.7269.

Gregoretti, C., and G. Dalla Fontana (2008), The triggering of debris flow due to channel-bed failure in some alpine headwater basins of the Dolomites: Analyses of critical runoff, *Hydrolog. Processes*, *22*, 2248–2263.

Guzzetti, F., S. Peruccacci, M. Rossi, and C. P. Stark (2007), Rainfall thresholds for the initiation of landslides in central and southern Europe, *Meteor. Atmos. Phys.*, *98*(3–4), 239–267; doi:10.1007/s00703-007-0262-7.

Guzzetti, F., S. Peruccacci, M. Rossi, and C. Stark (2008), The rainfall intensity-duration control of shallow landslides and debris flows: an update, *Landslides*, *5*(1), 3–17; doi:10.1007/s10346-007-0112-1.

Iverson, R. (2005), Debris-flow mechanics, 105–134, in *Debris-Flow Hazards and Related Phenomena*, edited by M. Jakob and O. Hungr, *Springer Praxis Books*, Springer Berlin Heidelberg, Berlin, Heidelberg.

Jakob, M., and H. Weatherly (2003), A hydroclimatic threshold for landslide initiation on the north shore mountains of Vancouver, British Columbia, *Geomorphology*, *54*(3–4), 137–156; doi:10.1016/S0169-555X(02)00339-2.

Marra, F., E. I. Nikolopoulos, J.-D. Creutin, and M. Borga (2014), Radar rainfall estimation for the identification of debris-flow occurrence thresholds, *J. Hydrol.*, *519, Part B*(0), 1607–1619; doi:10.1016/j.jhydrol.2014.09.039.

Nikolopoulos, E. I., F. Marra, J. D. Creutin, and M. Borga (2015), Estimation of debris flow triggering rainfall: Influence of rain gauge density and interpolation method, *Geomorphology*, *243*, 40–50.

Nikolopoulos, E. I., S. Crema, L. Marchi, F. Marra, F. Guzzetti, and M. Borga (2014), Impact of uncertainty in rainfall estimation on the identification of rainfall thresholds for debris flow occurrence, *Geomorphology*, *221*, 286–297; doi:10.1016/j.geomorph.2014.06.015.

Norbiato, D., M. Borga, R. Merz, G. Blöschl, and A. Carton (2009), Controls on event runoff coefficients in the eastern

Italian Alps, *J. Hydrol.*, *375*(3–4), 312–325; doi:10.1016/j.jhydrol.2009.06.044.

Petley, D. (2012), Global patterns of loss of life from landslides, *Geology*, *40*(10), 927–930; doi:10.1130/G33217.1.

Seo, D. J. (1998), Real-time estimation of rainfall fields using rain gage data under fractional coverage, *J. Hydrol.*, *208*(1), 25–36.

Seo, D.-J. (2013), Conditional bias-penalized kriging (CBPK), *Stochas. Environ. Res. Risk Assess.*, *27*(1), 43–58; doi:10.1007/s00477-012-0567-z.

Tabios, G. Q., and J. D. Salas (1985), A comparative analysis of techniques for spatial interpolation of precipitation, *J. Amer. Water Resour. Assoc.*, *21*(3), 365–380.

Tiranti D., R. Cremonini, F. Marco, A. R. Gaeta, and S. Barbero (2014), The DEFENSE (DEbris Flows triggEred by storms, Nowcasting SystEm): An early warning system for torrential processes by radar storm tracking using a Geographic Information System (GIS), *Comput. Geosci.*, *70*, 96–109.

Tiranti D., S. Bonetto, and G. Mandrone (2008), Quantitative basin characterization to refine debris-flow triggering criteria and processes: an example from the Italian Western Alps, *Landslides*, *5*(1), 45–57.

Wieczorek, G. F. (1996), Landslide triggering mechanisms, 76–90, in *Landslides: Investigation and Mitigation*, edited by A. K. Turner and L. R. Schuster, Transportation Research Board Special Report 247.

22

Prospects in Landslide Prediction: Confronting the Challenges of Precipitation Uncertainty

Natasha Markuzon,[1] Catherine Slesnick,[1] Erin Leidy,[1,2] John Regan,[1] Xiang Gao,[2] and Adam Schlosser[2]

ABSTRACT

Landslides are one of the most damaging geohazard phenomena in terms of historical death tolls and socioeconomic losses. Yet, understanding their underlying causes and predicting their occurrence remain a significant challenge. We discuss the effect of antecedent precipitation patterns on landslide activity and assess the extent to which uncertainty in precipitation measurements affects the accuracy of landslide prediction models. We compare modeling results while using precipitation data of different spatial resolution, and we introduce an alternative methodology that predicts regional precipitation-induced landslide activity using combined composites of atmospheric state and dynamic variables rather than using precipitation measurements directly. In particular, for the Swiss Flood and Landslide Database historical records, we derive models predicting landslides using (1) rain gauge data, (2) lower spatial resolution precipitation reanalysis information, and (3) composites of atmospheric variables taken from the same reanalysis product. All models are developed using data-driven methodologies and supervised learning techniques. At every stage, we compare the models' predictive results and assess uncertainties associated with each model's input data. We also discuss the atmospheric composite modeling approach as a potential for landslide prediction when no good precipitation data are available, as in forecasting landslide risk under global climate change.

22.1. INTRODUCTION

Every year landslides cost the United States $3.5 billion in damages and often result in fatalities [*Schuster and Highland*, 2001]. Knowing which areas are particularly vulnerable to high damage slides is a critical first step to reducing the risk under which many people consistently live, and should inform disaster response, urban, and transportation planning [*Alexander*, 2005]. The implications of this statement are underscored by the recent landslide in Washington that resulted in 41 casualties, one of the most fatal landslides ever to occur in the United States [*Berman*, 2014].

Because precipitation is one of the most common triggers of landslides, the danger these disasters pose to humans and the environment will increase with the expected increase in extreme precipitation events due to climate change [*O'Gorman and Schneider*, 2009]. In this chapter, we identify prospects of improving predictions of landslide activity by addressing the effects of uncertainty in precipitation measurements and predictions. Several methods exist for estimating rainfall conditions that could potentially trigger landslides [*Guzzetti et al.*, 2007]. The most common technique [*Martelloni et al.*, 2011] is to identify rainfall intensity-duration thresholds [*Dahal et al.*, 2009, 2008; *Guzzetti et al.*, 2007; *Giannecchini et al.*, 2012] consistent with historical landslide events. A number of studies concentrated on identifying

[1]*Charles Stark Draper Laboratory, Cambridge, Massachusetts, USA*
[2]*Massachusetts Institute of Technology, Cambridge, Massachusetts, USA*

Natural Hazard Uncertainty Assessment: Modeling and Decision Support, Geophysical Monograph 223, First Edition.
Edited by Karin Riley, Peter Webley, and Matthew Thompson.

antecedent rainfall [*Glade*, 2000], or cumulative rainfall [*Polemio and Sdao*, 1999] associated with triggering landslides, though determining the number of days to be used for optimal prediction presents a challenge [*Guzzetti et al.*, 2007]. A combination of antecedent and event rainfall as well as a combination of temperature, atmospheric pressure, and wind associated with landslides is important in reducing the uncertainty of their prediction [*Zezere et al.*, 2005; *Aleotti*, 2004].

We exemplify our analyses of differences in precipitation-driven model results with data collected in Switzerland from 1972 to 2012. We start by developing daily landslide predictions using rain gauge and ground weather station data. We focus on analysis of combinations of weather conditions associated with landslides, which can be helpful in reducing the uncertainty of landslide prediction. We use the same techniques and landslide dataset to develop landslide prediction models from lower resolution reanalysis data that combine both ground-based and satellite weather information. By comparing results of these two models, we can assess the effects that precipitation data resolution has on landslide predictions.

Extending precipitation-induced landslide models to predict future occurrence under different climate change scenarios is challenged by high levels of uncertainty and low resolutions associated with projected precipitation outputs. In particular, general circulation models (GCMs) that are used for estimating future climate parameters have extremely coarse resolution relative to the scale of landslide processes [*Crozier*, 2010]. As a result, GCMs spatially smooth and subsequently underestimate the localized heavy precipitation events that are most relevant to landslide occurrence. When applied to landslide models, this effect manifests as an underestimate of landslide occurrence. Multiple efforts have been made to identify the distinct large-scale dynamic conditions that induce local-scale precipitation extremes [*Rudari et al.*, 2004; *Rudari et al.*, 2005; *Grotjahn*, 2011; *DeAngelis et al.*, 2013]. These composite atmospheric conditions are typically achieved by conditioning atmospheric reanalysis synoptic flows and fluxes on the occurrence of extreme events identified from local surface station observations. This type of approach bridges the scale gap between large-scale, smoothed model precipitation outputs and observed heavy precipitation in localized regions smaller than the coarse resolution of the reanalysis data.

We present here a novel approach to future landslide prediction that uses this methodology to bypass the modeled rain that comes from GCMs. Building off of the work by *Gao et al.* [2014], we have developed composites of atmospheric conditions over all days with observed landslides. We demonstrate that the models developed with these data have predictive potential to successfully detect landslide events. The final stage of this approach will be completed in a future phase of this project. In this stage, we will repeat the composite analysis using output from GCMs, to show extension of the methodology to predictions of future landslides under different climate change scenarios.

All models discussed in this chapter are data driven. Thus, uncertainties in the data used to derive the models (e.g., inaccurate records, measurement error, etc.) are propagated to model results. For example, landslide records, while available both at state and country-specific levels, are hard to verify and often reflect the date the landslide was observed, not the date it occurred, making it difficult to analyze factors that preceded the landslide event. The most reliable records of landslide occurrence are typically found in newspaper articles, which are biased to reflect only the largest and most costly (to humans) events. These reports also tend to neglect important physical details, such as precise location or size of the landslide, or its run-out distance.

A second major source of uncertainty specific to precipitation-induced landslide modeling lies in the collection of precipitation information. Precipitation data have traditionally been collected using ground-based rain gauges such as tipping buckets and weighing, optical, and capacitance gauges [*Alexander*, 1989]. *Nikolopoulos et al.* [2014] find that the natural variability in rainfall can create large discrepancies between the rainfall volume and duration at a landslide site and that of the closest rain gauge. Such instruments are often sparsely placed or absent in many parts of the world, particularly in areas with mountainous terrains where differences in elevation over short distances can lead to dramatic changes in precipitation distribution due to the interaction between topography and atmospheric flows. Even in the continental United States, where rain gauges are relatively plentiful, the distance from a weather station or rain gauge to the site of a landslide often exceeds 50 or 100 km. Remote-sensing observations of weather and ground parameters provide consistent data at uniform resolution and can be used in areas where no in situ observations are available. However, precipitation variables are often measured remotely at spatial resolution that is lower than needed to accurately support landslide models [*Huffman et al.*, 2007; *Hong et al.*, 2007; *Tralli et al.*, 2005].

22.2. DATA

22.2.1. Switzerland Landslides

We use Switzerland as the study area because of the availability of a comprehensive database that contains records of landslides spanning 40 yr. The Swiss Flood and Landslide Database [*Hilker et al.*, 2009] contains

3366 records of landslides that have occurred between 1972 and 2012. Most of the records originate from news sources and are thus considered to be accurate reflections of a specific landslide date and location. We note however that, as is the case with most landslide records, the sample is biased toward large landslides in populated areas. Any dates that were indicated as uncertain were removed. The final dataset contains 2473 records.

22.2.2. High-Resolution Precipitation

The interpolated rain gauge data were obtained from the European Climate Assessment and Dataset project, which maintains daily observational data from weather stations throughout Europe. The E-OBS product is a daily gridded reanalysis dataset that includes mean temperature, minimum temperature, maximum temperature, sea level pressure, and total precipitation [*Haylock et al.*, 2008; *van den Besselaar et al.*, 2011]. The data resolution is 0.25°. Records are available for the entire time period covered in the Switzerland Landslide Database.

22.2.3. Low-Resolution Precipitation

NASA's Modern Era Retrospective Analysis for Research and Application (MERRA) is a reanalysis product of both remotely sensed and in situ weather observations [*Rienecker et al.*, 2011]. For this work, we used the MERRA-Land product derived from an improved set of land surface hydrological fields generated by rerunning a revised version of the land component of the MERRA system [*Reichle,* 2012]. The MERRA-Land estimates benefit from corrections to the precipitation forcing with the global gauge-based NOAA Climate Prediction Center Unified (CPCU) precipitation product and from revised parameter values in the rainfall interception model, changes that effectively correct for known limitations in the MERRA surface meteorological forcings. We used the IAU 2d Simulated land surface diagnostics (tavg1_2d_mld_Nx) dataset, which contains 50 parameters at a 0.5 by 0.66° resolution. Though in other studies we utilized a significant proportion of these parameters, for the current analysis we used only total surface precipitation (PRECTOT) values.

22.2.4. Low-Resolution Atmospheric Data

MERRA data were also used in development of the landslide models derived from composite atmospheric data because the low resolution of these data most closely proxies outputs from GCMs. Future work will extend the methodology to directly produce models from GCM atmospheric variables. The relevant atmospheric variables used to construct the composites associated with landslide events were based on MERRA products MAI3CPASM and MAI1NXINT. MAI3CPASM contains pressure variables on 42 levels in 3 hr increments. Geopotential height, sea level pressure, and vertical pressure velocity were drawn from this file. MAI1NXINT has a single level resolution and is measured in daily increments. These data are at a 1.25° resolution and are available for the entire time period covering events in the Landslide Database.

22.3. PRECIPITATION-BASED LANDSLIDE PREDICTION

To understand temporal weather patterns associated with landslide events, we developed models from the E-OBS and MERRA data for the occurrence of landslides based on the weather conditions in the days preceding the event. The spatial resolution of the prediction grid was defined by the data, with E-OBS weather data at 0.25° resolution, and MERRA defined by the raster, with 0.5 by 0.66° resolution. We applied a data-driven methodology, where each day and grid point was described by a set of precipitation features, and an associated target indicating presence or absence of landslides (Fig. 22.1). While it is well established that slope, land cover, and a number of other nonprecipitation-related variables are crucial in predicting landslides [*Terlien,* 1998; *Markuzon et al.*, 2012; *Regan et al.*, 2013], they are not considered in this study in order to amplify the contribution of precipitation, and other weather-related variables.

For each day in the observed period, the feature vector reflecting the following information was created using either E-OBS or MERRA data: cumulative precipitation in the preceding 1, 2, 3, 7, 14, 30, days, up to 182 days; average daily temperature for the same periods; and average surface atmospheric pressure values for the same periods. In total, there were 122 independent variables associated with the target reflecting presence or absence of landslides on that day. The created datasets were split into training and test sets, with evaluation of the model performed on the unseen data in the test set. We evaluated the models' performance by computing the total accuracy, and the associated confusion matrix. The results are reported using 10-fold cross validation. For evaluation purposes, the data were balanced between landslide and nonlandslide days, making the number of target and nontarget days in the dataset equal.

We developed models predicting days with landslides using either E-OBS or MERRA data and compared their predictive accuracy. In addition, we evaluated the change in predictive accuracy of models if only antecedent precipitation data were used for prediction while temperature and pressure were omitted. And finally, we demonstrated that the models' performance degraded if we reduced the

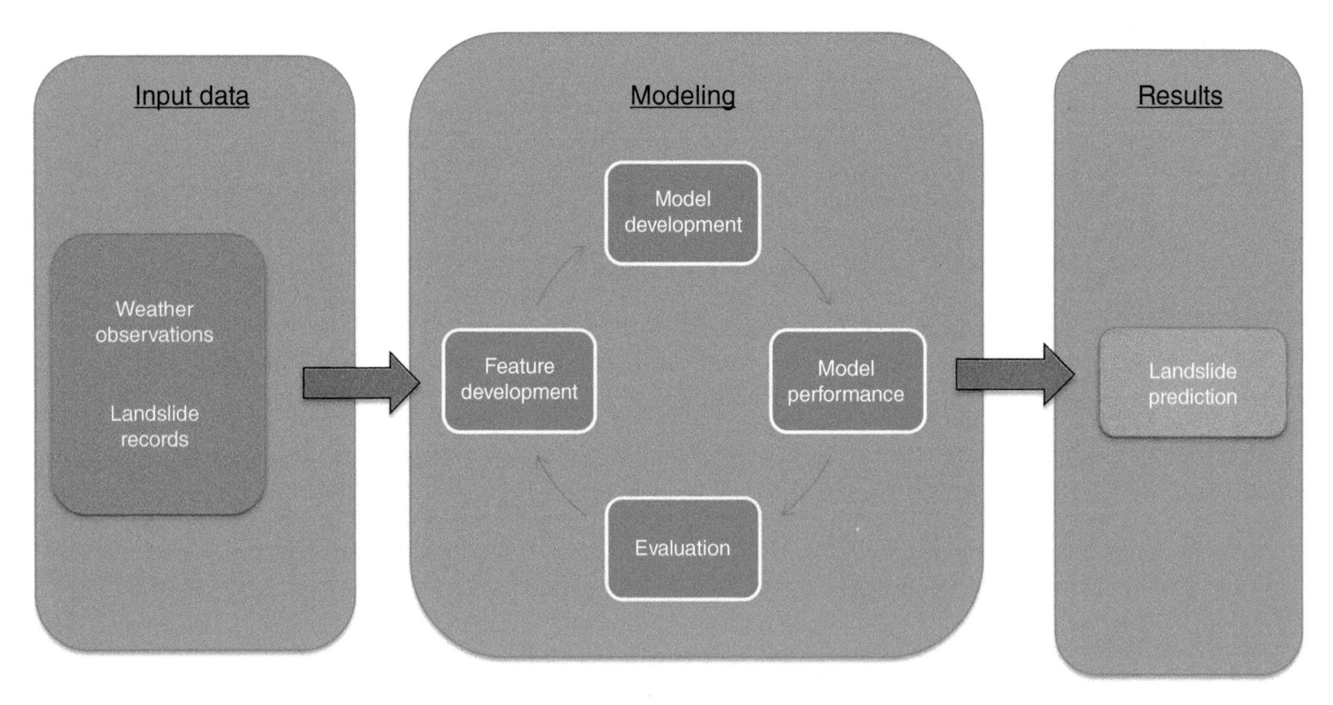

Figure 22.1 High level flow of the modeling approach used in this analysis. Input data in the form of landslide records and weather data are transformed into 1-dimensional feature vectors, each of which contains descriptive variables for a single landslide event. Supervised learning models are built and evaluated using training and test subsets of the data, respectively. Model parameters and feature definition are adjusted based on measured model performance, and the analytical cycle is repeated until predictive performance has been optimized, after which final landslide predictions are generated.

number of variables to include precipitation data about only the 3 days preceding a landslide. Both training and evaluation were performed using the same set of landslides, thus allowing comparison of differences in models' performance due to differences in the E-OBS and MERRA datasets, and selected subsets of variables.

Both sets of precipitation-based models were derived and tested using a variety of classifiers; the results are reported for models developed using a Decision Tree Random Forest Classifier [*Breiman,* 2001]. A decision tree is a recursive algorithm that partitions the input feature space with a set of classification rules. Each resulting region, or terminal node, is then labeled with a class. To classify a new instance represented by a feature vector, the instance is passed down through the decision tree and classified according to the terminal node it lands in. Random forests generate an ensemble of decision trees by randomly selecting a set of features and building a tree from a sample of training data, repeating this numerous times. Once a large number of trees are generated, the random forest classifies by unweighted voting.

22.4. EXTENDING LANDSLIDE PREDICTION TO CLIMATE CHANGE SCENARIOS

Extension of derived landslide models to future scenarios in a changing climate presents a great interest. However, as discussed in Section 22.1, current GCMs typically used for estimating future precipitation patterns are not appropriate for prediction of localized heavy rainfall events [*Gao et al.*, 2014]. We have developed a methodology to bypass the modeled rain that comes from GCMs by introducing a set of atmospheric composite measurements. Large-scale atmospheric conditions associated with landslide activity were determined from composites of atmospheric conditions over all days with observed landslides. These atmospheric variables have been shown to detect the occurrence of heavy precipitation events more consistently when compared with observations than model-simulated precipitation [*Gao et al.*, 2014].

As a first step to demonstrating this methodology, we developed models using historical data. The next phase of this work will repeat the analysis presented here using model outputs from GCMs, and compare the two sets of results. The ultimate goal is to show that we can develop robust projections of future landslide activities anticipated under global change. Historic atmospheric data associated with heavy precipitation were obtained from MERRA reanalysis, including 500 hpa geopotential height, 500 hpa vertical pressure velocity, sea level pressure, and total precipitable water. To standardize the resolution with climate models, all the data were linearly interpolated to 2.5° × 2°. All variables have been converted to a standardized anomaly, defined as the anomaly from the seasonal climatological mean over the time

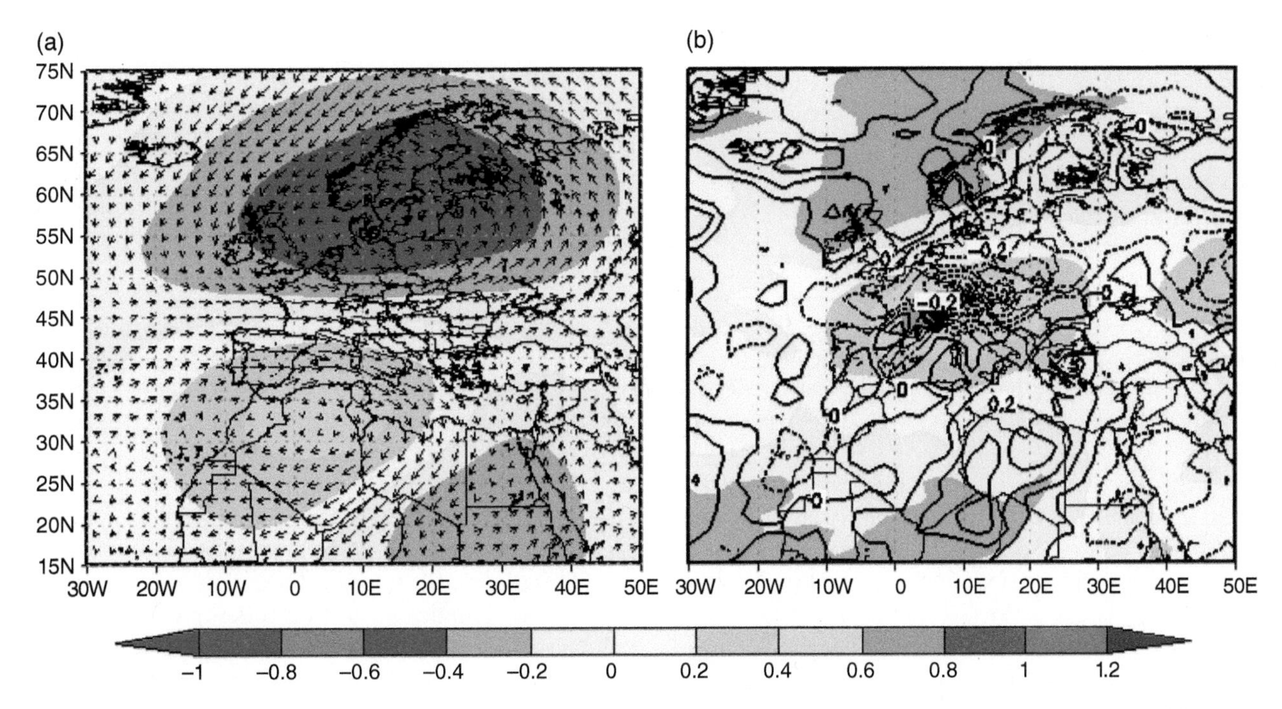

Figure 22.2 Composites of all Swiss winter slide dates. The colors in panel (a) correspond to the value of the standardized anomaly of 500-hpa geopotential height (Z500) and the arrows show vertical integral atmospheric vapor flux. The colors of panel (b) correspond to the level of total precipitable water (TPW) and the contour lines are 500-hpa vertical pressure velocity (ω500).

period under consideration, divided by its standard deviation. The atmospheric conditions of each landslide day were determined, and a composite was created that averaged all landslide days into one pattern. Because of the variation in weather patterns among seasons, averaging patterns over all landslide days together was not meaningful. While the winter months (December, January, and February) accounted for only ~13% of all landslides within out dataset, we found combining only the winter months produced the strongest composite patterns across all variables, and thus for this first proof-of-concept study, we chose to focus on composites only of winter events averaged across all years. Models for other seasons will be developed in future efforts.

The composites shown in Figure 22.2 are used to identify a pattern associated with the occurrence of landslide events. We examined the data that contributed to each composite and made a map, for each variable, of the number of days of data when the value in a particular 2.5° × 2° grid cell was either positive or negative. From these maps, we identified hotspots, defined as a cluster of grid cells that show strong consistency among many members. The grid(s) with the maximum consistent sign count serve as a lower threshold to the smallest hotspot that must be matched. We also defined a cutoff in the spatial correlation between the composite and an individual day. The cutoff is measured as the spatial correlation value above which a day has a statistically determined higher than random chance of being a landslide day. Of all the winter days in the 34 yr time span, 161 of them are defined as landslide days and 2908 are not. Days with a higher spatial correlation have a higher chance of being a landslide day.

Days with a high probability of a landslide event were thus defined as having:

1. Three or more variables with signs consistent with the hotspot grid cells.
2. At least one of four variables (Z500, ω500, vapor flux, and TPW) exceeding the cutoff value for spatial correlation.
3. Positive spatial correlations of all variables.

As indicated, the requirements to identify a landslide event (or day) do not require all of the variables to exceed the correlation cutoff or have sign consistency within the hotspots. These relaxed requirements reflect, in general, the complex and nonunique nature of the large-scale (precursory) atmospheric conditions associated with landslides. In the results presented in Section 22.5, the technique indicates predictive potential but also points to ongoing developments in these methods.

22.5. COMPARISON OF MODEL RESULTS

For both sets of data, E-OBS and MERRA, we found that long-term precipitation patterns were tightly correlated with landslide occurrence. Table 22.1 shows a summary of the models' performances using either E-OBS or MERRA data with different subsets of input variables. Reported values are the average of test set performance

Table 22.1 Average Models' Test Set Performance Using Long-Term Weather Patterns

Switzerland landslide prediction data Type	Average test set accuracy (%)	Average precision (%)	Average recall (%)	ROC area
E-OBS all variables	81.4	79.9	84.0	89.2
E-OBS precipitation only	75.4	73.3	79.7	82.8
E-OBS precipitation 3 days	69.2	68.0	72.5	75.4
MERRA precipitation only	67.2	66.2	70.4	72.8
MERRA precipitation 3 days	61.5	61.1	63.2	65.5

Note: Accuracy is the probability of making a correct prediction for both positive and negative targets; precision is the percent of true positives in all positive predictions made by the model; recall, or sensitivity, is the percent of positives correctly identified by the model; and ROC area reflects the sensitivity against the specificity for different discrimination thresholds, where specificity is a percent of true negatives in all negative predictions made by the model.

over 50 independent runs using 10-fold cross validation. All results are reported on the balanced data, a randomly selected subset with equal number of target and nontarget days.

Using the higher resolution E-OBS data, we achieved 81% total accuracy on the balanced evaluation set using all 122 temporal variables of precipitation, temperature, and atmospheric pressure. Models developed using precipitation variables only resulted in a decrease in predictive performance to 75%. Excluding the long-term variables (all data more than 3 days preceding the landslide event) further decreased performance to 69%. The false positive rate on the balanced test set increased from average 16% for models including all variables, to 22% for precipitation variables only, and to 30% for the 3 days model. Model results using lower resolution MERRA precipitation data showed a decrease in performance as compared with similar models using E-OBS data. Using precipitation data for the range of up to 182 days, we achieved 67% test set accuracy and 30% false alarm rate as compared with 75% test set accuracy and 22% false positive rate for E-OBS. That performance dropped to 61% if using only 3-day precipitation pattern computed with MERRA data, with 37% false alarm rate (Table 22.1).

For the composite analysis, the criterion of detection was evaluated on 34 yr of MERRA data, including all the winter days between 1979 and 2012. All days were compared against the composite constructed from the observed landslide days to determine hotspot sign consistency and spatial correlation. If a day met the criteria detailed in Section 22.4, it was considered a landslide day. Performance was measured by the success rate in identifying an observed landslide date, and the number of false positives that the criterion detects. The success rate is the fraction of days that the analogue detected and matched the observed landslide days over the total number of observed events. The success rate reached 20%–25% for exact-day matches, as compared with the random chance that a landslide would occur on any given day (=161 observed landslides/3069 days-of-record ≅ 5%). The success rate increased to 45%–50% for a 7 day window surrounding the landslide event. Overall, demonstration of this technique indicates, to first order, predictive potential. Nevertheless, the marginal improvement over success from a random draw suggests that a number of additional factors should be considered, such as additional atmospheric variables, preconditioning of soil conditions, local geography, and/or unobserved landslide events.

22.6. DISCUSSION AND CONCLUSIONS

The prediction of current and future landslides in a changing climate faces a number of uncertainties, and this chapter has presented the start of an approach to confront the challenges presented with the uncertainty of precipitation estimation. We have demonstrated that the use of machine learning can generate predictive models of landslide occurrence both from rain gauge precipitation data and from lower resolution reanalysis information. We find that assessment of both recent precipitation (1–3 days) and longer temporal patterns (up to several months) together is important for making an accurate landslide prediction. The relatively high false positive rate observed for all models could be due in part to the incompleteness of the data, and the fact that the Swiss Flood and Landslide Database is biased toward only the largest events that directly impact humans. Therefore, days identified as negative (no landslides) may in fact have had one or more events that were not recorded either because they occurred in remote areas or did not cause large monetary damages.

The performance of models using the E-OBS gauge-based weather data proved to be superior to that of MERRA-based precipitation. Because training and evaluation were performed using the same set of landslides, our analysis can diagnose the effect of different precipitation estimates on the performance of predictive models. Lower spatial resolution of the MERRA-based precipitation could have contributed to the difference in the performance as low-resolution data inherently miss many of the local extreme weather events that can trigger landslides.

In this work, we also show that atmospheric conditions present and prior to the occurrence of a landslide are a viable means of detection and prediction, and thus can be used to estimate the change in landslide activity according to predicted climate change patterns. As such, the model offers an alternative and complementary approach by removing sole reliance on modeled precipitation. Seasonal analysis proved to be the most useful, with winter months providing the most encouraging results. We find strong predictive potential at a 7 day window of atmospheric conditions detecting (an observed) landslide, which is consistent with our previous conclusion that the considerations of longer-term weather patterns hold the most promise for predictive potential. Ultimately, these techniques can be used to provide quantitative insights of the expected changes in landslide activity associated with climate change, and whether actions to mitigate human-altered climate change will lead to an associated reduction in landslide risk, or alternatively, are there adaptive measures that would be necessary to protect life, property, and infrastructure under a changing threat? Given the challenges presented by precipitation uncertainty, the scientific community continues to pursue novel predictive approaches, such as those presented here, to provide actionable information to the fullest extent possible.

The final stage of this project will be to repeat the atmospheric composite analysis using predictive outputs from GCMs. We will develop landslide models for the Swiss Landslide Database against either the precipitation GCM output or the atmospheric GCM output and compare results. Next steps for the analysis beyond this project include incorporation of additional variables such as land cover, slope, and drainage area to create a more comprehensive estimate based on a more complete set of conditions.

ACKNOWLEDGMENTS

This work is funded by NASA ROSES grant number 09-IDS09-0049. We offer special thanks to our collaborators from the Massachusetts Institute of Technology: Dr. Dino Bellugi, Prof. Taylor Perron, and Prof. Paul O'Gorman.

REFERENCES

Aleotti, P.(2004), A warning system for rainfall-induced shallow failures, *Eng. Geol.*, *73*(3), 247–265.

Alexander, D. (1989), Urban landslides, *Prog. Phys. Geog.*, *13*(2), 157–189.

Alexander, D. (2005), Vulnerability to landslides, *Landslide Hazard and Risk*, Wiley, Chichester, 175–198.

Berman, M. (2014), Active Search Through Washington Landslide Debris Ends, *Washington Post*, 29 April 2014.

Breiman, L. (2001), Random forests, *Mach. Learn.*, *45*(1), 5–32.

Crozier, M. (2010), Deciphering the effect of climate change on landslide activity: A review, *Geomorphology*, *124*, 260–267.

Dahal, R., S. Hasegawa, A. Nonomura, M. Yamanaka, T. Masuda, and K. Nishino (2009), Failure characteristics of rainfall-induced shallow landslides in granitic terrains of Shikoku Island of Japan, *Environ. Geol.*, *56*(7), 1295–1310.

Dahal, R. K., and S. Hasegawa (2008), Representative rainfall thresholds for landslides in the Nepal Himalaya, *Geomorphology*, *100*(3–4), 429–444.

DeAngelis, A. M., A. J. Broccoli, and S. G. Decker (2013), A comparison of CMIP3 simulations of precipitation over North America with observations: Daily statistics and circulation features accompanying extreme events, *J. Climate*, *26*, 3209–3230.

Gao, X., et al. (2014), An analogue approach to identify heavy precipitation events: Evaluation and application to CMIP5 climate models and the United States, *J. Climate*, *27*, 5941D5963.

Giannecchini, R., Y. Galanti, and G. D. Avanzi (2012), Critical rainfall thresholds for triggering shallow landslides in the Serchio River Valley (Tuscany, Italy), *Nat. Hazards Earth Syst. Sci.*, *12*(3), 829–842.

Glade, T. (2000), Applying probability determination to refine landslide-triggering rainfall thresholds using an empirical "antecedent daily rainfall model," *Pure Appl. Geophys.*, *157*(6), 1059–1079.

Grotjahn, R. (2011), Identifying extreme hottest days from large scale upper air data: A pilot scheme to find California Central Valley summertime maximum surface temperatures, *Climate Dyn.*, *37*, 587–604.

Guzzetti, F., et al. (2007), Rainfall thresholds for the initiation of landslides in central and southern Europe, *Meteor. Atmos. Phys.*, *98*(3–4), 239–267.

Haylock, M., A. Klein Tank, E. Klok, P. Jones, and M. New (2008), A European daily high-resolution gridded dataset of surface temperature and precipitation, *J. Geophys. Res. Atmos.*, *113*.

Hilker, N., A. Badoux, and C. Hegg (2009), The Swiss flood and landslide damage database, 1972–2007, *Nat. Hazards Earth Syst. Sci.*, *9*(3), 913–925.

Hong, Y., R. Adler, and G. Huffman (2007), Use of satellite remote sensing data in the mapping of global landslide susceptibility, *Nat. Hazards*, *43*(2), 245–256.

Huffman, G. J., et al. (2007), The TRMM multisatellite precipitation analysis (TMPA): Quasi-global, multiyear, combined-sensor precipitation estimates at fine scales, *J. Hydrometeor.*, *8*(1), 38–55.

Markuzon, N., J. Regan, and C. Slesnick (2012), Using a weather-based data-driven approach for landslide prediction, *Proceedings of the AIAA Infotech@Aerospace Conference*, Garden Grove, CA.

Martelloni, G., S. Segoni, R. Fanti, and F. Catani (2011), Rainfall thresholds for the forecasting of landslide occurrence at regional scale, *Landslides*, 1–11.

Nikolopoulos, E. I., S. Crema, L. Marchi, F. Marra, F. Guzzetti,, and M. Borga (2014), Impact of uncertainty in rainfall estimation on the identification of rainfall thresholds for debris flow occurrence, *Geomorphology*, *221*, 286–297.

O'Gorman, P., and T. Schneider (2009), The physical basis for increases in precipitation extremes in simulations of 21st century climate change, *Proc. Nat. Acad. Sci.*, *106*(35), 14773–7; doi: 10.1073/pnas.0907610106.

Polemio, M., and F. Sdao (1999), The role of rainfall in the landslide hazard: the case of the Avigliano urban area (Southern Apennines, Italy), *Eng. Geol.*, *53*(3–4), 297–309.

Regan, J., E. Leidy, N. Markuzon, C. Slesnick, and E. Vaisman (2013), Modeling and assessment of weather-induced landslide activity, *Proceedings of the AIAA Infotech@Aerospace 2013*, Boston, MA.

Reichle, R. H. (2012), The MERRA-Land Data Product,. GMAO Office Note No. *3* (Version 1.2), available at http://gmao.gsfc.nasa.gov/pubs/office_notes.

Rienecker, M., et al. (2011), MERRA: NASA's modern-era retrospective analysis for research applications, *J. Climate*, *24*(14), 3624–3648.

Rudari, R., D. Entekhabi, and G. Roth (2004), Terrain and multiple-scale interactions as factors in generating extreme precipitation events, *J. Hydrometeor.*, *5*, 390–404.

Rudari, R., D. Entekhabi, and G. Roth (2005), Large-scale atmospheric patterns associated with mesoscale features leading to extreme precipitation events in northwestern Italy, *Adv. Water Resour.*, *28*, 601–614.

Schuster, R., and L. Highland (2001), Socioeconomic and environmental impacts of landslides in the Western Hemisphere, USGS Open File Report 01-0276.

Terlien, M. T. J. (1998), The determination of statistical and deterministic hydrological landslide-triggering thresholds, *Environ. Geol.*, *35*(2–3), 124–130.

Tralli, D. M., et al. (2005), Satellite remote sensing of earthquake, volcano, flood, landslide and coastal inundation hazards, ISPRS, *J. Photogramm. Remote Sens.*, *59*(4), 185–198.

Van den Besselaar, E., E. Hayock, G. van der Schrier, and A. Lein Tank (2011), A European daily high-resolution observational gridded data set of sea level pressure, *J. Geophys. Res.*, *116*.

Zêzere, J. L., R. M. Trigo, and I. F. Trigo (2005), Shallow and deep landslides induced by rainfall in the Lisbon region (Portugal): Assessment of relationships with the North Atlantic Oscillation, *Nat. Hazards Earth Syst. Sci.*, *5*(3), 331–344.

INDEX

Natural Hazard Uncertainty Assessment: Modeling and Decision Support, Geophysical Monograph 223, First Edition.
Edited by Karin Riley, Peter Webley, and Matthew Thompson.